# Fungal Biology

**Series Editors**
Vijai Kumar Gupta, Biorefining and Advanced Materials Research Center
Scotland's Rural College (SRUC), SRUC Barony Campus, Parkgate
Dumfries, Scotland, UK

Maria G. Tuohy, School of Natural Sciences
National University of Ireland Galway
Galway, Ireland

Fungal biology has an integral role to play in the development of the biotechnology and biomedical sectors. It has become a subject of increasing importance as new fungi and their associated biomolecules are identified. The interaction between fungi and their environment is central to many natural processes that occur in the biosphere. The hosts and habitats of these eukaryotic microorganisms are very diverse; fungi are present in every ecosystem on Earth. The fungal kingdom is equally diverse, consisting of seven different known phyla. Yet detailed knowledge is limited to relatively few species. The relationship between fungi and humans has been characterized by the juxtaposed viewpoints of fungi as infectious agents of much dread and their exploitation as highly versatile systems for a range of economically important biotechnological applications. Understanding the biology of different fungi in diverse ecosystems as well as their interactions with living and non-living is essential to underpin effective and innovative technological developments. This series will provide a detailed compendium of methods and information used to investigate different aspects of mycology, including fungal biology and biochemistry, genetics, phylogenetics, genomics, proteomics, molecular enzymology, and biotechnological applications in a manner that reflects the many recent developments of relevance to researchers and scientists investigating the Kingdom Fungi. Rapid screening techniques based on screening specific regions in the DNA of fungi have been used in species comparison and identification, and are now being extended across fungal phyla. The majorities of fungi are multicellular eukaryotic systems and therefore may be excellent model systems by which to answer fundamental biological questions. A greater understanding of the cell biology of these versatile eukaryotes will underpin efforts to engineer certain fungal species to provide novel cell factories for production of proteins for pharmaceutical applications. Renewed interest in all aspects of the biology and biotechnology of fungi may also enable the development of "one pot" microbial cell factories to meet consumer energy needs in the 21st century. To realize this potential and to truly understand the diversity and biology of these eukaryotes, continued development of scientific tools and techniques is essential. As a professional reference, this series will be very helpful to all people who work with fungi and should be useful both to academic institutions and research teams, as well as to teachers, and graduate and postgraduate students with its information on the continuous developments in fungal biology with the publication of each volume.

R. Z. Sayyed • Anjana Singh • Noshin Ilyas
Editors

# Antifungal Metabolites of Rhizobacteria for Sustainable Agriculture

Springer

*Editors*
R. Z. Sayyed
Asian PGPR Society and Auburn Ventures
Auburn, AL, USA

Anjana Singh
Central Department of Microbiology
Tribhuvan University
Kirtipur, Kathmandu, Nepal

Noshin Ilyas
Department of Botany
PMAS-Arid Agriculture University
Rawalpindi, Pakistan

ISSN 2198-7777 ISSN 2198-7785 (electronic)
Fungal Biology
ISBN 978-3-031-04804-3 ISBN 978-3-031-04805-0 (eBook)
https://doi.org/10.1007/978-3-031-04805-0



This Springer imprint is published by the registered company Springer Nature Switzerland AG
The registered company address is: Gewerbestrasse 11, 6330 Cham, Switzerland

# Foreword

If I wander to see contents of the book, it shows good compilation to unfold "Antifungal Metabolites of PGPR" in form of a compendium. For the plant fitness in agro-ecosystem, vigorous foodstuff is the main concern to promote sustenance of agriculture. It is noteworthy that with the increasing population all around globe, the need of dietary sources is increasing under habitat-imposed stresses the farmer always seeks to produce a healthy crop to gain high yield. Among the different pathogens of plant, fungi of different classes are the main destructor on different stages of soybean life cycle, starting from seed germination to seed production. Most known fungal species are strictly saprophytic, and among the approximately $10^5$ saprophytic species of the fungi, around $10^4$ are known to cause disease in plants. The interactions between plants and phytopathogenic fungi are complex, and fierce battles of attack and counterattack result in either resistance or plant disease. In general, to cause infection in host plants, fungi need to grow near to the plants and have to enter into plant tissue to feed. The way of growth may be different such as to grow on plant surface or to enter in the plant via natural openings and/or wounds.

Plant growth promoting rhizobacteria (PGPR) are a group of bacteria that colonize plant roots, first described by Kloepper and Schroth. PGPR are mostly present near the roots due to presence of root exudates secreted by plants and used as nutrient source by them. The nature of root exudates and soil conditions affect diversity of PGPR and their interaction with host plants. Based on occurrence, PGPR are classified as extracellular that present in the rhizosphere, spermosphere, and on the rhizoplane, while intracellular exist inside the root cells. The use of biological products based on PGPR helps in amelioration of functional milieu wherein crop grows well under fungal stress. Diverse range of benign traits of PGPR have been reported that work on alleviation of fungal stress, for example, volatile organic compounds (VOCs), siderophores, hydrogen cyanide production, antifungals, lytic enzymes, biosurfactants, and secondary metabolites

For this book, the editors compiled contents pooled from diverse nature of deployment of PGPR and benign role of PGPR to ameliorate environmental milieu for successful establishment of plant under biotic stress. In the present book, the editors provide glimpses of the following topics:

- Interaction of mycorrhizal fungi with rhizospheric microbes and their mode of action
- PGPR in biofilm formation and antibiotic production
- Potential of Streptomyces and its secondary metabolites
- The potential of plant growth promoting rhizobacteria (PGPR) as biological tools in enhancing agricultural sustainability
- *Pseudomonas* as BCA for fungal alleviation in rice crop
- Induction of hydrolytic enzymes: a criterion for biological control candidates against fungal pathogen
- Exploring the crucial role of PGPR in diabetes management
- Biofungicidal properties of rhizobacteria for plant growth promotion and plant disease resistance
- Role of PGPR in food security
- PGPR a sustainable agricultural mitigator for stressed agro-environments
- Hydrolytic enzymes from PGPR against plant fungal pathogens
- Utilization of plant growth promoting bacteria (PGPB) against phytopathogens
- Metabolomic profiling of PGPR for biological control of phytopathogens
- Fungal hydrolytic enzymes produced by PGPR
- Antifungal secondary metabolites against blast fungus *Magnaporthe oryzae*

  - The book's theme is very benign and selective to scientific fraternity around globe to understand the effective application of PGPR and its implementation *in situ* and *ex situ*. The book sheds light on various technologies deployed for PGPR application whereby researchers apply and analyze the implemented PGPR in sustenance of ago-ecosystem. The specific outcome of the book: explains the views on PGPR application and occurrence from across the world.

In addition, the book will be fruitful for students in terms of developing PGPR research in lab-based network and translational research henceforth.

I congratulate editors of proposed book who exhibited their potential to compile chapters that provide insight on role of PGPR in alleviation of fungal phytopathogen whereby plant can sustain under such menaces.

Amity University Uttar Pradesh (AUUP) Ajit Varma
Noida, Uttar Pradesh, India

# Preface

The world is currently experiencing a reinvigoration of microbial biotechnologies that utilize microbial benefits to improve agroecosystem functioning through enhanced soil health, crop vigor, and pest protection. Global food safety and security is yet another issue threatened by the numerous plant diseases caused by a wide variety of microbial pathogens. These diseases are traditionally controlled by the wide-scale use of chemical fungicides that pour recalcitrant compounds into the soil and develop resistant pathogens. A complex, well-structured, and regulated microbial community, with constant partner relationships, is always associated with plants. Such a community is called the phytomicrobiome, which is also referred to as plant growth-promoting rhizobacteria (PGPR). PGPR is a highly diverse group of organisms that control plant pathogens through various mechanisms such as antagonism and induction of systemic resistance against pathogens. Antagonistic PGPR produce numerous substances that inhibit pathogens. These antagonistic substances include siderophores, antibiotics, hydrogen cyanide, and hydrolytic enzymes. In plants, induced systemic resistance (ISR) resembles pathogen-induced systemic acquired resistance (SAR) under environments where the inducing bacteria and the challenging pathogen remain spatially separated. ISR and SAR renders uninfected plant parts more resistant to pathogens in several plant species. Resistance-inducing and antagonistic rhizobacteria have been used in formulating new inoculants with combinations of different mechanisms of action, leading to more efficient use for biocontrol strategies to improve cropping systems.

PGPR that produce antifungal metabolites have emerged as bioactive natural products against the pathogenic fungus and are used for the sustainable and cost-effective control of plant diseases. Various classes of secondary metabolites, including lipopeptides, macrolides, alkaloids, terpenoids, and phenolics from microorganisms and plants, strongly suppress fungal growth and can also control plant diseases in vitro and in vivo. Eco-friendly PGPR and their metabolites are excellent agents to control various fungal and bacterial phytopathogens.

Knowledge of the regulation of various strategies for bio-fungicidal properties of rhizobacteria for plant disease management will lead to PGPR development with improved reliability and efficacy. The traditional and novel secondary metabolites produced by PGPR that can be utilized in integrated pest management of plants will provide an eco-friendly and economically viable alternative to chemical pesticides.

Making better use of crop microbiome functions is expected to boost agricultural production. This shall make the bio-economy of less developed countries stronger. The advanced tools of study like rDNA technology, mathematical modeling, and computer technology in combination with a continuation of the classical approaches, such as importation and release of natural enemies and improved germplasm, breeding, and field testing, should quickly move research and technology into a new era. Further, the secondary metabolites produced from the plants and PGPR combination can be used in the treatment of human diseases too.

This book contains 17 chapters that deal with the vivid aspects of the biocontrol role of antifungal metabolites such as antibiotics, siderophores, hydrogen cyanide (HCN), volatile compounds, and volatile hydrolytic enzymes (chitinase, cellulose, protease, glucanase, etc.) produced by PGPR.

We believe that this book will be helpful to researchers, scientists, faculties, and students harnessing the emerging information on crop microbiomes. It is expected to generate new avenues to revolutionize agriculture by improving soil structure, increasing ecosystem nutrient availability, boosting plant resilience to stress, and developing a new intervention for insect pests.

Shahada, India — R. Z. Sayyed
Kirtipur, Nepal — Anjana Singh
Rawalpindi, Pakistan — Noshin Ilyas

# Acknowledgment

This book, entitled ***Antifungal Metabolites of Rhizobacteria for Sustainable Agriculture***, is a result of the dedicated teamwork of all the editors. As a team, we realized, at every step, the significance of planning, execution, mutual respect to timelines, and deliverance. We greatly value the support from all our institutions, especially colleagues, friends, and family members for their encouragement and motivational support.

We are thankful to all contributing authors whose research works and vast experiences in the field of antifungal metabolites of PGPR have played a very significant role in enriching the content of this publication. It was a pleasure to edit this book due to stimulating and great cooperation from the contributors' part.

We extend our sincere gratitude to our series editor, Dr. Vijai Gupta, and Springer, especially Eric Stannard, Senior Editor, Botany, Springer, USA; Kritheka Elango (Ms.), Project Coordinator (Books), Springer Nature, India; and all the members of Springer's team for reposing their faith in us as an editorial team and entrusting us in disseminating PGPR research in the scientific community across the globe.

R. Z. Sayyed
Anjana Singh
Noshin Ilyas

# Contents

**Part I Antifungal Metabolites (AFM) for Sustainable Agriculture**

1 **Potential of *Streptomyces* and Its Secondary Metabolites for Biocontrol of Fungal Plant Pathogens** . . . . 3
Anand Dave and Sanjay Ingle

2 **Antifungal Secondary Metabolites Against Blast Fungus *Magnaporthe oryzae*** . . . . 23
Moutoshi Chakraborty and Tofazzal Islam

3 **Utilization of Plant Growth-Promoting Bacteria (PGPB) Against Phytopathogens** . . . . 53
Ali Talha Khalil and Zabta Khan Shinwari

4 **PGPR in Biofilm Formation and Antibiotic Production** . . . . 65
Dibyajit Lahiri, Moupriya Nag, R. Z. Sayyed, Abdul Gafur, Mohammad Javed Ansari, and Rina Rani Ray

5 **Plant Growth-Promoting Rhizobacteria as Antifungal Antibiotics Producers** . . . . 83
Anima Shrestha, Manju Shree Shakya Hada, Reshma Tuladhar, Sarita Manandhar, Noshin Ilyas, Abdel Rahman Mohammad Al Tawaha, and Anjana Singh

6 **Biofungicidal Properties of Rhizobacteria for Plant Growth Promotion and Plant Disease Resistance** . . . . 103
Rajashree B. Patwardhan, Pragati S. Abhyankar, Suneeti S. Gore, Saylee V. Kalekar, and Shriya P. Umrani

7 **Antifungal Compounds of Plant Growth-Promoting *Bacillus* Species** . . . . 135
Sunanda Chakraborty, Tofazzal Islam, and Sunita Mahapatra

xii Index 375

8 **Bioactive Biomolecules from Plant Growth-Promoting Rhizobacteria (PGPR)** ........ 157
Manila Bajimaya, Sunita Basnet, Sailesh Malla, and Laxmi Prasad Thapa

**Part II Volatile Compounds and Hydrolytic Enzymes of PGPR**

9 **Metabolomic Profiling of Plant Growth-Promoting Rhizobacteria for Biological Control of Phytopathogens** ........ 181
Zakira Naureen, Syed Abdullah Gilani, Binta Kondoor Benny, Hajra Sadia, Fauzia Y. Hafeez, and Azra Khanum

10 **Hydrolytic Enzymes from PGPR Against Plant Fungal Pathogens** ........ 211
Suneeta Panicker and R. Z. Sayyed

11 **Induction of Hydrolytic Enzymes: A Criterion for Biological Control Candidates against Fungal Pathogen** ........ 239
Nusaibah Syd Ali, Tuan Muhammad Syafiq, and Muniroh Md Saad

**Part III Plant-Microbe Interactions and Agriculture Sustenance**

12 ***Pseudomonas* as Biocontrol Agent for Fungal Disease Management in Rice Crop** ........ 253
Kausik Basu, Amit C. Kharkwal, and Ajit Varma

13 **Interaction of Mycorrhizal Fungi with Rhizospheric Microbes and Their Mode of Action** ........ 269
Jay Kant Raut, Kuber Baral, Mahesh Kumar Adhikari, and Pramod Kumar Jha

14 **The Potential of Plant Growth-Promoting Rhizobacteria (PGPR) as Biological Tools in Enhancing Agricultural Sustainability** ........ 295
Garima Malik, Samira Chugh, and Akanksha Madan

15 **Role of Plant Growth-Promoting Rhizobacteria (PGPR) in Food Security** ........ 311
Ram Awadh Ram, Ajit Kumar Dubedi Anal, G. S. Tiwari, Vishal Nath, and Ram Kripal Pathak

16 **Plant-Microbe-Insect Interactions: Concepts and Applications for Agricultural Sustainability** ........ 335
V. Prithiv Raj, S. Thangavel, N. Ponnusamy, and V. Prabakaran

17 **Exploring the Crucial Role of Plant Growth-Promoting Rhizobacteria (PGPR) in Plant Secondary Metabolite Production and Diabetes Management** ........ 351
Deegendra Khadka and Kishor Pandey

**Index** 375

# Contributors


**Pragati S. Abhyankar** Department of Microbiology, Haribhai V. Desai College of Arts, Science and Commerce, Pune, India

**Mahesh Kumar Adhikari** Nepal Academy of Science and Technology, Khumaltar, Lalitpur, Nepal

**Nusaibah Syd Ali** Department of Plant Protection, Faculty of Agriculture, Universiti Putra Malaysia, Serdang, Selangor, Malaysia

**Ajit Kumar Dubedi Anal** ICAR-National Research Centre on Litchi, Muzaffarpur, India

**Mohammad Javed Ansari** Department of Botany, Hindu College Moradabad (Mahatma Jyotiba Phule Rohikhand University Bareilly), Moradabad, India

**Manila Bajimaya** Department of Fungal Biology, Polytechnic Research Institute of Nepal (PORIN), Koteshwor, Nepal

**Kuber Baral** Nepal Academy of Science and Technology, Khumaltar, Lalitpur, Nepal

**Sunita Basnet** Department of Natural Product and Bioresources, Polytechnic Research Institute of Nepal (PORIN), Koteshwor, Nepal

**Kausik Basu** Biofertilizer and Biologicals, Former Manager R&D, IPL Biologicals, Dehradun, Uttarakhand, India

**Binta Kondoor Benny** Department of Biological Sciences and Chemistry, College of Arts and Science, University of Nizwa, Nizwa, Sultanate of Oman

**Moutoshi Chakraborty** Institute of Biotechnology and Genetic Engineering, Bangabandhu Sheikh Mujibur Rahman Agricultural University, Gazipur, Bangladesh

**Sunanda Chakraborty** Department of Plant Pathology, Bidhan Chandra Krishi Viswavidyala, Mohanpur, Nadia, India

**Samira Chugh** Gargi College, University of Delhi, New Delhi, India

**Anand Dave** Department of Microbiology and Biotechnology Centre, The Maharaja Sayajirao University of Baroda, Vadoadara, India

**Abdul Gafur** Sinarmas Forestry Corporate Research and Development, Perawang, Indonesia

**Syed Abdullah Gilani** Department of Biological Sciences and Chemistry, College of Arts and Science, University of Nizwa, Nizwa, Sultanate of Oman

**Suneeti S. Gore** Department of Microbiology, Fergusson College, Pune, Maharashtra, India

**Manju Shree Shakya Hada** Department of Microbiology, Tri-Chandra Multiple Campus, Tribhuvan University, Kathmandu, Nepal

**Fauzia Y. Hafeez** COMSATS University, Islamabad, Pakistan

**Noshin Ilyas** Department of Botany, PMAS-Arid Agriculture University, Rawalpindi, Pakistan

**Sanjay Ingle** Department of Microbiology and Biotechnology Centre, The Maharaja Sayajirao University of Baroda, Vadoadara, India

**Tofazzal Islam** Institute of Biotechnology and Genetic Engineering (IBGE), Bangabandhu Sheikh Mujibur Rahman Agricultural University, Gazipur, Bangladesh

**Pramod Kumar Jha** Nepal Academy of Science and Technology, Khumaltar, Lalitpur, Nepal

Central Department of Botany, Tribhuvan University, Kathmandu, Nepal

**Saylee V. Kalekar** Department of Microbiology, Fergusson College, Pune, Maharashtra, India

**Deegendra Khadka** Molecular Biotechnology Unit, Faculty of Science, Nepal Academy of Science and Technology, Khumaltar, Lalitpur, Nepal

**Ali Talha Khalil** Department of Pathology, Lady Reading Hospital Medical Teaching Institution, Peshawar, Khyber Pakhtunkhwa, Pakistan

**Azra Khanum** University Institute of Biochemistry and Biotechnology, PMAS University of Arid Agriculture Rawalpindi, Rawalpindi, Pakistan

**Amit C. Kharkwal** Amity Institute of Microbial Technology (AIMT), Amity University Uttar Pradesh (AUUP), Noida, Uttar Pradesh, India

**Dibyajit Lahiri** Department of Biotechnology, University of Engineering and Management, Kolkata, West Bengal, India

**Akanksha Madan** Gargi College, University of Delhi, New Delhi, India

**Sunita Mahapatra** Department of Plant Pathology, Bidhan Chandra Krishi Viswavidyala, Mohanpur, Nadia, India

**Garima Malik** Raghunath Girls' Post Graduate College, Chaudhary Charan Singh University, Meerut, India

**Sailesh Malla** Strain Development Department, Discovery, R&D, Chr. Hansen A/S, Hørshlom, Denmark

**Sarita Manandhar** Department of Microbiology, Tri-Chandra Multiple Campus, Tribhuvan University, Kirtipur, Kathmandu, Nepal

**Moupriya Nag** Department of Biotechnology, University of Engineering and Management, Kolkata, West Bengal, India

**Vishal Nath** ICAR-National Research Centre on Litchi, Muzaffarpur, India

**Zakira Naureen** Department of Biological Sciences and Chemistry, College of Arts and Science, University of Nizwa, Nizwa, Sultanate of Oman

**Kishor Pandey** Central Department of Zoology, Tribhuvan University, Kirtipur, Kathmandu, Nepal

**Suneeta Panicker** Department of Microbiology, Dr. D.Y. Patil Arts, Commerce and Science College, Pimpri, Pune, India

**Ram Kripal Pathak** ICAR-Central Institute for Subtropical Horticulture, Lucknow, India

**Rajashree B. Patwardhan** Department of Microbiology, Haribhai V. Desai College of Arts, Science and Commerce, Pune, India

**N. Ponnusamy** Department of Agricultural Entomology, Dr. Rajendra Prasad Central Agricultural University, Pusa, Bihar, India

**V. Prabakaran** Department of Agricultural Entomology, B. A. College of Agriculture, Anand Agricultural University, Anand, Gujarat, India

**V. Prithiv Raj** Department of Agricultural Entomology, Sardarkrushinagar Dantiwada Agricultural University, Dantiwada, Gujarat, India

**Ram Awadh Ram** ICAR-Central Institute for Subtropical Horticulture, Lucknow, India

**Jay Kant Raut** Nepal Academy of Science and Technology, Khumaltar, Lalitpur, Nepal

**Rina Rani Ray** Department of Biotechnology, Maulana Abul Kalam Azad University of Technology, Haringhata, West Bengal, India

**Muniroh Md Saad** Department of Plant Protection, Faculty of Agriculture, Universiti Putra Malaysia, Serdang, Selangor, Malaysia

**Hajra Sadia** National University of Science and Technology, Islamabad, Pakistan

**R. Z. Sayyed** Asian PGPR Society and Auburn Ventures, Auburn, AL, USA

**Zabta Khan Shinwari** Department of Plant Sciences, Quaid-i-Azam University, Islamabad, Pakistan

Pakistan Academy of Sciences, Islamabad, Pakistan

**Anima Shrestha** Department of Microbiology, Tri-Chandra Multiple Campus, Tribhuvan University, Kathmandu, Nepal

**Anjana Singh** Central Department of Microbiology, Tribhuvan University, Kirtipur, Kathmandu, Nepal

**Tuan Muhammad Syafiq** Department of Plant Protection, Faculty of Agriculture, Universiti Putra Malaysia, Serdang, Selangor, Malaysia

**Abdel Rahman Mohammad Al Tawaha** Department of Biological Sciences, Al - Hussein Bin Talal University, Maan, Jordan

**S. Thangavel** Department of Agricultural Entomology, B. A. College of Agriculture, Anand Agricultural University, Anand, Gujarat, India

**Laxmi Prasad Thapa** Department of Molecular Biotechnology, Nepal Academy of Science and Technology, Khumaltar, Lalitpur, Nepal

**G. S. Tiwari** Teerthanker Mahaveer University, Moradabad, India

**Reshma Tuladhar** Central Department of Microbiology, Tribhuvan University, Kirtipur, Kathmandu, Nepal

**Shriya P. Umrani** Department of Microbiology, Siddhivinayak Mahila Mahavidyalaya, Pune, India

**Ajit Varma** Amity Institute of Microbial Technology (AIMT), Amity University Uttar Pradesh (AUUP), Noida, Uttar Pradesh, India

# About the Editors

**R. Z. Sayyed** is a professor of Microbiology. He has over 25 years of teaching and 20 years of research experience in microbiology and biotechnology. Currently, he serves as the president of the India Chapter of the Asian PGPR Society for Sustainable Agriculture. To his credit, he has received the Fellow Indian Phytopathological Society (2021), Best Teachers Award (2016), Young Scientist Award (2005, 2008, and 2012), Prof. M M Sharma Award, Springer-Society Award (2020), and Award for Excellence in PGPR Research (2017, 2018, and 2019). He is associate editor of *Environmental Sustainability* (Springer) and guest editor of *Sustainability* (MDPI), Frontiers in Microbiology, Peer J and PLOS ONE. He has authored and co-authored over 250 research papers in high IF international journals; 27 books with Springer, Wiley, CRC, and Cambridge Press; and 43 book chapters in reputed edited volumes. Prof. Sayyed has trained several graduate, postgraduate, and research students and produced seven PhDs under his guidance. He has delivered many invited talks at several Southeast Asian and European countries.

**Anjana Singh** Prof. Dr. Anjana (Maharjan) Singh is a professor in Central Department of Microbiology (CDM), Tribhuvan University (TU), Nepal. She was Fulbright postdoctoral fellow at the University of Virginia in 2008–09 and is recipient of 12 national/international awards. She is an academician in the Nepal Academy of Science & Technology (NAST) and in the editorial board of *Nepal Science and Technology Journal*, reviewer of many reputed journals, and coordinator of Women Scientists Forum, Nepal, NAST. She has over 125 national/international publications and has supervised 105 master's theses and 5 PhDs. She is an executive board member of AASSA-WISE committee since 2015 and has attended various workshops in India, Turkey, Korea, and Bangladesh. Dr. Singh was an executive member of the Board of Studies at the South Asian University from 2016 to 2018. She has presented her research work in 74 national and international workshops/conferences/symposiums.

**Noshin Ilyas** is an associate professor in the Department of Botany at PMAS Arid Agriculture University Rawalpindi, Pakistan. She always had distinction in her academic carrier, with a first-class first position in her master's course, and completed her PhD studies under a fellowship from the Higher Education Commission, Pakistan. She has supervised dissertational research work of 7 PhD and 31 MPhil students. Dr. Ilyas has a good publication record in prestigious journals with an impact factor of above 200 and 1700 citations. She is an academic editor for *PLoS One* and guest editor of sustainability and Frontiers in Microbiology. She has earned several research grants from the Higher Education Commission, Pakistan, and Pakistan Science Foundation. She has received a research award in the female category of assistant professor from PMAS-Arid University, and 1st Prof. Dr. Abdul Hameed Gold Medal in Applied and Environmental Microbiology from Applied Zoological Society of Pakistan. She is an active member of the scientific community and has presented her research work in various conferences in the country and abroad. She is president of Pakistan Asian PGPR society and a member of various professional societies.

# Part I
# Antifungal Metabolites (AFM) for Sustainable Agriculture

# Chapter 1
# Potential of *Streptomyces* and Its Secondary Metabolites for Biocontrol of Fungal Plant Pathogens

**Anand Dave and Sanjay Ingle**

## Contents


1.1 Introduction .... 3
1.2 *Streptomyces* Production of Metabolites for Biological Control .... 5
1.3 Main Phytopathogenic Fungi .... 7
1.4 Production of Enzymes .... 11
1.5 Volatile Organic Compounds (VOC) .... 12
1.6 Phytohormone Production .... 12
1.7 Enhancement of Plant Growth .... 13
1.8 Commercialization, Environmental Effects, and Biosafety of *Streptomyces* Products... 15
1.9 Conclusion .... 17
References .... 17


## 1.1 Introduction

*Streptomyces* are Gram-positive, filamentous bacteria belonging to the group actinomycetes, a group that encompasses the majority of soil bacterial species. Plant growth-promoting actinomycetes (PGPA) are root colonizing microbes with beneficial effects including plant growth promotion and disease control (Misra et al. 2017). *Streptomyces* is the most abundant genus. Continuous search for eco-friendly management of plant disease and promotion of plant growth and crop yield has headed toward the use of plant growth-promoting rhizobacteria (PGPR) as an effective and eco-friendly means of controlling soil-borne phytopathogens and simultaneously promoting the plant growth (Sayyed et al. 2007). Phytopathogens represent a global threat to the production of food and crops, causing devastating diseases that


A. Dave · S. Ingle (✉)
Department of Microbiology and Biotechnology Centre, The Maharaja Sayajirao University of Baroda, Vadoadara, India


 
R. Z. Sayyed et al. (eds.), *Antifungal Metabolites of Rhizobacteria for Sustainable Agriculture*, Fungal Biology, https://doi.org/10.1007/978-3-031-04805-0_1

result in significant economic losses (Valdés et al. 2017). At least 30–40% of the crop losses are caused by phytopathogenic infections and represent $200 billion a year worldwide (Sawicka and Egbuna 2019; FAO 2020). Phytopathogenic bacteria and fungi affect all plant food products, colonizing their surface or tissues. These microorganisms cause symptoms such as spots, blight, cankers, tissue rot, and hormonal imbalances that lead to excessive growth, growth retardation, atrophy, and root branching, among others (Kannan et al. 2015). Thus, these phytosanitary problems negatively affect the world's food supply.

Chemical pesticides and antibiotics have been used to fight microbial infections. However, the use of these compounds has brought different problems such as pathogen resistance to treatment, soil salinization, and environmental contamination (Petriccione et al. 2017; Rodríguez et al. 2020). Phosphorus (P) is an essential element for all living organisms including plants and classified as a macronutrient. It is the second highest essential nutrient element followed by nitrogen, but only 0.2% of P concentration is available in plant organic materials (Sayyed et al. 2007). P plays a significant role in the natural development and proper maturity of plants, photosynthesis, respiration, energy storage and transfer for cell division and enlargement, and several other processes. But usually, this has not been found in pure elemental form except only available as phosphates in many minerals.

Antibiotics production (antibiosis) by PGPR plays an important role in the management of plant diseases. The process has been defined as the inhibition or suppression of pathogenic microorganisms via the production of low-molecular-weight compounds (antibiotic) by other microorganisms. *Bacillus* species and fluorescent pseudomonas are playing active roles in the suppression of pathogenic microorganisms by producing extracellular metabolites that have inhibitory and antagonistic effects against their competitors. Additionally, to the direct antagonistic action, antibiotics have a vital role in induced systemic resistance (ISR) mechanism in plants. Different microorganisms have the ability to produce different antibiotics, for example, PGPR (*Bacillus* species) produces several antibiotics that comprise iturins, mycosubtilin, bacillomycin D, surfactin, fengycin, and zwittermicin A, whereas antibiotics produced by fluorescent pseudomonads include 2,4-diacetylphloroglucinol (DAPG), pyoluteorin, phenazines, pyrrolnitrin, oomycin A, viscosin, and massetolide A (Reshma et al. 2018).

According to this background, it is necessary to look for new antimicrobials to control phytopathogenic bacteria and fungi as alternative methods of plant protection, respectful with the environment, and less dependent on chemicals. Waksman and Henrici first characterized the genus *Streptomyces* in 1943. It is classified into the family of Streptomycetaceae based on its morphology and cell wall chemotype. Previous studies have reported that more than 74% of current antibiotics have been produced by the genus *Streptomyces*. *Streptomyces* species are abundant in the soil and are important members of the plant growth-promoting rhizobacteria (PGPR). Some species have been used in agriculture as biocontrol agents (Wang et al. 2018). *Streptomyces* spp. are Gram-positive filamentous soil bacteria producers of various secondary metabolites, including many valuable compounds such as antibiotics, volatile organic compounds, and enzymes. Also, it induces plant resistance to pathogens (Zhao et al. 2013; Faheem et al. 2015; Nah et al. 2021). Due to the extensive production of secondary metabolites, *Streptomyces* species have attracted much attention to the biological control of soil pathogens. The production and release of secondary metabolites depend on several factors, such as nutritional, biological, and environmental conditions. Furthermore, they can be induced or activated by other organisms when they invade the soil space, competing for available nutrients (Sun et al. 2016).

Soilborne phytopathogens are responsible for causing infection of roots, stems, leaves, and fruits. These phytopathogens occur in a broad spectrum of susceptible hosts under favorable environmental conditions. Such diseases are known to cause severe economic losses to variety of food crops and are commonly controlled by using synthetic fungicides or nonspecific chemical fungicides. These plant diseases are known to cause a loss of 30% crop yield posing economic hardship to producers (Sayyed et al. 2007).

The genus *Streptomyces* has been widely studied due to its biotechnological potential as a promising candidate for use in the biocontrol of phytopathogenic microorganisms. However, more information is still needed to develop novel formulations that could increase the shelf life of streptomycetes, thus ensuring their long-term viability, sporulation activity, and efficacy of their application as a microbial agrochemical. In this sense, encapsulation offers an advantageous and environmentally friendly option for developing new biological control products.

## 1.2 *Streptomyces* Production of Metabolites for Biological Control

Several studies have reported the effects of beneficial microorganisms and secondary plant metabolites as a bio-alternative to control phytopathogenic bacteria and fungi. These bioactive compounds tend to affect several pathogenic species and are biodegradable and friendly to the environment (Valdés et al. 2017). The use of microbial metabolites or the microorganisms themselves to control plant diseases has received greater attention because they do not negatively affect human or animal health, are respectful of the environment, and do not affect other beneficial organisms (Syed Ab Rahman et al. 2018) (Fig. 1.1). Although biological methods to control plant pathogens have been studied for more than 70 years, biocontrol products represent only 3.5% of the global pesticide market, which is still dominated by synthetic pesticides (Carbu et al. 2016). *Streptomyces*' antagonistic activities against pathogens can be achieved through three main mechanisms: competition for nutrients and space, antibiosis, and parasitism (Boukaew and Prasertsan 2014). *Streptomyces* spp. can colonize the surfaces of plant roots, survive in various soil types, and produce spores that allow them to remain longer and in different extreme conditions (Law et al. 2017). *Streptomyces* strains have a broad spectrum of antibiotics, volatile organic compounds (VOC), which act against pathogens, disrupting

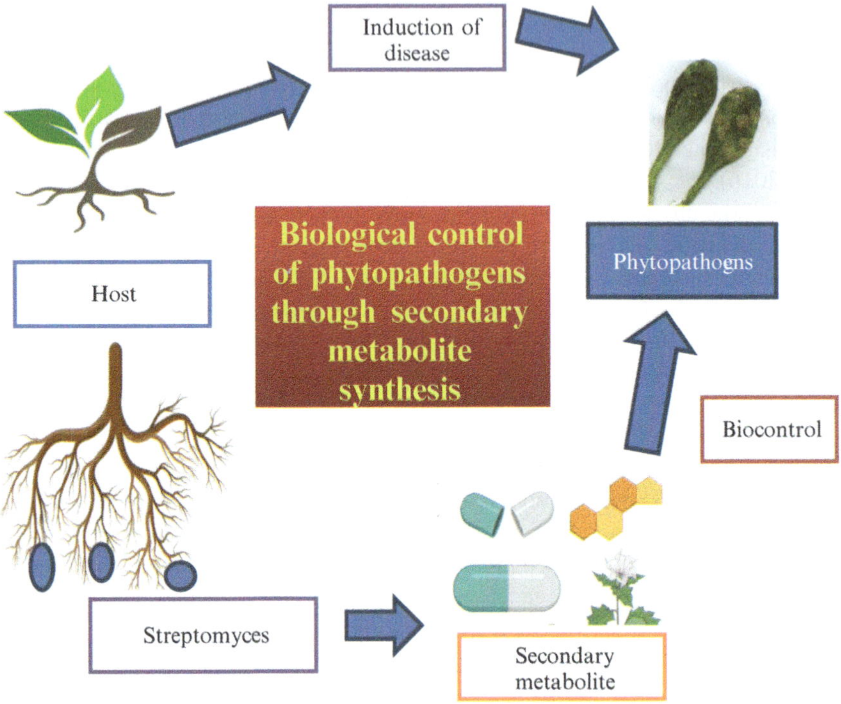


**Fig. 1.1** Biological control of phytopathogens through secondary metabolite synthesis

bacterial cell-cell communication (quorum sensing), and a variety of enzymes that degrade the cell wall of fungi (Dias et al. 2017; Wonglom et al. 2019; Song et al. 2020) (Table 1.1). The secondary metabolism of bacteria constitutes a rich source of bioactive compounds of potential pharmaceutical value, comprising biosynthetic pathways of many chemicals that have been and are being utilized as, e.g., antibiotics, cholesterol-lowering drugs, or antitumor drugs. Interestingly, the genes encoding the biosynthetic pathway responsible for the production of such a secondary metabolite are very often spatially clustered together at a certain position on the chromosome; such a compendium of genes is referred to as a "secondary metabolite biosynthesis gene cluster." This 24 genetic architecture has opened up the possibility for straightforward detection of secondary metabolite biosynthesis pathways by locating their gene clusters. antiSMASH is use for Secondary Metabolite screening. Based on profile hidden Markov models of genes that are specific for certain types of gene clusters, antiSMASH is able to accurately identify the gene clusters encoding secondary metabolites of all known broad chemical classes.

**Table 1.1** List of antibiotics produced by different Actinobacteria and their applications

| Metabolites | Type of metabolite | Pathogens | References |
|---|---|---|---|
| Oligomycin | Antibiotic | *Erwinia carotovora* | Doolotkeldieva et al. (2016) |
| Cycle (l-pro-l-Tyr) | Cyclic peptides | *Xanthomonas axonopodis* | Wattana-Amorn et al. (2016) |
| Cycle (d-pro-l-Tyr) | | *Ralstonia solanacearum*<br>*Clavibacter michiganensis* | |
| Proteases | Enzyme | *Colletotrichum dematium* | Song et al. (2020) |
| Blasticidin-S | Antibiotic | *Pyricularia oryzae* | Law et al. (2017) |
| Kasugamycin<br>Oligomycin A<br>Rapamycin<br>Pyrroles<br>b-1,3-Glucanases<br>2-Methyl-1-butanol<br>Acetic acid<br>Benzaldehyde<br>2-methylpropanoic acid<br>Butanoic acid<br>3-Methylbutanoic acid | Volatile compounds | *Colletotrichum* sp.<br>*Curvularia lunata* | Wonglom et al. (2019) |

Table 1.1 gives a list of antibiotics developed by Actinobacteria variations and how the use of antimicrobials has had a significant effect on the medical world where cancers, tumors, and even malaria have not previously been treated.

## 1.3 Main Phytopathogenic Fungi

Phytopathogenic fungi pose serious problems worldwide in cultivating economically important plants, especially in the subtropical and tropical regions. Phytopathogenic fungi are responsible for diseases in many vegetable and fruit products, appearing during crop development or postharvest. In addition, the phytopathogenic fungus produces toxic metabolites that can be carcinogenic and constitute a risk to public health. In contrast, the growth of specific fungi results in nutritional and chemical changes, poor appearance, and food flavor development (Vashistha and Chaudhary 2019). Among the phytopathogenic fungi with the highest incidence in agriculture and with the most significant economic impact are *Fusarium*, *Botrytis cinerea*, *Alternaria alternata*, and *Colletotrichum* spp. Plant diseases, especially those caused by plant pathogenic fungi, are key factors in the production and quality of crops and in the postharvest life of fresh fruit and vegetables (Sayyed et al. 2007). Among these pathogens, *Pyricularia oryzae* (*P. oryzae*) and *Botrytis cinerea* (*B. cinerea*) are two globally distributed fungi with a wide range of hosts, ranked as the first and second most important plant pathogens in the world based on their economic and scientific importance. These pathogens cause specific

**Table 1.2** Major phytopathogenic fungi with a critical impact in agriculture

| Microorganisms | Plant diseases | Infection mechanisms | Pathogenicity mechanisms | Reference |
|---|---|---|---|---|
| *Fusarium* | Bakanae disease of rice, corn ear rot, pine canker, contamination of corn and other cereals with mycotoxins | The pathogen directly invades the stems' base, near or below the soil surface, or enters the plant through the roots | They can produce molecules that suppress defense responses in plants, including the biosynthesis of phytoalexins and enzymes that destabilize plant membranes' integrity | Cook (2010); González et al. (2012); Aoki et al. (2014) |
| *Botrytis cinerea* | Gray mold disease | Infection occurs by direct penetration or by wounds that follow pruning or harvesting | Secretes enzymes and toxins that degrade and kill host cells | Dean et al. (2012); AbuQamar et al. (2017); Bolívar-Anillo et al. (2020) |
| *Alternaria alternata* | Necrotic lesions such as spots or brown/black spots. Leaf blight Dark lesions in roots and fruits | *A. alternata* could contaminate the plant through breaks in the skin or other weakened areas | Phytotoxins | Singh et al. (2015); Marin-Felix et al. (2019) |
| *Colletotrichum* spp. | Anthracnose strawberry crown rot, banana crown rot, red rot in coffee berries and sugar cane, blotch in cowpea | *Colletotrichum* infect the plants by attaching on the surface with the aspersorium squeezes itself through the plant cuticle into the epidermal cells | Toxins degradation enzymes | Joshi (2018); da Silva et al. (2020) |

diseases in plants using different infection mechanisms and pathogenicity mechanisms depending on the microorganism in question (Table 1.2).

### 1.3.1 Phytopathogenic Microorganism's Control

One of the traditional control methods is cultural control. Cultural control consists of standard agricultural practices to help prevent attacks by phytopathogens, make the environment less favorable for their development, destroy them, or reduce their damage (Valdés et al. 2017). The application of agricultural practices requires adequate knowledge about the physiology and phenology of cultivated plants and their agronomic characteristics, the modalities of suitable agricultural practices, and a good understanding of the biological behavior of local pests and their seasonal occurrence. Among the leading practices within cultural control, FAO (2021) recommends sowing healthy seeds, planting in soils without a history of the presence of these phytopathogens, using uncontaminated irrigation water, rotating crops with non-host crops such as garlic or onion, ensuring good soil drainage, control soil pests that can help spread and infection, and disinfect agricultural equipment with 2% sodium hypochlorite, agricultural iodine, or quaternary ammonia (Valdés et al. 2017). However, it is widespread that these practices are complemented with pesticides to increase their effectiveness.

The increasing use of synthetic chemicals to control phytopathogenic microorganisms has caused the development of pathogens' resistance to applied chemical products and the negative impact on the environment (Martins et al. 2018). The demand for chemical-free agrochemicals has increased due to the toxicity of the chemicals to humans. The control of diseases of microbial origin is based on the use of antibiotics and solutions whose main component is copper. The application of copper to the field has environmental impacts due to its toxicity in the soil and plants (Hippler et al. 2018). The use of cupric compounds as antimicrobial agents has provided the first evidence of cells resistant to this compound in *Xanthomonas* sp., *Pseudomonas* sp., and *Erwinia* sp. (Martins et al. 2018).

Conventional chemicals do not always provide adequate infection control. Some of the phytopathogenic microorganisms colonize the intercellular spaces of the leaves and invade other parts of the plant. In addition, frequent applications of synthetic agrochemicals increase the risk of microorganisms developing resistance. For example, *B. cinerea* is considered a high-risk pathogen in terms of resistance to fungicides (Bolívar-Anillo et al. 2020).

Chemical control by pesticides causes a lot of harm to humans and the environment. However, lack of access to effective pesticides and disease-resistant cultivars has motivated alternative strategies to control these diseases. For this reason, the development of complementary methods to chemical control, such as non-pathogenic microorganisms as biological control agents, new sources of natural antimicrobials that are more friendly to the environment and consumers, has been the focus of the scientific community.

If disease-resistant commercial cultivars are not available, the use of chemical control is widespread. The disadvantage of this method is the dangers to the environment and human health and its limited effectiveness. Crop rotation can prevent some diseases, but it is not always feasible; some pathogens can go undetected in asymptomatic hosts or survive for many years in the soil. This scenario justifies an urgent need for agents that effectively control phytopathogens.

The European Union and the USA are the primary consumers of chemical fungicides worldwide. However, since 2011 the use of these chemical agents has decreased, mainly in the USA, perhaps due to environmental protection and consumer health regulations (Carbu et al. 2016). The main phytosanitary companies have been investing in the field of Biocontrol in response to legal restrictions and consumer demand for pesticide-free foods (Bolívar-Anillo et al. 2020). In 2011, the global biocontrol market had a $2.1 billion reported value and was influenced by the growing demand for organic products (Velivelli et al. 2014). In 2019, organic products had a more significant market share, representing around 65.0%. The use of metabolites produced by microorganisms or the microorganisms themselves to control plant diseases has received increased attention, with a few exceptions. They have no adverse effects on human or animal health and are friendly to the environment (Bolívar-Anillo et al. 2020). Biological methods to control plant pathogens have been studied for more than 70 years. Biological control agents can directly inhibit pathogens through physical contact or through specific mechanisms to combat the pathogen (hyperparasitism, predation, etc.).

Furthermore, they can act indirectly through general mechanisms that do not target a particular type of pathogen (stimulation of plant defenses, competition by substrates, etc.). Otherwise, they can act by antagonism of diverse pathways (antibiotics, lytic enzymes, etc.), which are compatible with each other and can act simultaneously or synergistically (Bardin et al. 2015). However, biological control agents' effectiveness depends on climate variation, ecological competition, the intrinsic traits of the biological control agents, the exercise of selection pressure, and the quality of the product as formulated.

### 1.3.2 PGPA as Biocontrol Agents (BCAs)

The term biocontrol refers to the use of microbial antagonist to suppress the disease or pathogen. Biocontrol of pathogen involves application of beneficial rhizobia or their metabolites that neutralizes the negative effects caused by pathogens and thus promote positive responses by the plant (Junaid et al. 2013). Biological control of plant diseases using antagonist PGPR has emerged in recent years in agriculture as greater step toward sustainability, and public concern about the use of hazardous chemical fungicides and plant disease suppression by PGPR is the best possible alternative of reducing the doses of agrochemicals and its severity in agroecosystem. There are various mechanisms involved in biocontrol of plant diseases caused by PGPR. These mechanisms are generally classified as competition, lysis of cell components, antibiosis, and induction of host resistance. These biocontrol mechanisms indirectly promote the plant growth by reducing severity of diseases, hence, also known as indirect mechanisms of plant growth promotion.

### 1.3.3 Crop Diseases

Crop diseases are a major threat because they drastically reduce agricultural production and quality. Therefore, early detection and classification of diseases are important challenges for agricultural production (Bhadur and Rani 2020). The damage caused by agricultural pathogens is devastating. Losses in agriculture due to diseases worldwide amount to more than 220 billion dollars annually (FAO 2020). Plant diseases can be caused by different bacteria, fungi, and other agents. Diseased plant symptoms can include leaf spots, leaf blights, root rot, fruit rot, fruit spots, wilting, dieback, and decline. The significant impact of plant diseases is reducing the food available for humans by ultimately decreasing crop yields. This can result in unsuitable food for humans or lead to hunger in some areas (Loey et al. 2020). Therefore, recognizing plant diseases is of utmost importance in recommending and choosing the treatment for diseased plants and preventing infections of uninfected plants. The plant leaf is the most common way of detecting plant diseases as it shows different symptoms for different diseases.

Pigeon pea suffers high mortality because of serious seed and soil-borne fungal pathogens such as *F. udum* and *F. oxysporum*. Economically important crops such as pigeon pea need protection from various diseases as the demand for such crops is expected to steadily increase, particularly in the developing regions of the world. The fungi *Fusarium udum* are a soil-borne plant pathogen, and, therefore, chemical control is impractical in established cases. Moreover, extensive and inordinate use of synthetic and semi-synthetic compounds to improve pigeon pea productivity and disease control is a growing concern. In light of the above facts, it can be observed the application of PGPA can contribute to sustainable high yield and plant protection (Dave and Ingle 2021).

## 1.4 Production of Enzymes

*Streptomycetes* synthesize several lytic enzymes during their metabolic processes. They contribute substantially to carbon recycling by degrading different biopolymers that come from dead plants and animal material. These organic compounds, which include xylan, chitin, and cellulose, are degraded by exoenzymes (Shaikh et al. 2016; Jadhav et al. 2017). The products are channeled into the cell by specific carbohydrate importers that generally recognize mono- and disaccharides. Bentley et al. (2002), after sequencing all the genomes of the model organism *Streptomyces coelicolor* A3, reported a considerable number of 172 genes encoding secreted proteins such as hydrolases, chitinases, cellulases, lipases, nucleases, and proteases, as well as the appearance of 81 ATP-binding cassettes (ABC) permeases that are possibly used for the uptake of sugars, oligopeptides, nucleosides, and for drug export. Exoenzymes and ABC systems are 5- to 10-times higher than those of other bacteria, underscoring the general metabolic capacity of streptomycetes (Vurukonda et al. 2018).

Different types of proteins have been detected with the predicted signal peptides Sec or Tat and the antibiotic undecylprodigiosin and other proteins lacking signal peptides in vesicles within droplets found in *Streptomyces*. This has harmful effects on fungi and is a new way of providing proteins with potentially different applications (Gullón and Mellado 2018).

## 1.5 Volatile Organic Compounds (VOC)

The biological roles of VOCs are only now beginning to be analyzed, and initial studies show that they can have wide-ranging effects on both their producing organisms and their neighbors. VOCs can alter the antibiotic resistance profiles of bacteria, act as antifungal or antibiotic compounds, promote group behaviors such as motility and biofilm formation, and induce widespread changes in the gene expression of nearby microbes (Jones et al. 2018). VOCs can act as signaling molecules that influence gene expression, metabolism, and group behavior of responding microorganisms. *Bacillus* species use VOCs for biofilms production in response to volatile acetic acid (Chen et al. 2015). *Streptomyces* VOC production has been found to allow environmental conditions such as pH to be modulated. These compounds can act as signaling molecules, affecting gene expression and response group behavior within microbial communities. Most studies have explored the ecological roles of VOCs in communication and competition between unrelated species.

The antimicrobials production by *Streptomyces* strains can be significantly increased depending on different nutritional and cultivation conditions. A deficit of nutrients can influence the development of hyphae and, therefore, produce these antimicrobial compounds. These environmental signals are interpreted by cells as stress, so these compounds are considered stress metabolites due to their role in adaptability. The production of these metabolites involves some genes encoding phosphorylated guanosine nucleotide phosphate (ppGpp) (an important protein for metabolite synthesis) and regulatory proteins (Sivapragasam et al. 2017). Antibiotic production is triggered by alarmone ppGpp due to nutritional stress. In addition, some species-specific enzymes, such as lysosomal acid lipase and *Streptomyces* antibiotic regulatory proteins, are important in regulating metabolic pathways. The type of signals received and sent by *Streptomyces* determines the metabolites that will be produced. For example, the attack of pathogens activates a signal, the deficit of nutrients sends a signal, etc. The α-butyrolactones are among the main signaling molecules in PSPG cell-cell communication (Olanrewaju and Babalola 2019).The basal medium can play an essential role in secondary metabolites production through fermentation. The main parameters that can influence are the nutrients (nitrogen, source of phosphorus, and carbon), growth rate, enzymatic inactivation, and some variable conditions such as oxygen supply, temperature, light, and pH. Furthermore, the production of valuable metabolites by actinomycetes differs qualitatively and quantitatively according to the used strains (Oskay 2011).

## 1.6 Phytohormone Production

Phytohormones are also called as plant growth regulators (PGRs) and are important for plant growth and development. Various PGPR are known to produce phytohormone, namely, auxins, cytokinin, and gibberellins (Saharan and Nehra 2011). When plants encounter growth-limiting conditions, they often attempt to adjust the levels of their endogenous phytohormones in order to decrease the negative effects. Phytohormones are organic in nature which promote the growth of plants, even at very low concentrations, and help in tissue development. Plant growth promotion by phytohormone-producing PGPR has been reported by many workers (Joo et al. 2005; Sayyed et al. 2007; Spaepen et al. 2007; Akhtar and Siddiqui 2009; Kang et al. 2009; Ahemad and Khan 2012; Nithyapriya et al. 2021). Indoleacetic acid (IAA) is a most studied phytohormone produced by 80% rhizospheric bacteria (Patten and Glick 1996; Akhtar and Siddiqui 2009).

## 1.7 Enhancement of Plant Growth

*Bacillus* sp. is one of the most studied PGPRs. Members of this species produce a wide range of plant-growth-promoting (PGP) traits such as antibiotics (Ahmed et al. 2021; Enshasy et al. 2020; Zakaria et al. 2019; Sagar et al. 2022), hydrolytic enzymes (Jadhav et al. 2017), phytohormones, antioxidant enzymes (Khan et al. 2020; Kusale et al. 2021), siderophores, etc. (Patel et al. 2018; Sayyed et al. 2007). Plant growth promotion is the most important eco-friendly and sustainable aspect of PGPR. A large number of bacterial species including *Alcaligenes*, *Azospirillum*, *Arthrobacter*, *Acinetobacter*, *Bradyrhizobium*, *Bacillus*, *Burkholderia*, *Enterobacter*, *Erwinia*, *Flavobacterium*, *Pseudomonas*, and Rhizobium associated with the rhizosphere of plant and are able to exert many beneficial effects on plant growth (Saharan and Nehra 2011). There are several mechanisms by which PGPR can increase the plant growth; they are also known as direct mechanisms of plant growth promotion, and these include nitrogen fixation, production of phytohormones, lowering of ethylene concentration by producing ACC deaminase (Sagar et al. 2020), and solubilization of phosphorus (Sharma et al. 2016) and various other minerals. Co-inoculation with multifarious *Bradyrhizobium* sp. and *Pseudomonas* sp. improves plant growth, plant, and soil nutrients and enzymes through the production of siderophores, phytohormones enzymes, exopolysaccharide, stress tolerance, and phosphate solubilization (Nasab and Sayyed, 2019; Nasab et al. 2021 Basu et al. 2021; Jabborova et al. 2020a, b, c; Sayyed et al. 2007).

PGPR may have a direct or roundabout effect on the growth of plants in two natural ways. Aberrant development happens when PGPR diminishes or forestalls the destructive impacts of at least one harming microorganism. This is basically investigated through biocontrol or the enmity of soil plant microbes. Especially, the impacts of microorganism intrusion and foundation can be firmly forestalled by colonization or the biosynthesis of anti-microbials and other auxiliary metabolites. Direct development advances plant development by PGPR when the plant is provided with a bacterial integrated compound, or when PGPR in any case encourages plant take-up of soil supplements. The separate was initially chosen for the organic control of *Rhizoctonia solani*. It has been accounted for that *Streptomyces pulcher*, *Streptomyces canescens*, and *Streptomyces citreofluores* were utilized in the control of bacterial, *Fusarium*, and *Verticillium* withers, early scourge, and bacterial ulcer of tomato. Like most rhizobacteria, it appears to be exceptionally plausible that *Streptomyces* are prepared to do straightforwardly improving plant development. *Streptomyces* are promising microbial biocontrol living beings that can estrange and additionally slaughter contagious and bacterial plant microorganisms. Their biocontrol action is regularly performed before the microorganisms totally taint their particular host, for example, by treating seeds with endophytic *Streptomyces* spp. What's more, *Micromonospora* spp. preceding planting, this perception infers that the microbial rivals entered the seeds during their germination and colonized the seedlings. Bacon and Hinton detailed that shifting degrees of sickness concealment in the field were emphatically corresponded with comparative outcomes got from in vitro explores. In different examinations, a critical microbe restraint in vitro was not generally connected with disease assurance *in planta*. There is a need to develop and promote environment-friendly strategies to cope with drought stress. Inoculation of plant growth-promoting rhizobacteria (PGPR) in the rhizosphere of crops is one of the most suitable strategies in agriculture (Sayyed et al. 2007). Application of plant growth-promoting rhizobacteria (PGPR), mycorrhizae, and seaweed, along with supplementary irrigation, have been suggested as sustainable strategies to combat the drought stress and low fertility of soil (Sayyed et al. 2007).

### *1.7.1 Siderophore Production*

Microbial siderophores are well known for their ecological significance. Bacteria belonging to the genera *Azotobacter*, *Bacillus*, *Pseudomonas*, *Rhizobium*, *Serratia*, etc. produce siderophores (Patel et al. 2018; Kusale et al. 2021; Kloepper et al. 1980; Neilands 1995; Arora et al. 2001; West and Buckling 2002; Rajkumar et al. 2010; Saha et al. 2012). Siderophore production in iron-stress conditions confers upon these organisms an added advantage, resulting in exclusion of pathogens due to iron starvation. Siderophores contribute to disease suppression by conferring a competitive advantage to biocontrol agents for the limited supply of essential trace minerals in natural habitats. Siderophores thus are involved both in plant growth promotion and health protection (Nithyapriya et al. 2021). Siderophore- producing strains of *Penicillium chrysogenum* and *P. aeruginosa* assist in nodulation, nitrogen fixation, plant growth promotion, biological control, release of organic acids, and in managing abiotic stress tolerance (Mahmod and Allah 2001). Siderophores are thus one of the most diverse biomolecules secreted by PGPRs which perform a multitude of functions falling both under direct and indirect roles. A more collective approach is required to determine their complete role in the rhizosphere and for development of future inoculants.

### 1.7.2 Phosphate Solubilization

Phosphorus is the second important key element after nitrogen as a mineral nutrient in terms of quantitative plant requirement (Yadav et al. 2012; Sharma et al. 2013). It plays a vital role in most of the major metabolic reactions, viz., photosynthesis, signal transduction, energy transfer, biosynthesis of macromolecules, and respiration (Khan et al. 2010; Yadav et al. 2012). Despite this fact, the most of this phosphorus is insoluble which present in organic and inorganic form and therefore not available for plant growth (Glick 2012). This problem can be solved by phosphate solubilizing bacteria (PSB) which solubilize the insoluble phosphate into a soluble form and make available to the plant for its growth and development, and hence PSB are widely used in biofertilizer preparation. PSB plays an important role in the phosphate nutrition of plants in a more eco-friendly manner. The naturally abundant PSB solubilize insoluble phosphate and convert it into soluble forms for the crop plants.

## 1.8 Commercialization, Environmental Effects, and Biosafety of *Streptomyces* Products

*Streptomyces* producing antimicrobial secondary metabolites present an attractive alternative to chemical fertilizers, pesticides, and supplements, which may result in a significant increase in agricultural plant growth and pest and disease control (Sousa et al. 2008). While increasing our knowledge of the mechanisms triggered by actinomycetes for suppressing plant diseases, improving nutrient uptake by plants, and stimulating and/or increasing the production of phytohormones in planta, a great deal of research is being carried out worldwide for the development of correct formulations containing actinomycete inoculants as their active ingredients. Nevertheless, very few actinomycete-based products are currently commercialized. Although biocontrol with PGPR is an acceptable green approach, the proportion of registration of *Streptomyces* spp. as biocontrol agents for commercial availability is very low. Mycostop (Verdera Oy, Finland) is the only *Streptomyces*-based plant protection product registered in the EU; it is also registered in Canada and the USA. Actofit and Astur, based on *Streptomyces avermitilis*, are registered as insecticides in the Ukraine. Any formulation with an increased shelf life and a broad spectrum of actions, such as plant growth promotion and/or disease suppression under field conditions, could open the way for technological exploitation and marketing. Many reports suggest that commercial biocontrol agents are easy to deliver, induce plant growth and stress resistance, and, eventually, increase plant biomass and yield. More recently, Sharma et al. (2013) reported that avermectin B1b, a component of commercially available abamectin, was obtained as a fermentation product of *Streptomyces avermitilis*, which has frequently been used as an insecticidal agent. Very few *Streptomyces*-based commercial formulations are available in the worldwide market, compared with products based on *Streptomyces* metabolites; the former are mainly indicated for pest and disease control.

### 1.8.1 Formulation and Inoculation Methods

For any agricultural and pharmaceutical industry, "formulation" is an important challenge to the success or failure of commercial products made from experimentally effective biological pest control agents. The formulation should contain a calibrated amount of the new microorganism as the active ingredient, a collection of other inert ingredients (often not defined in detail by the manufacturer), and produce commercially available ingredients. Over-the-counter formulations must comply with local and national pesticide laws (especially the law on microbial inoculum, if applicable) and grower requirements. The standard storage conditions for growers are 5–30 °C (Bhardwaj et al. 2014). Four forms can be processed and made available on the market for industrial microbial formulations: powder, liquid suspension, granules, and slurry (Bashan et al. 2014). In addition, in both in vitro and in vivo studies, biocontrol agents do not produce the same results; this is considered to be a major problem in the industrial production of bio-inoculants. Soil organic matter, pH, nutrient levels, and moisture levels influence the effectiveness of biocontrol agents, including *Streptomyces*. Variations in agri-environmental conditions appear to affect the results of biocontrol agents that perform well in vitro: in greenhouse or field experiments, an experimentally excellent biocontrol agent can fail. Therefore, when choosing an effective biocontrol agent for a particular area, environmental variables should always be taken into careful consideration.

Every microbiological method can help achieve effective outcomes in commercial agriculture (open fields, nurseries, glasshouse development, and so on) (Suprapta 2012; Law et al. 2017). As discussed in earlier chapters, most *Streptomyces* are soil dwellers, invading the rhizosphere and frequently demonstrating the ability to enter plants, effectively colonizing plant tissues as endophytes. Pathogens require an endophytic state to commence infection, hence an effective endophytic biocontrol agent, such as a selected *Streptomyces* spp., should be able to quickly migrate from the rhizosphere into the roots and/or other plant parts. As a result, its antagonistic activity can be both preventive (competitive exclusion of plant pathogens in the rhizosphere and planta) and curative (competitive exclusion of plant pathogens in the rhizosphere and planta) in nature (killing plant pathogens post-infection). The collection and selection of prospective *Streptomyces* spp. from the industrial point of view. The aim should necessarily be regarded as a candidate for the production and implementation of creative and sustainable biopesticides.

Industrial manipulation of research findings is not feasible by definition. The majority of articles published around the world show *Streptomyces*' superior biocontrol activity in in vitro or in planta experiments conducted under stringent standardized circumstances. There are no field study findings that support the applicability of experimental data in commercial greenhouses/fields on a regular basis. This is clearly a flaw that stymies the commercialization of active biopesticides. Actinobacteria, particularly *Streptomyces*, are further hampered in their development since they are completely reliant on the substrate on which they grow (Bibb 2013) and on the natural microbial population around them to produce antimicrobial compounds.

## 1.9 Conclusion

The utilization of rhizo-microflora as bio-fertilizers and biocontrol agents has become a significant part of modern agricultural techniques, with huge potential to dominate agricultural markets in the coming decade. As filamentous spore-forming bacteria with exceptional antipathogen and nutrient cycling activity, actinomycetes are one of the most promising PGPRs. They can improve overall soil health and increase agricultural output. However, some unsolved issues must be addressed in order to transfer discoveries from controlled laboratory contexts to large-scale field trials and commercial commercialization. While actinomycetes can sporulate and have a high potential for survival, new formulation methods are needed to extend shelf life and improve transportation. More comprehensive studies of the complicated actinomycete-rhizosphere milieu, as well as PGP action mechanisms, are also needed. It may be able to utilize molecular tools and genetic engineering to promote colonization, add more PGP features to the already beneficial actinomycete strain, and increase crop compatibility. In addition, the symbiotic relationship of actinomycetes with other PGPR should be investigated in order to develop a highly effective and efficient bio-inoculant system that can be used in a variety of soil types and environmental situations.

**Acknowledgments** The authors are grateful to Department of Microbiology and Biotechnology Centre, The Maharaja Sayajirao University of Baroda, Vadodara, Gujarat, India, for their support.

## References

AbuQamar S, Moustafa K, Tran LSP (2017) Mechanisms and strategies of plant defense against Botrytis cinerea. Crit Rev Biotechnol 37:262–274. https://doi.org/10.1080/07388551.2016.1271767

Ahemad M, Khan MS (2012) Evaluation of plant-growth-promoting activities of rhizobacterium pseudomonas putida under herbicide stress. Ann Microbiol 62:1531–1540. https://doi.org/10.1007/s13213-011-0407-2

Ahmed S, ChoudhuryA R, Kumar S, Choi RJ, SayyedR Z, Sa TM (2021) Biomolecular painstaking utilization and assimilation of phosphorus under indigent stage in agricultural crops. In: Singh HB, Vaishnav A, Sayyed R (eds) Antioxidants in plant-microbe interaction. Springer, Singapore, pp 565–588.

Akhtar MS, Siddiqui ZA (2009) Use of plant growth promoting rhizobacteria for the biocontrol of root-rot disease complex of chickpea. Aust Plant Pathol 38:44–50 and its effect on plant growth. Biotechnol Lett 31:277–281

Aoki T, O'Donnell K, Geiser DM (2014) Systematics of key phytopathogenic fusarium species: current status and future challenges. J General Plant Pathol 80:189–201. https://doi.org/10.1007/s10327-014-0509-3

Arora NK, Kang SC, Maheshwari DK (2001) Isolation of siderophore producing strain of rhizobium meliloti and their biocontrol potential against Macrophomina phaseolina that causes charcoal rot of groundnut. Curr Sci 25:674–677

Bardin M, Ajouz S, Comby M, Lopez-Ferber M, Graillot B, Siegwart M et al (2015) Is the efficacy of biological control against plant diseases likely to be more durable than that of chemical pesticides? Front Plant Sci 6:566. https://doi.org/10.3389/fpls.2015.00566

Bashan Y, de-Bashan LE, Prabhu SR, Juan-Pablo H (2014) Advances in plant growth promoting bacterial inoculant technology: formulations and practical perspectives (1998–2013). Plant Soil 378:1–33

Basu A, Prasad P, Das SN, Kalam S, Sayyed RZ, Reddy MS, El Enshasy H (2021) Plant growth promoting Rhizobacteria (PGPR)as green bioinoculants: recent developments, constraints, and prospects. Sustainability 13:1140.https://doi.org/10.3390/su13031140

Bentley SD, Chater KF, Cerdeno-Tarraga AM, Challis GL, Thomson NR, James KD et al (2002) Complete genome sequence of the model actinomycete Streptomyces coelicolor A3(2). Nature 417:141–147. https://doi.org/10.1038/417141a. Berlin, pp 295–343

Bhadur G, Rani R (2020) Agricultural crops disease identification and classification through leaf images using machine learning and deep learning technique: a review. SSRN Electron J 2020:1–9. https://doi.org/10.2139/ssrn.3564973

Bhardwaj D, Ansari MW, Sahoo RK, Tuteja N (2014) Biofertilizers function as key player in sustainable agriculture by improving soil fertility, plant tolerance and crop productivity. Microb Cell Factories 13:66

Bibb MJ (2013) Understanding and manipulating antibiotic production in actinomycetes. Biochem Soc Trans 41:1355–1364

Bolívar-Anillo HJ, Garrido C, Collado IG (2020) Endophytic microorganisms for biocontrol of the phytopathogenic fungus Botrytis cinerea. Phytochem Rev 19:721–740. https://doi.org/10.1007/s11101-019-09603-5

Boukaew S, Prasertsan P (2014) Suppression of rice sheath blight disease using a heat stable culture filtrate from Streptomyces philanthi RM-1-138. Crop Protect 61:1–10. https://doi.org/10.1016/j.cropro.2014.02.012

Carbu M, Gonzalez-Rodriguez V, Garrido C (2016) New biocontrol strategies for strawberry fungal pathogens. In: Husaini A, Neri D (eds) Strawberry: growth, development and diseases. CABI, Boston

Chen Y, Gozz K, Yan F, Chai Y (2015) Acetic acid acts as a volatile signal to stimulate bacterial biofilm formation. MBio 6:e00392–e00315. https://doi.org/10.1128/mBio.00392-15

Cook RJ (2010) Fusarium root, crown and foot roots and associated seedling diseases. In: Bockus WW, Bowden RL, Hunger RM, Morrill WL, Murray TD, Smiley RW (eds) Compendium of wheat diseases and pests, 3rd edn. APS Press, St. Paul, pp 37–39

da Silva LL, Moreno HLA, Correia HLN, Santana MF, de Queiroz MV (2020) Colletotrichum: species complexes, lifestyle, and peculiarities of some sources of genetic variability. Appl Microbiol Biotechnol 104:1891–1904. https://doi.org/10.1007/s00253-020-10363-y

Dave A, Ingle S (2021) *Streptomyces* sp. S-9 promotes plant growth and confers resistance in pigeon pea (*Cajanus cajan*) against *Fusarium* wilt. 3Biotech 11:459. https://doi.org/10.1007/s13205-021-02989-0

Dean R, van Kan J, Pretorius ZA (2012) The top 10 fungal pathogens in molecular plant pathology. Mol Plant Pathol 13:1–17. https://doi.org/10.1111/j.1364-3703.2012.2011.00783.x

Dias MP, Bastos MS, Xavier VB, Cassel E, Astarita LV, Santarém ER (2017) Plant growth and resistance promoted by Streptomyces spp. in tomato. Plant Physiol Biochem 118:479–493. https://doi.org/10.1016/j.plaphy.2017.07.017

Doolotkeldieva T, Bobusheva S, Suleymankisi A (2016) Biological control of Erwinia carotovora ssp. by Streptomyces species. Adv Microbiol 06:104–114. https://doi.org/10.4236/aim.2016.62011
Enshasy HE, Ambehati KK, Ashraf F, Ramchuran S, Sayyed RZ, Amalin D, Dailin DJ, Hanapi SZ (2020). Trichoderma:biocontrol agents for promoting plant growth and soil health. In: Agriculturally important fungi for sustainable agriculture: vol 2:Functional annotation for crop protection, A N Yadav, S Mishra, D Kour, N Yadav, A Kumar (Eds), Switzerland, pp 239–259
Faheem M, Raza W, Zhong W, Nan Z, Shen Q, Xu Y (2015) Evaluation of the biocontrol potential of Streptomyces goshikiensis YCXU against fusarium oxysporum f. sp. niveum. Biol Control 81:101–110. https://doi.org/10.1016/j.biocontrol.2014.11.012
FAO (2020) Crop prospects and food situation—quarterly global report. Crop Prospects and Food. Available online at: http://www.fao.org. Accessed 26 March 2021
FAO (2021) Integrated Pest and pesticide management. Available online at: http://www.fao.org/pest-and-pesticide-management/ipm/principles-andpractices/
Glick BR (2012) Plant growth promoting bacteria: mechanisms and applications. Scientifica 2012:1–15
González I, Yailén A, Peteira B (2012) Aspectos generales de la interacción Fusarium oxysporum f. sp. lycopersici-tomate. Rev Protec Veg 27:1–7
Gullón S, Mellado RP (2018) The cellular mechanisms that ensure an efficient secretion in Streptomyces. Antibiotics 7:33. https://doi.org/10.3390/antibiotics7020033v
Hippler FWR, Boaretto RM, Dovis VL, Quaggio JA, Azevedo RA, Mattos-Jr D (2018) Oxidative stress induced by cu nutritional disorders in citrus depends on nitrogen and calcium availability. Sci Rep 8:1641. https://doi.org/10.1038/s41598-018-19735-x
Jabborova D, Kannepalli A, Fayzullaeva M, Sulaymonov K, Kadirova D, Jabbarova Z, Sayyed RZ (2020a) Isolation andcharacterization of endophytic bacteria from ginger (Zingiber officinale Rosc.). Ann Phytomedicine 9(1):116–121
Jabborova D, Wirth S, Kannepalli A, Narimanov A, Desouky S, Davranov K, Sayyed RZ, Enshasy HE, Malek RA, Syed A,Bahkali AH (2020b) Co-inoculation of rhizobacteria and biochar application improves growth and nutrient in soybean andenriches soil nutrients and enzymes. Agronomy 10:1142. https://doi.org/10.3390/agronomy10081142
Jabborova D, Wirth S, Kannepalli A, Narimanov A, Desouky S, Davranov K, Sayyed RZ, Enshasy H, AbdMalek R, Syed A,Bahkali AH (2020c) Co-Inoculation of Rhizobacteria and biochar application improves growth and nutrients in soybean andenriches soil nutrients and enzymes. Agronomy 10(8):1142
Jadhav HP, Shaikh SS, Sayyed RZ (2017) Role of hydrolytic enzymes of Rhizoflora in biocontrol of fungal Phytopathogens: anoverview in Rhizotrophs: plant growth promotion to bioremediation. Springer-Nature Springer, Singapore, pp 183–203
Jones S, Pham C, McKillip J, Zambri M, Carlson E, Elliot M (2018) Streptomyces volatile compounds influence exploration and microbial community dynamics by altering iron availability. BioRxiv 396606. https://doi.org/10.1101/396606
Joo GJ, Kim YM, Kim JT, Rhee IK, Kim JH, Lee IJ (2005) Gibberellins-producing rhizobacteria increase endogenous gibberellins content and promote growth of red peppers. J Microbiol 43(6):510–515
Joshi R (2018) A review on Colletotrichum spp. virulence mechanism against host plant defensive factors. J Med Plants Stud 6:64–67. https://doi.org/10.22271/plants.2018.v6.i6b.02
Junaid JM, Dar NA, Bhat TA, Bhat AH, Bhat MA (2013) Commercial biocontrol agents and their mechanism of action in the management of plant pathogens. Int J Mod Plant Ani Sci 1:39–57
Kang SM, Joo GJ, Hamayun M, Na CI, Shin DH, Kim HY, Hong JK, Lee IJ (2009) Burkholderia sp. KCTC 11096BP as a newly isolated gibberellin producing bacterium. J Microbiol 47:167–171
Kannan V, Bastas K, Antony R (2015) Plant pathogenic bacteria: an overview. In: Sustainable approaches to controlling plant pathogenic bacteria, 1st edn, pp 1–16. https://doi.org/10.1201/b18892-2

Khan MS, Zaidi A, Ahemad M, Oves M, Wani PA (2010) Plant growth promotion by phosphate solubilizing fungi – current perspective. Arch Agron Soil Sci 56(1):73–98. https://doi.org/10.1080/03650340902806469

Khan I, Awan SA, Ikram R, Rizwan M, Akhtar N, Yasmin H, Sayyed RZ, Ali S, Ilyas N (2020) 24-Epibrassinolide regulated antioxidants and osmolyte defense and endogenous hormones in two wheat varieties under drought stress. Physiologia Plantarum 1–11. https://doi.org/10.1111/ppl.13237

Kloepper JW, Leong J, Teintze M, Schroth MN (1980) Enhanced plant growth by Siderophores produced by plant growth promoting rhizobacteria. Nature 286:885–886

Kusale SP, Attar YC, Sayyed RZ, Malek RA, Ilyas N, Suriani NL, Khan N, Enshasy HE (2021) Production of plant beneficial and antioxidants metabolites by Klebsiella variicola under salinity Stress. Molecules 26:1894. https://doi.org/10.3390/molecules26071894

Law JWF, Ser HL, Khan TM, Chuah LH, Pusparajah P, Chan KG et al (2017) The potential of Streptomyces as biocontrol agents against the rice blast fungus, Magnaporthe oryzae (Pyricularia oryzae). Front Microbiol 8:3. https://doi.org/10.3389/fmicb.2017.00003

Loey M, ElSawy A, Afify M (2020) Deep learning in plant diseases detection for agricultural crops: a survey. Int J Serv Sci Manag Eng Technol 11:41–58. https://doi.org/10.4018/IJSSMET.2020040103

Mahmod ALE, Allah MH (2001) Siderophore production by some microorganisms and their effect on Bradyrhizobium mung bean symbiosis. Int J Agric Microbiol 3(2):158–162

Marin-Felix Y, Hernández-Restrepo M, Iturrieta-González I, García D, Gené J, Groenewald JZ et al (2019) Genera of phytopathogenic fungi: GOPHY 3. Stud Mycol 94:1–124. https://doi.org/10.1016/j.simyco.2019.05.001

Martins PMM, Merfa MV, Takita MA, De Souza AA (2018) Persistence in phytopathogenic bacteria: do we know enough? Front Microbiol 9:1099. https://doi.org/10.3389/fmicb.2018.01099

Misra S, Dixit VK, Khan MH, Kumar MS, Dviwedi G, Yadav S et al (2017) Exploitation of agro-climatic environment for selection of 1-aminocyclopropane-1-carboxylic acid (ACC) deaminase producing salt tolerant indigenous plant growth promoting rhizobacteria. Microbiol Res 205:25–34. https://doi.org/10.1016/j.micres.2017.08.007

Nah H-J, Park J, Choi S, Kim E-S (2021) WblA, a global regulator of antibiotic biosynthesis in Streptomyces. J Ind Microbiol Biotechnol 48:kuab007. https://doi.org/10.1093/jimb/kuab007

Nasab BF, Sayyed RZ (2019) Plant growth promoting rhizobacteria and salinity stress: a journey into the soil. In: Sayyed RZ,Arora NK, Reddy MS (eds) Plant growth promoting rhizobacteria for sustainable stress management vol 1 abiotic stressmanagement. Springer, Singapore, pp 21–34

Nasab BF, Sayyed RZ, Ahmad PR, Rahmani F (2021) Biopriming and Nanopriming: green revolution wings to increase plant yield, growth, and development under stress condition and forward dimensions. In: Singh HB, Vaishnav A, Sayyed R (eds) Antioxidants in plant-microbe interaction. Springer, Singapore, pp 623–655

Neilands JB (1995) Siderophores: structure and function of microbial iron transport compounds. J Biol Chem 270:26723–26726

Nithyapriya S, Lalitha S, Sayyed RZ, Reddy MS, Dailin DJ, Enshasy HE, Suriani NL, Herlambang S (2021) Production,purification, and characterization of bacillibactin siderophore of Bacillus subtilis and its application for improvement in plantgrowth and oil content in sesame. Sustainability 13:5394. https://doi.org/10.3390/su13105394

Olanrewaju OS, Babalola OO (2019) Streptomyces: implications and interactions in plant growth promotion. Appl Microbiol Biotechnol 103:1179–1188. https://doi.org/10.1007/s00253-018-09577-y

Oskay M (2011) Effects of some environmental conditions on biomass and antimicrobial metabolite production by Streptomyces sp., KGG32. Int J Agric Biol 13:317–324

Patel PR, Shaikh SS, Sayyed RZ (2018) Modified chrome azurol S method for detection and estimation of siderophores havingaffinity for metal ions other than iron. Environ Sustainab 1(1):81–87

Patten CL, Glick BR (1996) Bacterial biosynthesis of indole-3-acetic acid. Can J Microbiol 42:207–220

Petriccione M, Zampella L, Mastrobuoni F, Scortichini M (2017) Occurrence of copper-resistant pseudomonas syringae pv. Syringae strains isolated from rain and kiwifruit orchards also infected by P. s. Pv. Actinidiae. Eur J Plant Pathol 149:953–968. https://doi.org/10.1007/s10658-017-1246-1

Rajkumar M, Noriharu A, Prasad MNV, Helena F (2010) Potential of siderophore-producing bacteria for improving heavy metal phytoextraction. Trend Biotechnol 28(3):142–149

Reshma P, Naik MK, Aiyaz M, Niranjana SR, Chennappa G, Shaikh SS, Sayyed RZ (2018) Induced systemic resistance by 2,4 diacetylphloroglucinol positive fluorescent Pseudomonas strains against rice sheath blight. Indian J Exp Biology 56(3):207–212

Rodríguez M, Torres M, Blanco L, Béjar V, Sampedro I, Llamas I (2020) Plant growth-promoting activity and quorum quenching-mediated biocontrol of bacterial phytopathogens by Pseudomonas segetis strain P6. Sci Rep 10:1–12. https://doi.org/10.1038/s41598-020-61084-1

Sagar A, Sayyed RZ, Ramteke PW, Sharma S, Marraiki N, Elgorban AM, Syed S (2020) ACC deaminase and antioxidant enzymes producing halophilic Enterobacter sp. PR14 promotes the growth of rice and millets under salinity stress. Physiol Mol Biol Plants 26:1847–1854

Sagar A, Rai S, Ilyas N, Sayyed RZ, Al-Turki AI, Enshasy HE, Simarmata T (2022) Halotolerant rhizobacteria for salinity stress mitigation: diversity, mechanism and molecular approaches, sustainability, 14:490. https://doi.org/10.3390/su14010490

Saha R, Saha N, Donofrio RS, Besterbelt LL (2012) Microbial siderophores: a mini review. J Basic Microbiol 52:1–15

Saharan BS, Nehra V (2011) Plant growth promoting rhizobacteria: a critical review. Life Sci Med Res 21:1–30

Sawicka B, Egbuna C (2019) Pests of agricultural crops and control measures. In: Natural remedies for pest, disease and weed control, pp 1–16. https://doi.org/10.1016/B978-0-12-819304-4.00001-4

Sayyed RZ, Patel DC, Patel PR (2007) Plant growth promoting potential of P solubilizing *pseudomonas* sp. occurring in acidic soil of Jalgaon. Asian J Microbiol Biotechnol Environ Sci 4:925–928

Shaikh SS, Reddy MS, Sayyed RZ (2016) Plant growth promoting rhizobacteria: an eco-friendly approach for sustainable agroecosystemplant soil-microbes. Springer, Switzerland, pp 182–201

Sharma SB, Sayyed RZ, Trivedi MH, Gobi TA (2013) Phosphate solubilizing microbes: sustainable approach for managing phosphorus deficiency in agricultural soils. Springer Plus 2:587

Sharma S, Sayyed R, Sonawane M, Trivedi M, Thivakaran G (2016) Neurospora sp SR8, a novel phosphate solubiliser from rhizosphere of soil of Sorghum in Kachh, Gujarat, Indian J Exp Biology 54:644–649

Singh V, Shrivastava A, Jadon S (2015) Alternaria diseases of vegetable crops and its management control to reduce the low production. Int J Agric Sci 7:834–840

Sivapragasam S, Deochand DK, Meariman JK, Grove A (2017) The stringent response induced by phosphate limitation promotes purine salvage in agrobacterium fabrum. Biochemist 56:5831–5843. https://doi.org/10.1021/acs.biochem.7b00844

Song L, Jiang N, Wei S, Lan Z, Pan L (2020) Isolation, screening, and identification of actinomycetes with antifungal and enzyme activity assays against Colletotrichum dematium of Sarcandra glabra. Mycobiology 48:37–43. https://doi.org/10.1080/12298093.2020.1716604

Sousa SC, Soares FAC, Garrido SM (2008) Characterization of Streptomycetes with potential to promote plant growth and biocontrol. Sci Agric 65(1):50–55

Spaepen S, Vanderleyden J, Remans R (2007) Indole-3-acetic acid in microbial and microorganism plant signaling. FEMS Microbiol Rev 31:425–448

Sun J, Pei Y, Li E, Li W, Hyde KD, Yin WB et al (2016) A new species of Trichoderma hypoxy-lon harbours abundant secondary metabolites. Sci Rep 6:37369. https://doi.org/10.1038/srep37369

Suprapta DN (2012) Potential of microbial antagonists as biocontrol agents against plant fungal pathogens. J ISSAAS 18:1–8

Syed Ab Rahman S, Singh E, Pietersen C, Schenck P (2018) Emerging microbial biocontrol strategies for plant pathogens. Plant Sci 267:102–111. https://doi.org/10.1016/j.plantsci.2017.11.012

Valdés R, Castillo F, Cabello J (2017) Review of antibacterial activity of plant extracts and growth-promoting microorganism (Gpm) against phytopathogenic. Eur J 4:11–36

Vashistha A, Chaudhary A (2019) Effects of biocontrol agents and heavy metals in controlling soilborne phytopathogens. Int J Herb Med 7:41–45

Velivelli SL, De Vos P, Kromann P, Declerck S, Prestwich BD (2014) Biological control agents: from field to market, problems, and challenges. Trends Biotechnol 32:493–496. https://doi.org/10.1016/j.tibtech.2014.07.002

Vurukonda SSKP, Giovanardi D, Stefani E (2018) Plant growth promoting and biocontrol activity of Streptomyces spp. as endophytes. Int J Mol Sci 19:952. https://doi.org/10.3390/ijms19040952

Wang C, Wang Y, Ma J, Hou Q, Liu K, Ding Y et al (2018) Screening and whole-genome sequencing of two Streptomyces species from the rhizosphere soil of peony reveal their characteristics as plant growthpromoting rhizobacteria. Biomed Res Int 2018:2419686. https://doi.org/10.1155/2018/2419686

Wattana-Amorn P, Charoenwongsa W, Williams C, Crump MP, Apichaisataienchote B (2016) Antibacterial activity of cyclo(L-pro-L-Tyr) and cyclo(D-pro-L-Tyr) from Streptomyces sp. strain 22-4 against phytopathogenic bacteria. Nat Prod Res 30:1980–1983. https://doi.org/10.1080/14786419.2015.1095747

West SA, Buckling A (2002) Cooperation, virulence and siderophore production in bacterial parasites. Proc R Soc Lond 270:37–44

Wonglom P, Suwannarach N, Lumyong S, Ito S, Matsui K, Sunpapao A (2019) Streptomyces angustmyceticus NR8-2 as a potential microorganism for the biological control of leaf spots of Brassica rapa subsp. Pekinensis caused by Colletotrichum sp. and Curvularia lunata. Biol Control 138:104046. https://doi.org/10.1016/j.biocontrol.2019.104046

Yadav RS, Meena SC, Patel SI, Patel KI, Akhtar MS, Yadav BK, Panwar J (2012) Bioavailability of soil P for plant nutrition. In: Lichtfouse E (ed) Farming for food and water security, Sustainable agriculture reviews, vol 10. Springer, Dordrecht, pp 177–200

Zakaria AK, Sayyed RZ, Enshasy HE (2019) Biosynthesis of antibiotics by PGPR and their roles in biocontrol of plant diseases.In: Plant growth promoting Rhizobacteria for sustainable stress management vol II Rhizobacteria in biotic stress management.Springer, Singapore, pp 1–36

Zhao J, Xue QH, Niu GG, Xue L, Shen GH, Du JZ (2013) Extracellular enzyme production and fungal mycelia degradation of antagonistic Streptomyces induced by fungal mycelia preparation of cucurbit plant pathogens. Ann Microbiol 63:809–812. https://doi.org/10.1007/s13213-012-0507-7

# Chapter 2
# Antifungal Secondary Metabolites Against Blast Fungus *Magnaporthe oryzae*

**Moutoshi Chakraborty and Tofazzal Islam**

**Contents**

2.1 Introduction .... 23
2.2 Plant Infection Mechanism by *Magnaporthe oryzae* .... 25
2.3 Biological Control .... 26
2.4 Effects of Bioactive Natural Products Against Blast Fungus .... 27
2.5 Mechanisms of Important Bioactive Secondary Metabolites Against Cereal Blast Fungus .... 35
2.6 Conclusion and Future Perspectives .... 43
References .... 44

## 2.1 Introduction

Food security for the ever-increasing global population is being threatened on a continuous basis by numerous adverse biotic and abiotic factors that agricultural scientists need to address. Cereals including rice, wheat, maize, and barley are primary carbohydrate sources for the majority of the world's population. However, many abiotic and biotic stresses also affect the production of these cereals, including unfavorable soil and environmental factors, insect outbreaks, diseases, and weed invasion. Cereal yield loss due to pathogenic fungi and bacteria are one of the main food and nutritional security challenges in developing countries (Li et al. 2011). Among several fungal diseases, the most destructive in cereals is the blast caused by a heterothalic ascomycete *Magnaporthe oryzae* (Howard and Valent 1996; Islam et al. 2016a, b). Because of the economic importance of the disease it causes, this fungus is ranked top of the 10 most important pathogens of plants (Dean et al. 2012). Multiple strains of this fungal pathogen can infect at least 50 plant species of

M. Chakraborty · T. Islam (✉)
Institute of Biotechnology and Genetic Engineering, Bangabandhu Sheikh Mujibur Rahman Agricultural University, Gazipur, Bangladesh
e-mail: tofazzalislam@bsmrau.edu.bd



R. Z. Sayyed et al. (eds.), *Antifungal Metabolites of Rhizobacteria for Sustainable Agriculture*, Fungal Biology, https://doi.org/10.1007/978-3-031-04805-0_2

the Poaceae family including major food crops like rice, wheat, barley, oat, finger millet, *Setaria*, and ryegrass with crop losses up to 100% (Howard and Valent 1996; Islam et al. 2016a, b). *M. oryzae* consists of a number of morphologically similar but genetically distinct host-specific pathotypes infecting rice (*Oryza* pathotype-MoO), wheat (*Triticum* pathotype-MoT), ryegrass (*Lolium* pathotype-MoL), foxtail millet (*Setaria* pathotype-MoS), finger millet or ragi (*Eleusine* pathotype-MoE), oat (*Brachiaria* pathotype-MoB), as well as other graminaceous hosts (Valent et al. 2019). Because of climate change and intercontinental agricultural commodity trading, the fungus has been reported to spread widely with infected grains and adapted for colonization of new hosts. For example, *M. oryzae Triticum* (MoT) became a damaging pathogen of wheat through host jumping in Brazil in 1985, and then dispersed to neighboring Argentina, Bolivia, and Paraguay (Perelló et al. 2015). It was first spotted in an Asian country Bangladesh in February 2016 that destroyed more than 1500 hectares of wheat with crop losses of up to 100% (Islam et al. 2016a, b, 2019; Gupta et al. 2018). This deadly pathogen is now spreading to South Asia's neighboring wheat-growing regions, posing a significant threat to this heavily populated region's food and nutritional security (Islam et al. 2019, 2020). The MoT management has become very challenging owing to the sudden epidemic of wheat blast disease (Islam et al. 2020).

Usage of resistant cultivars and application of fungicides are considered to be effective control strategies for several phytopathogens. However, regulation of *M. oryzae* by commercial fungicides is always ineffective, and the possible source of resistance is also limited in the development of blast-resistant varieties (Dutta et al. 2018). On the other hand, the use of chemical fungicides could not be commercially feasible and is harmful to the environment (Chung et al. 2008; Todorova and Kozhuharova 2010). Non-judicial repetitive spraying can also contribute to the development of fungicide resistance leading to rapid mutation in the pathogen population (Hahn 2014). Isolated strains of wheat and other grasses have shown extensive distribution of cyt b, Pg, Pp, Pu, and Pygt mutants in Brazil arising from excessive application of triazole and strobilurin (QoI) fungicides. These strains displayed varying degrees of resistance to those commonly used fungicides (Castroagudín et al. 2015; Dorigan et al. 2019). While wheat blast fungus has recently invaded South East Asia, our current knowledge of its infection pathway and pathogen biology is limited. However, examples from Brazil and other neighboring countries suggest that novel eco-friendly methods are required for the sustainable management of blast diseases in cereal crops.

Biological control of phytopathogens using bioactive agents derived from microbes is an effective alternative and a sustainable plant protection method (Shoda 2000; Chan et al. 2003). Microorganisms including *Bacillus* spp., *Pseudomonas* spp., and other antagonistic bacteria, as well as Actinomycetes fungi that produce a range of antibiotics and enzyme inhibitors, have been used to control rice blast disease (Prabavathy et al. 2006; Karthikeyan and Gnanamanickam 2008). Furthermore, many plant secondary metabolites also exhibited inhibitory activity against *M. oryzae* (Chutrakul et al. 2009; Xiong et al. 2013) and possess high potential for future control of blast disease caused by new cereal pathotypes. However, few of these effective natural compounds were marketed for blast disease control mainly because of their inconsistent performance in field conditions. More importantly, the basic mechanisms of their mode of action against blast fungus are still little understood.

Many studies have been reported evaluating a number of potential natural products to combat blast diseases in rice, wheat, ragi, or millets (Chakraborty et al. 2021). Although numerous reviews have been published on the biological control of different plant pathogens on diverse crop products (Ganeshan and Manoj Kumar 2005; Nayak et al. 2017), no detailed review has been reported on the management of blast disease using bioactive compounds. For the first time, this book chapter discusses the infection mechanism of the wheat blast fungus and comprehensively reviews the biocontrol of cereal blast disease utilizing diverse antifungal, bioactive natural secondary metabolites. We also discuss and propose strategies to use current knowledge for the development of products for the control of biorational blast disease utilizing biotechnological methods for field application. Genetic modification of plants for producing or overexpressing bioactive natural products has been reviewed recently (Anderson et al. 2016), and will not be discussed in this chapter.

## 2.2 Plant Infection Mechanism by *Magnaporthe oryzae*

*M. oryzae* fungus can attack the host at any growth stage and organ including leaves, internodes, necks, spikes, and panicles (Howard and Valent 1996). Infection at the reproductive stage (panicle initiation and development) is highly damaging and can cause yield losses up to 100% in cereals such as rice and wheat (Islam et al. 2016a, b). *M. oryzae* begins the process of infection with the asexually formed, three-celled pyriform airborne conidium, which is adhered to the host surface by an adhesive. The germ tube is rapidly developed which ultimately results in an appressorium driven by cell cycle progression (Hamer et al. 1988). Emergence of an appressorium needs a hard-hydrophobic surface, which originates after 4–6 hours of germination from the germ tube by mitotic cell division of a migrant nucleus from the three-celled conidium (Yan and Talbot 2016). After mitosis, one nucleus returns to the conidium, and then the conidium with three nuclei is completely destroyed through a mechanism called autophagy. There remains another nucleus in the appressorium which gradually becomes mature. This nucleus undergoes a mitotic sequential division and inevitably develops the invasive fungal hyphae (Saunders et al. 2010). The matured appressorium rich in chitin and melanin creates turgor pressure by accumulating a high concentration of compatible solvents such as glycerol. This turgor pressure is converted into a mechanical force that helps appressorium/penetration peg to infiltrate the leaf epidermis or stalk cuticle. After penetration, the pathogen differentiates the main hypha into bulbous and branched infection hyphae (Kankanala et al. 2007). Infection hyphae have cap-like structures known as the biotrophic interfacial complex that are involved in releasing effectors into plant cells (Kankanala et al. 2007; Mosquera et al. 2009; Giraldo et al. 2013). The *M. oryzae* secreted effector proteins such as Avr-Pita, Avr-Pii, Avr-Piz-t, Pwl1, ACE1, Bas1–4, Slp1, and Mc69 inhibit the host immunity by binding chitin oligosaccharides in the apoplast and attacking multiple components of host immunity (Mentlak et al. 2012; Liu et al. 2013; Chen et al. 2014). In addition, at this stage, *M. oryzae* secretes antibiotic biosynthesis mono-oxygenase and mycotoxin tenuazonic acid, which allows the fungus to impair the plant defense signaling to promote successful biotrophic growth of the fungus (Kankanala et al. 2007; Patkar et al. 2015; Yun et al. 2015). Lesions of the disease generally become visible 72–96 hours after infection accompanied by profuse sporulation under favorable environmental conditions (El Refaei 1977). The fungal infection induces the accumulation of reactive oxygen species which eventually leads to death of host cells or necrotic lesions promoting further hyphal growth and nutrient absorption (Egan et al. 2007). Aerial conidiophores with sympodially arranged conidia are generally disseminated to new host plants by wind, dewdrop splash, infected seeds, and/or residues of infected crops (Urashima et al. 1999; Wilson and Talbot 2009).

Typical leaf symptoms occur as water-soaked areas of diamond shape that at the beginning are gray to green in color, turning tan into brown with a necrotic border at the mature stage, coalescing with other lesions in order to destroy the entire leaf. The symptoms are also found in panicles and necks of rice (Liberato et al. 2006), spikes of wheat (Islam et al. 2016a, b), finger millet or ragi (Gashaw et al. 2014) and in nodes as well as internodes of foxtail millet (Sharma et al. 2014) resulting in unfilled or partially filled grains with shrinkage, premature bleaching of spikelets, and entire heads (Igarashi 1990; Malaker et al. 2016). The white patchy symptoms are observed from a considerable distance from the field at the reproductive stage. High temperature (16–30 °C) and high relative humidity >80% are ideal environmental conditions for the development of blast disease (Cardoso et al. 2008; Gória et al. 2013).

## 2.3 Biological Control

Generally, biological control was defined as "the action of introduced or resident living species, or parts thereof, to maintain the population density of another species at a lower average than would occur in its absence" (Chernin and Chet 2002). The idea of using microbial metabolites as biocontrol agents dates back to the nineteenth century. It has now become an efficient and sustainable method for plant disease control to reduce the use of synthetic agrochemicals, their harmful environmental and human health effects and the growth of pathogens' resistance to chemical substances. Several natural compounds have been identified over the years for the management of cereal crop pathogens.

Different microorganisms generate nearly 23,000 secondary bioactive metabolites, and more than 10,000 of these compounds have been identified as antifungal compounds (Sayyed et al. 2015, 2016; Awla et al. 2016; Jadhav et al. 2020; Sagar et al. 2020; Baba et al. 2021; Kusale et al. 2021). For example, *Streptomyces* alone can produce almost 7600 compounds, including geldanamycin and nigericin, which inhibit plant pathogens' growth (Law et al. 2017). The mechanisms of action of various bioactive secondary metabolites include hydrolyzing pathogenic cell walls, disrupting the integrity of cell membranes, blocking chitin synthesis, inhibiting protein synthesis, disrupting cytoskeletal network, protein kinase C activities, and induced systemic resistance (Berdy 2005; Islam 2008; Islam et al. 2011; Gwinn 2018; Reshma et al. 2018). However, the mechanisms of antifungal activities of all bioactive substances have not yet been fully understood.

A large number of substances can successfully inhibit the development of various plant pathogens and are usually of plant origin and are called natural plant products or botanicals. They vary in composition and contents of the bioactive compounds that have high specificity for target pathogens with low toxicity to humans and the environment. Botanical products contain plant extracts, as well as essential oils. There are mixtures of 20 to 60 volatile and aromatic compounds like acids, acetones, alcohols, esters, alkaloids, phenols, and terpenoids (Bakkali et al. 2008; Negi 2012; Azmir et al. 2013). These occur at higher concentrations in glandular trichomes or secretory cell wall cavities, representing 1 to 2% of plant biomass. Since these particular cells can be found in most plants, theoretically botanical substances can be extracted from any part of the plant (Bakkali et al. 2008; Koul et al. 2008). However, the composition and contents of these natural products vary depending on the type of plant (e.g., species, subspecies, variation, and chemotype), the plant organ (e.g., leaves, branches, roots, flowers, and seeds), the stage of development, soil composition, season, and atmosphere (Licata et al. 2015; Moghaddam and Mehdizadeh 2015). Plant candidates are selected to search for potential sources of botanical compounds depending on their known medicinal or food value. For example, plants like *Camptotheca acuminate*, *Oryza sativa*, *Allium sativum*, *Aglaia edulis*, *Maesa japonica*, *Trevesia palmate*, and *Chelidonium majus* are often utilized to extract bioactive natural products in large amounts due to their recognized antagonistic activity against a group of plant pathogens (Barrera-Necha et al. 2008).

## 2.4 Effects of Bioactive Natural Products Against Blast Fungus

In recent years, application of bioactive compounds and botanicals has gained popularity in plant protection owing to health hazards and environmental pollution from the indiscriminate use of synthetic fungicides. Antagonistic microorganisms synthesize and secrete diverse metabolites, such as antibiotics (Vinay et al. 2016; Shaikh et al. 2018; Zaman et al. 2021). Most of these metabolites directly suppress mycelial growth, inhibit conidial germination, and the formation of appressoria of various plant pathogenic fungi, including the blast fungus of rice and wheat (Table 2.1). For example, various volatiles and diffusible compounds produced by various strains of *Bacillus* often possess significant inhibition effect against cereal blast fungi (Leelasuphakul et al. 2006; Arrebola et al. 2009; Hossain et al. 2016). For instance, *B. velezensis* G34 and *B. subtilis* BS155 synthesize bacillomycin L (1)

**Table 2.1** List of bioactive secondary metabolites, their sources and blast disease suppressing activities

| Products | Source | Inhibitory concentrations | Activities | Blast control (%) | Type of test | References |
|---|---|---|---|---|---|---|
| Bacillomycin L (**1**) | *Bacillus velezensis* strain G341 | 250 μg/ml | Reduces disease severity caused by MoO | 56 | In planta and | Lim et al. (2017) |
| Fengycin (**2**) | *B. subtilis* BS155 | 20 μg/ml | Induces cell<br>Membrane<br>Damage and dysfunction of organelles, disrupting MMP, oxidative stress, and chromatin<br>Condensation of MoO | 51 | In vitro | Zhang and Sun (2018) |
| Surfactin (**3**) | *B. Licheniformis* BC98 | 1 μg/ml | Inhibits mycelial growth of MoO | 97 | In vitro | Tendulkar et al. (2007) |
| Staurosporine (**4**) | *Streptomyces roseoflavus* strain LS-A24 | 3.125 μg/ml | Inhibits the activities of protein kinase C of MoO | 55 | In vitro | Sugahara et al. (2019) |
| Blasticidin S (**5**) | *S. Griseo-chromogene* | 10–20 μg/ml, 1–3 g/ha | Inhibited the protein synthesis of MoO | 95 | In vitro | Yamaguchi (1982) |
| Kasugamycin (**6**) | *S. kasugaensis* | 2 g/ha | Inhibited the protein synthesis of MoO | 95 | In vitro | |
| Kasugamycin (**6**) | *S. kasugaensis* | 400–500 ml/acre | Inhibits mycelial growth of MoO | 90–95 | In vitro | Umezawa et al. (1965) |
| Kasugamycin (**6**) | *S. kasugaensis* | 0.2% | Inhibits mycelial growth of MoO | 60 | In vitro | Magar et al. (2015) |
| Streptomycin + Thiophanate Methyl (**7**) | *S. griseus* | 0.15% | Inhibits mycelial growth of MoO | 62 | In vitro | |
| Oligomycin A (**8**) | *S. libani strain* As1 | 3 μg/ml | Inhibits mycelial growth of MoO | 80 | In vitro | Kim et al. (1999) |

| | | | | | | |
|---|---|---|---|---|---|---|
| Oligomycin A (**8**) | *Streptomyces* spp. AMA49 | 8 µg/ml | Inhibits conidial germination and appressorium formation of MoO | 100 | In vitro | Buatong et al. (2019) |
| Polyoxin A (**9**) | *S. Cacaoi asoensis* | 6.25 µg/ml | Inhibits germ tube formation and causes hyphal tip swelling of MoO | 60 | In vitro | Nguyen et al. (2018) |
| Polyoxin B (**10**) | | | | | | |
| Cyclic lipopeptides N3 (**11**) | *P. koreensis* strain COW3 | 10 µg/ml | Inhibits appressoria formation of MoO | 52 | In vitro | Omoboye et al. (2019) |
| Entolysin or Orfamide (**12**) | *P. Putida* strain COR5 | | | 48 | | |
| Lokisin (**13**) | *P. koreensis* strain COR10 | | | 54 | | |
| White line inducing principle (WLIP) (**14**) | *P. Putida* strain RW10S2 | | | 47 | | |
| Phenaminomethylacetic acid (**15**) | *Bacillus methylotrophicus* strain BC79 | 4000 g/667 $m^2$ | Inhibits mycelial growth of MoO | 85 | In vitro | Shan et al. (2013) |
| Antifungalmycin 702 (**16**) | *S. padanus* JAU4234 | 20 µg/ml | Inhibits mycelial growth, suppresses conidia germination and appressorium formation of MoO | 90 | In vitro | Xiong et al. (2013) |
| Cyclo (L-leucyl-L- prolyl) (**17**) | *Streptomyces* Sp. KH-614 | 2.5 µg/ml | Inhibits mycelial growth of MoO | 85 | In vitro | Rhee Hyeong (2003) |
| Dihydroxyviridio-fungin (C37H58N2O10) (**18**) | *S. Flavotricini* | 55 µg/ml | Inhibits mycelial growth of MoO | 34 | In vitro | Khalil et al. (2014) |
| Pyrrolo [1,2-a] pyra- zine-1,4-dione (**19**) | *Streptomyces* spp. isolate UPMRS4 | 100 µg/ml | Inhibits mycelial growth of MoO | 98 | In vitro | Awla et al. (2016) |

(continued)

**Table 2.1** (continued)

| Products | Source | Inhibitory concentrations | Activities | Blast control (%) | Type of test | References |
|---|---|---|---|---|---|---|
| Chitinase (**20**) | *Pseudomonas fluorescens* isolate 1 (Pfl) | 20 μg/ml | Inhibits mycelial growth of MoE | 73 | In vitro | Radjacommare et al. (2004) |
| Pyrrolnitrin (**21**) | *P. Cepacia strain* RB425 | 10 μg/ml | Inhibits mycelial growth of MoO | 81 | In vitro | Homma et al. (1989) |
| Ascherxanthone B (**22**) | *Aschersonia luteola* BCC 8774 | 0.58 μg/ml | Inhibits mycelial growth of MoO | 86 | In vitro | Chutrakul et al. (2009) |
| Alternariol monomethyl ether (**23**) | *Alternaria tenuissima* | 51.0 μg/ml | Inhibits appressoria formation of MoO | 80–100 | In vitro | Jeon et al. (2010) |
| (4S)-α,β-dehydrocurvularin (**24**) | *Alternaria* sp. | 10.6 μg/ml | | | | |
| HDFO (**25**) | *Biscogniauxia* spp. O821 | 5 μg/ml | Inhibits conidial germination and blast lesion formation of MoO | 99 | In vitro | Nguyen et al. (2018) |
| HDFO (**25**) | *Biscogniauxia* sp. strain O-811 | Fivefold | Inhibits conidial germination and appressorium formation of MoO | 93.5 | In vitro | Moriguchi et al. (2019) |
| Caripyrin (**26**) | *Caripia montagnei* | 10 μg/ml | Inhibit conidial germination and appressorium<br>Formation of MoO | 80 | In vitro | Rieger et al. (2010) |
| Cryptocin (**27**) | *Cryptosporiopsis* cf. *quercina* | 0.39 μg/ml | Inhibits mycelial growth of MoO | 70 | In vitro | Li et al. (2000) |
| Chaetoviridin A (**28**) | *Chaetomium globosum strain* F0142 | 62.5 lg/ml | Inhibits mycelial growth of MoO | 80 | In vitro | Park et al. (2005) |
| Griseofulvin (**29**) | *Xylaria* sp. strain F0010 | 50 μg/ml | | 95 | | |
| GKK1032A2 (**30**) | *Penicillium* spp. IBWF-029-96 | 2 μg/ml | Inhibits conidial germination of MoO | 100 | In vitro | Becker et al. (2011) |

| | | | | | | |
|---|---|---|---|---|---|---|
| Tanzawaic acid K (**31**) | *Penicillium* spp. IBWF104–06 | 25 μg/ml | Inhibits conidial germination of MoO | 90 | In vitro | Sandjo et al. (2014) |
| Tanzawaic acid L (**32**) | | 50 μg/ml | | | | |
| Tanzawaic acid A (**33**) | | 10 μg/ml | | | | |
| Tanzawaic acid E (**34**) | | 25 μg/ml | | | | |
| Arohynapene A (**35**) | | 25 μg/ml | | | | |
| Arohynapene B (**36**) | | 25 μg/ml | | | | |
| Gageopeptide A (**37**) | *Bacillus subtilis* strain 109GGC020 | 10.0 μg/disk | Inhibits mycelial growth of MoT | 50.5 | In vitro | Chakraborty et al. (2020a) |
| Gageopeptide B (**38**) | | 1.5 μg/disk | | 54.3 | | |
| Gageopeptide C (**39**) | | 2.5 μg/disk | | 58.1 | | |
| Gageopeptide D (**40**) | | 2.5 μg/disk | | 45.7 | | |
| Gageotetrin B (**41**) | | 10.0 μg/disk | | 60.9 | | |
| Oligomycin B (**42**) | *Streptomyces* spp. strain B8739 | 0.05 μg/disk | Inhibits mycelial growth of MoT | 57.1 | In vitro | Chakraborty et al. (2020b) |
| Oligomycin F (**43**) | *Streptomyces* spp. strains A171 | 0.005 μg/disk | | 73.9 | | |

and fengycin (2), respectively, which showed strong antifungal activities against rice blast fungus MoO (Lim et al. 2017; Zhang and Sun 2018). *Bacillus licheniformis* BC98 generates surfactin (3), a cyclic lipopeptide that has inhibited MoO mycelial growth at a minimum inhibitory concentration (MIC) of 1 µg / ml (Tendulkar et al. 2007). *B. tequilensis* GYLH001 is also capable of producing many bioactive macromolecular and small molecular compounds like cellulase, protease, gelatinase, 1-aminocyclopropane-1-carboxylate deaminase, and indole-3-actic acid. A recent study showed that the thermal and acid-stable sterilized filtrate of these *Bacillus* strain inhibited *M. oryzae*'s growth, indicating a potential application of this strain towards phytopathogens under extreme environmental conditions (Li et al. 2018). In an experiment, we found that five linear lipopeptides gageopeptide A-D (37–40) and gageotetrin B (41) synthesized by *B. subtilis* strain 109GGC020 inhibit the hyphal growth of wheat blast fungus MoT at 1.5, 2.5, 2.5, 10, and 10 µg/ml, respectively. The compounds also inhibit conidiogenesis, conidia germination, appressoria formation of MoT, and suppress blast disease development in detached wheat leaves (Chakraborty et al. 2020a).

*Streptomyces* species also generate many secondary metabolites which showed significant antifungal activities against MoO (Rhee Hyeong 2003; Prabavathy et al. 2006; Xiong et al. 2013; Awla et al. 2016). For instance, staurosporine (4) synthesized by *S. roseoflavus* strain LS-A24 inhibited conidial germination and appressorial formation of MoO at MIC 3.125 µg/ml (Sugahara et al. 2019). Blasticidin S (5) and kasugamycin (6) synthesized by *S. griseochromogenes* and *S. kasugaensis*, respectively, have inhibited MoO by 95% (Yamaguchi 1982). Oligomycin A (8), polyoxin A (9), polyoxin B (10), and cyclo (L-leucyl-L- prolyl) (19) produced by *Streptomyces* spp. showed strong antifungal activities against MoO at 8 µg/ml, 6.25 µg/ml, 6.25 µg/ml, and 2.5 µg/ml, respectively (Isono et al. 1965; Rhee Hyeong 2003; Buatong et al. 2019). Recently, we have found that oligomycin B (42) and F (43) synthesized by *Streptomyces* spp. strain B8739 and A171 inhibit the mycelial growth of MoT at 0.05 and 0.005 µg/ml. The compounds also inhibit conidiogenesis, conidia germination, and appressoria formation of MoT, and suppress blast disease development in detached wheat leaves (Chakraborty et al. 2020b).

Homma et al. (1989) found that pyrrolnitrin (23) produced by *Pseudomonas cepacia* RB425 inhibited rice blast disease by 81%. Chitinase (22) extracted from *P. fluorescens* isolate 1 (Pfl) strongly suppressed the mycelial growth of the finger millet blast fungus MoE (Radjacommare et al. 2004). *Pseudomonas* species also produced many lipopeptides such as cyclic lipopeptides N3 (11), entolysin or orfamide (12), lokisin (13), and WLIP (14) that were shown to inhibit the appressorial formation of MoO with a MIC of 10 µg/ml (Omoboye et al. 2019).

Various fungal species are also known to produce secondary metabolites with potential inhibitory activities against phytopathogens. Among the fungi, *Aschersonia luteola* BCC 8774 synthesizes ascherxanthone B (22) and *Cryptosporiopsis* cf. *quercina* produces cryptocin (27), which possesses excellent biocontrol activities against rice blast fungus (Chutrakul et al. 2009; Li et al. 2000). Compound 22 and 27 inhibited the mycelial growth of rice blast fungus at 0.39 µg/ml and 0.58 µg/ml (MIC), respectively. GKK1032A2 (30) isolated from *Penicillium* spp. IBWF-029-96 suppressed 100% conidial germination of MoO at 2 µg/ml (Becker et al. 2011).

Strains of *Alternaria* was reported to produce alternariol monomethyl ether (23) and (4S)-α,β-dehydrocurvularin (24) that could inhibit appressorial formation of MoO by 80–100% (Jeon et al. 2010). To date, several bioactive secondary metabolites have been reported that showed inhibitory activities against the cereal blast pathogen *Magnaporthe* are listed in Table 2.1 with their source of isolation, inhibitory activity (%), and mode of action.

Plant products or botanicals are also a useful source of antifungal metabolites that are used in effective plant disease management because of their biodegradability and low toxicity (Vu et al. 2015). Extracts and phytochemicals from several herbs, spices, and medicinal plants have been investigated for their antimicrobial activity on different plant pathogens including rice blast fungus (Table 2.2). For example, camptothecin (44) isolated from the leaves of *Camptotheca acuminate* showed strong antifungal activity against *M. oryzae*. It has inhibited 100% mycelial growth and conidial germination of MoO at MIC 10 μg/mL (Xu et al. 2019). Recently, Norvienyeku et al. (2019) reported that a pentacyclic triterpenoid saponin bayogenin 3-O-cellobioside (45) derived from the leaf extracts of *Oryza sativa* completely suppressed conidia germination and appressorium formation of MoO with a MIC of 5 nM/L. Other five new acylated triterpenoid saponins such as maejaposide A (46), maejaposides A-1 (47), maejaposide C-1 (48), maejaposide C-2 (49), and maejaposide C-3 (50) isolated from the methanol extract of *Maesa japonica* significantly reduced MoO by 85–98% (Ngo et al. 2019). Kim et al. (2018) also documented that four triterpenoid saponins, i.e., TPG1 (hederagenin-3-*O*-*β*-D--glucopyranosyl-(1 → 3)-*α*-L-rhamnopyranosyl-(1 → 2)-*α*-L-rhamnopyranosyl-(1 → 2)-*α*-L-arabinopyranoside) (51), TPG2 (macranthoside A) (52), TPG3 (*α*-hederin) (53), and TPG5 (3-*O*-*α*-L-rhamnopyranosyl asiatic acid) (54) extract from the methanol extract of *Trevesia palmate*, have strong antagonistic activities against MoO. These antifungal compounds completely inhibited the mycelial growth of MoO with MIC at 4 to 16 μg/ml. Pannellin (55) and rocaglaol (56) were flavaglines isolated from *Aglaia* spp. and suppressed 95% germ tube formation of MoO at 1.6 μg/ml and 3 μg/ml, respectively (Froyd et al. 1976; Engelmeier et al. 2000). Chelerythrine chloride (58) isolated from *Chelidonium majus* strongly inhibited germ tube formation and appressorial development (Penn et al. 2015). Allicin (59) also known as diallyl-thiosulfinate, extracted from garlic tissue, suppressed 90–99% spore germination of MoO with a MIC of 50 μg/ml and 84 μg/ml, respectively (Curtis et al. 2004; Fry et al. 2005). However, active ingredients in these extracts and their mechanisms of action against blast fungus still remain unclear. Botanicals, which have been reported as successful inhibitors of cereal blast fungus until now, are listed in Table 2.2 with their source, inhibitory activity, and mode of action. The structures of different bioactive natural products isolated from microorganisms and plants are given in Fig. 2.1.

In developing countries, farmers frequently use wide-spectrum fungicides to a certain pathogen without proper diagnosis of the problem and prior knowledge of fungicide specificity which resulted in reduced fungicide sensitivity and increased disease management costs. In these conditions the use of different microbes and natural products obtained from plants should be cost-effective and more sustainable.

**Table 2.2** List of bioactive plant products suppressing blast fungus

| Products | Source | Inhibitory concentrations | Activities | Blast control (%) | Type of test | References |
|---|---|---|---|---|---|---|
| Camptothecin (**44**) | *Camptotheca acuminata* | 10 μg/mL | Inhibits mycelial growth and conidial germination of MoO | 100 | In vitro | Xu et al. (2019) |
| Bayogenin 3-O-Cellobioside (**45**) | *Oryza sativa* | 5 nM/L | Inhibits conidia germination and appressorium formation of MoO | 100 | In vitro | Norvienyeku et al. (2019) |
| Maejaposide A (**46**) | *Maesa japonica* | 32 μg/ml | Suppresses disease development caused by MoO | 85–98 | In planta | Ngo et al. (2019) |
| Maejaposides A-1 (**47**) | | 16 μg/ml | | | | |
| Maejaposides C-1 (**48**) | | 4 μg/ml | | | | |
| Maejaposides C-2 (**49**) | | 4 μg/ml | | | | |
| Maejaposides C-3 (**50**) | | 4 μg/ml | | | | |
| Triterpene glycosides TPG1 (**51**) | *Trevesia palmata* | 8 μg/ml | Inhibits mycelial growth of MoO | 100 | In vitro | Kim et al. (2018) |
| Triterpene glycosides TPG2 (**52**) | | 8 μg/ml | | | | |
| Triterpene glycosides TPG3 (**53**) | | 4 μg/ml | | | | |
| Triterpene glycosides TPG5 (**54**) | | 16 μg/ml | | | | |
| Pannellin (**55**) | *Aglaia edulis* | 3 μg/ml | Inhibits germ tube formation of MoO | 95 | In vitro | Engelmeier et al. (2000) |
| Rocaglaol (**56**) | *A. odorata* | 1.6 μg/ml | | | | |
| Tricyclazole (**57**) | *Neostapfia colusana* | 0.56 kg/ha | Inhibits mycelial growth of MoO | 96 | In vitro | Froyd et al. (1976) |

(continued)

**Table 2.2** (continued)

| Products | Source | Inhibitory concentrations | Activities | Blast control (%) | Type of test | References |
|---|---|---|---|---|---|---|
| Tricyclazole (**57**) | Eli Lilly company, greenfield, Indiana | 0.1 μg/ml | Inhibits melanin biosynthesis Of MoO | 100 | In planta | Woloshuk et al. (1983) |
| Chelerythrine chloride (**58**) | *Chelidonium majus* | 25 μM | Inhibits the activities of protein kinase C of MoO | 43 | In vitro | Penn et al. (2015) |
| Allicin (**59**) | *Allium sativum* | 84 μg/ml | Inhibits conidia germination of MoO | 90 | In vitro | Curtis et al. (2004) |
| Allicin (**59**) | *A. sativum* | 50 μg/ml | Inhibits conidia germination of MoO | 99 | In vitro | Fry et al. (2005) |
| Isobavachalcone (IBC) (**60**) | *Psoralea corylifolia* L | 0.3 g a.i./ha | Inhibits mycelial growth by cell wall degradation of MoO | 92 | In vitro | Liu et al. (2018) |
| Trans-2-Hexenal (leaf aldehyde) (**61**) | Chisso petrochemical co., Chiba, Japan | 0.85 μg/ml | Reduces disease severity caused by MoO | 100 | In planta | Tajul et al. (2011) |
| Cis-3-hexenol (leaf alcohol) (**62**) | Nippon Zeon co., Kanagawa, Japan | | | 87 | | |
| Cis-3-hexenal (**63**) | | | | 100 | | |

## 2.5 Mechanisms of Important Bioactive Secondary Metabolites Against Cereal Blast Fungus

Biopesticides must have a unique mode of action which is non-toxic to non-target pests and shows limited human and environmental toxicity (Leahy et al. 2014). Between 1997 and 2010, most registrations of pesticides were centered on biological systems: natural products (36%), synthetics extracted from natural products (6%), and biological control agents (58%) (Cantrell et al. 2012). Biopesticide formulations based on natural resources are essential strategies for overcoming the agricultural challenge of supplying the world's population with adequate, sustainable, and nutritious food stocks. But focusing on natural or naturally produced

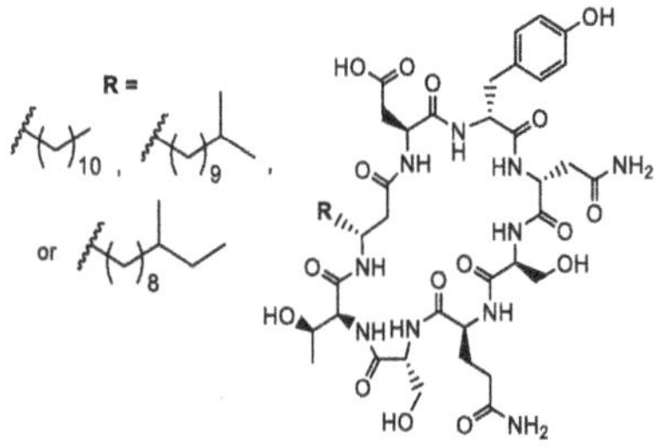

**Bacillomycin L (1)**

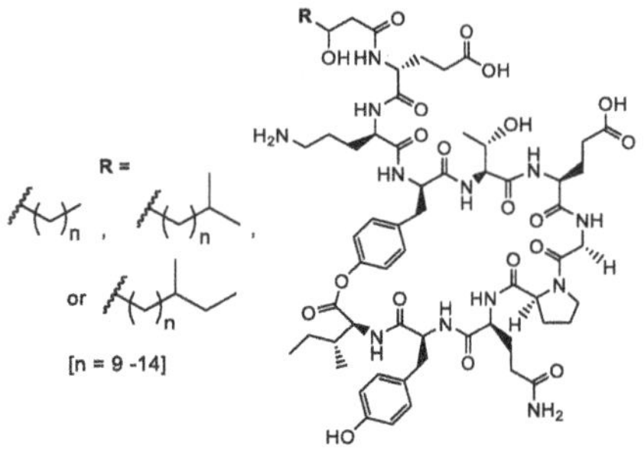

**Fengycin A (2)**

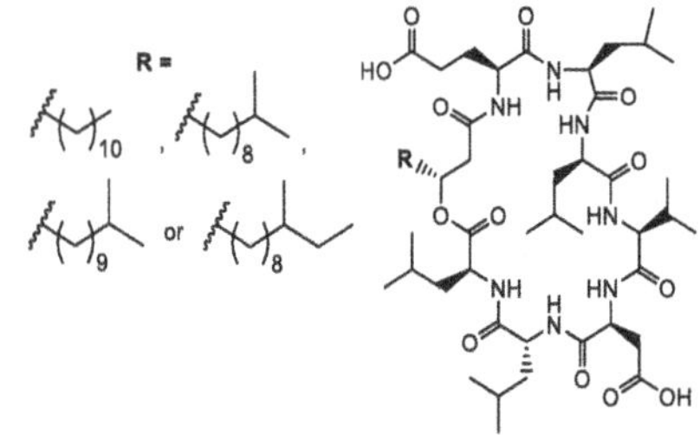

**Surfactin (3)**

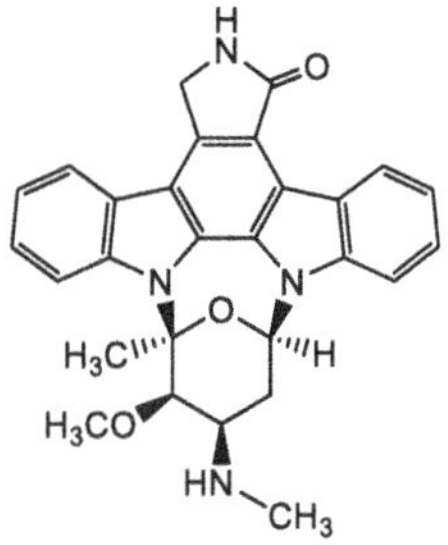

**Staurosporine (4)**

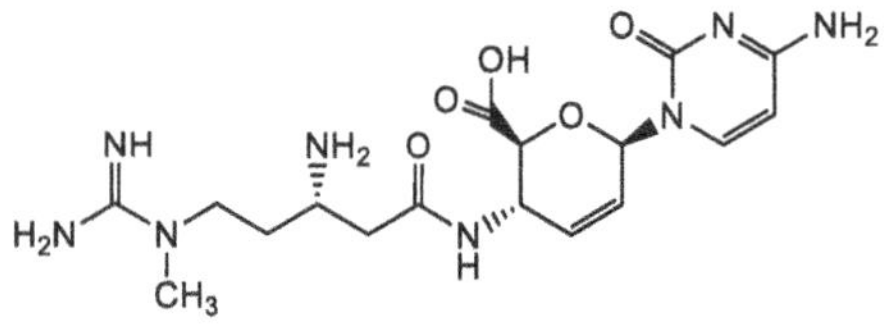

**Blasticidin S (5)**

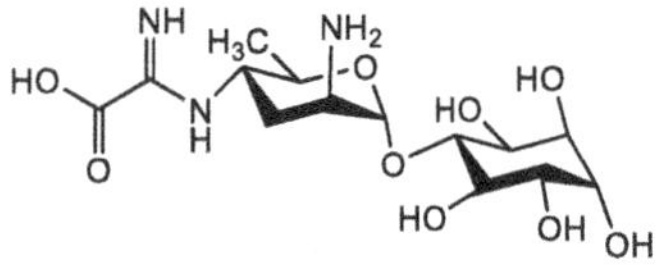

**Kasugamycin (6)**

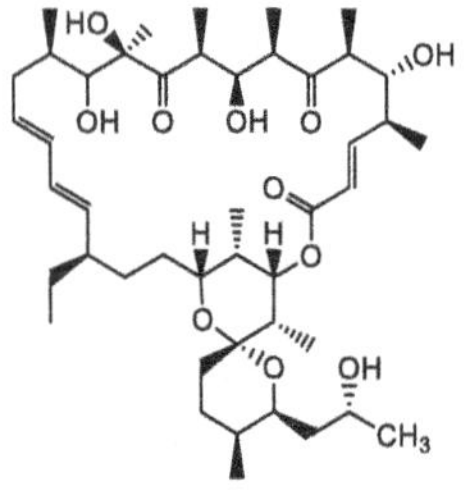

**Oligomycin A (7)**

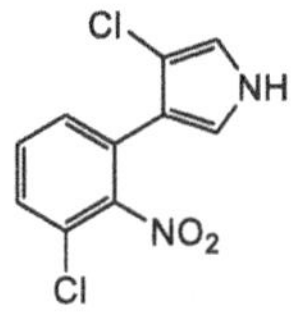

**Pyrrolnitrin (21)**

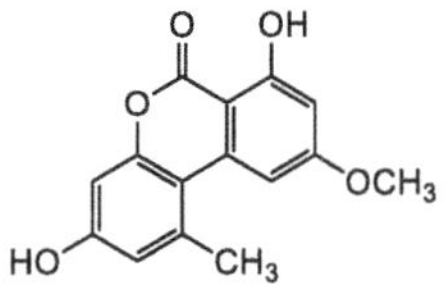

**Alternariol monomethyl ether (23)**

**Fig. 2.1** Structures of different bioactive natural compounds against *Magnaporthe oryzae*

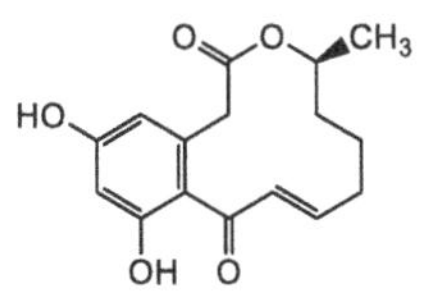

(4*S*)-α,β-dehydrocurvularin (24)

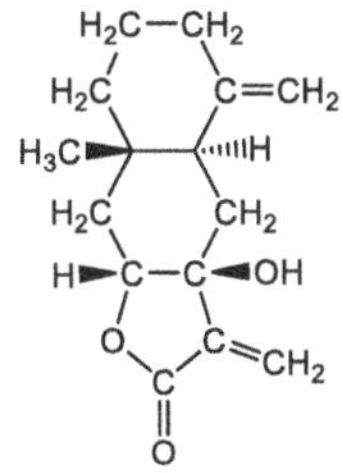

HDFO (25)

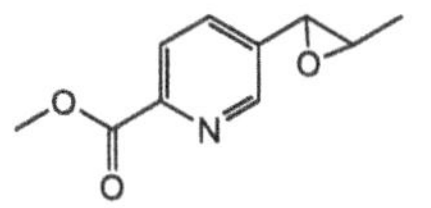

Caripyrin (26)

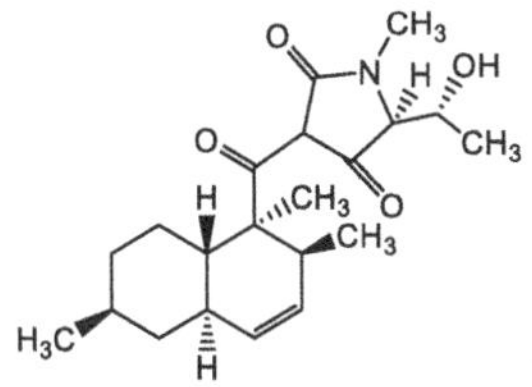

Cryptocin (27)

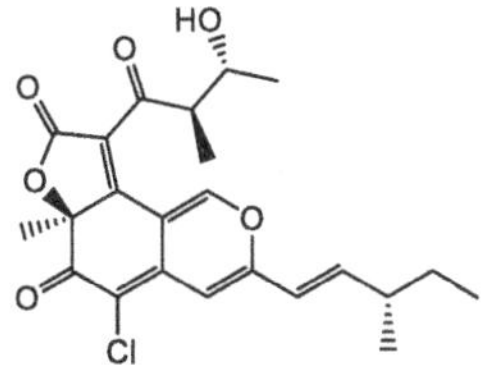

Chaetoviridin A (28)

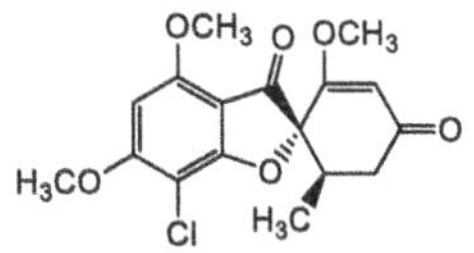

Griseofulvin (29)

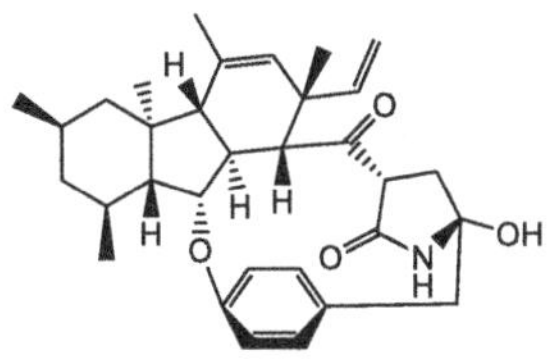

GKK1032A2 (30)

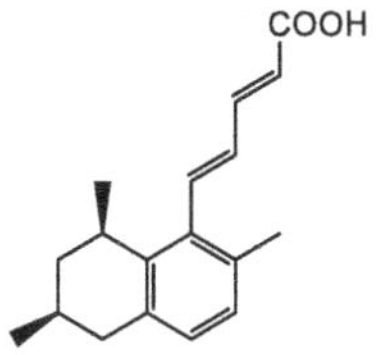

Tanzawaic acid A (31)

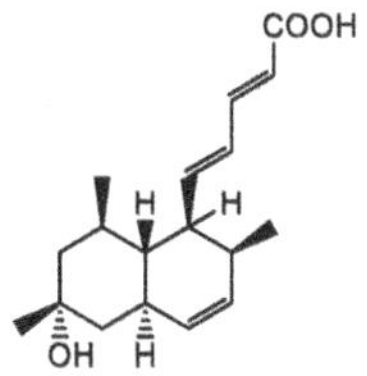

Tanzawaic acid E (32)

**Fig. 2.1** (continued)

synthetic products of a limited range will make disease control unsustainable. For instance, strobilurin was originally extracted from the *Strobilurus tenacellus*, a wood rotting fungus that is also generated by a number of related fungi. Strobilurin and its derivatives bind to the Qo site of cytochrome b, thereby blocking the flow of electrons between it and cytochrome c (US EPA 2019). Approximately 10% of global fungicides contain 6 strobilurins and therefore remain a significant part of plant disease management including cereal blast fungus, in many agricultural practices (Fungicide Resistance Action Committee 2016). The development of pathogen resistance to the synthetic QoI fungicides, however, produced an enormous problem (Kim et al. 2010). To manage the worrisome cereal blast fungus, new natural products with novel mode of action and reliance on a consortium of products are therefore required.

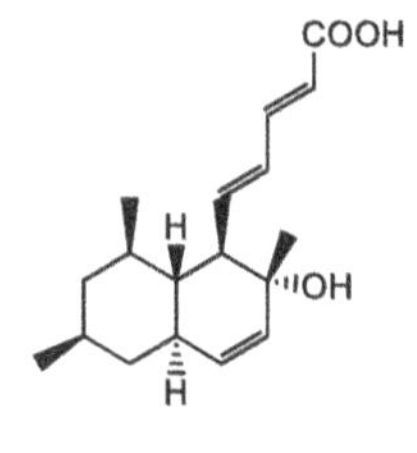

Tanzawaic acid K (33)

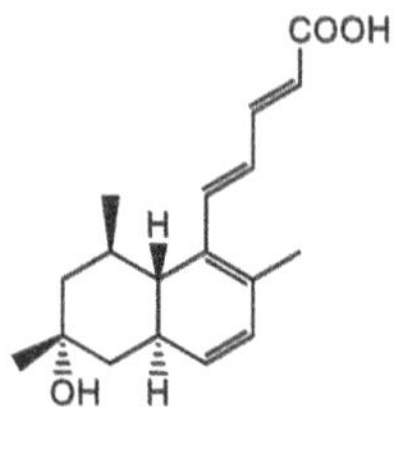

Tanzawaic acid L (34)

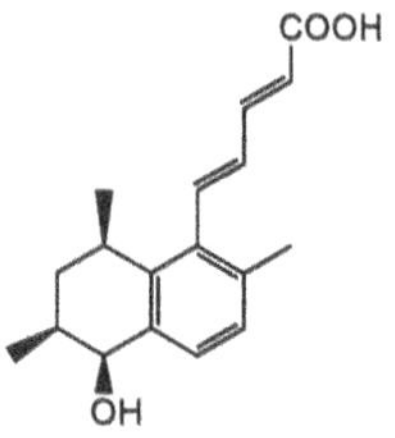

Arohynapene A (35)

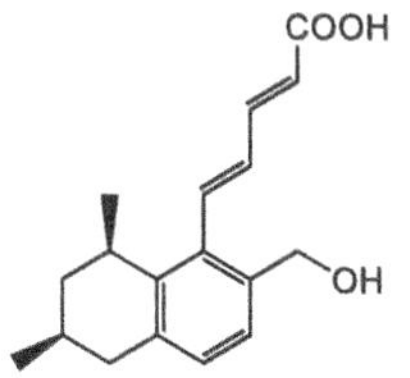

Arohynapene B (36)

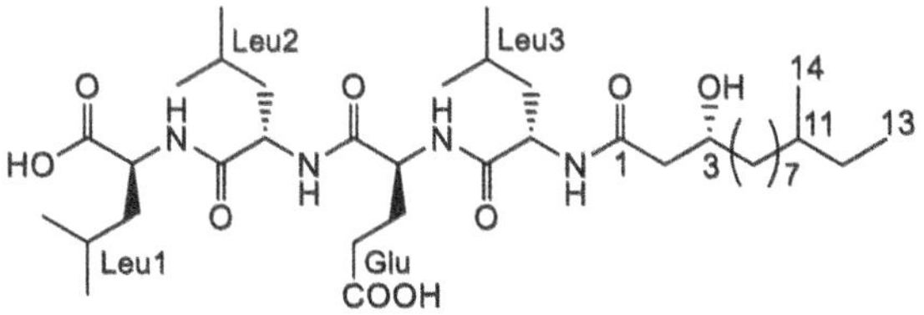

Gageopeptide A (37)

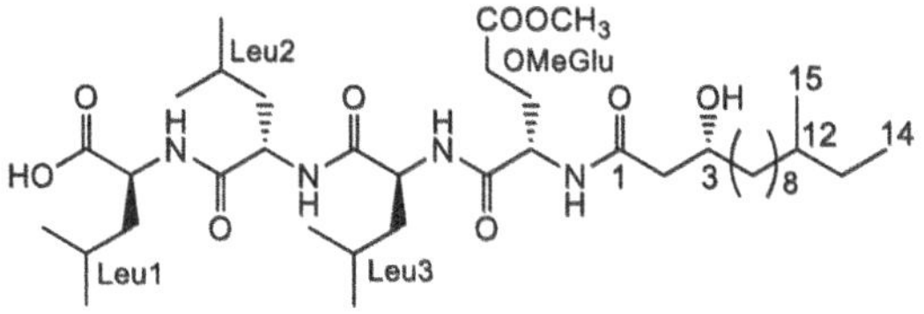

Gageopeptide B (38)

**Fig. 2.1** (continued)

Most species of soil bacteria can produce two to four antimicrobial lipopeptides, peptidyl antibiotics including surfactin, iturin, and fengycin which are considered to have high antagonistic activity and stability (Ishii 2015; He et al. 2019). These natural molecules are likely to interact with the surface tension of the fungal cell membranes, leading to the formation of micropores and leakage of essential ions, causing cell death (Ongena and Jacques 2008; Kim et al. 2010). In addition, lipopetides induce systemic resistance (Mihalache et al. 2018). Bacillomycin L (1), fengycin (2), surfactin (3), gageopeptide A-D (37–40), and gageotetrin B (41) are cyclic and linear lipopeptides extracted from *Bacillus* species showed strong antifungal activities against MoO and MoT, affecting the distribution of cytoskeletal elements (e.g., actin and tubulin) and leading to loss of the directional growth (Tendulkar et al. 2007; Lim et al. 2017; Zhang and Sun 2018; Chakraborty et al. 2020a). *Pseudomonas* species also produced several cyclic lipopeptides, such as N3 (11), entolysin or orfamide (12), lokisin (13), and WLIP (14) which were reported to inhibit the

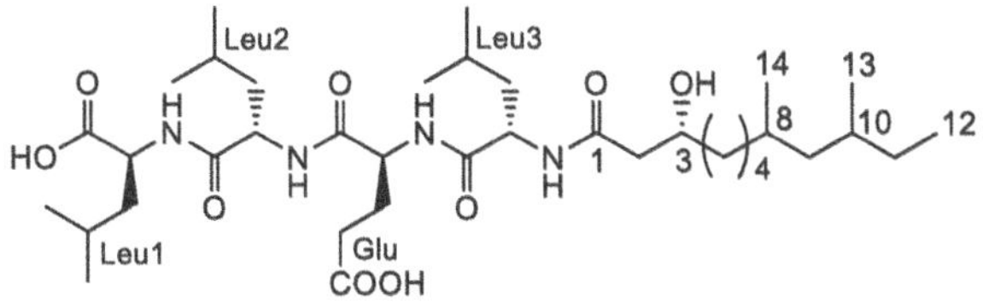


**Gageopeptide C (39)**

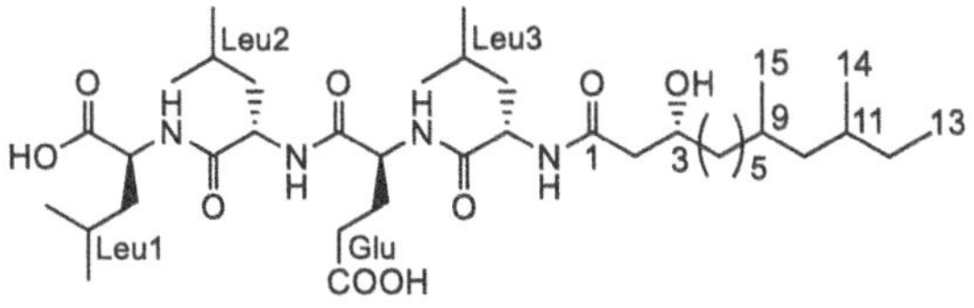


**Gageopeptide D (40)**

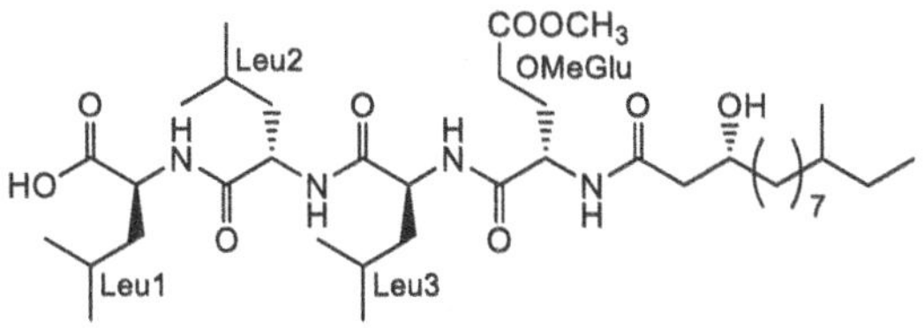


**Gageotetrin B (41)**

**Fig. 2.1** (continued)

appressorial formation of MoO, though their modes of action have yet to be investigated (Omoboye et al. 2019).

Blasticidin S (5) is a peptidyl nucleoside antibiotic produced by *Streptomyces griseochromogenes* that shows strong fungicidal activity against different fungi including *M. grisea* (Yamaguchi 1982; Barka et al. 2016; Lamberth 2016). This secondary metabolite blocks protein synthesis (D2) (Li et al. 2013) by interfering peptide bond formation and peptidyl-tRNA hydrolysis through deformed conformation of tRNA (Copping and Duke 2007; Svidritskiya et al. 2013). Compound **5** displayed strong fungicidal activity against MoO (Yamaguchi 1982). The practical application of this compound as commercial fungicide is still to be studied. *Streptomyces* species produce oligomycins that inhibit ATP synthase by blocking the proton channel needed for oxidative phosphorylation of ADP to ATP (Nakata et al. 1995). Oligomycin A (8), B (42), and F (43) have shown effective antifungal activities against MoO and MoT by suppressing its mycelial growth (Rhee Hyeong 2003; Buatong et al. 2019; Chakraborty et al. 2020b). Compound 8 isolated from

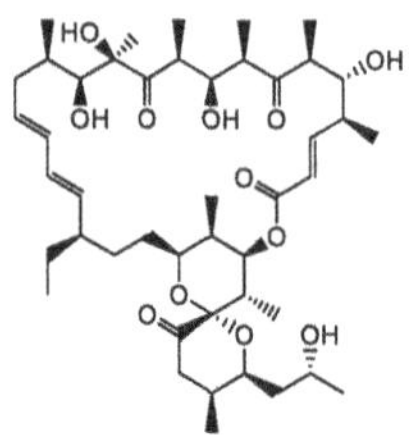


Oligomycin B (42)

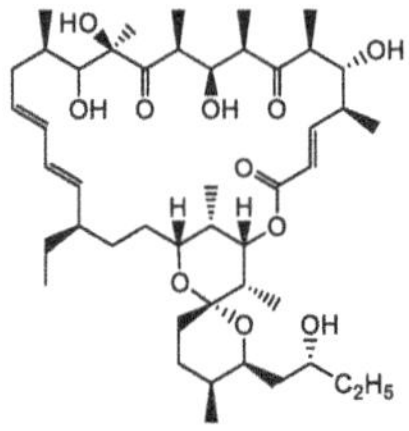


Oligomycin F (43)

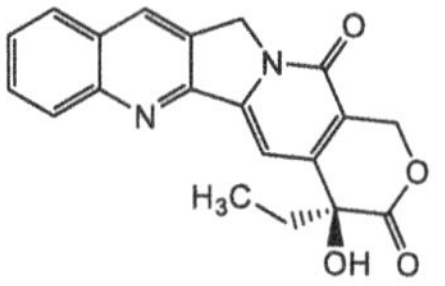


Camptothecin (44)

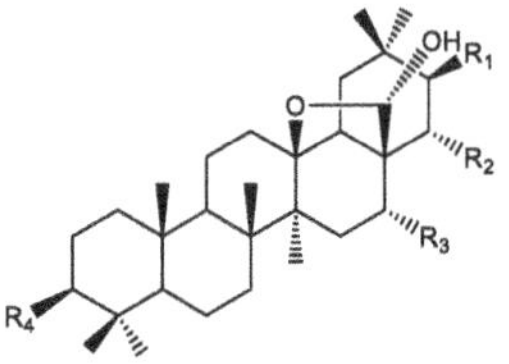


Maejaposide core structure

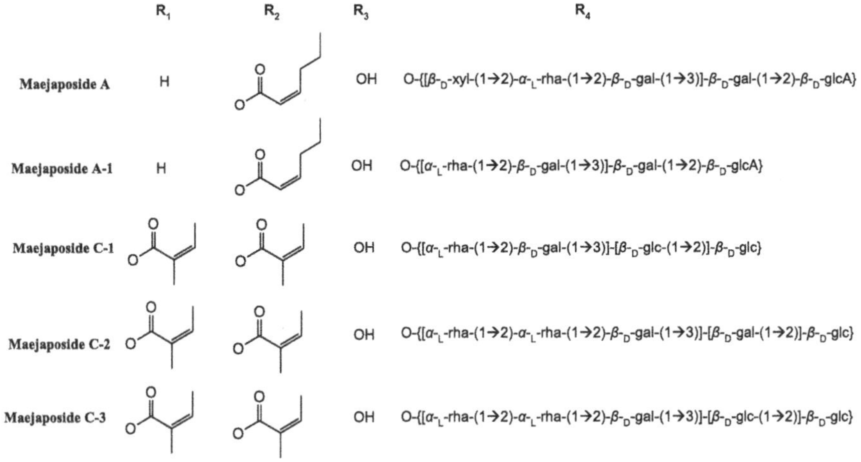


Compounds (46-50)

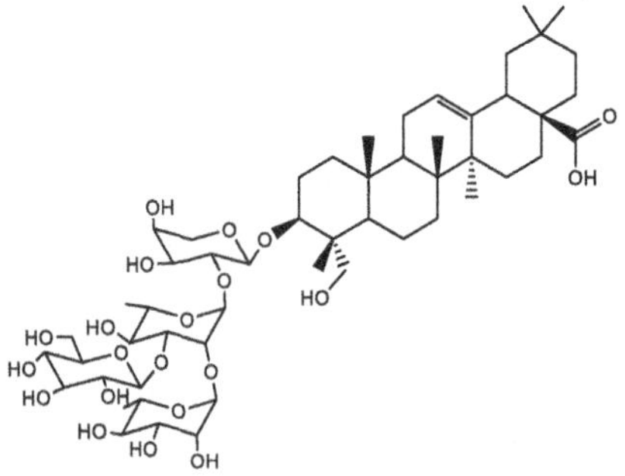

TPG1 (51)

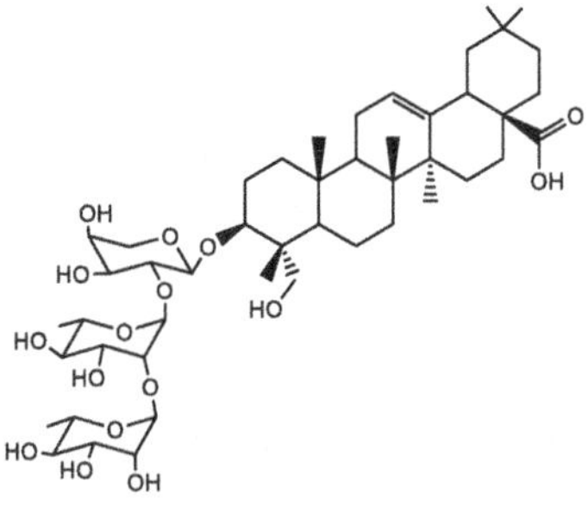

TPG2 (52)

**Fig. 2.1** (continued)

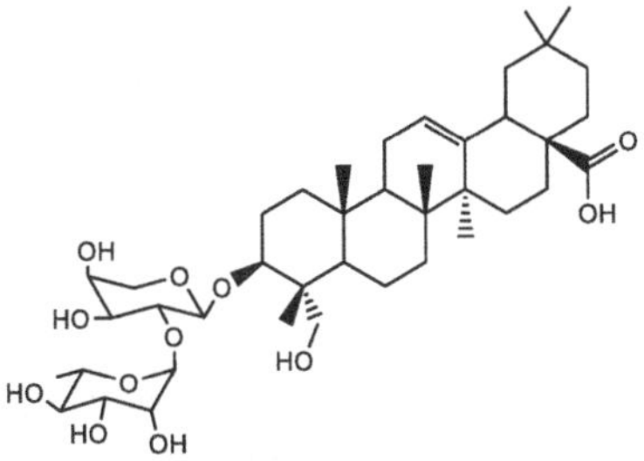

**TPG3 (53)**

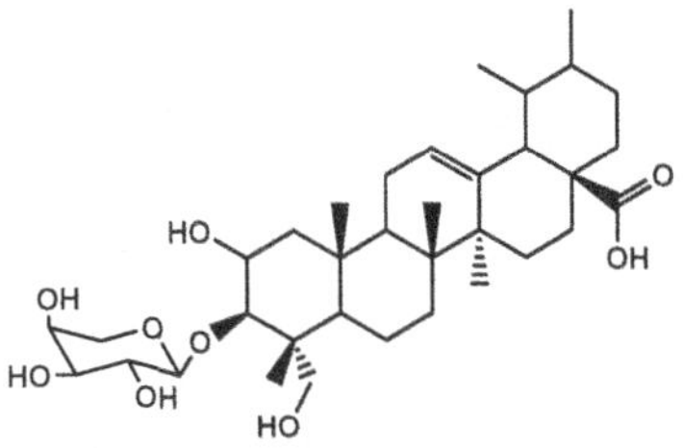

**TPG5 (54)**

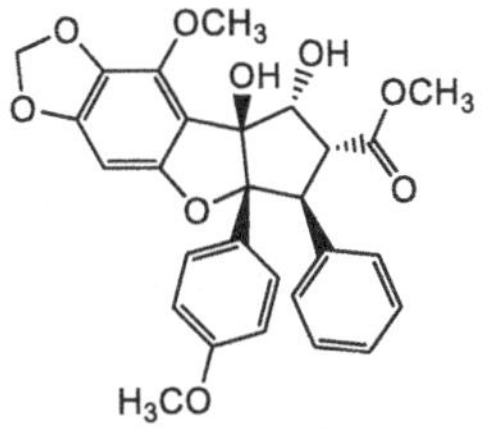

**Pannellin (55)**

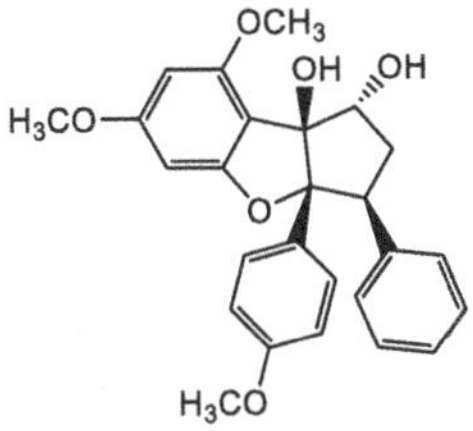

**Rocaglaol (56)**

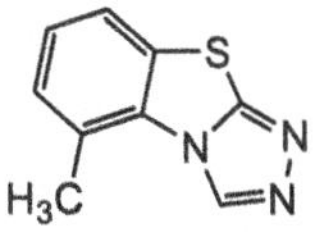

**Tricyclazole (57)**

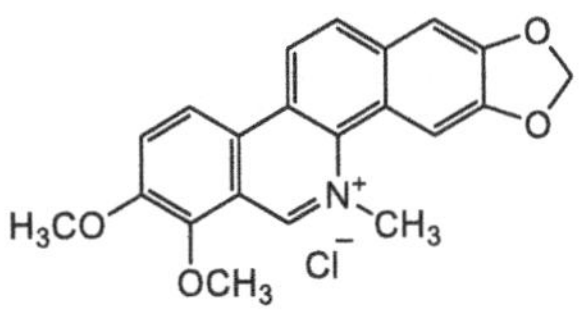

**Chelerithrine chloride (58)**

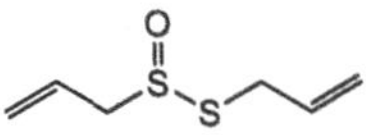

**Allicin (59)**

**Fig. 2.1** (continued)

marine *Streptomyces* has been reported as motility inhibitor and lysis inducer of peronosporomycete zoospores (Dame et al. 2016; Islam et al. 2016a, b).

Polyoxin A (9) and polyoxin B (10) extracted from *S. cacaoi asoensis* are efficient in controlling MoO (Isono et al. 1965). These natural products are known to block chitin synthesis in the cell wall, which disrupts germ tube formation and induces swelling of hyphal tip (Lamberth 2016; Mihalache et al. 2018). Although precise mechanisms of action of these bioactive compounds are yet to be elucidated, compounds directly targeting cell wall biosynthesis of *M. oryzae* may be useful lead compounds for producing new fungicide against cereal blast. On the other hand, Kasugamycin (6) was synthesized by *S. kasugaensis*, an inhibitor of protein synthesis (D3), and also showed inhibitory effects against MoO (Umezawa et al. 1965; Barka et al. 2016). The specificity of the activity of compound **6** against different pathogens has yet to be examined.

Staurosporine (4) and streptomycin (7) isolated from *Streptomyces* spp. and chelerythrine chloride (58) derived from a herbaceous perennial plant *Chelidonium majus* (Family Papavaraceae) are selective inhibitors of protein kinase C (PKC) and have been shown to be an effective inhibitor of *M. oryzae* and developmental transition of its asexual life stages (Magar et al. 2015; Penn et al. 2015; Sugahara et al. 2019). Inhibition of PKC by compound 4 and chelerythrine also suppressed asexual spore formation and motility of the phytopathogenic oomycetes, *Aphanomyces cochlioides* and *Plasmopara viticola* (Tamoki et al. 1986; Islam et al. 2011). In an in vitro experiment, we have found strong antifungal activities of staurosporine and chelerythrine chloride against wheat blast fungus MoT (data not shown). Another perennial woody plant *Camptotheca acuminata*-derived alkaloid camptothecin (44) also showed strong antifungal activities against *M. oryzae*. It inhibits protein synthesis, RNA synthesis, and carbohydrate metabolism of MoO and causes membrane damage (Xu et al. 2019).

Other class of phytochemicals recognized as saponins are isolated from various plants that have an antifungal mechanism as their ability to form complexes with sterols in fungal membranes, leading to pore formation and cell membrane disruption followed by cell component leakage (Ngo et al. 2019). Some triterpenoid saponins, such as bayogenin 3-O-Cellobioside (45), maejaposide A (46), maejaposide A-1 (47), maejaposide C-1 (48), maejaposide C-2 (49), maejaposide C-3 (50), triterpene glycosides TPG1 (51), triterpene glycosides TPG2 (52), triterpene glycosides TPG3 (53), and triterpene glycosides TPG5 (54), extracted from methanol extract of plant parts, significantly suppressed the hyphal growth, conidia germination, and appressorium formation of MoO (Kim et al. 2018; Ngo et al. 2019; Norvienyeku et al. 2019). Pannellin (55) and rocaglaol (56) are flavaglines, isolated from *Aglaia* spp., inhibited germ tube formation of MoO (Basmadjian et al. 2013). Flavaglines are usually referred to as natural potent anti-cancer drugs which display highly selective cytotoxic effects on cancer cells by preventing the progression of cell cycle (Li-Weber 2014). However, further investigation is needed for the antimicrobial mechanisms of flavaglines.

Tricyclazole (57) isolated from plant species is an antipenetrant that is systemic in rice and has the potential to suppress rice blast disease at any stage of plant development (Peterson 1990). It prevents rice plants from MoO fungal invasion by inhibiting penetration by the fungus into the epidermal cell wall. The compound acts by preventing biosynthesis of melanin, thereby inducing a loss of rigidity in the appressorial wall (Woloshuk et al. 1983). Its activity is defensive but not curative. Allicin (59) is an active component of garlic that prevents the vegetative and reproductive growth of a number of microorganisms, including antibiotic-resistant strains. It interferes with thiol groups, which can inactivate major enzymes. This compound's mechanism of action has been documented as a combination of a decrease in glutathione levels, unfolding stress, and inactivation of vital metabolic enzymes via S-allylmercapto modification of cysteines (Müller et al. 2016). Compound 59 significantly suppressed MoO by inhibiting the growth of fungal mycelium (Curtis et al. 2004; Fry et al. 2005). It also forms part of a natural and primitive antimicrobial defense system in plants. Due to the potential of disulfide-S-monoxides to inhibit a broad array of cysteine-containing enzymes, the interest in using allicin has increased in recent decades which decreases the risk of microbes developing resistance towards this agent (Ankri and Mirelman 1999).

Reviewing a broad range of literature, it appears that the major mechanism of action of natural biofungicides against blast fungus *M. oryzae* involves inhibition of plasma membrane synthesis, cell wall biosynthesis, mitochondrial respiration inhibition, and ion channel interference. In order to find a suitable alternative to strobilurin fungicides to combat blast disease in major cereal crops including rice and wheat, the specificity and commercial value of natural compounds that exhibited significant inhibitory activities against blast fungus need to be examined.

## 2.6 Conclusion and Future Perspectives

As an effective and sustainable approach of crop protection from plant pathogens, using bioactive agents is gaining popularity. Globally, a major endeavor is under progress to develop cost-effective and rational products for biocontrol. An enormous amount of beneficial secondary metabolites were screened for antimicrobial activities, with a fraction of these evaluated compounds displaying positive results in inhibiting phytopathogen growth in both laboratory and field conditions. However, studies have suggested that some potential antimicrobial products did not perform effectively in field conditions compared with the laboratory or in vitro. A more effective method of laboratory screening would also need to be developed that can reflect the field output of these agents from simulation modeling in controlled environment.

Integrating microbial and plant-derived biological control agents with conventional management methods can minimize fungicide use significantly and enable crop protection and production more sustainable. In order to formulate such microbial strains into products for field usage, more research is required. While a large amount of microbial and plant extracts was found to be very useful in controlling plant pathogens including blast fungus, the active ingredients and modes of action are still unknown. Engineering novel strains with improved bioactive molecule biosynthesis will increase the effectiveness of biocontrol agents. Unraveling the molecular and physiological targets of bioactive compounds towards blast fungus is of vital interest in designing a long-term program for disease management. Since the biology and pathogenicity of wheat blast fungus *M. oryzae Triticum* (MoT) is poorly understood in contrast with the rice blast fungus *M. oryzae Oryzae* (MoO), more study is needed to unravel host specificity, epidemiology, and pathogenicity factors to control blast diseases more effectively. The screening of naturally occurring wheat genotypes which are resistant to blast fungus should be another important aspect of research. These resistant genotypes can also be used as resistance source for the breeding of new resistant wheat cultivars. An emerging area of research may also be the elucidation of the fundamental resistance mechanism in wheat against blast disease.

Continued research and development on bio-pesticides against plant diseases would allow researchers to find more effective strains, considering these species' ubiquity and diversity. However, the large-scale use of biocontrol agents is still limited in field conditions due to the inconsistency and variability of biocontrol activity. Among the few challenges that need to be addressed are reducing the variability and increasing the consistency of disease inhibition. The activity of biocontrol agents relies on the relationships among themselves, the host plant, the pathogen, and the biotic and abiotic environmental factors in field conditions. As such, optimizing and documenting these variables and making the information accessible to end users would help to achieve reliable outcomes. Instead of using the organism itself, using metabolites from genetically modified species would greatly reduce the potential danger of introducing such organism in the ecosystem and the unintended impacts it may carry. In addition, this method would certainly be useful in eliminating complicated procedures for the registration and marketing of any biocontrol agents. Developing a viable blast disease control program would focus on filling our information gaps in wheat-blast interactions, and integrating biological and biorational products into the broad disease management framework.

**Acknowledgments** This work was funded by the Krishi Gobeshona Foundation (KGF), Bangladesh through KGF TF 50-C/17 to Tofazzal Islam of the Institute of Biotechnology and Genetic Engineering of BSMRAU, Bangladesh.

## References

Anderson MZ, Porman AM, Wang N, Mancera E, Huang D, Cuomo CA, Bennett RJ (2016) A multistate toggle switch defines fungal cell fates and is regulated by synergistic genetic cues. PLoS Genet 12(10):e1006353

Ankri S, Mirelman D (1999) Antimicrobial properties of allicin from garlic. Microbes Infect 1:125–129

Arrebola E, Sivakumar D, Korsten K (2009) Effect of volatile compounds by *bacillus* strains on post-harvest decay in citrus. Biol Control 53:122–128

Awla HK, Kadir J, Othman R, Rashid TS, Wong M-Y (2016) Bioactive compounds produced by *Streptomyces* spp. isolate UPMRS4 and antifungal activity against *Pyricularia oryzae*. Am J Plant Sci 7:1077–1085

Azmir J, Zaidul ISM, Rahman MM, Sharif KM, Mohamed A, Sahena F, Jahurul MHA, Ghafoor K, Norulaini NAN, Omar AKM (2013) Techniques for extraction of bioactive compounds from plant materials: a review. J Food Eng 117:426–436

Baba ZA, Hamid B, Sheikh TA, Alotaibi S, Enshasy HE, Ansari JA, Zuan ATK, Sayyed R Z (2021) Psychrotolerant *Mesorhizobium* sp. isolated from temperate and cold desert regions solubilize potassium and produces multiple plant growth promoting metabolites. Molecules 26:5758

Bakkali F, Averbeck S, Averbeck D, Idaomar M (2008) Biological effects of essential oils-a review. Food Chem Toxicol 46:446–475

Barka EA, Vatsa P, Sanchez L, Gaveau-Vaillant N, Jacquard C, Klenk H-P, Clement C, Ouhdouch Y, van Wezel GP (2016) Taxonomy, physiology, and natural products of actinobacteria. Microbiol Mol Biol Rev 80:1–43

Barrera-Necha LL, Garduno-Pizana C, Garcia-Barrera LJ, Garduño-Pizaña C, Garcia-Barrera LJ (2008) *In vitro* antifungal activity of essential oils and their compounds on mycelial growth of *fusarium oxysporum* f. spp. *gladioli* (Massey) Snyder and Hansen. Plant Pathol J 8:17–21

Basmadjian C, Thuaud F, Ribeiro N, Désaubry L (2013) Flavaglines: potent anticancer drugs that target prohibitins and the helicase eIF4A. Future Med Chem 5(18):2185–2197

Becker J, Liermann JC, Opatz T, Anke H, Thines E (2011) GKK1032A2, a secondary metabolite from *Penicillium* spp. IBWF-029-96, inhibits conidial germination in the rice blast fungus *Magnaporthe oryzae*. J Antibiot 65(2):99–102

Berdy J (2005) Bioactive microbial metabolites. J Antibiot 58:1–26

Buatong J, Rukachaisirikul V, Sangkanu S, Surup F, Phongpaichit S (2019) Antifungal metabolites from marine-derived *Streptomyces* spp. AMA49 against *Pyricularia oryzae*. J Pure Appl Microbiol 13(2):653–665

Cantrell CL, Dayan FE, Duke SO (2012) Natural products as sources for new pesticides. J Nat Prod 75(6):1231–1242

Cardoso CADA, Reis EM, Moreira EN (2008) Development of a warning system for wheat blast caused by *Pyricularia grisea*. Summa Phytopathol 34:216–221

Castroagudín VL, Ceresini PC, de Oliveira SC, Reges JTA, Maciel JLN, Bonato ALV, Dorigan AF, McDonald BA (2015) Resistance to QoI fungicides is widespread in Brazilian populations of the wheat blast pathogen *Magnaporthe oryzae*. Phytopathology 105:284–294

Chakraborty M, Mahmud NU, Gupta DR, Tareq FS, Shin HJ, Islam T (2020a) Inhibitory effects of linear lipopeptides from a marine *Bacillus subtilis* on the wheat blast fungus *Magnaporthe oryzae Triticum*. Front Microbiol 11:665

Chakraborty M, Mahmud NU, Muzahid ANM, Rabby SMF, Shin HJ, Islam T (2020b) Oligomycins inhibit *Magnaporthe oryzae Triticum* and suppress wheat blast disease. PLoS One 15(8):e0233665

Chakraborty M, Mahmud NU, Ullah C, Rahman M, Islam T (2021) Biological and biorational management of blast diseases in cereals caused by Magnaporthe oryzae. Crit Rev Biotechnol 41(7):994–1022.

Chan YK, McCormick WA, Seifert KA (2003) Characterization of an antifungal soil bacterium and its antagonistic activities against *fusarium* species. Can J Microbiol 49:253–262

Chen XL, Shi T, Yang J, Shi W, Gao X, Chen D, Xu X, Xu JR, Talbot NJ, Peng YL (2014) N-glycosylation of effector proteins by an alpha-1,3-mannosyltransferase is required for the rice blast fungus to evade host innate immunity. Plant Cell 26:1360–1376

Chernin L, Chet I (2002) Enzymes in the environment: activity, ecology, and applications. Marcel Dekker, NY:171–225

Chung S, Kong H, Buyer JS, Lakshman DK, Lydon J, Kim S-D, Roberts DP (2008) Isolation and partial characterization of *Bacillus subtilis* ME488 for suppression of soilborne pathogens of cucumber and pepper. Appl Microbiol Biotechnol 80:115–123

Chutrakul C, Boonruangprapa T, Suvannakad R, Isaka M, Sirithunya P, Toojinda T, Kirtikara K (2009) Ascherxanthone B from *Aschersonia luteola*, a new antifungal compound active against rice blast pathogen *Magnaporthe grisea*. J Appl Microbiol 107(5):1624–1631

Copping LG, Duke SO (2007) Natural products that have been used commercially as crop protection agents. Pest Manag Sci 63(6):524–554

Curtis H, Noll U, Störmann J, Slusarenko AJ (2004) Broad-spectrum activity of the volatile phytoanticipin allicin in extracts of garlic (*Allium sativum* L.) against plant pathogenic bacteria, fungi and oomycetes. Physiol Mol Plant P 65(2):79–89

Dame ZT, Islam MT, Helmke E, von Tiedemann A, Laatsch H (2016) Oligomycins and pamamycin homologs impair motility and induce lysis of zoospores of the grapevine downy mildew pathogen, *Plasmopara viticola*. FEMS Microbiol Lett 363:fnw167

Dean R, Van Kan JAL, Pretorius ZA, Hammond-Kosack KE, Di Pietro A, Spanu PD, Rudd JJ, Dickman M, Kahmann R, Ellis J, Foster GD (2012) The top 10 fungal pathogens in molecular plant pathology. Mol Plant Pathol 13(4):414–430

Dorigan AF, Carvalho GD, Poloni NM, Negrisoli MM, Maciel JLN, Ceresini PC (2019) Resistance to triazole fungicides in *Pyricularia* species associated with invasive plants from wheat fields in Brazil. Acta Scientia Agron 41(1):39332

Dutta S, Zahan SM, Gupta DR, Mahmud NU, Chanclud E, Win J, Kamoun S, Islam MT (2018) Genomic analyses reveal that biocontrol of wheat blast by *Bacillus* spp. may be linked with production of antimicrobial compounds and induced systemic resistance in host plants, Figshare

Egan MJ, Wang Z-Y, Jones MA, Smirnoff N, Talbot NJ (2007) Generation of reactive oxygen species by fungal NADPH oxidases is required for rice blast disease. Proc Natl Acad Sci 104(28):11772–11777

El Refaei MI (1977) Epidemiology of rice blast disease in tropics with special reference to the leaf wetness in relation to disease development. Indian Agricultural Research Institute, New Delhi, India

Engelmeier D, Hadacek F, Pacher T, Vajrodaya S, Greger H (2000) Cyclopenta[b] benzofurans from *Aglaia* species with pronounced antifungal activity against rice blast fungus (*Pyricularia grisea*). J Agr Food Chem 48(4):1400–1404

Froyd J, Paget C, Guse L, Dreikorn B, Pafford J (1976) Tricyclazole: a new systemic fungicide for control of *Pyricularia oryzae* on rice. Phytopathology 66:1135–1139

Fry FH, Okarter N, Baynton-Smith C, Kershaw MJ, Talbot NJ, Jacob C (2005) Use of a substrate/alliinase combination to generate antifungal activity *in situ*. J Agr Food Chem 53(3):574–580

Fungicide Resistance Action Committee., www.frac.com (2016)

Ganeshan G, Manoj Kumar A (2005) *Pseudomonas fluorescens*, a potential bacterial antagonist to control plant diseases. J Plant Interact 1(3):123–134

Gashaw G, Alemu T, Tesfaye K (2014) Morphological, physiological and biochemical studies on *Pyricularia grisea* isolates causing blast disease on finger millet in Ethiopia. J Appl Biosci 74(1):6059

Giraldo MC, Dagdas YF, Gupta YK, Mentlak TA, Yi M, Martinez-Rocha AL, Saitoh H, Terauchi R, Talbot NJ, Valent B (2013) Two distinct secretion systems facilitate tissue invasion by the rice blast fungus *Magnaporthe oryzae*. Nat Commun 4:1996

Gória MM, Ghini R, Bettiol W (2013) Elevated atmospheric CO2 concentration increases rice blast severity. Trop Plant Pathol 38:253–257

Gupta DR, Avila CR, Win J, Soanes DM, Ryder LS, Croll D, Bhattacharjee P, Hossain MS, Mahmud NU, Mehebub MS, Surovy MZ, Rahman MM, Talbot NJ, Kamoun S, Islam MT (2018) Cautionary notes on use of the MoT3 diagnostic assay for *Magnaporthe oryzae* wheat and rice blast isolates. Phytopathology 109(4):504–508

Gwinn KD (2018) Bioactive natural products in plant disease control. Stud Nat Prod Chem:1229–1246

Hahn M (2014) The rising threat of fungicide resistance in plant pathogenic fungi: *botrytis* as a case study. J Chem Biol 7(4):133–141

Hamer JE, Howard RJ, Chumley FG, Valent B (1988) A mechanism for surface attachment in spores of a plant pathogenic fungus. Science 239:288–290

He Y, Zhu M, Huang J, Hsiang T, Zheng L (2019) Biocontrol potential of a *Bacillus subtilis* strain BJ-1 against the rice blast fungus *Magnaporthe oryzae*. Can J Plant Pathol 41(1):47–59

Homma Y, Kato Z, Hirayama F, Konno K, Shirahama H, Suzui T (1989) Production of antibiotics by *pseudomonas cepacia* as an agent for biological control of soilborne plant pathogens. Soil Biol Biochem 21:723–728

Hossain MA, Claggett JM, Edwards SR, Shi A, Pennebaker SL, Cheng MY, Hasty J, Johnson TL (2016) Post-transcriptional regulation of Gcr1 expression and activity is crucial for metabolic adjustment in response to glucose availability. Mol Cell 62(3):346–358

Howard RJ, Valent B (1996) Breaking and entering: host penetration by the fungal rice blast pathogen *Pyricularia grisea*. Annu Rev Microbiol 50:491–512

Igarashi S (1990) Update on wheat blast (*Pyricularia oryzae*) in Brazil. In: Saunders D (ed) Proceedings of the international conference – wheat for the nontraditional warm areas. CIMMYT, Mexico DF, Mexico, pp 480–483

Ishii H (2015) Fungicide resistance in plant pathogens—principles and a guide to practical management. In: Hollomon DW (ed) Ishii H. Springer, New York, pp 35–48

Islam MT (2008) Disruption of ultrastructure and cytoskeletal network is involved with biocontrol of damping-off pathogen *Aphanomyces cochlioides* by *Lysobacter* spp. strain SB-K88. Biol Control 46(3):312–321

Islam MT, Tiedemann AV, Laatsch H (2011) Protein kinase C is likely to be involved in zoosporogenesis and maintenance of flagellar motility in the peronosporomycete zoospores. Mol Plant Microbe 24(8):938–947

Islam MT, Croll D, Gladieux P, Soanes DM, Persoons A, Bhattacharjee P, Hossain MS, Gupta DR, Rahman MR, Mahboob GM, Cook N, Salam MU, Surovy MZ, Sancho VB, Maciel JLN, Nhani Júnior A, Castroagudín VL, de Assis Reges JT, Ceresini PC, Ravel S, Kellner R, Fournier E, Tharreau D, Lebrun M-H, McDonald BA, Stitt T, Swan D, Talbot NJ, Saunders DGO, Win J, Kamoun S (2016a) Emergence of wheat blast in Bangladesh was caused by a south American lineage of *Magnaporthe oryzae*. BMC Biol 14:84

Islam MT, Laatsch H, von Tiedemann A (2016b) Inhibitory effects of macrotetrolides from *Streptomyces* spp. on zoosporogenesis and motility of peronosporomycete zoospores are likely linked with enhanced atpase activity in mitochondria. Front Microbiol 7:1824

Islam MT, Kim KH, Choi J (2019) Wheat blast in Bangladesh: the current situation and future impacts. Plant Pathol J 35(10):1–10

Islam, M.T., Gupta, D.R., Hossain, A., Roy, K.K., He, X., Kabir, M.R., Singh, P.K., Khan, M.A.R., Rahman, M., Wang, G.L. (2020) Wheat blast: a new threat to food security. Phytopathology Research 2(1). https://doi.org/10.1186/s42483-020-00067-6

Isono K, Nagatsu J, Kawashima Y, Suzuki S (1965) Studies on polyoxins, antifungal antibiotics. Agric Biol Chem 29(9):848–854

Jadhav HP, Sonawane MS, Khairnar MH, Sayyed RZ (2020) Production of alkaline protease by rhizospheric *Bacillus cereus* HP_RZ17 & *Paenibacillus xylanilyticus* HP_RZ19. Environ Sustain 3:5–13

Jeon YT, Ryu KH, Kang MK, Park S-H, Yun H, Pham QT, Kim S-U (2010) Alternariol monomethyl ether and α, β-dehydrocurvularin from endophytic fungi *Alternaria* spp. inhibit appressorium formation of *Magnaporthe grisea*. J Korean Soc Appl Bi 53(1):39–42

Kankanala P, Czymmek K, Valent B (2007) Roles for rice membrane dynamics and plasmodesmata during biotrophic invasion by the blast fungus. Plant Cell 19:706–724

Karthikeyan V, Gnanamanickam S (2008) Biological control of Setaria blast (*Magnaporthe grisea*) with bacterial strains. Crop Prot 27:263–267

Khalil MS, Moubasher H, Hasan FF (2014) Biological control of rice blast disease by *Streptomyces flavotricini*. Res J Pharm Biol Chem Sci 5:1453–1461

Kim BS, Moon SS, Hwang BK (1999) Isolation, identification, and antifungal activity of a macrolide antibiotic, oligomycin a, produced by *Streptomyces libani*. Can J Bot 77:850–858

Kim PI, Ryu J, Kim YH, Chi YT (2010) Production of biosurfactant lipopeptides iturin a, fengycin and surfactin a from *Bacillus subtilis* CMB32 for control of *Colletotrichum gloeosporioides*. J Microbiol Biotechnol 20:138–145

Kim BR, Han JW, Ngo TM, Le DQ, Kim JC, Kim H, Choi GJ (2018) Identification of novel compounds, oleanane- and ursane-type triterpene glycosides, from *Trevesia palmata*: their biocontrol activity against phytopathogenic fungi. Sci Rep 8:14522

Koul O, Walia S, Dhaliwal SG (2008) Essential oils as green pesticides: potential and contraints. Biopestic Int 4:68–84

Kusale SP, Attar YC, Sayyed RZ, Malek RA, Ilyas N, Suriani NL, Khan N, Enshasy HE (2021) Production of plant beneficial and antioxidants metabolites by *Klebsiella variicola* under salinity stress. Molecules 26:1894

Lamberth C (2016) Naturally occurring amino acid derivatives with herbicidal, fungicidal or insecticidal activity. Amino Acids 48940:929–940

Law JW, Ser H, Khan TM, Chuah L, Pusparajah P, Chan K, Goh B, Lee L (2017) The potential of *Streptomyces* as biocontrol agents against the rice blast fungus, *Magnaporthe oryzae* (*Pyricularia oryzae*). Front Microbiol 8:3

Leahy J, Mendelsohn M, Kough J, Jones R, Berckes N (2014) Biopesticides: state of the art and future opportunities. In: Gross A, Coats JR, Duke SO, Seiber JN (eds) American Chemical Society, Washington, DC, pp 3–18

Leelasuphakul W, Sivanunsakul P, Phongpaichit S (2006) Purification, characterization and synergistic activity of β-1,3-glucanase and antibiotic extract from an antagonistic *Bacillus subtilis* NSRS 89-24 against rice blast and sheath blight. Enzym Microb Technol 38(7):990–997

Li JY, Strobel GA, Harper JK, Lobkovsky E, Clardy J (2000) Cryptocin, a potent tetramic acid antimycotic from the endophytic fungus *Cryptosporiopsis cf. quercina*. Org Lett 2:767–770

Li Q, Jiang Y, Ning P, Zheng L, Huang J, Li G, Jiang D, Hsiang T (2011) Suppression of *Magnaporthe oryzae* by culture filtrates of *Streptomyces globisporus* JK-1. Biol Control 58(2):139–148

Li L, Wu J, Deng Z, Zabriskie TM, He X (2013) *Streptomyces lividans* Blasticidin S deaminase and its application in engineering a Blasticidin S-producing strain for ease of genetic manipulation. Appl Environ Microbiol 79(7):2349–2357

Li H, Guan Y, Dong Y, Zhao L, Rong S, Chen W, Miaomiao Lv XH, Gao X, Chen R, Li L, Xu Z (2018) Isolation and evaluation of endophytic *bacillus tequilensis* GYLH001 with potential application for biological control of *Magnaporthe oryzae*. PLoS One 13(10):e0203505

Liberato JR, Rodrigues FA, Cother EJ, Bonman JM, Shivas RG (2006) Rice blast disease (*Magnaporthe oryzae*). Available online: PaDIL - http://www.padil.gov.au

Licata M, Tuttolomondo T, Dugo G, Ruberto G, Leto C, Napoli EM, Rando R, Rita Fede M, Virga G, Leone R, La Bella S (2015) Study of quantitative and qualitative variations in essential oils of *Sicilian oregano* biotypes. J Essent Oil Res 27:293–306

Lim SM, Yoon MY, Choi GJ, Choi YH, Jang KS, Shin TS, Park HW, Yu NH, Kim YH, Kim J-C (2017) Diffusible and volatile antifungal compounds produced by an antagonistic *bacillus velezensis* G341 against various phytopathogenic fungi. Plant Pathol J 33(5):488–498

Liu W, Liu J, Ning Y, Ding B, Wang X, Wang Z, Wang GL (2013) Recent progress in understanding PAMP- and effector-triggered immunity against the rice blast fungus *Magnaporthe oryzae*. Mol Plant 6:605–620

Liu X, Li W, Hu B, Wang M, Wang J, Guan L (2018) Identification of isobavachalcone as a potential drug for Rice blast disease caused by the fungus *Magnaporthe grisea*. J Biomol Struct Dyn 1–17

Li-Weber M (2014) Molecular mechanisms and anti-cancer aspects of the medicinal phytochemicals rocaglamides (=flavaglines). Int J Cancer 137(8):1791–1799

Magar PB, Acharya B, Pandey B (2015) Use of chemical fungicides for the management of rice blast (*Pyricularia Grisea*) disease at Jyotinagar, Chitwan, Nepal. Int J Appl Sci Biotechnol 3(3):474

Malaker PK, Barma NC, Tewari TP, Collis WJ, Duveiller E, Singh PK, Joshi AK, Singh RP, Braun HJ, Peterson GL, Pedley KF, Farman ML, Valent B (2016) First report of wheat blast caused by *Magnaporthe oryzae* pathotype *Triticum* in Bangladesh. Plant Dis 100(11):2330

Mentlak TA, Kombrink A, Shinya T, Ryder LS, Otomo I, Saitoh H, Terauchi R, Nishizawa Y, Shibuya N, Thomma BPHJ, Talbot NJ (2012) Effector-mediated suppression of chitin-triggered immunity by *Magnaporthe oryzae* is necessary for rice blast disease. Plant Cell 24(1):322–335

Mihalache G, Balaes T, Gostin I, Stefan M, Coutte F, Krier F (2018) Lipopeptides produced by *Bacillus subtilis* as new biocontrol products against fusariosis in ornamental plants. Environ Sci Pollut Res Int 25(30):29784–29793

Moghaddam M, Mehdizadeh L (2015) Variability of total phenolic, flavonoid and rosmarinic acid content among Iranian basil accessions. LWT - Food Sci Technol 63:535–540

Moriguchi Y, Kihara J, Ueno M (2019) Suppression effects of a secondary metabolite of *Biscogniauxia* spp. strain O-811 obtained from mushrooms against the rice blast fungus *Magnaporthe oryzae*. Bull Fac Life Env Sci Shimane Univ 24:14–18

Mosquera G, Giraldo MC, Khang CH, Coughlan S, Valent B (2009) Interaction transcriptome analysis identifies *Magnaporthe oryzae* BAS1-4 as biotrophy-associated secreted proteins in rice blast disease. Plant Cell 21:1273–1290

Müller A, Eller J, Albrecht F, Prochnow P, Kuhlmann K, Bandow JE, Slusarenko AJ, Leichert LIO (2016) Allicin induces thiol stress in bacteria through S-allylmercapto modification of protein cysteines. J Biol Chem 291(22):11477–11490

Nakata M, Ishiyama T, Akamatsu S, Hirose Y, Maruoka H, Suzuki R, Tatsuta K (1995) Synthetic studies on oligomycins. Synthesis of the oligomycin B spiroketal and polypropionate portions. Bull Chem Soc Jpn 68(3):967–989

Nayak SK, Nayak S, Mishra BB (2017) Antimycotic role of soil *bacillus* spp. against rice pathogens: a biocontrol prospective. Microb Biotechnol 1:29–60

Negi PS (2012) Plant extracts for the control of bacterial growth: efficacy, stability and safety issues for food application. Int J Food Microbiol 156:7–17

Ngo MT, Han JW, Yoon S, Bae S, Kim S-Y, Kim H, Choi GJ (2019) Discovery of new triterpenoid saponins isolated from *Maesa japonica* with antifungal activity against rice blast fungus *Magnaporthe oryzae*. J Agri Food Chem 67(27):7706–7715

Nguyen QT, Ueda K, Tamura T, Kihara J, Itoh K, Yoshikiyo K, Ueno M (2018) Antifungal activity of a novel compound purified from the culture filtrate of *Biscogniauxia* spp. O821 against the rice blast fungus *Magnaporthe oryzae*. J Gen Plant Pathol 84(2):142–147

Norvienyeku J, Lin L, Waheed A, Chen X, Bao J, Shabbir A, Lin L, Zhong Z, Batool W, Aliyu SR, Zhou J, Lu G, Wang Z (2019) Bayogenin 3-O-Cellobioside is a novel non-cultivar specific anti-blast metabolite produced in rice in response to *Pyricularia oryzae* infection. Biorxiv 647636

Omoboye OO, Oni FE, Batool H, Yimer HZ, De Mot R, Höfte M (2019) *Pseudomonas* cyclic lipopeptides suppress the rice blast fungus *Magnaporthe oryzae* by induced resistance and direct antagonism. Front Plant Sci 10:901

Ongena M, Jacques P (2008) *Bacillus* lipopeptides: versatile weapons for plant disease biocontrol. Trends Microbiol 16:115–125

Park J-H, Choi GJ, Jang KS, Lim HK, Kim HT, Cho KY, Kim J-C (2005) Antifungal activity against plant pathogenic fungi of chaetoviridins isolated from *Chaetomium globosum*. FEMS Microbiol Lett 252(2):309–313

Patkar RN, Benke PI, Qu Z, Constance CYY, Yang F, Swarup S, Naqvi NI (2015) A fungal monooxygenase-derived jasmonate attenuates host innate immunity. Nat Chem Biol 11:733–740

Penn TJ, Wood ME, Soanes DM, Csukai M, Corran AJ, Talbot NJ (2015) Protein kinase C is essential for viability of the rice blast fungus *Magnaporthe oryzae*. Mol Microbiol 98(3):403–419

Perelló A, Martinez I, Molina M (2015) First report of virulence and effects of *Magnaporthe oryzae* isolates causing wheat blast in Argentina. Plant Dis 99(8):1177

Peterson LG (1990) Tricyclazole for control of *Pyricularia Oryzae* on Rice: the relationship of the mode of action and disease occurrence and development. In: Grayson BT, Green MB, Copping LG (eds) Pest management in rice. Springer, Dordrecht, pp 122–130

Prabavathy VR, Mathivanan N, Murugesan K (2006) Control of blast and sheath blight diseases of rice using antifungal metabolites produced by *Streptomyces* spp. PM5. Biol Control 39(3):313–319

Radjacommare R, Ramanathan A, Kandan A, Sible GV, Harish S, Samiyappan R (2004) Purification and antifungal activity of chitinase against *Pyricularia grisea* in finger millet. World J Microbiol Biotechnol 20(3):251–256

Reshma P, Naik MK, Aiyaz M, Niranjana SR, Chennappa G, Shaikh SS, Sayyed RZ (2018) Induced systemic resistance by 2,4diacetylphloroglucinol positive fluorescent *pseudomonas* strains against rice sheath blight. Indian J Exp Biol 56(3):207–212

Rhee Hyeong KI (2003) Purification and identification of an antifungal agent from *Streptomyces* spp. KH-614 antagonistic to rice blast fungus, *Pyricularia oryzae*. J Microbiol Biotechnol 13:984–988

Rieger PH, Liermann JC, Opatz T, Anke H, Thines E (2010) Caripyrin, a new inhibitor of infection-related morphogenesis in the rice blast fungus *Magnaporthe oryzae*. J Antibiot 63(6):285–289

Sagar A, Sayyed RZ, Ramteke PW, Sharma S, Marraiki N, Elgorban AM, Syed A (2020) ACC deaminase and antioxidant enzymes producing halophilic *Enterobacter* sp. PR14 promotes the growth of rice and millets under salinity stress. Physiol Mol Biol Plants 26:1847–1854

Sandjo LP, Thines E, Opatz T, Schüffler A (2014) Tanzawaic acids I–L: four new polyketides from *Penicillium* sp. IBWF104-06. Beilstein J Org Chem 10:251–258
Saunders DGO, Aves SJ, Talbot NJ (2010) Cell cycle-mediated regulation of plant infection by the rice blast fungus. Plant Cell 22(2):497–507
Sayyed RZ, Patel PR, Shaikh SS (2015) Plant growth promotion and root colonization by EPS producing *Enterobacter* sp. RZS5 under heavy metal contaminated soil. Indian J Exp Biol 53:116–123
Sayyed RZ, Patel PR, Shaikh SS (2016) Dynamism of PGPR in bioremediation and plant growth promotion in heavy metal contaminated soil. Indian J Exp Biol 54:286–290
Shaikh SS, Wani SJ, Sayyed RZ, Thakur R, Gulati A (2018) Production, purification and kinetics of chitinase of *Stenotrophomonas maltophilia* isolated from rhizospheric soil. Indian J Exp Biol 56(4):274–278
Shan H, Zhao M, Chen D, Cheng J, Li J, Feng Z, An D (2013) Biocontrol of rice blast by the phenaminomethylacetic acid producer of *bacillus methylotrophicus* strain BC79. Crop Prot 44:29–37
Sharma R, Girish AG, Upadhyaya HD, Humayun P, Babu TK, Rao VP, Thakur RP (2014) Identification of blast resistance in a core collection of foxtail millet germplasm. Plant Dis 98(4):519–524
Shoda M (2000) Bacterial control of plant diseases. J Biosci Bioeng 6:515–521
Sugahara A, Yoshimi A, Shoji F, Fujioka T, Kawai K, Umeyama H, Komatsu K, Enomoto M, Kuwahara S, Hagiwara D, Katayama T, Horiuchi H, Miyazawa K, Nakayama M, Abe K (2019) A novel antifungal compound Z-705 specifically inhibits protein kinase C of filamentous fungi. Appl Environ Microbiol 85:e02923–e02918
Svidritskiya E, Lingb C, Ermolenkob DN, Korosteleva AA (2013) Blasticidin S inhibits translation by trapping deformed tRNA on the ribosome. Proc Natl Acad Sci U S A 110:12283–12288
Tajul MI, Motoyama T, Hatanaka A, Sariah M, Osada H (2011) Green-odour compounds have antifungal activity against the rice blast fungus *Magnaporthe oryzae*. Eur J Plant Pathol 132(1):91–100
Tamoki T, Nomoto H, Takahashi I, Kato Y, Morimoto M, Tomita F (1986) Staurosporine, a potent inhibitor of phospholipid calcium dependent protein kinase. Biochem Biophys Res Commun 135:397–402
Tendulkar SR, Saikumari YK, Patel V, Raghotama S, Munshi TK, Balaram P, Chattoo BB (2007) Isolation, purification and characterization of an antifungal molecule produced by *bacillus licheniformis* BC98, and its effect on phytopathogen *Magnaporthe grisea*. J Appl Microbiol 103(6):2331–2339
Todorova S, Kozhuharova L (2010) Characteristics and antimicrobial activity of *Bacillus subtilis* strains isolated from soil. World J Microbiol Biotechnol 26:1207–1216
Umezawa H, Okami Y, Hashimoto T, Suhara Y, Hamada M, Takeuchi T (1965) A new antibiotic, kasugamycin. J Antibiot Ser A 18:101–103
Urashima AS, Hashimoto Y, Le Don D, Kusaba M, Tosa Y, Nakayashiki H, Mayama S (1999) Molecular analysis of the wheat blast population in Brazil with a homolog of retrotransposon MGR583. Jpn J Phytopathol 65:429–436
US Environmental Protection Agency (EPA) (2004) https://www3.epa.gov/pesticides/chem_search/reg_actions/registration/decision_PC-100053_2-Feb-04.pdf. Accessed 25 Nov 2019
Valent B, Farman M, Tosa Y, Begerow D, Fournier E, Gladieux P, Islam MT, Kamoun S, Kemler M, Kohn LM, Lebrun MH, Stajich JE, Talbot NJ, Terauchi R, Tharreau D, Zhang N (2019) *Pyricularia graminis-tritici* is not the correct species name for the wheat blast fungus: response to Ceresini et al., (MPP 20:2). Mol Plant Pathol 20(2):173–179
Vinay JU, Naik MK, Rangeshwaran R, Chennappa G, Shaikh SS, Sayyed RZ (2016) Detection of antimicrobial traits in fluorescent pseudomonads and molecular characterization of an antibiotic pyoluteorin. 3. Biotech 6(227):1–11
Vu TT, Kim H, Tran VK, Le Dang Q, Nguyen HT, Kim H, Kim IS, Choi GJ, Kim JC (2015) *In vitro* antibacterial activity of selected medicinal plants traditionally used in Vietnam against human pathogenic bacteria. BMC Complem Altern M 16(1):32

Wilson RA, Talbot NJ (2009) Under pressure: investigating the biology of plant infection by *Magnaporthe oryzae*. Nat Rev Microbiol 3:185–195

Woloshuk CP, Sisler HD, Vigil EL (1983) Action of the antipenetrant, tricyclazole, on appressoria of *Pyricularia oryzae*. Physiol Plant Pathol 22(2):245–259

Xiong Z-Q, Tu X-R, Wei S-J, Huang L, Li X-H, Lu H, Tu GQ (2013) *In vitro* antifungal activity of antifungalmycin 702, a new polyene macrolide antibiotic, against the rice blast fungus *Magnaporthe grisea*. Biotechnol Lett 35(9):1475–1479

Xu S, Wang B, Li L, Zhou Q, Tian M, Zhao X, Peng J, Liu F, Cheng Y, Xu Y, Feng X (2019) Effects of camptothecin on the rice blast fungus *Magnaporthe oryzae*. Pestic Biochem Physiol 163:108–116

Yamaguchi I (1982) Fungicides for control of rice blast disease. J Pest Sci 7:307–316

Yan X, Talbot NJ (2016) Investigating the cell biology of plant infection by the rice blast fungus *Magnaporthe oryzae*. Curr Opin Microbiol 34:147–153

Yun CS, Motoyama T, Osada H (2015) Biosynthesis of the mycotoxin tenuazonic acid by a fungal NRPS-PKS hybrid enzyme. Nat Commun 6:8758

Zaman HB, Farooq S, Fatima S, Sayyed RZ, Baba ZA, Sheikh TA, Reddy MS, El Enshasy H, Gafur A, Ni Luh S (2021) Bacterial plant biostimulants: a sustainable way towards improving growth, productivity, and health of crops. Sustainability 13(5):2856

Zhang L, Sun C (2018) Fengycins, cyclic lipopeptides from marine *Bacillus subtilis* strains, kill the plant-pathogenic fungus *Magnaporthe grisea* by inducing reactive oxygen species production and chromatin condensation. Appl Environ Microbiol 84:e00445–e00418

# Chapter 3
# Utilization of Plant Growth-Promoting Bacteria (PGPB) Against Phytopathogens

**Ali Talha Khalil and Zabta Khan Shinwari**

**Contents**

3.1 Introduction ........ 53
3.2 Phytopathogens ........ 55
3.3 Antibiotic Used on Plants to Control Phytopathogens ........ 56
3.4 PGPB for Control of Phytopathogens ........ 57
3.5 Conclusions ........ 60
References ........ 60

## 3.1 Introduction

Plants in the fields may appear as an individual organism to an unaided eye, whereas these plants are actually a representation of a complex community that harness and maintain a constant, well-structured and dynamic relationship with other partners, like the microorganisms (Kumar et al. 2020; Backer et al. 2018). This dynamic relationship lays foundation of the phytomicrobiome. This relationship of microorganisms and plants dates back to the earlier days after the colonization of land by the terrestrial plants. The phytomicrobiome that constitutes of bacteria and fungi are in association with the major plant structures, like flowers, leaves, roots, etc. in a specialized manner (Backer et al. 2018), that perform different functions that promote the plant growth (Fig. 3.1). With the advent of latest cutting-edge tools used in metagenomics and microbiology, it's not possible to develop a deep understanding

A. T. Khalil
Department of Pathology, Lady Reading Hospital Medical Teaching Institution, Peshawar, Khyber Pakhtunkhwa, Pakistan
e-mail: alitalha.khalil@lrh.edu.pk

Z. K. Shinwari (✉)
Department of Plant Sciences, Quaid-i-Azam University, Islamabad, Pakistan

Pakistan Academy of Sciences, Islamabad, Pakistan



R. Z. Sayyed et al. (eds.), *Antifungal Metabolites of Rhizobacteria for Sustainable Agriculture*, Fungal Biology, https://doi.org/10.1007/978-3-031-04805-0_3

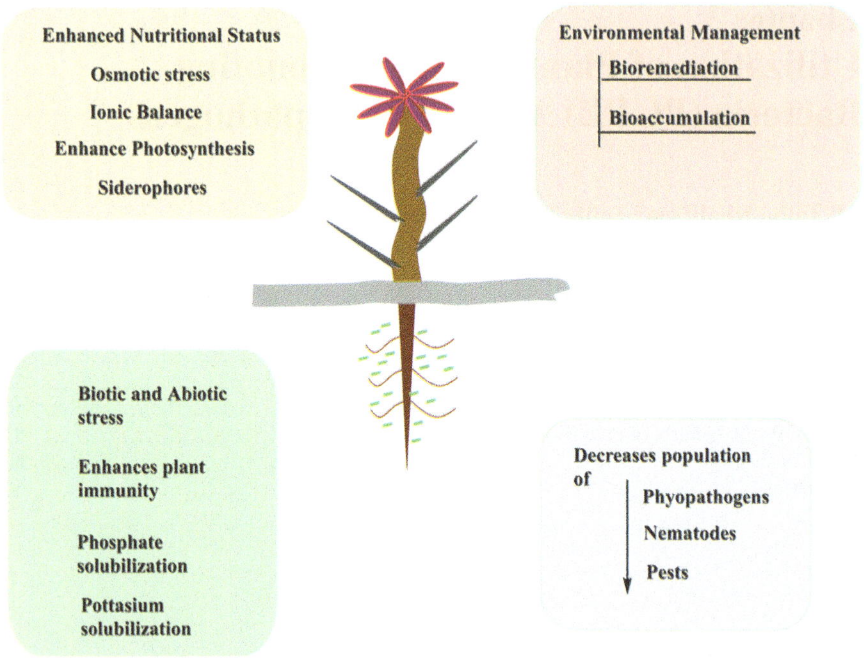


**Fig. 3.1** Applications of plant growth-promoting bacteria

of this relation alongside the different factors affecting it like conditions, genotype, etc. These PGPB, for example, include the bacteria that inhabit the rhizosphere region (PGPR) have the ability to facilitate plant growth, or they are present in the interior tissues in the form of endophytes (Santoyo et al. 2016, 2021). PGPR are among the widely studied for the plant growth promotion potential. It is now established that the PGPR can establish endophytic colonies in the root's interior, may also transcend to other parts of the plants, and adapt to the new ecological niches in the stem, leave, or other organs of the plant (Tharek et al. 2021; Rai et al. 2020). Plants can regulate the composition of the associated PGPB in different ways like secreting signal compounds that dictate the genetic and biochemical regulation of the microbial flora. Recent studies have also indicated the regulation of specific PGPB species, referred as a hub species or core species, which further regulates the entire PGPR community. It is proposed that the core species have remained in association with the plants for long, therefore, occupies a central position in the hierarchy (Toju et al. 2018).

Plant growth-promoting bacteria or PGPB are found around plant tissues that facilitate the growth promotion of the plants through direct or indirect mechanisms (Fig. 3.2). The direct manner in which the bacteria promote the overall well-being of the plant to which they are associated can be through modulation of the plant growth hormones auxin, cytokinin, gibberellin, etc. It is estimated that around 80% of the PGPR has the ability to synthesize and release the auxin as secondary

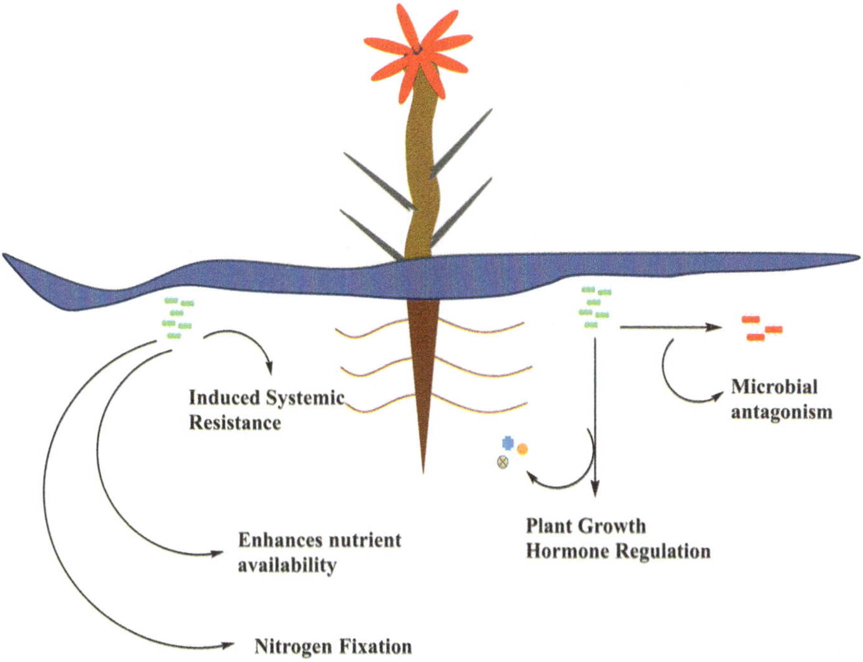


**Fig. 3.2** Different mechanisms of plant growth promotion through plant growth-promoting bacteria

metabolite (Olanrewaju et al. 2017; Shah et al. 2021). Other direct mechanisms include, for example, iron sequestration, phosphorus solubilization, nitrogen fixation, etc. Besides direct modulation from the PGPB, there are indirect mechanisms through which the associated bacteria manifest the plant growth promotion. PGPB are reported to protect the plants from oxidative stress (Kusale et al. 2021). Plant growth-promoting bacteria are reported to enhance the uptake of nutrients (Jabborova et al. 2021; Manasa et al. 2021). These include, for example, biosynthesis of antibiotics and biocidal compounds to counteract the invading pathogens, secretion of enzymes that modulate cell wall degradation, creating a competition for resources, etc. (Olanrewaju et al. 2017).

## 3.2 Phytopathogens

Plant pathogens (phytopathogens) represent an alarming threat to the plant health, overall plant-based agricultural productivities, and the stability of the ecosystem. Phytopathogens rely on their host plants to fulfill their requirements like nutrients and shelter. The physiological dysfunction of host plants manifested by the phytopathogens causes a diseased condition in plants, therefore, are considered as a constant threat. Host plants have evolved intricate mechanisms to counteract the effects of the invading pathogenic microorganisms (Shittu and Obiazikwor 2021). As the human dependence on plants for food, medicinal, and industrially important products has been increasing, the producers have become more reliant on synthetic chemicals as a means of protection of the plants (Compant et al. 2005). These pathogenic microorganisms outsmart the current interventions as they evolve and adapt rapidly (Savary et al. 2019). However, the ever-increasing synthetic means possess myriad complexities and negative effects like contributing to the pathogen resistance and their nontarget environmental impacts (Alsendi et al. 2021).

## 3.3 Antibiotic Used on Plants to Control Phytopathogens

According to the global estimates by CABI (Centre for Agriculture and Bioscience International), 20–40% of the yielded crops is wasted due to pests and diseases (CABI 2021). Losses of staple and commodity crops can have a direct negative impact on the food security, nutrition, and global agriculture-based economies. With the climate fluctuations worldwide, the threat of the phytopathogens has been amplified several folds that can significantly hinder the progress towards achieving sustainable development. The extreme weather conditions influence the phytopathogens and plants. For example, stressed conditions like drought can cause changes to the stomatal aperture which are considered as main gates of entry of the pathogen in leaf disease (Bostock et al. 2014; Juroszek et al. 2020; Qi et al. 2018). Besides the use of pesticides, farmers and producers have now amplified the use of antibiotics to ensure protection of plants from phytopathogens. Use of antibiotics on crops further raises concerns on the development of resistance to antibiotics that are also of medical relevance to human. Resistance to antibiotics has already been developed by phytopathogenic bacteria and can spread across the food chain (Srichamnong et al. 2021; Taylor and Reeder 2020). It is estimated that the use of antibiotics in agriculture is around 63–240 k tons annually, in which the application on crops accounts for 0.26 and 0.5%. However, in some regions the use of antibiotics on crops is high by magnitudes. In response to the citrus greening caused by *Candidatus* Liberibacter spp., the US-EPA proposed streptomycin (292 tons / year) for treating the citrus crops while in the USA, the annual human use of the streptomycin is around 6.3 tons (Jacobs 2019). Oxytetracycline has been applied on plant disease where resistance to the streptomycin has been developed (Sundin and Wang 2018). Other antibiotics like gentamycin, kasugamycin, and oxolinic acid have been used in different countries (Taylor and Reeder 2020). Resistance to the fungicides like azoxystrobin is reported (Kutman et al. 2013; Yin et al. 2020). Considering the potential adverse impacts in the form of antibiotic resistance, alternative ways need to be sought to counteract the phytopathogens for ensuring sustainable production through plant resources.

## 3.4 PGPB for Control of Phytopathogens

Control of plant diseases can be achieved through the applications to different chemical means; however, in view of their long-term negative consequences, chemical means are discouraged. Instead, there is a shift towards biological means or their derived components for controlling the plant infections, which provide eco-friendly as well as a cheap platform to control plant diseases (Wankhade et al. 2019). In this regard, the PGPB has been revealing promising results which through different mechanisms promote the plant health (Verma et al. 2019). Many studies have documented the protection of plant against the phytopathogens. Seedling damping off is considered to significantly hamper the productivity of soybean. Causes of damping off are complex; however, the causative agents that are responsible for the seedling damping off are *Rhizoctonia*, *Pythium*, *Fusarium*, and *Phytophthora*. Studies revealed potential application of PGPR on controlling damping off as well as root rot disease besides other beneficial outcomes in soybean, by inoculation of *Bradyrhizobium japonicum*, *Bacillus cereus*, and *Trichoderma viride* strains. Significant inhibition of *Rhizoctonia* solani, which causes root disease, was achieved by the application of PGPR strains *like Bradyrhizobium japonicum*, *Bacillus megaterium*, *Methylobacterium aminovorans*, *Methylobacterium rhodinum*, and *Trichoderma viride*. More than 50% inhibition of the *Rhizoctonia solani* was obtained using dual plate method (Omara et al. 2018). Inoculation with PGPR strains is reported to enhance the overall health of soybeans in *Macrophomina phaseolina* infested soil, reduced damping off, rotting, and wilting (El-Barougy et al. 2009).

Among the phytopathogens, *Verticillium dahlia* that causes *Verticillium* wilt is of considerable concern as it effects around 160 agronomically important crops (Guenoun et al. 2019). While antifungal antibiotics like aureofungin and benomyl are used to control the infection, however, fungal resistance to these antibiotics has rendered them ineffective (Goicoechea 2009). On the other hand, the biological control of *Verticillium dahlia* has revealed promising results, for example, in olive plants through PGPR *Pseudomonas putida* and *Pseudomonas fluorescens*. Studies revealed in vitro antagonism between *Verticillium dahlia* and *Pseudomonas* spp. *Serratia plymuthica* also revealed in vitro antagonism with fungal strains including *V. dahlia* (Arias-Calderón et al. 2015). Recently, *Bacillus* species also revealed promising results in the biocontrol of *Verticillium* wilt in pepper. Among the *Bacillus* species, *Bacillus subtilis* was found as most. Table 3.1 Updated review of literature (2019–2020) regarding PGPB for controlling phytopathogens effective as compared to the *B. amyloliquefaciens* (Guenoun et al. 2019). *Bacillus zanthoxyli has* emerged as another potential biocontrol agents for *Phytophthora capsica* and *Ralstonia olanacearum* in paprika and tomato (Usmonov et al. 2021).

**Table 3.1** Summarizes recent studies on the application of different biocontrol agents

| PGPB as BCA | Isolated from | Biocontrol applications | Reference |
|---|---|---|---|
| *Pseudomonas segetis* | *Salicornia europaea* | Reduced soft rot symptoms caused by *Dickeya solani*, *Pectobacterium atrosepticum*, and *P. carotovorum* on potato and carrot<br>Protected tomato plants against manifestations of *pseudomonas syringae* infection | (Rodríguez et al. 2020) |
| *Bacillus velezensis* | *Olea europaea* | Inhibited *Verticillium dahlia* up to 92% in vitro that causes *Verticillium* wilt<br>In field conditions, it decreased the percentage of dead plants and disease severity index | (Azabou et al. 2020) |
| *Pseudomonas mediterranea* | *Nicotiana tabacum* | Revealed broad-spectrum antagonism against the causative agents of different plant diseases like:<br>*Rhizoctonia solani.* (wheat sharp eyespot)<br>*Phytophthora nicotianae* (tobacco black shank)<br>*Pseudomonas syringae* (tobacco wildfire disease)<br>*Fusarium graminearum* (corn stalk rot) | (Gu et al. 2020) |
| *Bacillus* spp. | *Zea mays* | *B. Licheniformis* protected pepper from gray mold<br>*B. Licheniformis* and *B. pumilus* inhibited *B. cinerea* and *F. solani* | (Márquez et al. 2020) |
| *Bacillus velezensis* | *Lilium leucanthum* | Inhibited phytopathogens *F. fujikuroi*, *B. dothidea*, *F. oxysporum*, *B. cinerea* | (Khan et al. 2020) |
| *Bacillus*, *pseudomonas*, *Sphingomonas* | *Dendrobium officinale* | Revealed antagonistic activity against *Sclerotium rolfsii* which causes southern blight on *Dendrobium officinale* | (Shen et al. 2020) |
| *Bacillus subtilis*, *bacillus flexus* | *S. surattense*<br>*A. indicum* | Revealed inhibition of the wilt causing pathogens like *Ralstonia solanacearum* and *fusarium oxysporum* | (Jinal and Amaresan 2020) |

(continued)

**Table 3.1** (continued)

| PGPB as BCA | Isolated from | Biocontrol applications | Reference |
|---|---|---|---|
| *B. subtilis, B. velezensis, B. pumilus, B. altitudinis B. Laterosporus* | Ryegrass | Revealed pathogen antagonism on different bacterial and fungal strains | (Li et al. 2020) |
| *B. subtilis, B. amyloliquefaciens, P. fluorescens, P. aeruginosa* | *Lycopersicon esculentum* | Revealed significant potential in reduction of the bacterial wilt disease caused by *Clavibacter michiganensis* | (Abo-Elyousr et al. 2019) |
| *Streptomyces* isolates | *Oryza sativa* | Revealed promising results in the biocontrol of bacterial panicle blight caused by *Burkholderia glumae* | (Suárez-Moreno et al. 2019) |
| Isolates of *bacillus, staphylococcus, Ochrobactrum, pseudomonas, Lysinibacillus, micrococcus, Leifsonia, Exiguobacterium, Arthrobacter* | *Camellia sinensis* | Conferred resistance to rice plants from *R. solani* | (Bhattacharyya et al. 2020) |
| *Bacillus xiamenensis* | *Saccharum officinarum* | Suppression of red rot and enhanced sugarcane growth Inhibited multiple fungal strains | (Xia et al. 2020) |

### 3.4.1 Mechanistic Insights

The PGPB enhances protection of plants against pathogenic organisms through diverse mechanisms that mainly comprise of the synthesis of antibiotics, secondary metabolites, induced systemic resistance, and synthesis of siderophore (Reshma et al. 2018; Nithyapriya et al. 2021). The PGPB has the ability to secrete lytic enzymes that brings about degradation of the harmful organisms. These enzymes include, for example, extracellular secretion of chitinases, proteases, etc. It is established that these enzymes are secreted to degrade the invading pathogenic organisms (Meena et al. 2020).

Production of siderophores is yet another mechanism under debate, by which the PGPR respond to the invading pathogens. These siderophores are low molecular weight compounds with affinity for iron. Through the siderophores, PGPR scavenges the iron and deprives the invading pathogens from iron availability (Tirry et al. 2021; Sultana et al. 2021). Synthesis of antibiotics by the PGPB against the phytopathogens is well established. Antibiotics are small chemical structures that through diverse mechanisms act on the competing organisms. For example, *Bacillus subtilis*, a PGPR, has been known to synthesize more than 20 different compounds that can kill the invading pathogens. *Pseudomonas fluorescens* is known for the production of Pyrrolnitrin, an antifungal antibiotic. Bacillomycin, fengycin, phenazine, polymyxin, mycosubtilin, kanosamine, etc. are some of the antibiotics that have been known to be produced by the PGPB (Meena et al. 2020).

## 3.5 Conclusions

There is immense potential of the plant growth-promoting bacteria for enhancing agriculture produce, ensuring a sustainable food supply, and reducing the environmental burden of the synthetic compounds used for plant protection. Engineering the phytomicrobiome would suppress the invading phytopathogens yielding significant economic outcomes. However, the selection of the PGPR strains for applications in field should be done carefully by considering the biosafety aspects. Research in the field of PGPB is still in infancy and requires research to be done in the basic and applied areas; however, their potential applications has attracted scientists to explore this exciting plant-bacteria relationship for extensive exploration. Expanded research on mechanism of growth promotion would help in prioritizing the PGPB in large scale commercial applications. This intimate relationship between plants and microbes can be source of great interest in the search of bioactive compounds and antibiotics.

**Acknowledgments** We are thankful to the editor for allowing us opportunity to contribute.

## References

Abo-Elyousr KA, Bagy HMK, Hashem M, Alamri SA, Mostafa YS (2019) Biological control of the tomato wilt caused by Clavibacter michiganensis subsp. michiganensis using formulated plant growth-promoting bacteria. Egyptian Journal of Biological Pest Control 29:1–8

Alsendi A, Havasi M, Golmohammadi GR, Farahani S, Kheradmand K (2021) Efficacy of microcapsule essential oil of peppermint (Mentha piperita L.) on biological parameters of two-spotted spider mite, Tetranychus urticae Koch (Acari: Tetranychidae). Plant Prot 44:1–21

Arias-Calderón R, Rodríguez-Jurado D, Bejarano-Alcázar J, Belaj A, De La Rosa R, León L (2015) Evaluation of Verticillium wilt resistance in selections from olive breeding crosses. Euphytica 206:619–629

Azabou MC, Gharbi Y, Medhioub I, Ennouri K, Barham H, Tounsi S, Triki MA (2020) The endophytic strain bacillus velezensis OEE1: an efficient biocontrol agent against Verticillium wilt of olive and a potential plant growth promoting bacteria. Biol Control 142:104168

Backer R, Rokem JS, Ilangumaran G, Lamont J, Praslickova D, Ricci E, Subramanian S, Smith DL (2018) Plant growth-promoting rhizobacteria: context, mechanisms of action, and roadmap to commercialization of biostimulants for sustainable agriculture. Front Plant Sci 9:1473

Bhattacharyya C, Banerjee S, Acharya U, Mitra A, Mallick I, Haldar A, Haldar S, Ghosh A, Ghosh A (2020) Evaluation of plant growth promotion properties and induction of antioxidative defense mechanism by tea rhizobacteria of Darjeeling, India. Sci Rep 10:1–19

Bostock RM, Pye MF, Roubtsova TV (2014) Predisposition in plant disease: exploiting the nexus in abiotic and biotic stress perception and response. Annu Rev Phytopathol 52:517–549

CABI Centre for Agriculture and Bioscience International [Online]. Accessed 15 Feb 2021

Compant S, Duffy B, Nowak J, Clément C, Barka EA (2005) Use of plant growth-promoting bacteria for biocontrol of plant diseases: principles, mechanisms of action, and future prospects. Appl Environ Microbiol 71:4951–4959

El-Barougy E, Awad NM, Turky AS, Hamed HA (2009) Antagonistic activity of selected strains of rhizobacteria against Macrophomina phaseolina of soybean plants. Am Euras J Agr Environ Sci 5:337–347

Goicoechea N (2009) To what extent are soil amendments useful to control Verticillium wilt? Pest Management Science: Formerly Pesticide Science 65:831–839
Gu Y, Wang J, Xia Z, Wei H-L (2020) Characterization of a versatile plant growth-promoting rhizobacterium Pseudomonas mediterranea strain S58. Microorganisms 8:334
Guenoun K, Chattaoui M, Bouri M, Rhouma A, Naghmouchi K, Raies A (2019) Biological control of growth promoting rhizobacteria against verticillium wilt of pepper plant. Biologia 74:237–250
Jabborova D, Kannepalli A, Davranov K, Narimanov A, Enakiev Y, Syed A, Elgorban AM, Bahkali AH, Wirth S, Sayyed R (2021) Co-inoculation of rhizobacteria promotes growth, yield, and nutrient contents in soybean and improves soil enzymes and nutrients under drought conditions. Sci Rep 11:1–9
Jacobs A (2019) Citrus farmers facing deadly bacteria turn to antibiotics, alarming health officials—the New York times. The New York Times
Jinal NH, Amaresan N (2020) Evaluation of biocontrol bacillus species on plant growth promotion and systemic-induced resistant potential against bacterial and fungal wilt-causing pathogens. Arch Microbiol 202:1785–1794
Juroszek P, Racca P, Link S, Farhumand J, Kleinhenz B (2020) Overview on the review articles published during the past 30 years relating to the potential climate change effects on plant pathogens and crop disease risks. Plant Pathol 69:179–193
Khan MS, Gao J, Chen X, Zhang M, Yang F, Du Y, Munir I, Xue J, Zhang X (2020) The endophytic bacteria bacillus velezensis Lle-9, isolated from Lilium leucanthum, harbors antifungal activity and plant growth-promoting effects. J Microbiol Biotechnol 30:668–680
Kumar A, Singh S, Gaurav AK, Srivastava S, Verma JP (2020) Plant growth-promoting bacteria: biological tools for the mitigation of salinity stress in plants. Front Microbiol 11:1216
Kusale SP, Attar YC, Sayyed R, Malek RA, Ilyas N, Suriani NL, Khan N, El Enshasy HA (2021) Production of plant beneficial and antioxidants metabolites by Klebsiellavariicola under salinity stress. Molecules 26:1894
Kutman BY, Kutman UB, Cakmak I (2013) Foliar nickel application alleviates detrimental effects of glyphosate drift on yield and seed quality of wheat. J Agric Food Chem 61:8364–8372
Li Z, Song C, Yi Y, Kuipers OP (2020) Characterization of plant growth-promoting rhizobacteria from perennial ryegrass and genome mining of novel antimicrobial gene clusters. BMC Genomics 21:1–11
Manasa M, Ravinder P, Gopalakrishnan S, Srinivas V, Sayyed R, El Enshasy HA, Yahayu M, Kee Zuan AT, Kassem HS, Hameeda B (2021) Co-inoculation of bacillus spp for growth promotion and iron fortification in sorghum. Sustainability 13:12091
Márquez R, Blanco EL, Aranguren Y (2020) Bacillus strain selection with plant growth-promoting mechanisms as potential elicitors of systemic resistance to gray mold in pepper plants. Saudi journal of biological sciences 27:1913–1922
Meena M, Swapnil P, Divyanshu K, Kumar S, Tripathi YN, Zehra A, Marwal A, Upadhyay RS (2020) PGPR-mediated induction of systemic resistance and physiochemical alterations in plants against the pathogens: current perspectives. J Basic Microbiol 60:828–861
Nithyapriya S, Lalitha S, Sayyed R, Reddy M, Dailin DJ, El Enshasy HA, Luh Suriani N, Herlambang S (2021) Production, purification, and characterization of Bacillibactin Siderophore of Bacillus subtilis and its application for improvement in plant growth and oil content in sesame. Sustainability 13:5394
Olanrewaju OS, Glick BR, Babalola OO (2017) Mechanisms of action of plant growth promoting bacteria. World J Microbiol Biotechnol 33:1–16
Omara AE-D, Nour El-Din M, Hauka F, Hafez A, El-Nahrawy S, Ghazi A, Elsakhawy T, Fusco V (2018) Suppression of Rhizoctonia solani damping-off in soybean (Glycine max L.) by plant growth promoting rhizobacteria strains. *Environment*. Biodiversit Soil Sec 2:39–49
Qi J, Song CP, Wang B, Zhou J, Kangasjärvi J, Zhu JK, Gong Z (2018) Reactive oxygen species signaling and stomatal movement in plant responses to drought stress and pathogen attack. J Integr Plant Biol 60:805–826

Rai PK, Singh M, Anand K, Saurabh S, Kaur T, Kour D, Yadav AN, Kumar M (2020) Role and potential applications of plant growth-promoting rhizobacteria for sustainable agriculture. Elsevier, New and Future Developments in Microbial Biotechnology and Bioengineering

Reshma, P., Naik, M., Aiyaz, M., Niranjana, S., Chennappa, G., Shaikh, S. & Sayyed, R. 2018. Induced systemic resistance by 2, 4-diacetylphloroglucinol positive fluorescent pseudomonas strains against rice sheath blight

Rodríguez M, Torres M, Blanco L, Béjar V, Sampedro I, Llamas I (2020) Plant growth-promoting activity and quorum quenching-mediated biocontrol of bacterial phytopathogens by Pseudomonas segetis strain P6. Sci Rep 10:1–12

Santoyo G, Moreno-Hagelsieb G, Del Carmen Orozco-Mosqueda M, Glick BR (2016) Plant growth-promoting bacterial endophytes. Microbiol Res 183:92–99

Santoyo G, Urtis-Flores CA, Loeza-Lara PD, Orozco-Mosqueda M, Glick BR (2021) Rhizosphere colonization determinants by plant growth-promoting Rhizobacteria (PGPR). Biology 10:475

Savary S, Willocquet L, Pethybridge SJ, Esker P, Mcroberts N, Nelson A (2019) The global burden of pathogens and pests on major food crops. Nature ecology & evolution 3:430–439

Shah A, Nazari M, Antar M, Msimbira LA, Naamala J, Lyu D, Rabileh M, Zajonc J, Smith DL (2021) PGPR in agriculture: A sustainable approach to increasing climate change resilience. Front Sustain Food Syst 5:667546. https://doi.org/10.3389/fsufs

Shen Y-C, Korkor NL, Xiao R, Pu Q, Hu M, Zhang S-S, Kong D-D, Zeng G, Hu X-F (2020) Antagonistic activity of combined bacteria strains against southern blight pathogen of Dendrobium officinale. Biol Control 151:104291

Shittu HO, Obiazikwor OH (2021) Arms race between phytopathogens and host plants: A similitude of two nations at war. J Microbiol Biotechnol Food Sci 2021:646–650

Srichamnong W, Kalambaheti N, Woskie S, Kongtip P, Sirivarasai J, Matthews KR (2021) Occurrence of antibiotic-resistant bacteria on hydroponically grown butterhead lettuce (Lactuca sativa var. capitata). Food Sci Nutrit

Suárez-Moreno ZR, Vinchira-Villarraga DM, Vergara-Morales DI, Castellanos L, Ramos FA, Guarnaccia C, Degrassi G, Venturi V, Moreno-Sarmiento N (2019) Plant-growth promotion and biocontrol properties of three Streptomyces spp isolates to control bacterial rice pathogens. *Frontiers in microbiology* 10:290

Sultana S, Alam S, Karim MM (2021) Screening of siderophore-producing salt-tolerant rhizobacteria suitable for supporting plant growth in saline soils with iron limitation. J Agr Food Res 4:100150

Sundin GW, Wang N (2018) Antibiotic resistance in plant-pathogenic bacteria. Annu Rev Phytopathol 56:161–180

Taylor P, Reeder R (2020) Antibiotic use on crops in low and middle-income countries based on recommendations made by agricultural advisors. CABI Agr Biosci 1:1–14

Tharek M, Khairuddin D, Najimudin N, Ghazali AH (2021) Plant growth promoting potentials of beneficial endophytic Escherichia coli USML2 in association with Rice seedlings. Tropical Life Sci Res 32:119

Tirry N, Kouchou A, El Omari B, Ferioun M, El Ghachtouli N (2021) Improved chromium tolerance of Medicago sativa by plant growth-promoting rhizobacteria (PGPR). J Genetic Eng Biotechnol 19:1–14

Toju H, Peay KG, Yamamichi M, Narisawa K, Hiruma K, Naito K, Fukuda S, Ushio M, Nakaoka S, Onoda Y (2018) Core microbiomes for sustainable agroecosystems. Nature Plants 4:247–257

Usmonov A, Yoo S-J, Kim ST, Yang JS, Sang MK, Jung HW (2021) The bacillus zanthoxyli HS1 strain renders vegetable plants resistant and tolerant against pathogen infection and high salinity stress. Plant Pathol J 37:72

Verma PP, Shelake RM, Das S, Sharma P, Kim J-Y (2019) Plant growth-promoting rhizobacteria (PGPR) and fungi (PGPF): potential biological control agents of diseases and pests. Springer, Microbial Interventions in Agriculture and Environment

Wankhade S, Patil C, Padghan P, Pardey V (2019) Efficacy of fungicides, plant extracts and bioagents against sclerotium rolfsii incitant. Collar Rot Betel Vine

Xia Y, Farooq MA, Javed MT, Kamran MA, Mukhtar T, Ali J, Tabassum T, Ur Rehman S, Munis MFH, Sultan T (2020) Multi-stress tolerant PGPR bacillus xiamenensis PM14 activating sugarcane (Saccharum officinarum L.) red rot disease resistance. Plant Physiol Biochem 151:640–649

Yin X-D, Ma K-Y, Wang Y-L, Sun Y, Shang X-F, Zhao Z-M, Wang R-X, Chen Y-J, Zhu J-K, Liu Y-Q (2020) Design, synthesis, and antifungal evaluation of 8-Hydroxyquinoline metal complexes against Phytopathogenic fungi. J Agric Food Chem 68:11096–11104

# Chapter 4
# PGPR in Biofilm Formation and Antibiotic Production

**Dibyajit Lahiri, Moupriya Nag, R. Z. Sayyed, Abdul Gafur, Mohammad Javed Ansari, and Rina Rani Ray**

**Contents**

4.1 Introduction ........ 65
4.2 Plant-Microbiome Interaction by the Biofilm Formed by PGPR ........ 67
4.3 PGPR and Its Role in Agriculture ........ 68
4.4 Antibiotic Production by PGPR ........ 69
4.5 Conclusion and Future Prospects ........ 76
References ........ 76

## 4.1 Introduction

Diverse groups of microbes are found to be thriving within the soil at favorable conditions. The plants which are found growing on fields comprise of complex community having a symbiotic relationship with the organisms. The plants that grow within the field are not solitary but comprise of complex community (Lundberg et al. 2012) and keep maintaining a stable relationship with the other communities.

D. Lahiri · M. Nag
Department of Biotechnology, University of Engineering and Management, Kolkata, West Bengal, India

R. Z. Sayyed
Department of Microbiology, PSGVP Mandal's S I Patil Arts, G B Patel Science and SKTV Sangh Commerce College, Shahada, India

A. Gafur
Sinarmas Forestry Corporate Research and Development, Perawang, Indonesia

M. J. Ansari
Department of Botany, Hindu College Moradabad (Mahatma Jyotiba Phule Rohikhand University Bareilly), Moradabad, India

R. R. Ray (✉)
Department of Biotechnology, Maulana Abul Kalam Azad University of Technology, Haringhata, West Bengal, India



R. Z. Sayyed et al. (eds.), *Antifungal Metabolites of Rhizobacteria for Sustainable Agriculture*, Fungal Biology, https://doi.org/10.1007/978-3-031-04805-0_4

Thus a plant comprises well-coordinated microbial species that remain associated with plants (Turner et al. 2013; Chaparro et al. 2014). The soil is a favorable habitat comprising of diverse microbial populations that made the researchers to find their activities and functions. The microbes play a major role in promoting the growth of the plants and the quality of the soil. They possess the ability of converting the nutrients and help in binding with the soil particle thus helping in the growth of the plants. Various types of microbes remain associated with the soil comprising of beneficial bacterial cells like rhizobacteria that colonize upon the root surface termed as plant growth-promoting rhizobacteria (PGPR) (Lugtenberg and Kamilova 2009) that play a major role in promoting the growth of the plants. This mechanism of promoting plant growth can be achieved by various processes that include nitrogen fixation (Kusale et al. 2021), enhancement of nutrients like phosphates in soil (Sharma et al. 2013, 2016; Jabborova et al. 2020; Kannepalli et al. 2021), enhancing the synthesis of antibiotics (Zakaria et al., 2019; Vinay et al. 2016; Reshma et al. 2018), suppression of various plant ailments (Sukmawati et al. 2021; Sharma et al. 2020), synthesis of plant hormones (Khan et al. 2020), phytostimulation (Kalam et al. 2020; Basu et al. 2021; Hamid et al. 2021; Khan et al. 2021; Sivan and Chet 1992; Zehnder et al. 2001), and various enzymes (Sagar et al. 2020, 2022; Kusale et al. 2021). The microbial species mainly comprise of bacteria and fungi that remain associated with major plant parts comprising of stems, fruits, flowers, and roots (Berg et al. 2016). It has been observed that the microbial colonies that are associated with the roots are largest in number mainly those that remain associated with higher plants. As most of the microbial communities that are associated with plants cannot be cultured, they are studied by metagenomic analysis (Hirsch and Mauchline 2012) and associated methodologies that help in assessing the change in association with the change in genotype (Delaplace et al. 2015). The rhizomicrobiome is greatly controlled by the plant (Zhang et al. 2017). The exudates that come out from the plant roots are of various compositions which act as carbon source for the microbes existing in association with the plants (Chaparro et al. 2012). It has also been observed that biochemical and genetic expressions of some microbial species are regulated by the signal compounds that are produced by the plants (Nelson and Sadowsky 2015). The microbial species that colonize undergo density-dependent communication mechanism known as quorum sensing at conditions requiring the mechanism of physiological shift (Chauhan et al. 2015). The analogs being produced by the plants help in interacting with the quorum sensing (QS) compounds thus regulating rhizomicrobiome (Ortiz-Castro et al. 2009).

On the other hand, biofilm associated microbiota play the pivotal role in PGPR synthesis. Although not much studies have been performed in determining the existence of PGPR in the form of biofilm, it was opined by scientists that the existence of PGPR in the form of biofilm provides greater ability to guard from the microbial competitions and environmental stresses (Sayyed et al. 2015; Ilyas et al. 2020; Ramey et al. 2004; Seneviratne et al. 2008a, b). The significance of biofilm construction in the plant-microbe interactions cannot be ignored as it promotes plant growth. Since from the perspective of agricultural improvement, the biofilm associated PGPR and their positive role in plant nutrition and production of beneficial secondary metabolites like antibiotics are of immense importance, this domain needs more research and development. The present chapter highlights the role of biofilm in regulation and promotion of plant growth and the production of antibiotics by PGPR having efficacy against wide range of plant pathogens.

## 4.2 Plant-Microbiome Interaction by the Biofilm Formed by PGPR

The microbial species present at the rhizospheric region of the plant exist as sessile and biofilm associated, not in the planktonic form (Watnick and Kolter 1999; Davey and O'Toole 2000). Growth of the plant and quality of crop and yield are regulated by the existence of biofilm at the rhizospheric region of the plant (Saleh-Lakha and Glick 2006). Bacterial species like *Pseudomonas putida*, *P. fluorescens*, and other groups of pseudomonads are found to form effective biofilm having an important role in promoting plant growth (Ude et al. 2006) (Table 4.1).

It has been observed that bacterial species like *P. fluorescens* are observed to grow discretely and colonize on the surface of the roots and develop biofilm along the epidermal tissues. The *Pseudomonas* spp. possess the ability to undergo transition from small multicellular clusters to extensive biofilm. The microbial species existing within the biofilm derive its nutrition from the exudates of the roots (Bais et al. 2006). The carbon content present within the root exudates accounts for 0.4% of that total carbon fixed during photosynthesis but plays a significant role in

**Table 4.1** PGPR synthesizing plant associated bacteria

| Name of the bacteria | Characteristics | Reference |
|---|---|---|
| *Rhizobium* | Colonizes within the roots of the legumes and also colonizes upon the roots of cowpea | Krysciak et al. (2011) |
| *Rhizobium leguminosarum* | Colonizes within the roots of legumes | Vanderlinde et al. (2009); Janczarek and Skorupska (2011) |
| *Pantoea agglomerans* | Colonizes within the roots of chickpea and wheat | Chauhan and Nautiyal (2010) |
| *Klebsiella pneumoniae* | Colonizes upon the roots of wheat | Iniguez et al. (2004) |
| *Cyanobacteria* spp. | Helps in the development of heterogeneous biofilm along with *pseudomonas*, *Azotobacter*, *rhizobium* spp. | Prasanna et al. (2011) |
| *Bacillus polymyxa* | Colonizes upon the roots of cucumber | Yang et al. (2004) |
| *Bacillus amyloliquefaciens* | Colonizes upon the roots of maize, tomato, and *Arabidopsis thaliana* | Fan et al. (2011) |
| *Azorhizobium caulinodans* | Colonizes upon the roots of rice plants | Van Nieuwenhove et al. (2000) |
| *Acinetobacter calcoaceticus* | Colonizes upon the roots of duckweeds | Yamaga et al. (2010) |

maintaining the soil microbiota (Rovira 1969). The organic compounds that are present within the exudates of the roots play a significant role in providing various factors derived from the plants and thus helping in the mechanism of colonization (Lugtenberg et al. 1999) and development of biofilm upon the root surface (Walker and Bais 2004). The formation of biofilm at the sites of the roots is stimulated in the presence of plant-derived factors which are profoundly similar to the interactions between Rhizomes-legumes and interactions that are taking place between the bacterial cells (de Ruijter et al. 1999). Studies showed that the exudates from the roots help in the growth of *P. putida* which results in the development of biofilm. The mechanism of QS plays an important role by the bacterial cells in expressing and coordinating the expression of genes in accordance to the available cellular densities. The signaling cascades are responsible in coordinating the interactions performed by the bacterial cells with plants by bringing about its control in the production of antibiotics, maceration of tissues, release of toxins, and the process of horizontal gene transfer (HGT) (von Bodman et al. 2003). Thus QS helps in the phenotypic regulation of PGPR and the formation of the biofilm (Loh et al. 2002). In PGPR the mechanism of QS is associated with various types of signaling cascades comprising of gamma-butyrolactones within the *Streptomyces* spp., acylated homoserine lactones (AHL) within the groups of proteobacteria, cis-11-methyl-2-dodecanoic acid (DSF) within *Xylella*, *Xanthomonas*, and associated species and oligopeptides produced by the groups of Gram-positive microbes (Danhorn and Fuqua 2007). The rhizospheric bacterial cells use AHL in the process of cell-to-cell communication mechanism. AHLs' help is assessing the cellular density and enhances the communication between the cells (Pierson et al. 1998a; b). The common types of proteins that are associated with the mechanism of AHL-associated QS mechanism are LuxI, LuxR, and AHL synthase, but variations are observed in bacterial species like *P. fluorescens* and *Vibrio harveyi* which are associated with proteins comprising of LuxM AHL and Hdt AHL synthase which are the types of LuxI (Case et al. 2008). Various types of bacterial cells colonizing upon the roots and *Pseudomonas* spp. implement AHL-dependent QS mechanisms (Juhas et al. 2005). Thus, biofilm-associated PGPR has various roles and can be exploited in the process of green agriculture (Fig. 4.1).

## 4.3 PGPR and Its Role in Agriculture

The biggest concern in the present scenario is the side effects of various types of agro-chemicals that have resulted in new mechanisms of enhancing the productivity of crops. Thus, the implementations of various types of beneficial microorganisms (Table 4.2) help in enhancing the growth of the plants by providing various nutrient supplements without bringing about any damage to the environment and enhancing the productivity of the soil (O'Connell 1992). The limitations of availability of the inocula of these beneficial microbial species have prevented overall commercialization of this process (van Elsas et al. 1986).

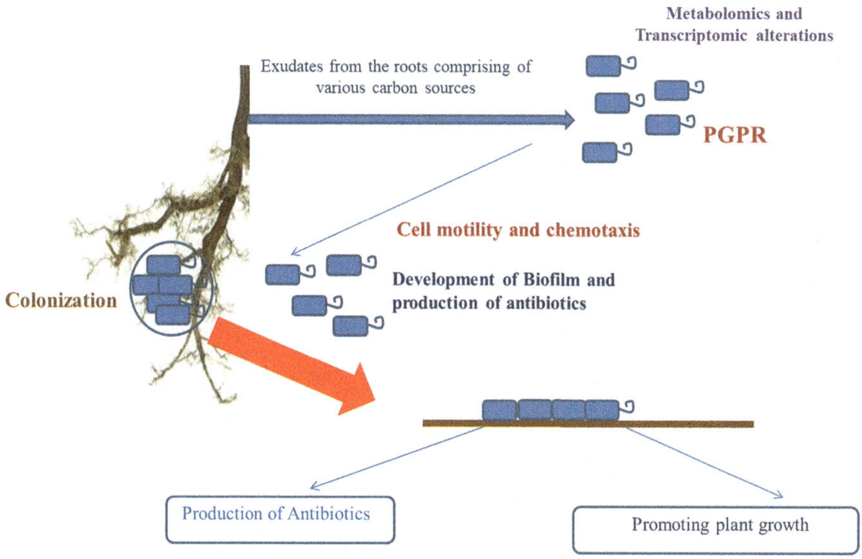


**Fig. 4.1** Development of biofilm and its role

## 4.4 Antibiotic Production by PGPR

The growth of the plant pathogens within the plants can be prevented by the microbial antagonists and has been a potential alternative to chemical pesticides. The microbial colonies prevent the growth of plant pathogens with the help of various types of extracellular metabolites at its inhibitory concentrations. The bacterial species that are associated in preventing the plant diseases are referred to as biocontrol agents. The bacterial cells associated with PGPR play vital role in inhibiting the pathogenic groups of microbial cells by the production of various types of hydrolytic enzymes comprising of glucanases, proteases, chitinases, and lipases possessing the ability of killing the pathogenic groups of fungal species (Neeraja et al. 2010), competing with the nutrients, preventing the colonization of harmful microbes upon root surface (Stephens et al. 1993), and production of antibiotics. The antibiotics that are produced by the bacterial cells comprise of volatile substances like aldehydes, hydrogen cyanide, ketones, aldehydes, and alcohols, whereas the non-volatile groups of antibiotics comprise of polyketides like 2,4-diacetylphloroglucinol (DAPG), diacetylphloroglucinol, and mupirocin; heterocyclic nitrogenous compounds like pyocyanin, phenazine-1-carboxylic acid, hydroxyphenazines, and phenazine derivatives; and phenylpyrrole antibiotics (Table 4.3) (De Souza et al. 2003). The antibiotics that are synthesized by PGPR comprise of 2,4-diacetylphloroglucinol, oligomycin A, butyrolactones, xanthobaccin, phenazine-1-carboxylic acid, zwittermicin A, pyrrolnitrin, and viscosinamide (Viveros et al. 2010). *Bacillus* spp. and pseudomonads play an important role in inhibiting the growth of the pathogenic microorganisms. The various types of antibiotics that are produced include phenazine-1-carboxyclic

**Table 4.2** Role of PGPR biofilm in different domains of agriculture

| Functions performed by PGPR biofilm | Name of microbes | Mechanism | References |
|---|---|---|---|
| Plant protection | *Paenibacillus polymyxa* | The development of biofilm provides advantage to the PGPR for colonizing easily upon the root surface resulting in the establishment of competition with the existing indigenous microflora resulting in the enhancement of plant growth. The biofilm formed by *Paenibacillus polymyxa* upon the surface of the roots thus protecting the roots from attack of other bacterial cells and drought | Timmusk et al. (2005); Seneviratne et al. (2008a, b) |
| Biofertilizers | *Bradyrhizobium elkanii*<br>*Pleurotus* spp.<br>*Pseudomonas fluorescens*, *Pleurotus ostreatus biofilm* | The PGPR biofilm helps in the mechanism of nitrogen fixation, enhances the accumulation of nitrogen and nitrogenase, and thus enhances the growth of the plants | Jayasinghearachchi and Seneviratne (2004) |
| Plant growth-promoting agent | *Azoarcus* spp., *pseudomonas* spp., *Herbaspirillum seropedicae*, *Serratia* spp., *Burkholderia* spp., *Rhizobium leguminosarum*, *Bradyrhizobium japonicum*, *Klebsiella* sp. and *Azorhizobium caulinodans*<br>Non-pathogenic fungal endophytes<br>*Epicoccum purpurascens*, *Alternaria alternata*, and *Cladosporium tenuissimum*<br>*Streptomyces* spp. | PGPR stimulates the production of indoleacetic acid (IAA) thus stimulates the growth of the plant | Bandara et al. (2006) |

**Table 4.3** Chemistry and mode of action of antibiotics produced by PGPR

| PGPR | Antibiotic produced | Chemical structure | Activity | Reference |
|---|---|---|---|---|
| *Bacillus* spp. | Plipastatins A and B | They possess well-defined structure comprising of one β-turn as well as 6–9 residues along with hydrogen bonds | They are the groups of antifungal antibiotics that prevent the growth of the fungal cells by bringing about inhibition to phospholipase $A_2$ | Volpon et al. (2000) |
| | Bacillomycin | They are chemically the groups of lipopeptide compounds | They possess the ability of inhibiting various types of pathogens | Xu et al. (2013); Volpon et al. (1999) |
| | | They are the group of lipopeptide compounds having identical amino acid sequences | The hemolytic and antifungal properties make it to be an efficient bio-control agent | Dang et al. (2019); Constantinescu (2001) |
| | Zwittermicin A | Chemically these compounds comprise of α-D amino acids, ethanolamine, and glycosyl moieties. They also possess terminal amide group that is produced by the virtue of modification of non-proteinogenic amino acid, i.e., β-ureidoalanine | It checks the growth of fungi, bacteria, and protists | Kevany et al. (2009) |
| | Kanosamine | Chemically it is 3-amino-3-deoxy-D-glucose | It has potent role in inhibiting the growth of plant pathogens | Milner et al. (1996) |
| *Pseudomonas* spp. | Azomycin | – | They have potential antibacterial properties | Shoji et al. (1989) |
| | Pseudomonic acid | – | They have potential antibacterial properties | Banks et al. (1998) |

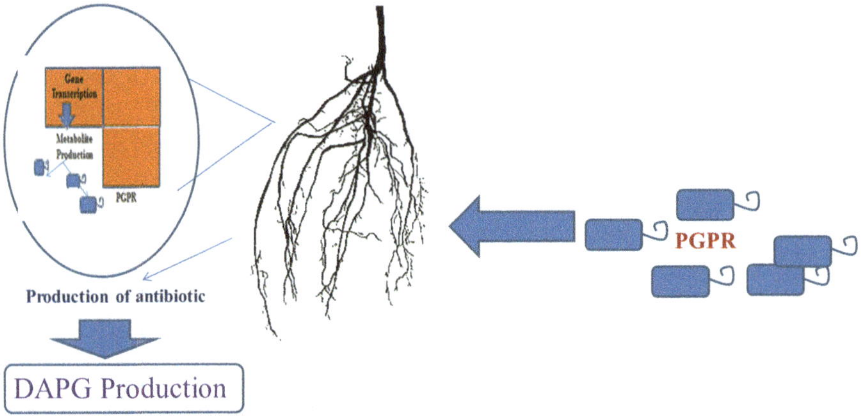


**Fig. 4.2** Production of antibiotics by PGPR

acid, zwittermicin A, phenazine-1-carboxamide, ecomycins, pyrrolnitrin, pseudomonic acid, azomycin, and many more. These compounds possess various properties that include antimicrobial, antiviral, antitumor, cytotoxic, antioxidant, phytotoxic, antihelminthic, and plant growth enhancing activities. Pyrrolnitrin is the antibiotic produced by *P. fluorescens* which is responsible for the destruction of *Rhizoctonia solani.* The destruction of the membrane of *Pythium* spp. can be achieved by the use of DAPG which is one of the most extensively studied antibiotic from PGPR (de Souza and Raaijmakers 2003). It has also been observed that phenazine produced by *Pseudomonas* spp. possess antagonistic activity against *Fusarium oxysporum* (Beneduzi et al. 2012). The antibiotics that are produced by the *Bacillus* spp. comprise of polymyxin, colistin, and circulin that possess the ability of inhibiting the growth of Gram-positive, Gram-negative bacterial cells and fungi. The antibiotics that are produced by the *Bacillus* spp. are chemically lipopeptide in nature and comprise of surfactin, iturin, and fengycin. The six major groups of antibiotics like phloroglucinols, pyoluteorin, cyclic lipopeptides, pyrrolnitrin, hydrogen cyanide, and lipopeptides play an important role in preventing diseases associated with roots (Beneduzi et al. 2012). Lipopeptide and biosurfactants are the two major groups of antibiotics that have developed immense importance due to their ability to act against wide range of pathogenic organisms comprising of bacteria, fungi, protozoa, and nematodes (De Bruijn et al. 2007) (Fig. 4.2).

### *4.4.1 Biosynthesis of Polyketide Group of Antibiotics by PGPR*

PGPR is associated with the production of various types of antibiotics, but polyketide group of antibiotics like mupirocin, pyoluteorin, and 2,4-diacetylphloroglucinol plays an important role in inhibiting the growth of pathogens that are associated with plants. Diacetylphloroglucinol (DAPG) are majorly produced by fluorescent bacterial species like pseudomonads that play an important role in inhibiting nematodes, bacteria, and fungi (Keel et al. 1992). Among various types of extracellular metabolites produced by the microbial cells associated with PGPR, DAPG shows highest level of protection in plants. The evidential mechanisms by which DAPG brings about protection in plants involve mutation within the cluster of genes that are responsible for the synthesis of DAPG causing reduction in the biocontrol activity performed by the antagonistic bacterial species (Keel et al. 1992). The amount of production of DAPG has direct effect on the inhibition of the disease causing pathogens (Raaijmakers et al. 1999), and presence of DAPG producing organisms along with PGPR results is better protection of the crops.

### 4.4.2 Biosynthesis of DAPG by PGPR

The synthesis of DAPG is associated with *phl* gene cluster consisting of *phlACBDE* operon and genes *phlF*, *phlG*, *phlH*, and *phlI*, respectively. The first step involves the cyclization of malonyl-CoA to phloroglucinol by type III polyketide synthase (encoded by gene *phlD*) (Achkar et al. 2005). This is followed by the acetylation of phloroglucinol to MAPG by a monoacetylphloroglucinol (MAPG) acetyltransferase (encoded by *phlACB*) and then DAPG (Kidarsa et al. 2011). Studies have shown that the biosynthesis of DAPG which is a type of polyketide antibiotic is produced by the condensation of three acetyl coenzyme A along with malonyl coenzyme A resulting in the formation of precursor known as monoacetylphloroglucinol (MAPG). Transacetylation of the precursor results in the synthesis of Phloroglucinol (PHL) using a chalcone synthase (CHS) type enzyme (Shanahan et al. 1992; Keel et al. 2000). The last gene of the phl operon (*phlE*) is responsible for the outflow of DAPG (Abbas et al., 2004). Genes *phlF* and *phlH* are involved in the subjugation and stimulation of DAPG synthesis (Keel et al. 2000). Gene *phlG* codes a hydrolase helping the back synthesis of DAPG to MAPG (Bottiglieri and Keel 2006), whereas the role of *phlI* (encoding a hypothetical protein) in DAPG synthesis and regulation is yet to be known (Almario et al. 2017a, b).

### 4.4.3 Genetic Regulation of DAPG Production by PGPR

*Pseudomonas* species are well known for its interaction with eukaryotic hosts such as plants, fungi, or human. In some cases, interaction with plants leads to some advantageous effects (Mirza et al., 2006) such as providing defense against plant pathogens (Mazurier et al. 2009; Almario et al. 2017a, b) via production of secondary metabolites such as DAPG. The mechanism of action of DAPG involves induction of plant defenses by induced systemic resistance (ISR) and directly modulating plant hormonal balance by acting as an auxin-mimetic compound (Almario et al. 2017a, b). The production of DAPG by pseudomonads marks its ability to combat against diseases. The pseudomonads that are responsible for the production of

DAPG can be taxonomically differentiated on the basis of amplified ribosomal DNA restriction analysis fingerprints (ARDRA) (McSpadden Gardener et al. 2000). The communication existing between the rhizospheric bacterial cells and pseudomonads by AHL plays a significant role in the regulation of the expression of the antibiotic gene (Pierson et al. 1998a, b).

### 4.4.4 Factors Regulating the Production of DAPG

The performance of the fluorescent pseudomonads is dependent on the biotic and abiotic factors associated with the environment of the crop (Thomashow and Weller 1995). Alteration on the expression of the gene *phlA* is associated with various types of biotic factors like species of plants, age of the plant, pathogens, and cultivars (Notz et al. 2001). Various types of abiotic resources like various types of minerals and carbon sources also influence the production of DAPG within *P. fluorescens* (Nowak-Thompson et al. 1999).

### 4.4.5 Biosynthesis of Pyoluteorin by PGPR

Pyoluteorin (Plt) is a type of phenolic polyketide possessing a resorcinol ring. The ring is associated with bichlorinated pyrrole moiety (Kitten et al. 1998). First isolation of antibiotic was from *P. fluorescens* and *P. aeruginosa* that exhibited fungicidal with bactericidal properties. The damping off in the cotton plants can be prevented by the addition of *Plt* gene within the cotton plants (Howell and Stipanovic 1980).

### 4.4.6 Biosynthesis of Mupirocin by PGPR

Mupirocin is another type of antibiotic produced by *P. fluorescens* which shows various degrees of antimicrobial activities. Chemically it is pseudomonic acid which exhibits bactericidal activities (Fuller et al. 1971). It has the ability of inhibiting translational machinery like isoleucyl-tRNA synthetase thus preventing the incorporation of isoleucine within the newly synthesized proteins (Hughes and Mellows 1980). Mupirocin also results in the alteration of the target binding site of isoleucyl-tRNA synthetase.

### 4.4.7 Biosynthesis of Heterocyclic Group of Nitrogenous Compounds by PGPR

Rhizobacteria possess the ability of producing heterocyclic nitrogenous compounds having potential antimicrobial efficacy. One such most potent secondary metabolite having antimicrobial properties is phenazine (Chin-A-Woeng et al. 1998) that is produced by various pigmented bacterial species like *Brevibacterium*, *Burkholderia*, *Streptomyces*, and *Pseudomonas* (Turner and Messenger 1986) (Table 4.3).

### 4.4.8 Antibiotic Synthesized by PGPR Preventing Plant Diseases

The antibiotics that are synthesized (Table 4.4) and associated with PGPR act as a potent biocontrol agent for preventing various types of soil associated diseases in plants (Williams and Vickers 1986). Various types of antibiotics have been used potentially in eradicating various types of diseases that are associated with the plants.

**Table 4.4** Antibiotics synthesized by PGPR

| Types of antibiotics produced by PGPR | Organism from which it is produced | Pathogen on which it acts | Types of diseases | Reference |
|---|---|---|---|---|
| Pyrrolnitrin Pyoluteorin | *P. fluorescens* Pf-5 | *R. solani* and *Pythium ultimum* | Damping off | Howell et al. (1988) |
| Pyrrolnitrin, pseudane | *Burkholderia cepacia* | *Pyricularia oryzae* and *R. solani* | Damping off and rice blast | Homma et al. (1989) |
| Zwittermicin A | *Bacillus cereus* UW85 | *P. aphanidermatum* and *Phytophthora medicaginis* | Damping off | Smith et al. (1993) |
| Bacillomycin D | *Bacillus subtilis* AU195 | *Aspergillus flavus* | Aflatoxin contamination | Moyne et al. (2001) |
| Bacillomycin, fengycin | *Bacillus amyloliquefaciens* FZB42 | *Fusarium oxysporum* | Wilt | Koumoutsi et al. (2004) |
| Xanthobaccin A | *Lysobacter* spp. strain SB-K88 | *Aphanomyces cochlioides* | Damping off | Islam et al. (2005) |
| Gliotoxin | *Trichoderma virens* | *Rhizoctonia solani* | Rots on roots | Vargas et al. (2014) |
| Herbicolin | *Pantoea agglomerans* C9–1 | *Erwinia amylovora* | Fire blight | Wright et al. (2001) |
| DAPG | *Pseudomonas* spp. | *P. ultimum* | Sugar beet | Shanahan et al. (1992) |
| DAPG | *P. fluorescens* | *Theilaviopsis basicola* | Tobacco | Keel et al. (2000) |
| Aerugine | *P. fluorescens* | *C. orbiculare* | Cucumber | Lee et al. (2003) |
| Phenazine | *Pseudomonas* spp. | *F. Oxysporum* | Tomato | Chin-A-Woeng et al. (1998) |

## 4.5 Conclusion and Future Prospects

Biofilm research has been a central focus for various biotechnological outcomes. But still there is dearth of literatures in the field of biofilm associated with PGPR. It has been observed that plants benefit profusely by harboring endophytic microorganisms. The colonization of PGPR upon the plant roots results in various beneficial activities in association with the microbes. The mechanism of QS during biofilm formation helps in building the association of PGPR stronger and develops useful activities. The antibiotics produced by PGPR confer enhanced resistance to various pathogens and promote plant growth. Interactions with these endophytes also influence the plant productivity leading towards an improved agro-ecosystem. Moreover, use of natural product and designing of some future biocontrolling agents out of these PGPR will reduce the toxic effect of chemicals in nature to make the environment greener.

## References

Achkar J, Xian M, Zhao H, Frost JW (2005) Biosynthesis of phloroglucinol. J Am Chem Soc Apr 20 127(15):5332–5333

Almario J, Bruto M, Vacheron J, Prigent-Combaret C, Moënne-Loccoz Y, Muller D (2017a) Distribution of 2,4-Diacetylphloroglucinol biosynthetic genes among the *pseudomonas* spp. Reveals Unexpected Polyphyletism Front Microbiol 8:1218

Almario J, Bruto M, Vacheron J, Prigent-Combaret C, Moënne-Loccoz Y, Muller D (2017b) Distribution of 2,4-Diacetylphloroglucinol biosynthetic genes among the *pseudomonas* spp. Reveals Unexpected Polyphyletism. Front Microbiol 8:1218

Bais HP, Weir TL, Perry LG, Gilroy S, Vivanco JM (2006) The role of root exudates in rhizosphere interactions with plants and other organisms. Annu Rev Plant Biol 57:233–266

Bandara WMMS, Seneviratne G, Kulasooriya SA (2006) Interactions among endophytic bacteria and fungi: effects and potentials. J Biosci 31:645–650

Banks RM, Donald AC, Hannan PC, O'Hanlon PJ, Ragers NH (1998) Antimycoplasmal activities of the pseudomonic acids and structure-activity relationships of monic acid A derivatives. J Antibiot 41:609–613

Basu A, Prasad P, Das SN, Kalam S, Sayyed RZ, Reddy MS, El Enshasy HE (2021) Plant growth promoting Rhizobacteria (PGPR) as green bioinoculants: recent developments, constraints, and prospects. Sustainability 13:1140. https://doi.org/10.3390/su13031140

Beneduzi A, Ambrosini A, Passaglia LMP (2012) Plant growth-promoting rhizobacteria (PGPR): their potential as antagonists and biocontrol agents. Genet Mol Biol 35:1044–1051

Berg G, Rybakova D, Grube M, Koberl M (2016) The plant microbiome explored: implications for experimental botany. J Exp Bot 67:995–1002. https://doi.org/10.1093/jxb/erv466

Bottiglieri M, Keel C (2006) Characterization of PhlG, a hydrolase that specifically degrades the antifungal compound 2,4-diacetylphloroglucinol in the biocontrol agent *Pseudomonas fluorescens* CHA0. Appl Environ Microbiol 72(1):418–427

Case RJ, Labbate M, Kjelleberg S (2008) AHL-driven quorum-sensing circuits: their frequency and function among the Proteobacteria. ISME J 2:345–349

Chaparro JM, Sheflin AM, Manter DK, Vivanco JM (2012) Manipulating the soil microbiome to increase soil health and plant fertility. Biol Fertil Soils 48:489–499. https://doi.org/10.1007/s00374-012-0691-4

Chaparro JM, Badri DV, Vivanco JM (2014) Rhizosphere microbiome assemblage is affected by plant development. ISME J 8:790. https://doi.org/10.1038/ismej.2013.196

Chauhan PS, Nautiyal CS (2010) The purB gene controls rhizosphere colonization by *Pantoea agglomerans*. Lett Appl Microbiol 50:205–210

Chauhan H, Bagyaraj D, Selvakumar G, Sundaram S (2015) Novel plant growth promoting rhizobacteria—prospects and potential. Appl. Soil Ecol 95:38–53. https://doi.org/10.1016/j.apsoil.2015.05.011

Chin-A-Woeng TFC, Bloemberg GV, van der Bij AJ, van der Drift KMGM, Schripsema J, Kroon B, Scheffer RJ, Keel C, Bakker PAHM, De Bruijn FJ, Thomas-Oates JE, Lugtenberg BJJ (1998) Biocontrol by phenazine-1-carboxamide producing *pseudomonas chlororaphis* PCL1391 of tomato root rot caused by *fusarium oxysporum f.* sp. radicis-lycopersici. Mol Plant-Microbe Interact 11(11):1069–1077

Constantinescu F (2001) Extraction and identification of antifungal metabolites produced by some *B. subtilis* strains. Analele Institutului de Cercetari Pentru Cereale Protectia Plantelor 31:17–23

Dang Y, Zhao F, Liu X (2019) Enhanced production of antifungal lipopeptide iturin A by *bacillus amyloliquefaciens* LL3 through metabolic engineering and culture conditions optimization. Microb Cell Factories 18:68

Danhorn T, Fuqua C (2007) Biofilm formation by plant-associated bacteria. Annu Rev Microbiol 61:401–422

Davey ME, O'Toole AG (2000) Microbial biofilms: from ecology to molecular genetics. Microbiol Mol Biol Rev 64:847–867

De Bruijn I, de Kock MJD, Yang M, de Waard P, van Beek TA, Raaijmakers JM (2007) Genome-based discovery, structure prediction and functional analysis of cyclic lipopeptide antibiotics

de Ruijter NCA, Bisseling T, Emons AMC (1999) Rhizobium nod factors induce an increase in sub-apical fine bundles of actin filaments in Vicia sativa root hairs within minutes. Mol Plant-Microbe Interact 12:829–832

de Souza J, Raaijmakers JM (2003) Polymorphisms within the prnD and pltC genes from pyrrolnitrin and pyoluteorin-producing *pseudomonas* and *Burkholderia* spp. FEMS Microbiol Ecol 43:21–34

De Souza JT, Arnould C, Deulvot C, Lemanceau P, Gianinazzi-Pearson V (2003) Effect of 2, 4- diacetylphloroglucinol on Pythium: cellular responses and variation in sensitivity among propagules and species. Phytopathology 93:966–975

Delaplace P, Delory BM, Baudson C, Mendaluk-Saunier De Cazenave M, Spaepen S (2015) Influence of rhizobacterial volatiles on the root system architecture and the production and allocation of biomass in the model grass *Brachypodium distachyon* (L.) P. Beauv. BMC Plant Biol 15(195):10.1186/ s12870-015-0585-3

Fan B, Chen XH, Budiharjo A, Bleiss W, Vater J, Borriss R (2011) Efficient colonization of plant roots by the plant growth promoting bacterium *bacillus amyloliquefaciens* FZB42, engineered to express green florescent protein. J Biotechnol 151:303–311

Fuller AT, Mellows G, Woolford M, Banks GT, Barrow KD, Chain EB (1971) Pseudomonic acid: an antibiotic produced by *Pseudomonas fluorescens*. Nature 234416:417

Hamid B, Zaman M, Farooq S, Fatima S, Sayyed RZ, Baba ZA, Sheikh TA, Reddy MS, El Enshasy H, Gafur A, Suriani NL (2021) Bacterial plant biostimulants: A sustainable way towards improving growth, productivity, and health of crops. Sustainability 21(13):2856. https://doi.org/10.3390/su13052856

Hirsch PR, Mauchline TH (2012) Who's who in the plant root microbiome? Nat Biotechnol 30:961–962. https://doi.org/10.1038/nbt.2387

Homma Y, Sato Z, Hirayama F, Konno K, Shirahama H, Suzui T (1989) Production of antibiotics by *pseudomonas cepacia* as an agent for biological control of soilborne plant pathogens. Soil Biol Biochem 21:723–728

Howell CR, Stipanovic RD (1980) Suppression of Pythium ultimum-induced damping-off of cotton seedlings by *Pseudomonas fluorescens* and its antibiotic, pyoluteorin. Phytopathology 70:712–715

Howell CR, Beier RC, Stipanovic RD (1988) Production of ammonia by *Enterobacter cloacae* and its possible role in the biological control of pythiumm pre-emergence damping –off by the bacterium. Phatopathology 78:1075–1078

Hughes J, Mellows G (1980) Interaction of pseudomonic acid A with *Escherichia coli* B isoleucyl-tRNA synthetase. Biochem J 191:209–219

Ilyas N, Mumtaz K, Akhtar N, Yasmin H, Sayyed RZ, Khan W, Enshasy HE, Dailin DJ, Elsayed AE, Ali Z (2020) Exopolysaccharides producing bacteria for the amelioration of drought stress in wheat. Sustainability 2020(12):8876. https://doi.org/10.3390/su12218876

Iniguez AL, Dong Y, Triplett EW (2004) Nitrogen fixation in wheat provided by *Klebsiella pneumoniae* 342. Mol Plant-Microbe Interact 17:1078–1085

Islam MT, Hashidoko Y, Deora A, Ito T, Tahara S (2005) Suppression of damping-off disease in host plants by the rhizoplane bacterium Lysobacter spp. strain SB-K88 is linked to plant colonization and antibiosis against soilborne *Peronosporomycetes*. Appl Environ Microbiol Jul 71(7):3786–3796

Jabborova D, Wirth S, Kannepalli A, Narimanov A, Desouky S, Davranov K, Sayyed RZ, Enshasy HE, Malek RA, Syed A, Bahkali AH (2020) Co-inoculation of rhizobacteria and biochar application improves growth and nutrient in soybean and enriches soil nutrients and enzymes. Agronomy 10:1142. https://doi.org/10.3390/agronomy10081142

Janczarek M, Skorupska A (2011) Modulation of rosR expression and exopolysaccharide production in *rhizobium leguminosarum* bv. Trifolii by phosphate and clover root exudates. Int J Mol Sci 12:4132–4155

Jayasinghearachchi HS, Seneviratne G (2004) Can mushrooms fix atmospheric nitrogen? J Biosci 23:293–296

Juhas M, Eberl L, Tummler B (2005) Quorum sensing: the power of cooperation in the world of *pseudomonas*. Environ Microbiol 7:459–471

Kalam S, Basu A, Ahmad I, Sayyed RZ, Enshasy HE, Dailin DJ, Suriani NL (2020) Recent understanding of soil Acidobacteria and their ecological significance: A critical review. Front Microbiol 11:580024. https://doi.org/10.3389/fmicb.2020.580024

Kannepalli A, Davranov K, Narimanov A, Enakiev Y, Syed A, Elgorban AM, Bahkali AH, Wirth S, Sayyed RZ (2021) Gafur A (2021) co-inoculation of rhizobacteria promotes growth, yield, and nutrient contents in soybean and improves soil enzymes and nutrients under drought conditions. Scientific Report 11:22081. https://doi.org/10.1038/s41598-021-01337-9

Keel C, Schiner U, Maurhofer M, Voisard C, Laville J, Burger U, Wirthner P, Haas D, Defago G (1992) Suppression of root diseases by *Pseudomonas fluorescens* CHA0: importance of the bacterial secondary metabolite 2,4- Diacetylphloroglucinol. Mol Plant-Microbe Interact 5:4–13

Keel US, Seematter A, Maurhofer M, Blumer C, Duffy B, Bonnefoy CG, Reimmann C, Notz R, Défago G, Haas D, Keel C (2000) Autoinduction of 2, 4- Diacetylphloroglucinol biosynthesis in the biocontrol agent *Pseudomonas fluorescens* CHA0 and repression by the bacterial metabolites salicylate and pyoluteorin. J Bacteriol 182:1215–1225

Kevany BM, Rasko DA, Thomas MG (2009) Characterization of the complete zwittermicin A biosynthesis gene cluster from Bacillus cereus. Appl Environ Microbiol 75(4):1144–1155. https://doi.org/10.1128/AEM.02518-08

Khan I, Awan SA, Ikram R, Rizwan M, Akhtar N, Yasmin H, Sayyed RZ, Ali S, Ilyas N (2020) 24-Epibrassinolide regulated antioxidants and osmolyte defense and endogenous hormones in two wheat varieties under drought stress, *Physiologia Plantarum*. Oct 2020:1–11. https://doi.org/10.1111/ppl.13237

Khan N, Al S, Shahi MA, Mustafa A, Sayyed RZ, Curaá JA (2021) Insights into the interactions among roots, Rhizosphere and Rhizobacteria for improving plant growth and tolerance to abiotic stresses: a review. Cells 10(6):1551. https://doi.org/10.3390/cells10061551

Kidarsa TA, Goebel NC, Zabriskie TM, Loper JE (2011) Phloroglucinol mediates cross-talk between the pyoluteorin and 2,4-diacetylphloroglucinol biosynthetic pathways in *Pseudomonas fluorescens* Pf-5. Mol Microbiol 81(2):395–414

Kitten T, Kinscherf T, McEvoy G, Willis DK (1998) A newly identified regulator is required for virulence and toxin production in *pseudomonas syringae*. Mol Microbiol 28:917–929

Koumoutsi A, Chen X-H, Henne A, Liesegang H, Hitzeroth G, Franke P (2004) Structural and functional characterization of gene clusters directing non ribosomal synthesis of bioactive cyclic lipopeptides in *bacillus amyloliquefaciens* strain FZB42. J Bacteriol 186:1084–1096

Krysciak D, Schmeisser C, Preuss S, Riethausen J, Quitschau M, Grond S (2011) Involvement of multiple loci in quorum quenching of autoinducer I molecules in the nitrogen-fixing symbiont *rhizobium (Sinorhizobium)* spp. strain NGR234. Appl Environ Microbiol 77:5089–5099

Kusale SP, Attar YC, Sayyed RZ, Malek RA, Ilyas N, Suriani NL, Khan N, Enshasy HE (2021) Production of plant beneficial and antioxidants metabolites by *Klebsiella variicola* under salinity stress. Molecules 26:1894. https://doi.org/10.3390/molecules26071894

Lee JY, Moon SS, Hwang BK (2003) Isolation and antifungal and Antioomycete activities of Aerugine produced by *Pseudomonas fluorescens* strain MM-B16. Appl Environ Microbiol 69:2023–2021

Loh J, Pierson EA, Pierson LS III, Stacey G, Chatterjee A (2002) Quorum sensing in plantassociated bacteria. Curr Opin Plant Biol 5:285–290

Lugtenberg B, Kamilova F (2009) Plant-growth-promoting rhizobacteria. Annu Rev Microbiol 63:541–556

Lugtenberg BJJ, Kravchenko LV, Simons M (1999) Tomato seed and root exudate sugars: composition, utilization by *pseudomonas* biocontrol strains and role in rhizosphere colonization. Environ Microbiol 1:439–446

Lundberg DS, Lebeis SL, Paredes SH, Yourstone S, Gehring J, Malfatti S, Tremblay J, Engelbrektson A, Kunin V, Rio T, Edgar RC, Eickhors T, Ley RE, Hugenholtz PE, Tringe SG, Dangl JL (2012) Defining the core *Arabidopsis thaliana* root microbiome. Nature 488:86–90

Mazurier S, Corberand T, Lemanceau P, Raaijmakers JM (2009) Phenazine antibiotics produced by fluorescent pseudomonads contribute to natural soil suppressiveness to *fusarium* wilt. ISME J Aug 3(8):977–991

McSpadden Gardener BB, Schroeder KL, Kalloger SE, Raaijmakers JM, Thomashow LS, Weller DM (2000) Genotypic and phenotypic diversity of phlDcontaining *pseudomonas* strains isolated from the rhizosphere of wheat. Appl Environ Microbiol 66:1939–1946

Milner JL, Silo-Suh L, Lee JC, He H, Clardy J, Handelsman J (1996) Production of kanosamine by *Bacillus cereus* UW85. Appl Environ Microbiol 62(8):3061–3065

Moyne AL, Shalby R, Cleveland TE, Tuzun S (2001) Bacillomycin D: an iturin with antifungal activity against *aspergillus flavus*. J Appl Microbiol 90:622–629

Neeraja C, Anil K, Purushotham P, Suma K, Sarma P, Moerschbacher BM, Podile AR (2010) Biotechnological approaches to develop bacterial chitinases as a bioshield against fungal diseases of plants. Crit Rev Biotechnol 30:231–241

Nelson MS, Sadowsky MJ (2015) Secretion systems and signal exchange between nitrogen-fixing rhizobia and legumes. Front Plant Sci 6:491. https://doi.org/10.3389/fpls.2015.00491

Notz R, Maurhofer M, Schnider-Keel U, Duffy B, Haas D, Défago G (2001) Biotic factors affecting expression of the 2,4-diacetylpholoroglucinol biosynthesis gene phlA in *Pseudomonas fluorescens* biocontrol strain CHA0 in the rhizosphere. Phytopathology 91:873–881

Nowak-Thompson B, Chaney N, Wing JS, Gould SJ, Loper JE (1999) Characterization of the pyoluteorin biosynthetic gene cluster of *Pseudomonas fluorescens* Pf-5. J Bacteriol 181:2166–2174

O'Connell PF (1992) Sustainable agriculture – a valid alternative. Outlook Agric 21:5–12

Ortiz-Castro R, Contreras-Cornejo HA, Macias-Rodriguez L, LopezBucio J (2009) The role of microbial signals in plant growth and development. Plant Signal Behav 4:701–712. https://doi.org/10.4161/psb.4.8.9047

Pierson LS, Wood DW, Pierson EA, Chancey ST (1998a) N-acyl homoserine lactone-mediated gene regulation in biological control by fluorescent pseudomonads: current knowledge and future work. Eur J Plant Pathol 104:1–9

Pierson EA, Wood DW, Cannon JA, Blachere FM, Pierson LS III (1998b) Interpopulation signaling via N-acyl-Homoserine lactones among bacteria in the wheat rhizosphere. Mol Plant-Microbe Interact 11:1078–1108

Prasanna R, Pattnaik S, Sugitha TC, Nain L, Saxena AK (2011) Development of cyanobacterium-based biofilms and their in vitro evaluation for agriculturally useful traits. Folia Microbiol (Praha) 56:49–58

Raaijmakers J, Bonsall RF, Weller DM (1999) Effect pf population density of *Pseudomonas fluorescens* on production of 2, 4-Diacetylphloroglucinol in the rhizosphere of wheat. Phytopathology 89:470–475

Ramey BE, Matthysse AG, Fuqua C (2004) The FNR-type transcriptional regulator SinR controls maturation of *agrobacterium tumefaciens* biofilms. Mol Microbiol 52:1495–1511

Reshma P, Naik MK, Aiyaz M, Niranjana SR, Chennappa G, Shaikh SS, Sayyed RZ (2018) Induced systemic resistance by 2,4diacetylphloroglucinol positive fluorescent pseudomonas strains against rice sheath blight. Indian J Exp Biology 56(3):207–212

Rovira AD (1969) Plant root exudates. Bot Rev 35:35–57

Sagar A, Sayyed RZ, Ramteke PW, Sharma S, Marraiki N, Elgorban AM, Syed A (2020) ACC deaminase and antioxidant enzymes producing halophilic *Enterobacter* sp. PR14 promotes the growth of rice and millets under salinity stress. Physiol Mol Biology Plants 26:1847–1854

Sagar A, Rai S, Ilyas N, Sayyed RZ, Al-Turki AI, Enshasy HE, Simarmata T (2022) Halotolerant Rhizobacteria for salinity stress mitigation: diversity, mechanism and molecular approaches. Sustainability 14:490. https://doi.org/10.3390/su14010490

Saleh-Lakha S, Glick BR (2006) Plant growth-promoting bacteria. In: van Elsas JD, Jansson JK, Trevors JT (eds) Modern soil microbiology. CRC/Thomson Publishing, Boca Raton, FL/UK, pp 503–520

Sayyed RZ, Patel PR, Shaikh SS (2015) Plant growth promotion and root colonization by EPS producing *Enterobacter* sp. RZS5 under heavy metal contaminated soil. Indian J Exp Biology 53:116–123

Seneviratne G, Kecske's ML, Kennedy IR (2008a) Biofilmed biofertilisers: novel inoculants for efficient nutrient use in plants. In: Kennedy IR, ATMA C, Kecske's ML, Rose MT (eds) Efficient nutrient use in rice production in Vietnam achieved using inoculants biofertilisers. Proceedings of a project (SMCN/2002/073) workshop held in Hanoi, Vietnam, 12–13 October 2007. ACIAR Proceeding No. 130, ACIAR, Canberra, pp 126–130

Seneviratne G, Zavahir JS, Bandara WMMS, Weerasekara MLMAW (2008b) Fungal–bacterial biofilms: their development for novel biotechnological applications. World J Microbiol Biotechnol 24:739–743

Shanahan P, O'Sullivan DJ, Simpson P, Glennon JD, O'Gara F (1992) Isolation of 2,4-diacetylphloroglucinol from a fluorescent pseudomonad and investigation of physiological parameters influencing its production. Appl Environ Microbiol 58:353–358

Sharma SB, Sayyed RZ, Trivedi MH, Gobi T (2013) Phosphate solubilizing microbes: sustainable approach for managing phosphorus deficiency in agricultural soils. Springer Plus 2:587

Sharma S, Sayyed R, Sonawane M, Trivedi M, Thivakaran G (2016) *Neurospora* sp SR8, a novel phosphate solubiliser from rhizosphere of soil of sorghum in Kachh, Gujarat. Indian J Exp Biol 54:644–649

Sharma S, Gupta A, Dalela M, Sharma S, Sayyed RZ, Enshasy HE, Elsayed AE (2020) Linking organic metabolites as produced by *Purpureocillium lilacinum* 6029 cultured on Karanja deoiled cake medium for the sustainable management of root-knot nematodes. *Sustainability.* 12(9):8276. https://doi.org/10.3390/su12198276

Shoji J, Hinoo H, Terui Y, Kikuchi J, Hattori T, Ishii K, Matsumoto K, Yoshida T (1989) Isolation of azomycin from *Pseudomonas fluorescens*. J Antibiot 42:1513–1514

Sivan A, Chet I (1992) Microbial control of plant diseases. In: Mitchell R (ed) Environmental microbiology. Wiley-Liss, New York, pp 335–354

Smith KP, Havey MJ, Handelsman J (1993) Suppression of cottony leak of cucumber with *Bacillus cereus* strain UW85. Plant Dis 77:139–142

Stephens PM, Crowley JJ, O'Connell C (1993) Selection of pseudomonad strains inhibiting *Pythium ultimum* on sugarbeet seeds in soil. Soil Biol Biochem 25:1283–1288

Sukmawati D, Family N, Hidayat I, Sayyed RZ, Elsayed AE, Dailin DJ, Hanapi SZ, Wadaan MA, Enshasy HE (2021) Biocontrol Activity of *Aureubasidium pullulans* and Candida orthopsilosis isolated from *Tectona grandis* L Phylloplane against *Aspergillus* sp In Post-Harvested Citrus Fruit. Sustainability 13:7479. https://doi.org/10.3390/su13137479

Thomashow LS, Weller DM (1995) Current concepts in the use of introduced bacteria for biological disease control: mechanisms and antifungal metabolites. In: Stacey G, Keen NT (eds) Plant– microbe interactions, chapman and hall, New York, pp 187–235

Timmusk S, Grantcharova N, Gerhart E, Wagner H (2005) *Paenibacillus polymyxa* invades plant roots and forms biofilms. Appl Environ Microbiol 71:7292–7300

Turner JM, Messenger AJ (1986) Occurrence, biochemistry and physiology of phenazine pigment production. Adv Microbial Physiol 27:211–275

Turner TR, James EK, Poole PS (2013) The plant microbiome. Genome Biol 14:209. https://doi.org/10.1186/gb-2013-14-6-209

Ude S, Arnold DL, Moon CD, Timms-Wilson T, Spiers AJ (2006) Biofilm formation and cellulose expression among diverse environmental *pseudomonas* isolates. Environ Microbiol 8:1997–2011

van Elsas JD, Dijkstra AF, Govarert JM, van Veen JA (1986) Survival of *Pseudomonas fluorescens* and *Bacillus subtilis* introduced into soils of different texture in field microplots. FEMS Microbiol Ecol 38:150–160

Van Nieuwenhove C, Holm V, Kulasooriya SA, Vlassak K (2000) Establishment of *Azorhizobium caulinodans* in the rhizosphere of wetland rice (*Oryza sativa* L.). Biol Fertil Soils 31:143–149

Vanderlinde EM, Muszynski A, Harrison JJ, Koval SF, Foreman DL, Ceri H (2009) *Rhizobium leguminosarum* biovar viciae 3841, deficient in 27-hydroxyoctacosanoate- modified lipopolysaccharide, is impaired in desiccation tolerance, biofilm formation and motility. Microbiol 155:3055–3069

Vargas WA, Mukherjee PK, Laughlin D, Wiest A, Moran-Diez ME, Kenerley CM (2014) Role of gliotoxin in the symbiotic and pathogenic interactions of *Trichoderma virens*. Microbiology (Reading) 160(Pt 10):2319–2330

Vinay JU, Naik MK, Rangeshwaran R, Chennappa G, Shaikh SS, Sayyed RZ (2016) Detection of antimicrobial traits in fluorescent pseudomonads and molecular characterization of an antibiotic pyoluteorin, *3*. Biotech 6(227):1–11

Viveros OM, Jorquera MA, Crowley DE, Gajardo G, Mora ML (2010) Mechanisms and practical considerations involved in plant growth promotion by rhizobacteria. J Soil Sci Plant Nutr 10:293–319

Volpon L, Besson F, Lancelin JM (1999) NMR structure of active and inactive forms of the sterol dependent antibiotic bacillomycin L. Eur J Bioche 264:200–210

Volpon H, Besson F, Lancelin JM (2000) NMR structure of antibiotics plipastations A and B from *Bacillus subtilis* inhibitors of phospholipase A2. FEBS 485:76–80

von Bodman SB, Bauer WD, Coplin DL (2003) Quorum sensing in plant-pathogenic bacteria. Annu Rev Phytopathol 41:455–482

Walker TS, Bais HP, De'ziel E, Schweizer HP, Rahme LG, Fall R, Vivanco JM (2004) *Pseudomonas aeruginosa*-plant root interactions. Pathogenicity, biofilm formation, and root exudation. Plant Physiol 134:320–331

Watnick PI, Kolter R (1999) Steps in the development of a *vibrio cholerae* El tor biofilm. Mol Microbiol 34:586–595

Williams ST, Vickers JC (1986) The ecology of antibiotic production. Microbial Ecol 12:43–52

Wright SA, Zumoff CH, Schneider L, Beer SV (2001) *Pantoea agglomerans* strain EH318 produces two antibiotics that inhibit *Erwinia amylovora* in vitro. Appl Environ Microbiol 67(1):284–292

Xu Z, Shao J, Li B, Yan X, Shen Q, Zhang R (2013) Contribution of bacillomycin D in *bacillus amyloliquefaciens* SQR9 to antifungal activity and biofilm formation. Appl Environ Microbiol 79(3):808–815

Yamaga F, Washio K, Morikawa M (2010) Sustainable biodegradation of phenol by *Acinetobacter calcoaceticus* P23 isolated from the rhizosphere of duckweed *Lemna aoukikusa*. Environ Sci Technol 44:6470–6474

Yang J, Kharbanda PD, Mirza M (2004) Evaluation of *Paenibacillus polymyxa* PKB1 for biocontrol of Pythium disease of cucumber in a hydroponic system. Acta Hortic (ISHS) 635:59–66

Zehnder GW, Murphy IF, Sikora EJ, Kloepper JW (2001) Application to rhizobacteria for induced resistance. Eur J Plant Pathol 107:39–50

Zhang R, Vivanco JM, Shen Q (2017) The unseen rhizosphere rootsoil-microbe interactions for crop production. Curr Opin Microbiol 37:8–14. https://doi.org/10.1016/j.mib.2017.03.008

# Chapter 5
# Plant Growth-Promoting Rhizobacteria as Antifungal Antibiotics Producers

**Anima Shrestha, Manju Shree Shakya Hada, Reshma Tuladhar, Sarita Manandhar, Noshin Ilyas, Abdel Rahman Mohammad Al Tawaha, and Anjana Singh**

**Contents**

5.1 Introduction.......... 83
5.2 PGPR as Biocontrol Agent (BCA).......... 84
5.3 Mechanisms of PGPR in Biological Control.......... 85
5.4 Antibiotics Production by Antifungal PGPR.......... 86
5.5 Potential Antifungal PGPR and Their Applications.......... 92
5.6 Conclusion and Future Prospects.......... 92
References.......... 96

## 5.1 Introduction

Soil in close association with plant roots is highly influenced by the root system (Dobbelaere et al. 2003), especially root exudates which provide nutrients for bacteria in the rhizosphere. Rhizospheric soil is thus colonized by a higher number of bacteria than non-rhizospheric soil (Basu et al. 2021). A diverse group of bacteria that colonize the rhizosphere and rhizoplane is termed as rhizobacteria, and these

A. Shrestha (✉) · M. S. S. Hada · S. Manandhar
Department of Microbiology, Tri-Chandra Multiple Campus, Tribhuvan University, Kirtipur, Kirtipur, Kathmandu, Nepal

R. Tuladhar · A. Singh
Central Department of Microbiology, Tribhuvan University, Kirtipur, Kathmandu, Nepal

N. Ilyas
Department of Botany, PMAS-Arid Agriculture University, Rawalpindi, Pakistan

A. R. M. Al Tawaha
Department of Biological Sciences, Al - Hussein Bin Talal University, Maan, Jordan

 
R. Z. Sayyed et al. (eds.), *Antifungal Metabolites of Rhizobacteria for Sustainable Agriculture*, Fungal Biology, https://doi.org/10.1007/978-3-031-04805-0_5

bacteria can be both beneficial and detrimental to plants. The beneficial rhizobacteria in and around the influential area of roots that promote the growth of plants are referred to as plant growth-promoting rhizobacteria (PGPR). Only 2–5% of rhizobacteria promote plant growth (Kloepper and Schroth 1978).

Genera of *Bacillus*, *Azospirillum*, *Rhizobium*, *Serratia*, *Azotobacter*, *Xanthomonas*, *Paenibacillus*, and *Pseudomonas* (Qessaoui et al. 2019) are common PGPR in the rhizosphere. Among these, *Bacillus* spp. (Cao et al. 2018) and *Pseudomonas* spp. (Qessaoui et al. 2019) represent the core PGPR for many crops. *Bacillus* is considered as an attractive PGPR (Zeriouh et al. 2011) due to ease of storage as being spore-formers (Cao et al. 2018).

Depending on the degree of bacterial proximity to the root and intimacy of association, PGPR can be classified as extracellular PGPR (ePGPR) and intracellular PGPR (iPGPR) (Bhattacharyya and Jha 2012; Gray and Smith 2005). ePGPR, for example, genera of *Agrobacterium*, *Arthrobacter*, *Bacillus*, *Caulobacter*, *Erwinia*, *Micrococcus*, *Pseudomonas*, and *Serratia*, exist in the rhizosphere, on the rhizoplane, and in the spaces between cells of the root cortex, whereas iPGPR, for example, genera of *Allorhizobium*, *Bradyrhizobium*, *Mesorhizobium*, and *Frankia* exist inside root cells, generally in specialized nodular structures (Bhattacharyya and Jha 2012; Gray and Smith 2005).

## 5.2 PGPR as Biocontrol Agent (BCA)

PGPR directly or indirectly benefits several plants (Ali et al. 2020). As the name suggests, the primary role of PGPR is to promote the growth of associated plants by various mechanisms such as production of plant growth hormones (Kusale et al. 2021a) and enzymes (Sagar et al. 2020), siderophore production (Nithyapriya et al. 2021; Sayyed et al. 2019), nitrogen fixation (Kusale et al. 2021b), and phosphate solubilization (Sharma and Sayyed 2013; Sharma et al. 2016; Meena et al. 2016; Beneduzi et al. 2012). Besides these direct benefits to associated plants, they also provide indirect benefits by being antagonist to various phytopathogens including bacteria, fungi, nematodes, as well as viruses (Sivasakthi et al. 2014). This antagonistic property of PGPR leads to the biocontrol activity against these diverse phytopathogens.

Numerous studies have been carried out all over the world to explore the biocontrol activity of root colonizing bacteria. Researchers have identified various root associated bacteria that can help in the biocontrol of plant disease-causing microorganisms (Sharma et al. 2020; Arora et al. 2021; Ali et al. 2020; Bjelić et al. 2018; Fernando et al. 2006; Sivasakthi et al. 2014). They can suppress a broad spectrum of bacterial, fungal, and nematode diseases (Sukmawati et al. 2021). The use of PGPR has become a common practice in many regions of the world.

### *5.2.1 Advantages of PGPR over Chemical Biocides*

Plant health is affected by pathogenic microorganisms creating threat to the food production (Compant et al. 2005). The pressure to meet increasing demand for food has led to the use of chemical pesticides and fertilizers (Carvalho 2017) and development of genetically modified varieties of plants. But both strategies have limitations (Cao et al. 2018). Regarding agrochemicals, they are not effective against all the diseases and can pose a health and environmental risk as toxic residues can accumulate in the soil and enter the food supply.

In vitro studies of microbial agents show considerable promise for achieving good performance in the field where there is heterogeneity of abiotic and biotic factors and competition with organisms because chemical fungicides and synthetic fertilizers affect human health and environment (Singh et al. 2017). Despite disease-resistant genetically modified plants being developed, public acceptance of foods from them is low. Moreover, disease resistance of genetically modified plants is often overcome by pathogens within a few years (Cao et al. 2018).

Effective strategies are now in use for biological fertilizers like *Rhizobium* spp., *Bacillus* spp., *Pseudomonas* spp., *Trichoderma* spp., etc. for enhancing the growth of plants and eradicating disease caused by bacteria and fungi (Singh et al. 2017; Suriani et al. 2020). Thus PGPR as biocontrol agents help in promoting plant growth, increase crop production, and are also less expensive than chemical fertilizers (Olanrewaju et al. 2017).

## 5.3 Mechanisms of PGPR in Biological Control

There are various mechanisms by which PGPR acts as biocontrol agents. Production of cell wall degrading enzymes and hydrolytic enzymes (chitinase, glucanase, protease, and cellulase) (Jadhav et al. 2017, 2020a, b; Shaikh et al. 2018), production of hydrogen cyanide (HCN), competitive exclusion, induced systemic resistance (ISR) (Patel et al. 2016), and antibiotics production are the common mechanisms in biocontrol (Reshma et al. 2018).

Production of antibiotics is the major mechanism of PGPR that helps them to control the plant pathogens. The genera *Bacillus* and *Pseudomonas*, common PGPR, are reported to be potential antibiotic producers (Gouda et al. 2018). They produce a variety of metabolites which serve as antimicrobial, antifungal, antibacterial, anti-helminthic, antiviral, phytotoxic, antioxidant, cytotoxic, and antitumor agents (Olanrewaju et al. 2017), so suppress plant pathogens even at low concentration (Goswami et al. 2016). The strains of *Bacillus* produce a wide variety of antibacterial and antifungal antibiotics. Some of the compounds produced by *Bacillus* that are derived from ribosomal origin include subtilin, subtilosin A, TasA, and sublancin, whereas those formed by non-ribosomal peptide synthetases (NRPSs) and/or polyketide synthases (PKS) are bacilysin, chlorotetain, mycobacillin, rhizocticins, bacillaene, difficidin, and lipopeptides belonging to the surfactin, iturin, and fengycin families (Leclère et al. 2005). Antibiotics produced by *P. fluorescens* and *P. aeruginosa* include 2,4-diacetylphloroglucinol (DAPG), phenazine-1-carboxylic acid (PCA), phenazine-1-carboxamide (PCN), pyoluteorin (Plt), pyrrolnitrin (Prn), oomycinA, viscosinamide, butyrolactones, kanosamine, zwittermicin A, aerugine, rhamnolipids, cepaciamide A, ecomycins, pseudomonic acid, azomycin, antitumor antibiotics FR901463, cepafungins, and antibiotic karalicin (Olanrewaju et al. 2017).

## 5.4 Antibiotics Production by Antifungal PGPR

Antibiosis by PGPR plays a major role in the control of plant pathogens. PGPR can act as BCA against plant pathogens by producing one or more antibiotics. Besides direct antagonistic action, antibiotics also play a vital role in induced systemic resistance (ISR) mechanisms in plants (Vinay et al. 2016; Zakaria et al. 2019; Meena et al. 2020). The production of antifungal antibiotics by PGPR depends on biocontrol ability and the degree of fungal colonization in plant roots (Haas and Keel 2003). Antibiotics that suppress the growth of plant pathogens are grouped into volatile and non-volatile complexes (Table 5.1) (Fernando et al. 2006). The major antibiotics that play a vital role in the suppression of plant pathogens are majorly grouped into non-volatile compounds of antibiotics (Saraf et al. 2014), and is discussed in the chapter.

### *5.4.1 Non-volatile Antibiotics*

#### 5.4.1.1 Polyketides

Polyketides, usually produced by actinomycetes and plant associated pseudomonads, constitute antifungal antibiotics pyoluteorin (Plt) and 2,4-diacetylphloroglucinol (DAPG) and are highly effective in suppression of fungal plant pathogens (Bender et al. 1999).

**Table 5.1** Volatile and non-volatile antibiotics

| Volatile antibiotics | Non-volatile antibiotics |
|---|---|
| Hydrogen cyanide | Polyketides |
| Alcohols | Heterocyclic nitrogenous compounds |
| Aldehydes | Phenylpyrrole |
| Ketones | Lipopeptides |
| Sulfides | Cyclic lipopeptides |
| | Aminopolyols |

#### 5.4.1.1.1 Pyoluteorin (Plt)

Plt is a phenolic polyketide with a resorcinol ring coupled to a bichlorinated pyrrole moiety. Besides fungicidal, it also has bactericidal and herbicidal properties (Fernando et al. 2006). Several strains of *Pseudomonas* spp. that produce Plt suppressed plant diseases caused by phytopathogenic fungi (Andreolli et al. 2019; Bajpai et al. 2018; Cordero et al. 2012; Nowak-Thompson et al. 1999). Howell and Stipanovic (1980) mentioned that it was found to show its toxicity against oomycete *Pythium ultimum*, a plant pathogen causing harm to varieties of plants.

Similarly, genetic analysis of *Pseudomonas protegens* MP12 isolated from a soil of Northern Italy showed the gene *pltB* responsible for the synthesis of the antifungal compound pyoluteorin and exhibited inhibitory effects on in vitro mycelial growth of prominent grapevine (*Vitis vinifera*) phytopathogens *Botrytis cinerea*, *Alternaria alternata*, *Aspergillus niger*, *Penicillium expansum*, and *Neofusicoccum parvum*, and also in vivo antifungal activity against *B. cinerea* on grapevine leaves (Andreolli et al. 2019). Ten genes (*plt LABCDEFGRM*) required for the biosynthesis of plt were also identified within a 24-kb genomic region of *Pseudomonas fluorescens* Pf-5 (Nowak-Thompson et al. 1999).

Genetic study of *Pseudomonas protegens* strain W45, which inhibited the growth of *Sclerotinia sclerotiorum*, recovered from rhizosphere of wheat revealed the presence of *PltB* loci, responsible for the synthesis of plt (Bajpai et al. 2018).

#### 5.4.1.1.2 2, 4-Diacetylphloroglucinol (DAPG)

2,4-DAPG is a broad-spectrum antifungal and antibacterial antibiotic active against a wide range of rhizospheric plant pathogens (Chae et al. 2020; Kwak and Weller 2013; Zhao et al. 2020). It is a phenolic compound, a phloroglucinol derivative that protects the plant either by direct inhibition of the pathogen or by inducing resistance mechanisms in the plant (Kwak and Weller 2013). DAPG is produced by several *Pseudomonas* spp. (Zhao et al. 2020) and is considered the major antibiotic compound produced by *P. protegens* FD6 (Fernando et al. 2006; Zhang et al. 2020a, b). Genetic analysis of fluorescent *Pseudomonas* spp. showed that at least 22 genotypes of 2, 4-DAPG produced fluorescent *Pseudomonas* spp. The gene *phlD* is the key gene in *phlACBD* locus for the synthesis of 2, 4-DAPG (Kwak and Weller 2013; Zhao et al. 2020). Ali et al. (2020) detected 2, 4-DAPG in all tested *Pseudomonas* strains (*P. putida* MS6, *P. aeruginosa* FB1, *P. aeruginosa* FB2, *P. aeruginosa* FB3, *P. aeruginosa* FB5, *P. fluorescens* LB1, and *P. stutzeri* RMB6) included in their study through direct infusion analysis in mass spectroscopy. 2,4-DAPG produced by *Pseudomonas fluorescens* Pf-5 also inhibited growth of the plant pathogenic fungi *Pythium ultimum* and *Rhizoctonia solani* (Nowak-Thompson et al. 1994).

#### 5.4.1.2 Heterocyclic Nitrogenous Compounds

Rhizobacteria produce various extracellular heterocyclic nitrogen-containing antimicrobials (Fernando et al. 2006). Phenazines come under a large family of heterocyclic nitrogen-containing brightly colored pigments (Chin-A-Woeng et al. 2003). Phenazine is predominantly synthesized by fluorescent *Pseudomonas*. Its derivatives such as phenazine-1-carboxylic acid (PCA) and phenazine-1-carboxamide (PCN) are secreted by bacterial genera *Brevibacterium*, *Burkholderia*, *Streptomyces*, and *Pseudomonas* (Harikrishnan et al. 2016). Phenazines and their derivatives have broad-spectrum antimicrobial activity (Chin-A-Woeng et al. 2003; Powell et al. 2000). The phenazine compound has been found to be active against *Rhizoctonia solani* that cause sheath blight of rice (Karmegham et al. 2020). Powell et al. (2000) obtained a single active component phenazine-1-carboxylic acid from the cell extract of *Pseudomonas aureofaciens* Tx-1 (ATCC 55670) that exhibited greatest antifungal activity against turfgrass pathogens. The extracted PCA was also found to be effective against the fungal pathogen *Sclerotinia homoeocarpa* that cause dollar spot on creeping bentgrass and was found to be equally effective as commercial fungicides triadimefon and chlorothalonil (Powell et al. 2000).

#### 5.4.1.3 Phenylpyrrole

The antibiotic pyrrolnitrin (PRN), a chlorinated phenylpyrrole, belongs to the phenylpyrrole group (Fernando et al. 2006; Saraf et al. 2014). Chemically, PRN is 3-(2-nitro-3-chlorophenyl)-4-chloropyrrole and is a secondary metabolite derived from L-tryptophan (Saraf et al. 2014). PRN has a broad-spectrum antifungal activity and is active against a wide range of deuteromycete, basidiomycete, and ascomycete fungi (Fernando et al. 2006; Saraf et al. 2014). PRN is produced by fluorescent and non-fluorescent bacterial species, most common being *Pseudomonas*. However, it was first isolated from the bacteria *Burkholderia pyrrocinia*. PRN extracted from *Burkholderia cepacia* K87 is reported to be effective against *Rhizoctonia solani*, *Colletotrichum gloeosporioides*, and *Sclerotinia sclerotiorum* (Sultan et al. 2008). PRN also shows antifungal activity against *Saccharomyces cerevisiae*, *Penicillium atrovenetum*, and *P. oxalicum* (Tripathi and Gottlieb 1969).

#### 5.4.1.4 Lipopeptides

Lipopeptides secreted by several species of *Bacillus* are low-molecular-weight cyclic amphiphilic oligopeptides synthesized by multi-enzyme complexes called non-ribosomal peptide synthetases (NRPSs). The lipopeptides are categorized into three different families (iturin, fengycin, and surfactin) on the basis of their amino acid sequence and fatty acid branching (Chen et al. 2008; Stein 2005; Zeriouh et al. 2011). These three lipopeptide families are produced by *B. subtilis* and *B. amylofaciens* under in vitro conditions at high rates and are also secreted in natural condition in the rhizosphere as the main antimicrobials (Cawoy et al. 2015). Among three famillies of lipopeptides, surfactins show potential antibacterial, antiviral, and antimycoplasmal activity, but no marked antifungal activity. However, iturins and fengycins possess strong antifungal potential (Chen et al. 2008; Gond et al. 2015; Zeriouh et al. 2011).

Iturins are heptapeptides with a β-amino fatty acid, and the family includes iturin A, mycosubtilin, and bacillomycin (Romero et al. 2007; Saraf et al. 2014). They exhibit strong antifungal activity (Dang et al. 2019; Shi et al. 2018; Youyou Wang et al. 2020).

*B. subtilis* AU195 produces bacillomycin which is an inhibitor of *Aspergillus flavus* (Moyne et al. 2004). *B. subtilis* WL-2 that produces Iturin A can be used as a potential inhibitor of *Phytophthora infestans* affecting mycelium growth and controlling potato late blight (Pérez-García et al. 2011; Wang et al. 2020). *B. subtilis* strains, UMAF6614, UMAF6619, UMAF6639, and UMAF8561, were detected as antagonist against *Podosphaera fusca*, the main causal agent of cucurbit powdery mildew in Spain. They showed strong inhibitory effects due to the effect of bacillomycin, fengycin, and iturin A (Romero et al. 2007). Besides *B. subtilis*, other species of *Bacillus* including *B. amyloliquefaciens*, *B. licheniformis*, *B. thuringiensis*, and *B. methyltrophicus* have also been reported for iturin production (Dang et al. 2019). In a study of Arrebola et al., iturin A produced by *B. amyloliquefaciens* PPCB004 affected the fungal pathogens *Alternaria citri*, *Botryosphaeria* spp., *Colletotrichum gloeosporioides*, *Fusicoccum aromaticum*, *Lasiodiplodia theobromae*, *Penicillium crustosum*, and *Phomopsis persea* (Arrebola et al. 2010).

#### 5.4.1.5 Cyclic Lipopeptides (CLPs)

CLPs are composed of a fatty acid tail linked to a short oligopeptide, which is cyclized to form a lactone ring between two amino acids in the peptide chain (Raaijmakers et al. 2006). CLPs are biosurfactants produced by Gram-positive and Gram-negative bacteria such as *Bacillus* spp., *Pseudomonas* spp., and *Streptomyces* spp. (Raaijmakers et al. 2010).

Around hundred CLPs produced by various species of *Pseudomonas* have been identified. Geudens and Martins (2018) classified CLPs into viscosin, orfamide, amphisin, syringopeptin, and tolaasin groups. Based on the length of oligopeptide, at least 14 different groups of *Pseudomonas* CLPs have been identified (Geudens and Martins 2018; Omoboye et al. 2019). However, new CLPs are being known, and more new groups are being formed (Geudens and Martins 2018). The production of different CLPs (viscosinamide, tensin, and amphisin) is a common trait among many *P. fluorescens* strains in the soil environment, especially in the rhizosphere (Nielsen and Sørensen 2003). *P. koreensis* S150 synthesizes a cyclic lipopeptide lokisin that is dominant against plant fungal pathogens (Gu et al. 2020). *Pseudomonas* spp. CMR12a produces CLPs orfamides and sessilins that were able to control damping-off disease on Chinese cabbage (*Brassica chinensis*) caused by *Rhizoctonia solani* AG 2–1 and root rot disease on bean (*Phaseolus vulgaris* L.) caused by

*R. solani* AG 4-HGI (Olorunleke et al. 2015). The study of Omoboye et al. demonstrated that white line inducing principle (WLIP)-, lokisin-, and entolysin-producing *Pseudomonas* strains were capable to control the rice blast disease caused by *Magnaporthe oryzae* by ISR and direct effects (Omoboye et al. 2019). The gene cluster c-xtl, essential for inhibiting the development of *Magnaporthe oryzae*, was detected in *P. mosselii* BS011 which can be used as a promising biocontrol agent (Wu et al. 2018). Several species of plant associated *Pseudomonas* including *P. protogens* (Zhang et al. 2020a, b), *P. mosselii* (Wu et al. 2018), *P. syringae*, *P. tolaasii*, *P. fuscovaginae*, *P. corrugata*, *P. putida*, and *P. fluorescens* produce CLPs (Raaijmakers et al. 2006) capable of inhibiting fungal plant pathogens *Pythium ultimum* (Nielsen and Sørensen 2003), *Pythium myriotylum* (Oni et al. 2019), *Rhizoctonia solani* (Nielsen and Sørensen 2003), *Botrytis cinerea*, and *Monilinia fructicola* (Zhang et al. 2020a, b).

#### 5.4.1.6 Aminopolyols (Zwittermicin A)

Zwittermicin A (ZmA) is a highly polar, water-soluble aminopolyol antibiotic produced by *Bacillus* spp. and was first discovered from *B. cereus* UW85 (Milner et al. 1996; Sansinenea and Ortiz 2012; Silo-Suh et al. 1998). ZmA is a hybrid polyketide-nonribosomal peptide and displays broad-spectrum antimicrobial activity (Luo et al. 2011). It shows activity against phytopathogenic fungi *Alternaria*, *Fusarium*, *Helminthosporium*, and *Utilago* (Silo-Suh et al. 1998).

### 5.4.2 Mode of Action of Non-volatile Antifungal Antibiotics

#### 5.4.2.1 Polyketides

*Plt*, a polyketide antibiotic, consisting of a resorcinol ring linked to bichlorinated pyrrole moiety, has been found to be associated with the suppression of a series of plant diseases caused by pathogenic fungi, besides possessing bactericidal and herbicidal properties as well (Jan et al. 2011).

The phenolic compound, 2,4- DAPG, acts as an ISR elicitor, but the mechanism of ISR is unclear (Chae et al. 2020). DAPG, produced by pseudomonads, disrupt the membrane and inhibit the zoospore of *Pythium* (De Souza et al. 2003).

#### 5.4.2.2 Heterocyclic Nitrogenous Compounds (Phenazines)

Over 50 naturally occurring phenazines have been identified, among which commonly identified derivatives are pyocyanin, PCA, PCN, and hydroxy phenazines (Turner and Messenger 1986). The mechanisms of the action of phenazines in antifungal interactions are poorly understood (Chin-A-Woeng et al. 2003). Various studies assume they diffuse across or insert into the membrane and act as a reducing agent, resulting in the uncoupling of oxidative phosphorylation and the generation of toxic intracellular superoxide radicals and hydrogen peroxide which are harmful to the organism (Turner and Messenger 1986). A study showed that PCA can work together with the siderophore pyoverdine in promoting biofilm development, while phenazine pyocyanin (PYO) can facilitate biofilm formation via an iron-independent pathway even in the absence of siderophore. These mechanisms of the phenazines will make iron more bioavailable at later stages of infection by shifting the redox equilibrium between Fe(III) and Fe(II) (Wang et al. 2011).

#### 5.4.2.3 Phenylpyrrole

The study done on *S. cerevisiae* showed that pyrrolnitrin inhibits fungal growth by inhibiting the respiratory electron transport system specifically by inhibiting succinate oxidase, NADH oxidase, succinate-cytochrome c reductase, NADH-cytochrome c reductase, and succinate-coenzyme Q6 reductase. Phenylpyrrole also showed inhibitory effect on whole cells of *S. cerevisiae* by inhibiting the incorporation of 4C-glucose into nucleic acids and proteins. It also inhibited the incorporation of nucleotide bases into ribonucleic acid, deoxyribonucleic acid, and amino acids into protein (Tripathi and Gottlieb 1969).

#### 5.4.2.4 Lipopeptides

Lipopeptides iturins and fengycins produced by species of *Bacillus* exhibit strong antifungal activity (Romero et al. 2007). **Iturin A** causes cell membrane disruption leading to irregular internal structure of cell, triggers oxidative stress reactions similarly to reactive oxygen species (ROS), and causes mitochondrial damage leading to ATP production decline (Wang et al. 2020).

**Fengycin** acts on fungal pathogens by two ways: inducing systemic resistance in the plants by perturbing the cell membrane of the root cells and by directly attacking fungi by binding to their cell membrane, causing leakage and lysis (Sur et al. 2018).

#### 5.4.2.5 Cyclic Lipopeptides

Cyclic lipopeptides (CLPs) are versatile molecules with antimicrobial, cytotoxic, and surfactant properties (Raaijmakers et al. 2006). The study of Omoboye et al. (2019) showed that a CLP lokisin isolated from *Pseudomonas* spp. COR10 controlled rice blast disease caused by *Magnaporthe oryzae* by ISR and direct antagonism against the pathogen.

Most of the CLPs produced by *Pseudomonas* exhibit antifungal activity against a number of fungal pathogens by ISR and antagonism, but few either disrupt the fungal plasma membrane or lyse zoospore (Olorunleke et al. 2015).

#### 5.4.2.6 Aminopolyols (Zwittermicin A)

Zwittermicin A showed the strongest inhibition of germination of the cysts and elongation of germ tubes in *Pythium torulosum* (Shang et al. 1999).

## 5.5 Potential Antifungal PGPR and Their Applications

Though varieties of bacterial species such as *Azotobacter* spp. (Bjelić et al. 2018; Fatima et al. 2009), *Burkholderia cepacia* (Sultan et al. 2008), *Burkholderia contaminans* (Tagele et al. 2018), *Paenibacillis jamilae* (Wang et al. 2019), different species of *Bacillus* (Table 5.2), and *Pseudomonas* (Table 5.3) are capable of producing antifungal antibiotics and can be used in the biocontrol of phytopathogens, the potential PGPR that have antifungal properties are the species of *Bacillus* (Table 5.2) and *Pseudomonas* (Table 5.3). Various strains of *Bacillus* and *Pseudomonas* produce varieties of antifungal antibiotics and other antifungal metabolites. The exploration of these characteristics of the antifungal substances produced by PGPR and their applications help in the management of plant diseases caused by fungal pathogens (Tables 5.2 and 5.3).

## 5.6 Conclusion and Future Prospects

Microorganisms, specifically PGPR, have the potential to enhance growth of plants and to biologically control plant pathogens present in the rhizosphere through various mechanisms. This potential of PGPR may help reduce the use of chemical fertilizers and pesticides. The use of BCAs in soil helps to maintain the natural soil properties. PGPR in the rhizosphere can be used as BCAs that have antibacterial, antiparasitic, antiviral, or antifungal activity. Fungal diseases of plants have been a problem in agricultural and hence economic loss worldwide. However, researchers have identified specific microorganisms antagonist to the fungal plant pathogens, with the potential antifungal PGPR being the strains of *Bacillus* spp. and *Pseudomonas* spp. Novel strains of PGPR with increasing efficacy are being identified, and a potential area for the exploration of newer and more efficient rhizobacteria that would help not only in the enhancement of plant growth, but also as efficient BCAs. These PGPR could be used alone against the specific fungal pathogen, but the use of PGPR in combination of two or more strains of bacteria for an effective control of phytopathogens could be a new area for research.

The ability of rhizobacteria to act against the fungal pathogens is due to the production of various antifungal metabolites, especially antibiotics. Many antifungal antibiotics produced by PGPR, previously unidentified, are being identified by advanced techniques such as MALDI-TOF mass spectroscopy, solid-phase microextraction-gas chromatography, etc. Types of lipopeptide antibiotics produced

**Table 5.2** *Bacillus* spp. as potential antifungal PGPR and their applications in the management of fungal plant diseases

| Species and strains of *Bacillus* | Activity against | Host plant (Disease) | Antifungal metabolites produced | Study type | Reference |
|---|---|---|---|---|---|
| *Bacillus* spp. | *Helminthosporium* sp., *Macrophomina* sp., and *Fusarium* sp. | Maize (maize root disease) | Hydrolytic enzymes and siderophores | In vitro | Bjelić et al. 2018 |
| *B. velezensis* ZSY-1 | *Alternaria solani*, *Botrytis cinerea*, *Valsa mali*, *Monilinia fructicola*, *Fusarium oxysporum* f. sp. *capsicum*, *Colletotrichum lindemuthianum* | Tomato (early blight and gray mold | Pyrazine (2,5-dimethyl), benzothiazole, phenol (4-chloro-3-methyl), and phenol-2,4-bis (1,1-dimethylethyl) | In vitro | Gao et al. 2017 |
| *B. amyloliquefaciens* | *F. oxysporum* f.sp. *niveum* | Watermelon (*fusarium* wilt) | VOCs (2-nonanone and 2-heptanone) | In vitro and in vivo | Wu et al. 2019 |
| *B. methylotrophicus* XT1 CECT 8661 | *Botrytis cinerea* | Tomato, grapes, and strawberries | Lipopeptides (fengycin, surfactin, bacillomycin, and iturin) | In vitro and in vivo | Toral et al. 2018 |
| *Bacillus* spp. | *Rhizoctonia solani*, *Botrytis cinerea*, *Verticillium dahlia*, *Phytophthora infestans* | Tomato | Surfactin, fengycin, bacillibactin, petrobactin, lichenysin, and bacillaene | In vitro | Zhou et al. 2020 |
| *B. subtilis* | *Helminthosporium sativum* | Chickpea | Not studied | In vitro | Khan et al. 2019 |
| *B. subtilis* 30VD-1 | *Fusarium* spp. | Pea seeds | Volatiles and other antifungal molecules | In vitro and in vivo | Khan et al. 2018 |
| *B. subtilis* DZSY21 | *Curvularia lunata* | Maize | VOCs (2-Methylbutyric acid, 2-heptanone, and isopentyl acetate) | In vitro and in vivo | Xie et al. 2020 |
| *B. subtilis* KFP-5, KFP-7, KFP-17 | *Pyricularia oryzae* | Rice (rice blast disease) | Hydrolytic enzymes, siderophores | In vitro and in vivo | Rais et al. 2017 |

(continued)

**Table 5.2** (continued)

| Species and strains of *Bacillus* | Activity against | Host plant (Disease) | Antifungal metabolites produced | Study type | Reference |
|---|---|---|---|---|---|
| *B. subtilis* RB14 | *Rhizoctonia solani* K1 | Tomato (damping-off diseases) | Lipopeptide iturin A | In vitro and in vivo | Zohora et al. 2016 |
| *B. subtilis* subsp. *subtilis* MG27, *B. velezensis* MG33 *and* MG43, *B. pumilus* MG52 *and* MG84, *B. altitudinis* MG75, and *B. laterosporus* MG64 | *Magnaporthe oryzae* | Perennial ryegrass (severe blast disease) | Bacteriocins and terpenes | In vivo | Li et al. 2020 |
| *B. thuringiensis* | *Fusarium solani* | Chickpea | Not studied | In vitro | Khan et al. 2018 |
| *B. velezensis* NKG-2 | *Fusarium oxysporum* | Tomato | Not studied | In vitro and in vivo | Myo et al. 2019 |

**Table 5.3** *Pseudomonas* spp. as potential antifungal PGPR and their applications in the management of fungal plant diseases

| Species and strains of *Pseudomonas* | Activity against | Host plant (disease) | Antifungal metabolites produced | Study type | Reference |
|---|---|---|---|---|---|
| *P. aeruginosa* | *Fusarium oxysporum* f. sp. *cucumerinum* (*Foc*) | Cucumber (*Fusarium* wilt) | Siderophores and volatile compounds | In vitro | Islam et al. 2018 |
| *P. aeruginosa* | *Rhizoctonia solani* | Rice | Hydrolytic enzymes, siderophore | In vitro | Sheeba 2017 |
| *P. aeruginosa* FB2 | *Fusarium oxysporum*, *Fusarium moniliforme*, *Rhizoctonia solani*, *Colletotrichum gloeosporioides*, *Colletotrichum falcatum*, *Aspergillus niger*, *Aspergillus flavus* | Not mentioned | Antibiotics (iturin, surfactins, fengycin, DAPG, Phenazine, etc.), cell wall degrading enzymes (protease, chitinase, and cellulase), and siderophores | In vitro and in vivo | Ali et al. 2020 |

(continued)

**Table 5.3** (continued)

| Species and strains of *Pseudomonas* | Activity against | Host plant (disease) | Antifungal metabolites produced | Study type | Reference |
|---|---|---|---|---|---|
| *P. fluorescens* | *Pythium aphanidermatum* | Turmeric | Phenazine 1- carboxylic acid, 2, 4- DAPG, pyoluteorin, and pyrrolnitrin | In vitro | Prabhukarthikeyan and Raguchander 2016 |
| *P. fluorescens* HC1–07 | *Rhizoctonia solani AG-8* | Wheat (*Rhizoctonia* root rot) | Cyclic lipopeptide | In vitro and in vivo | Yang et al. 2017 |
| *P. protegens* FD6 | *Botrytis cinerea Monilinia fructicola* | Tomato (gray mold disease) | 2,4-DAPG, pyoluteorin, and pyrrolnitrin | In vitro and in vivo | Zhang et al. 2020a, b |
| *P. protegens* MP12 | *Botrytis cinerea* | Grapevine leaves | 2,4-DAPG, pyoluteorin, and pyrrolnitrin | In vitro and in vivo | Andreolli et al. 2019 |
| *P. protegens* RS-9 | *Ralstonia solanacearum* | Tomato | 2,4-DAPG, hydrogen cyanide (HCN), pyrrolnitrin, and pyoluteorin | In vitro and in vivo | Rai et al. 2017 |
| *P. chlororaphis* G05 | *Fusarium graminearum* | Cereal crops (*fusarium* head blight) | Pyrrolnitrin | In vitro | Huang et al. 2018 |
| *P. fluorescens* EP-5 | *Rhizoctonia solani* | Rice (rice sheath blight) | 2,4-diacetyl phloroglucinol (2,4-DAPG) | In vitro and in vivo | Reshma et al. 2018 |
| *P. putida* | *Rhizoctonia solani* | Sesamum | Hydrogen cyanide, hydrogen sulfide, siderophores, and pyoluteorin | In vitro | Vinay et al. 2016 |

by potential antifungal PGPR are being identified. Moreover, modes of action of known antibiotics pyoluteorin and DAPG are still unclear. The area of research on the mechanism of antibiotics is open for future researchers. Understanding of the basic underlying mechanisms of action will definitely help to exploit antifungal PGPR and their metabolites to be used as BCAs in the integrated management of fungal plant pathogens.

# References


Ali S, Hameed S, Shahid M, Iqbal M, Lazarovits G, Imran A (2020) Functional characterization of potential PGPR exhibiting broad-spectrum antifungal activity. Microbiol Res 232. https://doi.org/10.1016/j.micres.2019.126389

Andreolli M, Zapparoli G, Angelini E, Lucchetta G, Lampis S, Vallini G (2019) *Pseudomonas protegens* MP12: A plant growth-promoting endophytic bacterium with broad-spectrum antifungal activity against grapevine phytopathogens. Microbiol Res 219:123–131. https://doi.org/10.1016/j.micres.2018.11.003

Arora H, Sharma A, Sharma S, Haron FF, Gafur A, Sayyed RZ, Datta R (2021) *Pythium* damping-off and root rot of *Capsicum annuum* L.: Impacts, Diagnosis, and Management, Microorganisms 9:823. https://doi.org/10.3390/microorganisms9040823

Arrebola E, Jacobs R, Korsten L (2010) Iturin A is the principal inhibitor in the biocontrol activity of *Bacillus amyloliquefaciens* PPCB004 against postharvest fungal pathogens. J Appl Microbiol 108(2):386–395. https://doi.org/10.1111/j.1365-2672.2009.04438.x

Bajpai A, Singh B, Joshi S, Johri BN (2018) Production and characterization of an antifungal compound from *Pseudomonas protegens* strain W45. Proc Natl Acad Sci India Sect B - Biol Sci 88(3):1081–1089. https://doi.org/10.1007/s40011-017-0844-1

Basu A, Prasad P, Das SN, Kalam S, Sayyed RZ, Reddy MS, El Enshasy H (2021) Plant growth promoting Rhizobacteria (PGPR) as green bioinoculants: recent developments, constraints, and prospects. Sustainability 13:1140. https://doi.org/10.3390/su13031140

Bender C, Rangaswamy V, Loper J (1999) Polyketide production by plant associated pseudomonads. Annu Rev Phytopathol 37:175–196

Beneduzi A, Ambrosini A, Passaglia LMP (2012) Plant growth promoting rhizobacteria (PGPR): their potential as antagonists and biocontrol agents. Genet Mol Biol 35:1044–1051

Bhattacharyya PN, Jha DK (2012) Plant growth-promoting rhizobacteria (PGPR): emergence in agriculture. World J Microbiol Biotechnol 28(4):1327–1350. https://doi.org/10.1007/s11274-011-0979-9

Bjelić D, Marinković J, Tintor B, Mrkovački N (2018) Antifungal and plant growth promoting activities of indigenous Rhizobacteria isolated from maize (*Zea mays* L.) rhizosphere. Commun Soil Sci Plant Anal 49(1):88–98. https://doi.org/10.1080/00103624.2017.1421650

Cao Y, Pi H, Chandrangsu P, Li Y, Wang Y, Zhou H, Xiong H, Helmann JD, Cai Y (2018) Antagonism of two plant-growth promoting Bacillus velezensis isolates against *Ralstonia solanacearum* and *Fusarium oxysporum*. Sci Rep 8(1):1–14. https://doi.org/10.1038/s41598-018-22782-z

Cawoy H, Debois D, Franzil L, De Pauw E, Thonart P, Ongena M (2015) Lipopeptides as main ingredients for inhibition of fungal phytopathogens by *Bacillus subtilis/amyloliquefaciens*. Microb Biotechnol 8(2):281–295. https://doi.org/10.1111/1751-7915.12238

Chae DH, Kim DR, Cheong MS, Lee YB, Kwak YS (2020) Investigating the induced systemic resistance mechanism of 2,4-diacetylphloroglucinol (Dapg) using dapg hydrolase-transgenic arabidopsis. Plant Pathol J 36(3):255–266. https://doi.org/10.5423/PPJ.OA.02.2020.0031

Chen H, Wang L, Su CX, Gong GH, Wang P, Yu ZL (2008) Isolation and characterization of lipopeptide antibiotics produced by *Bacillus subtilis*. Lett Appl Microbiol 47(3):180–186. https://doi.org/10.1111/j.1472-765X.2008.02412.x

Chin-A-Woeng TFC, Bloemberg GV, Lugtenberg BJJ (2003) Phenazines and their role in biocontrol by *Pseudomonas* bacteria. New Phytol 157(3):503–523. https://doi.org/10.1046/j.1469-8137.2003.00686.x

Compant S, Duffy B, Nowak J, Christophe C, Barka AE (2005) PAS 100:2011 specification for composted materials. Appl Environ Microbiol 71(9):1–68. https://doi.org/10.1128/AEM.71.9.4951

Cordero P, Cavigliasso A, Príncipe A, Godino A, Jofré E, Mori G, Fischer S (2012) Genetic diversity and antifungal activity of native *Pseudomonas* isolated from maize plants grown in a central region of Argentina. Syst Appl Microbiol 35(5):342–351. https://doi.org/10.1016/j.syapm.2012.04.005

Dang Y, Zhao F, Liu X, Fan X, Huang R, Gao W, Wang S, Yang C (2019) Enhanced production of antifungal lipopeptide iturin A by *Bacillus amyloliquefaciens* LL3 through metabolic engineering and culture conditions optimization. Microb Cell Factories 18(1):1–14. https://doi.org/10.1186/s12934-019-1121-1

De Souza JT, Arnould C, Deulvot C, Lemanceau P, Gianinazzi-Pearson V, Raaijmakers JM (2003) Effect of 2,4-diacetylphloroglucinol on Pythium: cellular responses and variation in sensitivity among propagules and species. Phytopathology 93(8):966–975. https://doi.org/10.1094/PHYTO.2003.93.8.966

Dobbelaere S, Vanderleyden J, Okon Y (2003) Critical reviews in plant sciences plant growth-promoting effects of *Diazotrophs* in the rhizosphere plant growth. CRC Crit Rev Plant Sci 22(2):107–149

Fatima Z, Saleemi M, Zia M, Sultan T, Aslam M, Riaz-Ur-Rehman CMF (2009) Antifungal activity of plant growth-promoting rhizobacteria isolates against *Rhizoctonia solani* in wheat. African J Biotechnol 8(2):219–225. https://doi.org/10.5897/AJB2009.000-9040

Fernando WGD, Nakkeeran S, Zhang Y (2006) Biosynthesis of antibiotics by PGPR and its relation in biocontrol of plant diseases. PGPR Biocontrol Biofertilization: 67–109. https://doi.org/10.1007/1-4020-4152-7_3

Gao Z, Zhang B, Liu H, Han J, Zhang Y (2017) Identification of endophytic *Bacillus velezensis* ZSY-1 strain and antifungal activity of its volatile compounds against *Alternaria solani* and *Botrytis cinerea*. Biol Control 105:27–39. https://doi.org/10.1016/j.biocontrol.2016.11.007

Geudens N, Martins JC (2018) Cyclic lipodepsipeptides from *Pseudomonas* spp. - biological Swiss-Army knives. Front Microbiol 9:1–18. https://doi.org/10.3389/fmicb.2018.01867

Gond SK, Bergen MS, Torres MS, White JF (2015) Endophytic *Bacillus* spp. produce antifungal lipopeptides and induce host defence gene expression in maize. Microbiol Res 172:79–87. https://doi.org/10.1016/j.micres.2014.11.004

Goswami D, Thakker JN, Dhandhukia PC (2016) Portraying mechanics of plant growth promoting rhizobacteria (PGPR): A review. Cogent Food Agric 2(1). https://doi.org/10.1080/23311932.2015.1127500

Gouda S, Kerry RG, Das G, Paramithiotis S, Shin HS, Patra JK (2018) Revitalization of plant growth promoting rhizobacteria for sustainable development in agriculture. Microbiol Res 206:131–140. https://doi.org/10.1016/j.micres.2017.08.016

Gray EJ, Smith DL (2005) Intracellular and extracellular PGPR: commonalities and distinctions in the plant-bacterium signaling processes. Soil Biol Biochem 37(3):395–412. https://doi.org/10.1016/j.soilbio.2004.08.030

Gu Y, Ma YN, Wang J, Xia Z, Wei HL (2020) Genomic insights into a plant growth-promoting *Pseudomonas koreensis* strain with cyclic lipopeptide-mediated antifungal activity. Microbiology 9(9):1–15. https://doi.org/10.1002/mbo3.1092

Haas D, Keel C (2003) Regulation of antibiotic production in root-colonizing *Pseudomonas* spp. and relevance for biological Control of plant Disease. Annu Rev Phytopathol 41(63):117–153. https://doi.org/10.1146/annurev.phyto.41.052002.095656

Harikrishnan H, Shanmugaiah V, Nithya K, Balasubramanian N, Sharma MP, Gachomo EW, Kotchoni SO (2016) Enhanced production of phenazine-like metabolite produced by *Streptomyces aurantiogriseus* VSMGT1014 against rice pathogen, *Rhizoctonia solani*. J Basic Microbiol 56(2):153–161. https://doi.org/10.1002/jobm.201500362

Huang R, Feng Z, Chi X, Sun X, Lu Y, Zhang B, Lu R, Luo W, Wang Y, Miao J, Ge Y (2018) Pyrrolnitrin is more essential than phenazines for *Pseudomonas chlororaphis* G05 in its suppression of *Fusarium graminearum*. Microbiol Res 215(February):55–64. https://doi.org/10.1016/j.micres.2018.06.008

Islam MA, Nain Z, Alam MK, Banu NA, Islam MR (2018) In vitro study of biocontrol potential of rhizospheric *Pseudomonas aeruginosa* against *Fusarium oxysporum* f. sp. *cucumerinum*. Egypt J Biol Pest Control 28(1):1–11. https://doi.org/10.1186/s41938-018-0097-1

Jadhav HP, Shaikh SS, Sayyed RZ (2017) Role of hydrolytic enzymes of Rhizoflora in biocontrol of fungal Phytopathogens: an overview in Rhizotrophs : plant growth promotion to bioremediation. Springer-Nature Springer:183–203

Jadhav HP, Sayyed RZ, Shaikh SS, Bhamre HM, Sunita S, Enshasy HE (2020a) Statistically designed bioprocess for enhanced production of alkaline protease in *Bacillus cereus* HP_RZ17. J Sci Indust Res 79:491–498

Jadhav HP, Sonawane MS, Khairnar MH, Sayyed RZ (2020b) Production of alkaline protease by rhizospheric *Bacillus cereus* HP_RZ17 & *Paenibacillus xylanilyticus* HP_RZ19. Environment Sustainability 3:5–13

Jan AT, Azam M, Ali A, Rizwanul Haq QM (2011) Novel approaches of beneficial *Pseudomonas* in mitigation of plant diseases - an appraisal. J Plant Interact 6(4):195–205. https://doi.org/10.1080/17429145.2010.541944

Karmegham N, Vellasamy S, Nateesan B, Sharma MP, Al Farraj DA, Elshikh MS (2020) Characterization of antifungal metabolite phenazine from rice rhizosphere fluorescent pseudomonads (FPs) and their effect on sheath blight of rice. Saudi J Biol Sci. https://doi.org/10.1016/j.sjbs.2020.10.007

Khan N, Martínez-Hidalgo P, Ice TA, Maymon M, Humm EA, Nejat N, Sanders ER, Kaplan D, Hirsch AM (2018) Antifungal activity of *Bacillus* species against *Fusarium* and analysis of the potential mechanisms used in biocontrol. Front Microbiol 9:1–12. https://doi.org/10.3389/fmicb.2018.02363

Khan N, Bano A, Rahman MA, Guo J, Kang Z, Babar MA (2019) Comparative physiological and metabolic analysis reveals a complex mechanism involved in drought tolerance in chickpea (*Cicer arietinum* L.) induced by PGPR and PGRs. Sci Rep 9(1):1–19. https://doi.org/10.1038/s41598-019-38702-8

Kloepper JW, Schroth M (1978) Plant growth-promoting rhizobacteria on radishes. Proc. 4th Int. Conf. Plant Pathog. Bact 2:879–882

Kusale SP, Attar YC, Sayyed RZ, Malek RA, Ilyas N, Suriani NL, Khan N, Enshasy HA (2021a) Production of plant beneficial and antioxidants metabolites by *Klebsiella variicola* under salinity stress. Molecules 26:1894. https://doi.org/10.3390/molecules26071894

Kusale SP, Attar YC, Sayyed RZ, Enshasy HE, Hanapi Z, Ilyas N, Elgorban AE, Bahkali AH, Marraiki N (2021b) Inoculation of *Klebsiella variicola* alleviated slat stress salinity and improved growth and nutrients in wheat and maize. Agronomy 11:927. https://doi.org/10.3390/agronomy11050927

Kwak YS, Weller DM (2013) Take-all of wheat and natural disease suppression: A review. Plant Pathol J 29(2):125–135. https://doi.org/10.5423/PPJ.SI.07.2012.0112

Leclère V, Béchet M, Adam A, Guez JS, Wathelet B, Ongena M, Thonart P, Gancel F, Chollet-Imbert M, Jacques P (2005) Mycosubtilin overproduction by *Bacillus subtilis* BBG100 enhances the organism's antagonistic and biocontrol activities. Appl Environ Microbiol 71(8):4577–4584. https://doi.org/10.1128/AEM.71.8.4577-4584.2005

Li Z, Song C, Yi Y, Kuipers OP (2020) Characterization of plant growth-promoting rhizobacteria from perennial ryegrass and genome mining of novel antimicrobial gene clusters. BMC Genomics 21(1):1–11. https://doi.org/10.1186/s12864-020-6563-7

Luo Y, Ruan LF, Zhao CM, Wang CX, Peng DH, Sun M (2011) Validation of the intact zwittermicin a biosynthetic gene cluster and discovery of a complementary resistance mechanism in *Bacillus thuringiensis*. Antimicrob Agents Chemother 55(9):4161–4169. https://doi.org/10.1128/AAC.00111-11

Meena MK, Gupta S, Datta S (2016) Antifungal potential of PGPR, their growth promoting activity on seed germination and seedling growth of winter wheat and genetic variabilities among bacterial isolates. Int J Curr Microbiol Appl Sci 5(1):235–243. https://doi.org/10.20546/ijcmas.2016.501.022

Meena M, Swapnil P, Divyanshu K, Kumar S, Harish TYN, Zehra A, Marwal A, Upadhyay RS (2020) PGPR-mediated induction of systemic resistance and physiochemical alterations in plants against the pathogens: current perspectives. J Basic Microbiol 60(10):828–861. https://doi.org/10.1002/jobm.202000370

Milner JL, Stohl EA, Handelsman JO (1996) Zwittermicin A resistance gene from *Bacillus cereus*. J Bacteriol 178(14):4266–4272. https://doi.org/10.1128/jb.178.14.4266-4272.1996

Moyne AL, Cleveland TE, Tuzun S (2004) Molecular characterization and analysis of the operon encoding the antifungal lipopeptide bacillomycin D. FEMS Microbiol Lett 234(1):43–49. https://doi.org/10.1016/j.femsle.2004.03.011

Myo EM, Liu B, Ma J, Shi L, Jiang M, Zhang K, Ge B (2019) Evaluation of *Bacillus velezensis* NKG-2 for bio-control activities against fungal diseases and potential plant growth promotion. Biol Control 134(March):23–31. https://doi.org/10.1016/j.biocontrol.2019.03.017

Nielsen TH, Sørensen J (2003) Production of cyclic lipopeptides by *Pseudomonas fluorescens* strains in bulk soil and in the sugar beet rhizosphere. Appl Environ Microbiol 69(2):861–868. https://doi.org/10.1128/AEM.69.2.861-868.2003

Nithyapriya S, Sundaram L, Sayyed RZ, Reddy MS, Dailin DJ, Enshasy HE, Suriani NL, Herlambang S (2021) Production, purification, and characterization of bacillibactin siderophore of *Bacillus subtilis* and its application for improvement in plant growth and oil content in sesame. Sustainability 13:5394. https://doi.org/10.3390/su13105394

Nowak-Thompson B, Gould SJ, Kraus J, Loper JE (1994) Production of 2,4-diacetylphloroglucinol by the biocontrol agent *Pseudomonas fluorescens* Pf-5. Can J Microbiol 40(12):1064–1066. https://doi.org/10.1139/m94-168

Nowak-Thompson B, Chaney N, Wing JS, Gould SJ, Loper JE (1999) Characterization of the pyoluteorin biosynthetic gene cluster of *Pseudomonas fluorescens* Pf-5. J Bacteriol 181(7):2166–2174. https://doi.org/10.1128/jb.181.7.2166-2174.1999

Olanrewaju OS, Glick BR, Babalola OO (2017) Mechanisms of action of plant growth promoting bacteria. World J Microbiol Biotechnol 33(11):1–16. https://doi.org/10.1007/s11274-017-2364-9

Olorunleke FE, Hua GKH, Kieu NP, Ma Z, Höfte M (2015) Interplay between orfamides, sessilins and phenazines in the control of *Rhizoctonia* diseases by *Pseudomonas* sp. CMR12a. Environ Microbiol Rep 7(5):774–781. https://doi.org/10.1111/1758-2229.12310

Omoboye OO, Oni FE, Batool H, Yimer HZ, De Mot R, Höfte M (2019) *Pseudomonas* cyclic lipopeptides suppress the rice blast fungus *Magnaporthe oryzae* by induced resistance and direct antagonism. Front Plant Sci 10(July):1–17. https://doi.org/10.3389/fpls.2019.00901

Oni FE, Olorunleke OF, Höfte M (2019) Phenazines and cyclic lipopeptides produced by *Pseudomonas* sp. CMR12a are involved in the biological control of *Pythium myriotylum* on cocoyam (*Xanthosoma sagittifolium*). Biol Control 129(July 2018):109–114. https://doi.org/10.1016/j.biocontrol.2018.10.005

Patel S, Sayyed R, Saraf M (2016) Bacterial determinants and plant defense induction : their role as bio-control agent in agriculture plant soil microbes, plant soil microbes. Springer, Switzerland, pp 187–204

Pérez-García A, Romero D, de Vicente A (2011) Plant protection and growth stimulation by microorganisms: biotechnological applications of bacilli in agriculture. Curr Opin Biotechnol 22(2):187–193. https://doi.org/10.1016/j.copbio.2010.12.003

Powell JF, Vargas JM, Nair MG, Detweiler AR, Chandra A (2000) Management of dollar spot on creeping bentgrass with metabolites of *Pseudomonas aureofaciens* (TX-1). Plant Dis 84(1):19–24. https://doi.org/10.1094/PDIS.2000.84.1.19

Prabhukarthikeyan SR, Raguchander T (2016) Antifungal metabolites of *Pseudomonas fluorescens* against *Pythium aphanidermatum*. J Pure Appl Microbiol 10(1):579–584

Qessaoui R, Bouharroud R, Furze JN, El Aalaoui M, Akroud H, Amarraque A, Van Vaerenbergh J, Tahzima R, Mayad EH, Chebli B (2019) Applications of new Rhizobacteria *Pseudomonas* isolates in agroecology via fundamental processes complementing plant growth. Sci Rep 9(1):1–10. https://doi.org/10.1038/s41598-019-49216-8

Raaijmakers JM, De Bruijn I, De Kock MJD (2006) Cyclic lipopeptide production by plant-associated *Pseudomonas* spp.: diversity, activity, biosynthesis, and regulation. Mol Plant-Microbe Interact 19(7):699–710. https://doi.org/10.1094/MPMI-19-0699

Raaijmakers JM, de Bruijn I, Nybroe O, Ongena M (2010) Natural functions of lipopeptides from *Bacillus* and *Pseudomonas*: more than surfactants and antibiotics. FEMS Microbiol Rev 34(6):1037–1062. https://doi.org/10.1111/j.1574-6976.2010.00221.x

Rai R, Srinivasamurthy R, Dash PK, Gupta P (2017) Isolation, characterization and evaluation of the biocontrol potential of *Pseudomonas protegens* RS-9 against *Ralstonia solanacearum* in tomato. Indian J Exp Biol 55(9):595–603

Rais A, Jabeen Z, Shair F, Hafeez FY, Hassan MN (2017) *Bacillus* spp., a bio-control agent enhances the activity of antioxidant defense enzymes in rice against *Pyricularia oryzae*. PLoS One 12(11):1–17. https://doi.org/10.1371/journal.pone.0187412

Reshma P, Naik MK, Aiyaz M, Niranjana SR, Chennappa G, Shaikh SS, Sayyed RZ (2018) Induced systemic resistance by 2,4-diacetylphloroglucinol positive fluorescent *Pseudomonas* strains against rice sheath blight. Indian J Exp Biol 56(3):207–212

Romero D, De Vicente A, Rakotoaly RH, Dufour SE, Veening JW, Arrebola E, Cazorla FM, Kuipers OP, Paquot M, Pérez-García A (2007) The iturin and fengycin families of lipopeptides are key factors in antagonism of *Bacillus subtilis* toward *Podosphaera fusca*. Mol Plant-Microbe Interact 20(4):430–440. https://doi.org/10.1094/MPMI-20-4-0430

Sagar A, Sayyed RZ, Ramteke PW, Sharma S, Marraiki N, Elgorban AE, Syed A (2020) ACC deaminase and antioxidant enzymes producing halophilic *Enterobacter* sp. PR14 promotes the growth of rice and millets under salinity stress. Physiol Mol Biol Plants 26:1847–1854

Sansinenea E, Ortiz A (2012) Zwittermicin A: A promising Aminopolyol antibiotic from biocontrol bacteria. Curr Org Chem 16(8):978–987. https://doi.org/10.2174/138527212800194737

Saraf M, Pandya U, Thakkar A (2014) Role of allelochemicals in plant growth promoting rhizobacteria for biocontrol of phytopathogens. Microbiol Res 169(1):18–29. https://doi.org/10.1016/j.micres.2013.08.009

Sayyed RZ, Seifi S, Patel PR, Shaikh SS, Jadhav HP, Enshasy HE (2019) Siderophore production in groundnut rhizosphere isolate, *Achromobacter* sp. RZS2 influenced by physicochemical factors and metal ions. Environment Sustainability 1(3):295–301

Shaikh SS, Wani SJ, Sayyed RZ, Thakur R, Gulati A (2018) Production, purification and kinetics of chitinase of *Stenotrophomonas maltophilia* isolated from rhizospheric soil. Indian J Exp Biology 56(4):274–278

Shang H, Chen J, Handelsman J, Goodman RM (1999) Behavior of *Pythium torulosum* zoospores during their interaction with tobacco roots and *Bacillus cereus*. Curr Microbiol 38(4):199–204. https://doi.org/10.1007/PL00006787

Sharma SB, Sayyed RZ (2013) Mrugesh H Trivedi & Thivakaran Gobi, Phosphate solubilizing microbes: sustainable approach for managing phosphorus deficiency in agricultural soils. Springer Plus 2:587

Sharma S, Sayyed R, Sonawane M, Trivedi M, Thivakaran G (2016) *Neurospora* sp SR8, a novel phosphate solubiliser from rhizosphere of soil of sorghum in Kachh, Gujarat. Indian J Exp Biol 54:644–649

Sharma A, Gupta A, Dalela M, Sharma S, Sayyed RZ, Enshasy HE, Elsayed EA (2020) Linking organic metabolites as produced by *Purpureocillium lilacinum* 6029 cultured on Karanja deoiled cake medium for the sustainable management of root-knot nematodes. Sustainability 12(9):8276. https://doi.org/10.3390/su12198276

Sheeba J (2017) Isolation, screening and characterization of plant growth promoting bacteria and their antifungal effect on *Rhizoctonia solani* (J.G. Kühn 1858). Adv Plants Agric Res 7(5):369–375. https://doi.org/10.15406/apar.2017.07.00269

Shi J, Zhu X, Lu Y, Zhao H, Lu F, Lu Z (2018) Improving iturin A production of *Bacillus amyloliquefaciens* by genome shuffling and its inhibition against *Saccharomyces cerevisiae* in orange juice. Front Microbiol 9:1–8. https://doi.org/10.3389/fmicb.2018.02683

Silo-Suh LA, Stabb EV, Raffel SJ, Handelsman J (1998) Target range of Zwittermicin A, an aminopolyol antibiotic from *Bacillus cereus*. Curr Microbiol 37(1):6–11. https://doi.org/10.1007/s002849900328

Singh A, Singh M, Bhareti P, Amitabh Singh C, Singh K, Singh U (2017) Antifungal activity of some strains of plant growth-promoting rhizobacteria. ~ 577 ~. J Pharmacogn Phytochem 6(6):577–582

Sivasakthi S, Usharani G, Saranraj P (2014) Biocontrol potentiality of plant growth promoting bacteria (PGPR) - *Pseudomonas fluorescens* and *Bacillus subtilis*: A review. African J Agric Res 9(16):1265–1277. https://doi.org/10.5897/AJAR2013.7914

Stein T (2005) Bacillus subtilis antibiotics: structures, syntheses and specific functions. Mol Microbiol 56(4):845–857. https://doi.org/10.1111/j.1365-2958.2005.04587.x

Sukmawati D, Family N, Hidayat I, Sayyed RZ, Elsayed EA, Dailin DJ, Hanapi SZ, Wadaan MA, Enshasy HE (2021) Biocontrol Activity of *Aureubasidium pullulans* and *Candida orthopsilosis* isolated from *Tectona grandis* L Phylloplane against *Aspergillus* sp. In Post-Harvested Citrus Fruit Sustainability 13:7479. https://doi.org/10.3390/su13137479

Sultan MZ, Park K, Lee SY, Park JK, Varughese T, Moon SS (2008) Novel oxidized derivatives of antifungal pyrrolnitrin from the bacterium *Burkholderia cepacia* K87. J Antibiot (Tokyo) 61(7):420–425. https://doi.org/10.1038/ja.2008.58

Sur S, Romo TD, Grossfield A (2018) Selectivity and mechanism of Fengycin, an antimicrobial Lipopeptide from molecular dynamics. J Phys Chem B 122(8):2219–2226. https://doi.org/10.1021/acs.jpcb.7b11889.Selectivity

Suriani NL, Suprapta DN, Nazir N, Parwanayoni NMS, Darmadi AAK, Dewi DA, Sudatri NW, Fudholi A, Sayyed RZ, Syed A, Elgorban AM, Bahkali AH, El Enshasy HA, Dailin DJ (2020) A mixture of piper leaves extracts and rhizobacteria for sustainable plant growth promotion and bio-control of blast pathogen of organic Bali rice. Sustain 12(20):1–18. https://doi.org/10.3390/su12208490

Tagele SB, Kim SW, Lee HG, Kim HS, Lee YS (2018) Effectiveness of multi-trait *Burkholderia contaminans* KNU17BI1 in growth promotion and management of banded leaf and sheath blight in maize seedling. Microbiol Res 214:8–18. https://doi.org/10.1016/j.micres.2018.05.004

Toral L, Rodríguez M, Béjar V, Sampedro I (2018) Antifungal activity of lipopeptides from *Bacillus* XT1 CECT 8661 against *Botrytis cinerea*. Front Microbiol 9:1–12. https://doi.org/10.3389/fmicb.2018.01315

Tripathi RK, Gottlieb D (1969) Mechanism of action of the antifungal antibiotic pyrrolnitrin. J Bacteriol 100(1):310–318. https://doi.org/10.1128/jb.100.1.310-318.1969

Turner JM, Messenger AJ (1986) Occurrence, biochemistry and physiology of Phenazine pigment production. Adv Microb Physiol 27(C):211–275. https://doi.org/10.1016/S0065-2911(08)60306-9

Vinay JU, Naik MK, Rangeshwaran R, Chennappa G, Shaikh SS, Sayyed RZ (2016) Detection of antimicrobial traits in fluorescent pseudomonads and molecular characterization of an antibiotic pyoluteorin. 3. Biotech 6(2):1–11. https://doi.org/10.1007/s13205-016-0538-z

Wang Y, Wilks JC, Danhorn T, Ramos I, Croal L, Newman DK (2011) Phenazine-1-carboxylic acid promotes bacterial biofilm development via ferrous iron acquisition. J Bacteriol 193(14):3606–3617. https://doi.org/10.1128/JB.00396-11

Wang X, Li Q, Sui J, Zhang J, Liu Z, Du J, Xu R, Zhou Y (2019) Liu X (2019) Isolation and characterization of antagonistic bacteria *Paenibacillus jamilae* HS-26 and their effects on plant growth. Biomed Res Int. https://doi.org/10.1155/2019/3638926

Wang Y, Zhang C, Liang J, Wu L, Gao W, Jiang J (2020) Iturin A extracted from *Bacillus subtilis* WL-2 affects *Phytophthora infestans* via cell structure disruption, oxidative stress, and energy supply dysfunction. Front Microbiol 11:1–12. https://doi.org/10.3389/fmicb.2020.536083

Wu L, Xiao W, Chen G, Song D, Khaskheli MA, Li P, Zhang S, Feng G (2018) Identification of *Pseudomonas mosselii* BS011 gene clusters required for suppression of rice blast fungus *Magnaporthe oryzae*. J Biotechnol 282(April):1–9. https://doi.org/10.1016/j.jbiotec.2018.04.016

Wu Y, Zhou J, Li C, Ma Y (2019) Antifungal and plant growth promotion activity of volatile organic compounds produced by *Bacillus amyloliquefaciens*. Microbiology 8(8):1–14. https://doi.org/10.1002/mbo3.813

Xie S, Liu J, Gu S, Chen X, Jiang H, Ding T (2020) Antifungal activity of volatile compounds produced by endophytic *Bacillus subtilis* DZSY21 against *Curvularia lunata*. Ann Microbiol 70(1). https://doi.org/10.1186/s13213-020-01553-0

Yang M, Insects P, Wen S, Mavrodi DV, Mavrodi OV, Thomashow LS, Disease R, Control B, Guo J, Insects P, Weller DM, Disease R, Control B (2017) HHS. Public Access 104(3):248–256. https://doi.org/10.1094/PHYTO-05-13-0142-R.Biological

Zakaria AK, Sayyed RZ, Enshasy HE, Biosynthesis of antibiotics by PGPR and their roles in biocontrol of plant diseases (2019) Plant growth promoting Rhizobacteria for sustainable stress management Vol II Rhizobacteria in biotic stress management. Springer, Singapore, pp 1–36

Zeriouh H, Romero D, García-Gutiérrez L, Cazorla FM, De Vicente A, Pérez-García A (2011) The iturin-like lipopeptides are essential components in the biological control arsenal of *Bacillus subtilis* against bacterial diseases of cucurbits. Mol Plant-Microbe Interact 24(12):1540–1552. https://doi.org/10.1094/MPMI-06-11-0162

Zhang B, Zhao H, Wu X, Zhang LQ (2020a) The oxidoreductase DsbA1 negatively influences 2,4-diacetylphloroglucinol biosynthesis by interfering the function of Gcd in *Pseudomonas fluorescens* 2P24. BMC Microbiol 20(1):1–9. https://doi.org/10.1186/s12866-020-1714-1

Zhang QX, Kong XW, Li SY, Chen XJ, Chen XJ (2020b) Antibiotics of *Pseudomonas protegens* FD6 are essential for biocontrol activity. Australas Plant Pathol 49(3):307–317. https://doi.org/10.1007/s13313-020-00696-7

Zhao MM, Lyu N, Wang D, Wu XG, Zhao YZ, Zhang LQ, Zhou HY (2020) PhlG mediates the conversion of DAPG to MAPG in *Pseudomonas fluorescens* 2P24. Sci Rep 10(1):1–7. https://doi.org/10.1038/s41598-020-60555-9

Zhou L, Song C, Li Z, Kuipers O (2020) Antimicrobial activity screening of rhizosphere soil bacteria from tomato and genome-based analysis of their antimicrobial biosynthetic potential:1–23. https://doi.org/10.21203/rs.3.rs-26812/v1

Zohora US, Ano T, Rahman MS (2016) Biocontrol of *Rhizoctonia solani* K1 by iturin a producer *Bacillus subtilis* RB14 seed treatment in tomato plants. Adv Microbiol 6(May):424–431

# Chapter 6
# Biofungicidal Properties of Rhizobacteria for Plant Growth Promotion and Plant Disease Resistance

**Rajashree B. Patwardhan, Pragati S. Abhyankar, Suneeti S. Gore, Saylee V. Kalekar, and Shriya P. Umrani**

**Contents**

6.1 Introduction .... 103
6.2 Biodiversity of Rhizobacteria .... 105
6.3 Mechanisms of Plant Growth Promotion by Rhizobacteria .... 107
6.4 Fungal Diseases Associated with Plants .... 122
6.5 Production of Antifungal Metabolites by PGPR .... 124
6.6 Conclusion .... 125
References .... 126

## 6.1 Introduction

Microbes are an important part of living soil not only in transforming nutrients in the soil but also with several functions in influencing soil health. There are certain microbes which assist the plant to grow well in their presence by a variety of mechanisms (Basu et al. 2021). The rhizosphere is the narrow region of soil that is directly affected by root secretions and associated soil microorganisms known as the root microbiome. The plant growth-promoting rhizobacteria (PGPR) are a valuable group of soil bacteria that can reside plant roots and enhance plant growth and development. The concentration of microbes in the rhizosphere is 10 to 1000 times

R. B. Patwardhan (✉) · P. S. Abhyankar
Department of Microbiology, Haribhai V. Desai College of Arts, Science and Commerce, Pune, India

S. S. Gore · S. V. Kalekar
Department of Microbiology, Fergusson College, Pune, Maharashtra, India

S. P. Umrani
Department of Microbiology, Siddhivinayak Mahila Mahavidyalaya, Pune, India



R. Z. Sayyed et al. (eds.), *Antifungal Metabolites of Rhizobacteria for Sustainable Agriculture*, Fungal Biology, https://doi.org/10.1007/978-3-031-04805-0_6

greater than that in soil. Hence, such plant growth-promoting rhizobacteria (PGPR) should be exploited and utilized for sustainable agriculture (Sagar et al. 2020). The PGPR include *Azotobacter*, *Azospirillum*, *Acinetobacter*, *Agrobacterium*, *Arthrobacter*, *Bacillus*, *Burkholderia*, *Pseudomonas*, *Serratia*, *Streptomyces*, *Rhizobium*, *Bradyrhizobium*, *Mesorhizobium*, *Frankia*, *Thiobacillus*, *Actinobacteria*, *Bacteroidetes*, *Firmicutes*, and *Proteobacteria* belonging to different genera *Acetobacter*, *Achromobacter*, *Exiguobacterium*, *Flavobacterium*, *Gluconacetobacter*, *Herbaspirillum*, *Methylobacterium*, *Paenibacillus*, and *Staphylococcus* (Kour et al. 2019).

Rhizosphere and soil bacteria are particularly important in nearly all biochemical cycles in terrestrial ecosystems and participate in maintaining health and productivity of soil in agriculturally managed systems. A considerable variation in relative abundances of bacterial communities at both phylum and genus level has been observed among different crop systems. Compared to conventional farming systems, organic farming system shows higher percentage of phylum Proteobacteria (many PGPR) and lower percentage of phylum *Actinobacteria* (Reddy et al. 2019). Over the last decades, world agriculture has experienced high increase in crop yield. This is a result of high input of inorganic fertilizers and pesticides, and mechanization driven by fossil fuel. Repeated use over the years has led to serious environmental problems such as depletion of soil quality and health, ocean and ground water pollution, and emergence of resistant pathogens. It is a big challenge to feed the increasing world population on decreasing farmland areas without damaging environment. It is well known that rhizosphere and PGPR play an important role in maintaining crop and soil health through versatile mechanisms like nutrient cycling and uptake, suppression of plant pathogens, induction of resistance in plant host, and direct stimulation of plant growth (Anith et al. 2004) (Haas and Défago 2005). To act efficiently, microbes should remain active under a large range of conditions, such as varying pH, temperature, and concentrations of different ions. Preserving biodiversity of PGPR in soil could be an essential component of environment-friendly sustainable agriculture policies (Kour et al. 2020). Some studies have demonstrated that agricultural practices affect the diversity and function of rhizosphere and soil microorganisms (Raut et al. 2017). Organic farming differs from conventional agriculture in the production process, and it relies on techniques such as crop rotation, green manure, and biological pest control to maintain the soil productivity instead of chemical fertilizer and pesticides (Liu et al. 2017).

Plants lack adaptive immunity. Instead, plants are dependent on a heritable, innate immunity based on the recognition by receptors of the presence of microbial triggers (cues) including effector proteins and microbe-associated molecular patterns (Jones and Dangl 2006). The perception of microbial cues leads to the induction of broad spectrum of plant defenses called systemic acquired resistance (SAR) (Patel et al. 2016; Bostock 2005). Until recently, SAR was thought to be limited to induction of plant defenses against foliar microbial pathogens. However, recent results suggest that plants can activate signal exchanges between aboveground (AG) and belowground (BG) responses (Raut et al. 2017).

Root colonization by beneficial microbe is a process which is required for all mechanisms of biocontrol. Microbes are attracted chemotactically by certain components secreted by root. Weapons used by beneficial microbes to attack the pathogen include lytic enzymes such as chitinase, protease, cellulose, and glucanase (Shaikh et al. 2018; Jadhav et al. 2017, 2020a, b) and the antibiotics (Zakaria et al. 2019) such as phenazines, 2,4-diacetylphloroglucinol, pyoluteorin, pyrrolnitrin, hydrogen cyanide, cyclic lipopeptides, 2-hexyl-5-propyl resorcinol, and d-gluconic acid (Lugtenberg and Kamilova 2009). Based on the growth promotion activity of selected bacteria and their abilities to produce siderophores (Nithyapriya et al. 2021; Sayyed et al. 2019), phytohormones, and phosphate solubilizing activity (Sharma and Sayyed 2013; Sharma et al. 2016), the rhizobacteria may have great potential to increase the yield, growth, and nutrition of various vegetable crops under greenhouse and field conditions (Zaman et al. 2021). Therefore, these can be utilized as biofertilizer and biological control agents for fruit and vegetable production in sustainable and ecological agricultural systems (Lamsal et al. 2012).

The use of PGPR supports plant nutrition, and this has led to the development of a specific research area and to applications in agriculture. More recent advancements in knowledge of biology of a variety of soil-dwelling species are leading to an increased interest for their potential application for both plant growth promotion and protection against pathogens and parasites. These can be inhibited through direct and indirect mechanisms, thus helping to maintain crop health and productivity. Studies targeting at screening bacterial strains with such characteristics not only rise our understanding of the ecological significance of bacterial community related with plant roots but also create concrete prospects of application in agricultural contexts, in line with the principles of integrated pest management and eco-sustainability (Ruiu 2020).

The various types of fungal infections include the black spot, downy mildew, powdery mildew, blight, rust, wilt, club root, and anthracnose. Due to harm caused by chemical pesticides, use of PGPR to inhibit fungal diseases has been boosting interest (Kumar et al. 2019; Sharma et al. 2020; Arora et al. 2021). This review is an update concerning the bio-fungicidal properties of Rhizobacteria for plant growth promotion and plant disease resistance.

## 6.2 Biodiversity of Rhizobacteria

Biodiversity is an important element of environmental protection and is vital to agriculture production. Most microbial diversity of soil ecosystem is confined to rhizosphere. Depositions made into soil through plant root exudates play a major role in defining resident microflora, which differs from that in remaining bulk of soil. Rhizobacterial diversity is influenced by both, plant type and soil type. Soil factors, plant root exudates, and agricultural management are the factors that determine the microbial community composition within rhizosphere.

Based on their relationship with plants, PGPR have been divided into two major groups: symbiotic and free-living (Mubeen et al. 2006; Dominguez et al. 2017). PGPR have three main features: (i) root colonization ability, (ii) high survivability and multiplicity in root surroundings helping in plant growth promotion, and (iii) inhibition of phytopathogens (Bloemberg et al. 2000; Sagar et al. 2022; Sukmawati et al. 2021). PGPR strains occur in various taxonomic groups, which may be present concurrently in each soil. This suggests that taxonomically different PGPR strains may coexist in soil and colonize same rhizosphere, along with all non-PGPR members of the bacterial community. The taxonomic status of bacterial isolates based on their positive effect on plant growth or health, their ability to inhibit phytopathogens, or occurrence of particular gene or property of relevance for PGPR effect is characterized (Belimov et al. 2001)

Functional group approaches can be implemented when specific genes are recognized. For instance, nitrogen fixers can be screened using the *nifH* gene, which encodes dinitrogenase subunit of the nitrogenase. Its sequence is well conserved within the functional group, and it is commonly used as marker to monitor the size and diversity of diazotrophic community (Oldroyd and Dixon 2014). Some of these PGPR functional groups are taxonomically confined, such as the *Pseudomonas* 2,4-diacetylphloroglucinol (DAPG) producers (Vinay et al. 2016a, b; Reshma et al. 2018). In contrast, others show high diversity; certain bacterial functional groups may also comprise both PGPR and non-PGPR strains. For example, nitrogen fixers comprise PGPR as well as mutualistic symbionts and even a few pathogens.

Mutation is another vital component creating microbial variations that can interact diversely with plant host. Mutations influencing virulence-associated genes have striking results for advancement of destructiveness in wide evolution of pathogens (Sokurenko et al. 1999; McCann and Guttman 2008). Comparative phylogenetic analysis of DNA sequences of cloned 16S rRNA genes has shown that members of four major phylogenetic groups are ubiquitous to almost all soil types: class α-proteobacteria and phyla *Actinobacteria*, *Acidobacteria*, and *Verrucomicrobia*. These four groups are represented in >7% of 16S rRNA gene clone library studies of soil bacterial communities (Hugenholtz et al. 1998). Other classes of phylum Proteobacteria and phyla Firmicutes and Planctomycetes are detected in 25–75% of studies (Hugenholtz et al. 1998). Phyla *Proteobacteria*, *Cytophagales*, *Actinobacteria*, and Firmicutes are properly exemplified by cultivated microorganisms, and these four phyla account for 90% of all cultivated bacteria characterized by16S rRNA sequences from cultivated organisms in ARB database. The development of nif D and nif H specific primers has also proved extremely useful in screening diazotrophic strains.

The lack of information about the diversity of bacteria specifically isolated from rhizosphere of various plants needs to be filled up for our understanding of an important niche in microbial ecology of grasses such as rice. Some of standard rhizospheric and putative rhizospheric diazotrophs have been noticed to be naturally present in rhizosphere of the graminaceous plants like rice, sugarcane, wheat, kallar grass, etc. *Herbaspirillum* was first reported by (Olivares et al. 1996) as N2-fixing bacterium associated with roots of rice, maize, and sorghum. *Herbaspirillum seropedicae* as well as *Herbaspirillum rubrisubalbicans* (formerly identified as *Pseudomonas rubrisabalbicans*) are the confirmed diazotrophic bacteria. Another diazotrophic species *Herbaspirillum frisingense* has been isolated from C-4 fiber plants (Kirchhof et al. 2001)

*Azospirillum* spp. are known to be the most efficient diazotrophic bacteria isolated from rhizosphere of various plants. It is generally regarded as a rhizospheric bacterium and has often been reported to give best results upon inoculation to crop plants. Certain strains penetrate the roots suggesting that some strains of *Azospirillum* may also colonize within wheat tissues (Naiman et al. 2009). Till date several rhizospheric diazotrophic *Azospirillum* species including *Azospirillum lipoferum*, *A. amazonense*, *A. halopreferens*, *A. irakense*, and *A. doebereinerae* spp. have been reported (Khammas et al. 1989). *Klebsiella* spp., *Pseudomonas* spp., *Serratia* spp., *Azotobacter* spp., *Burkholderia* spp., *Alcaligenes* spp., *Enterobacter* spp., *Bacillus polymyxa*, *Gluconacetobacter* spp., *Azoarcus* spp., *Paenibacillus* spp., *Bacillus subtilis*, *B. licheniformis*, *B. pumilus*, *Brevibacterium halotolerans*, and *Pseudomonas putida* are plant associated bacteria reported from different plants.

## 6.3 Mechanisms of Plant Growth Promotion by Rhizobacteria

The primary goal of farming is production of high-quality, reliable, and reasonably priced food for an ever-increasing worldwide population. The widespread use of chemicals has harmful effects on environment, adverse effects on non-target organisms, carcinogenicity on living beings, and number of ecological problems. PGPR can promote plant growth by number of different mechanisms (Zope et al. 2019). They carry out the alteration of the whole microbial community in rhizosphere niche through production of secondary metabolites like hydrogen cyanide (Table 6.1),

**Table 6.1** Role of secondary metabolites secreted by PGPR

| Secondary metabolite | Inhibitory effect | Additional role |
|---|---|---|
| Hydrogen cyanide | Inhibits cytochrome C oxidase | Chelation of metals and cellular signalling |
| Volatile organic compounds | Antifungal and antibacterial | Plant growth promotion |
| Phenazines | Inhibition of conidial germination and mycelial growth, formation of reactive oxygen species (ROS) and reactive oxygen species (RNS) | Reduction in exopolysaccharide formation |
| Pyrrolnitrin | Affects electron transport and oxidative phosphorylation | Interferes with osmotic signal transduction |
| Pyoluteorin | Cell membrane destruction | Autoinduction and acts as a signal molecule |
| 2,4-diacetylphloroglucinol | Cell membrane destruction and reduction on disease symptoms | Intra and intracellular signals and co-PLT, antiglycation activity, and beneficial effects on plant growth |
| Cyclic Lipopeptides | Formation of ion channel in cell membrane, disruption of cell wall and cell membrane | Swarming movement and biosurfactants properties |

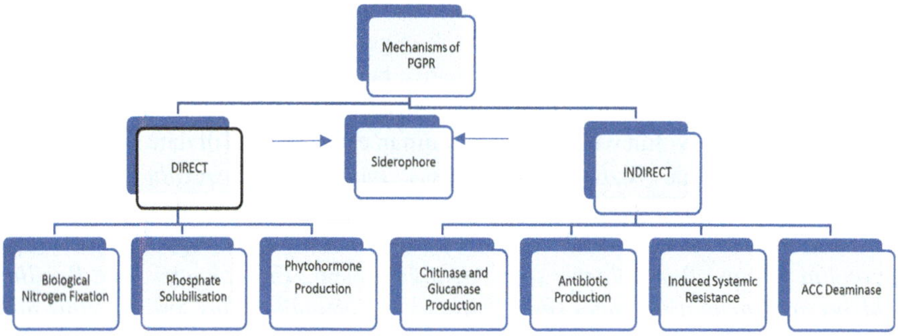


**Fig. 6.1** Mechanisms of plant growth promotion

lytic enzymes, antibiotics, siderophores, auxins like indole acetic acid, gibberellins, ACC deaminase, etc. (Kusale et al. 2021a). Directly PGPR promote the plant growth by enabling resource gain (nitrogen, phosphorus, and essential minerals) (Kusale et al. 2021b) or regulating plant hormone levels (Fig. 6.1).

## 6.3.1 *Alteration of the Whole Microbial Community in Rhizosphere Niche Through the Production of Various Substances*

### 6.3.1.1 Hydrogen Cyanide Production

Several PGPR release hydrogen cyanide, a volatile compound that employs moderate biocontrol activity and improve effect of bacterial antibiotics acting synergistically (Beneduzi et al. 2012). PGPB that can produce HCN also synthesize some antibiotics or cell wall degrading enzymes (Ramette et al. 2006). The low level of HCN synthesized by bacterium improves effectiveness of antifungal compound directed against fungal pathogens thereby ensuring that fungi do not develop resistance to antifungal in question. Thus, HCN synthesized by PGPB act synergistically with antibiotics.

HCN toxicity is attributed to its ability to inhibit cytochrome-C oxidase as well as other important metallo-enzymes (Nandi et al. 2017). Many bacterial genera such as *Rhizobium*, *Pseudomonas*, *Alcaligenes*, *Bacillus*, and *Aeromonas* have shown to be HCN producers (Ahmad et al. 2008; Zachow et al. 2017). Inhibition of tomato root knot disease caused by *Meloidogyne javanica* have been attributed to the effect of HCN (Siddiqui et al. 2006) as well as control of *Odontoter mesobesus*, a crop pest in India.

### 6.3.1.2 Production of Lytic Enzymes

Several PGPR produce enzymes chitinases, cellulases, −1,3 glucanases, proteases, and lipases that can lyse a portion of the cell walls of several pathogenic fungi. PGPB that synthesize one or more of these enzymes have been observed to have biocontrol activity against a range of pathogenic fungi including *Botrytis cinerea*, *Sclerotium rolfsii*, *Fusarium oxysporum*, *Phytophthora* spp., *Rhizoctonia solani*, and *Pythium ultimum* (Kim et al. 2008). A wide variety of microbial lytic enzymes have been studied, some of which are cellulases, glucanases, proteases, and chitinases. These enzymes efficiently prevent the proliferation of pathogenic fungi by hydrolyzing different components of their cell walls. PGPR produce extracellular hydrolytic enzymes that are engaged in hydrolysis of fungal cell wall components such as chitin, proteins, cellulose, hemicellulose, and DNA; these hydrolytic enzymes have ability of hindering fungal pathogens (Pal 2006). The antagonistic properties of hydrolytic enzymes against various phytopathogens play a major role in biocontrol (Kim et al. 2003) (Shaikh and Sayyed 2015). Hydrolytic enzymes can break down glycosidic bonds in chitin. *Stenotrophomonas*, *Pseudomonas*, and *Alcaligenes* inhabiting in soil produce chitinase to lyse the cell wall of phytopathogenic fungi (Shaikh et al. 2018). Proteases play a substantial role in cell wall lysis of phytopathogenic fungi because chitin and/or fibrils of β-glucan are inserted into protein matrix. Several *Bacillus* species are known to produce protease (Beg and Gupta 2003; Gerze et al. 2005). Some of the proteases produced by *Trichoderma* spp. are involved in inactivating extracellular enzymes of phytopathogenic fungi (Kredics et al. 2005). Application of efficient rhizobacterial strains secreting various hydrolytic enzymes helps to reduce the abundant use and doses of agrochemicals which is the most important prospect in PGPR research.

#### 6.3.1.3 Production of Antibiotics

It is one of the major biocontrol mechanisms of PGPR in nature. The diffusible compounds produced by PGPR are known to inhibit plant pathogens. A broad spectrum of antibiotics such as polyenes, macrolides, aminoglycosides, nucleosides, and benzoquinones are stated to be produced by PGPR. *Actinobacteria* are the major producer of antibiotics. An antibiotic that is recognized to control a pathogen to prevent plant damage from that pathogen might not be as effective against another pathogen on the same plant, and the antibiotic synthesizing PGPB may show varying differences in its actions under different field conditions. Many antibiotics have been derived from bacteria of the genera *Bacillus* and *Pseudomonas*. They produce a variety of metabolites which serve as antifungal, antibacterial, anti-helminthic, antiviral, antimicrobial, phytotoxic, antioxidant, cytotoxic, and antitumor agents. Fluorescent pseudomonads have been reported for pyoluteorin antibiotic production (Vinay et al. 2016a, b). At minimal concentrations, PGPR-secreted antibiotics are efficient at inhibiting the growth of other bacteria and fungi.

The problem on use of antibiotic-producing bacteria as biocontrol agents is, with increased use of these strains, some phytopathogens may develop resistance to specific antibiotics. Biocontrol strains synthesizing hydrogen cyanide and antibiotics can be used to overcome this problem. While hydrogen cyanide may not have much biocontrol activity by itself, it appears to act synergistically with bacterially encoded antibiotics (Raaijmakers et al. 2002).

#### 6.3.1.4 Siderophore Production

Siderophores are low-molecular-weight, iron-scavenging ligands produced by a wide variety of microorganisms to combat iron deficiency (Sayyed et al. 2013). They are tiny peptide molecules that have side chains and functional groups to which ferric ions can attach. Deficiency of iron in crop results in iron chlorosis, making them micronutrient deficient and hence susceptible to microbial diseases. Siderophore-producing rhizobacteria have been recognized as potential biocontrol agents for controlling plant diseases (Sairam et al. 2013). The ability of *Achromobacter* obtained from groundnut rhizosphere to produce siderophore in the presence of moderate/high levels of various metal ions can be exploited in bioremediation of metal contaminated soils (Sayyed et al. 2019). PGPR present in close vicinity to plant roots or its surface play a vital role in transporting iron nutrition to the crops, thereby endorsing plant health/growth as well as suppressing major phytopathogen, and have been seen as sustainable and eco-friendly substitute to chemical manures and chemical insecticides. PGPR serve as first defense against root-invading organisms and aid in removing toxic metals from polluted soil.

Siderophore-producing microbes can prevent pathogen proliferation by reducing amount of iron that is available to pathogen (Shen et al. 2013). PGPB synthesizing siderophores prevent proliferation of phytopathogens by secreting siderophores with an extremely high affinity for iron. These siderophores bind tightly to most of the Fe3 PLUS_SPI that is present in the rhizosphere of the host plant taking up bound iron either into PGPB or host plant. This avoids any fungal and bacterial pathogens in the host plant rhizosphere, where the biocontrol PGPB is bound, from acquiring sufficient iron for their growth. Thus, the pathogens are unable to proliferate because of lack of iron, causing them to lose ability to act as pathogens. Siderophore-producing rhizobacteria signify a promising option to chemical fertilizers by simultaneously tackling salt-stress impacts and Fe limitation in saline soils (Ferreira et al. 2019). Statistical optimization of siderophore production by *P. aeruginosa* RZS9 by applying Plackett–Burman design and response surface methodology (RSM) using central composite design (CCD) has been reported (Wani et al. 2016).

#### 6.3.1.5 Production of Auxins

The most described mechanism primarily used to explain the positive PGPB effects on plant growth is their capability to produce auxin. Microbial auxin production is the major factor not only responsible for strengthening the plant-microbe relationship, but it also promotes plant growth and development (Ahmed and Hasnain 2014). The compounds that have been reported to have auxin activity include IAA and its derivatives (Fig. 6.2) like indole-3-acetamide, indole-3-pyruvate, indole-3-acetaldehyde, etc. The scientific literature considers auxin and IAA to be interchangeable terms (Spaepen et al. 2007).

IAA is a common product of L- tryptophan metabolism. IAA assists in the production of longer roots with enhanced number of root hairs and root laterals which

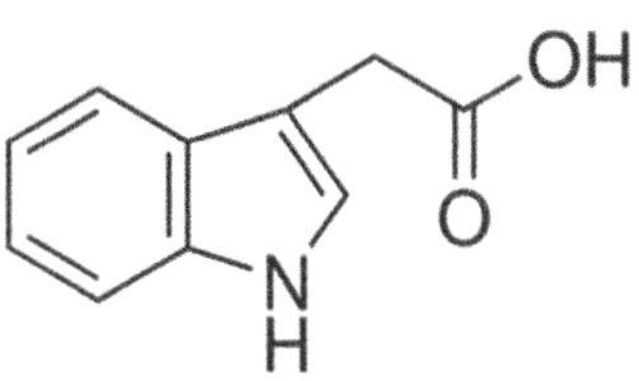


Indole- 3- acetic acid

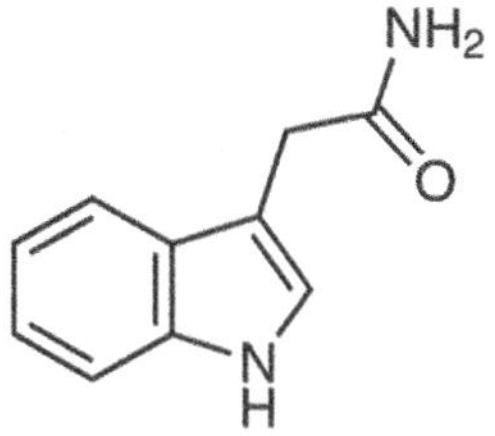


Indole -3-aetamide

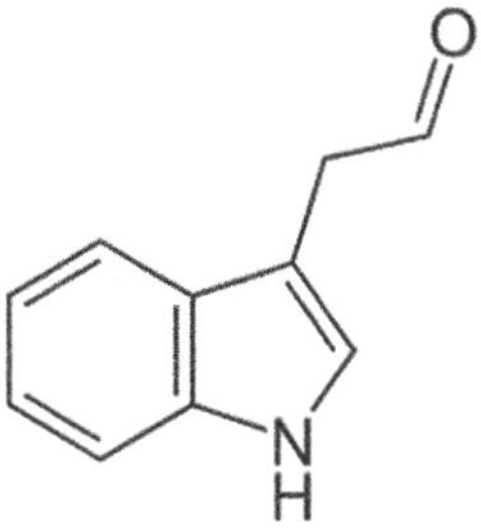


Indole-3-acetaldehyde

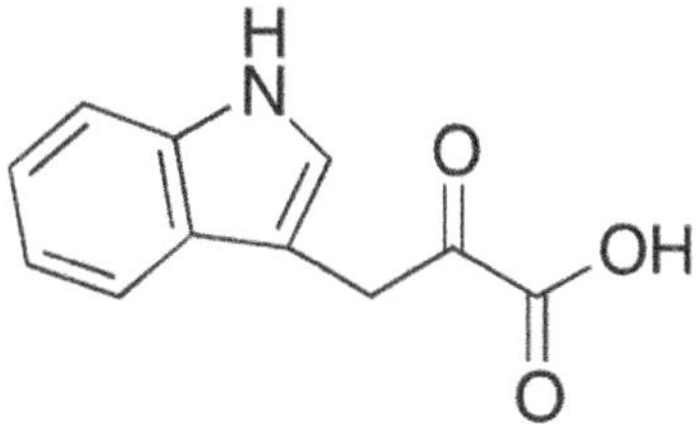


Indole-3-Pyruvic acid

**Fig. 6.2** Some known derivatives of IAA

are engaged in nutrient uptake (Datta and Basu 2000). IAA accelerates cell elongation by modifying certain conditions like rise in osmotic contents of the cell, rise in permeability of water into cell, reduction in wall pressure, increase in cell wall synthesis, and inducing protein synthesis. Rhizobacterial IAA changes plant auxin pools, eventually increasing root length as well as surface area, and in the process increasing the amount of root exudates available for uptake by plants as demonstrated by Ali et al. (2020). IAA affects cell division in plant, extension, and differentiation; stimulates seed and tuber germination; rises the rate of xylem and root growth; controls processes of vegetative growth; initiates lateral and adventitious root development; mediates responses to light, gravity, and florescence; and affects photosynthesis, pigment formation, biosynthesis of various metabolites, and resistance to stressful conditions (Tsavkelova et al. 2006)

IAA released by rhizobacteria interferes with many plant developmental processes because the endogenous pool of plant IAA may be altered by acquisition of IAA that has been secreted by soil bacteria. In plant roots, endogenous IAA may be suboptimal or optimal for growth (Lucas García et al. 2004), and additional IAA that is taken up from bacteria could alter the IAA level to either optimal or supraoptimal, resulting in plant growth promotion or inhibition, respectively. PGPB can use any functional IAA biosynthesis pathways. Synthesis of IAA is important in life and functioning of bacterium. IAA synthesized by PGPR is required at different levels in plant-bacterial interactions. Plant growth promotion and root nodulation both require IAA (Hynes et al. 2008). Bacterial IAA enhances root surface area and length and delivers the plant greater access to soil nutrients. Additionally, bacterial IAA loosens plant cell walls and enables an increasing amount of root exudation which offers additional nutrient to support growth of rhizosphere bacteria.

#### 6.3.1.6 Production of Gibberellins

Gibberellins are compounds which include a large group of tetracyclic diterpenoid carboxylic acids having either C20 or C19 carbon skeletons. Only 4 GAs have been identified in bacteria; GA1, GA3, GA4, and GA20 (Gupta et al. 2017), with GA1 and GA4 being most active (Sanghi et al. 2016). Gibberellins cause seed germination by breaking seed's dormancy and acting as chemical messenger. Gibberellins stimulate growth and activate important growth processes of plants including stem elongation, seed germination, flowering, fruit setting, improved photosynthesis rate, and chlorophyll content. PGPB production of GAs has been detected in *Achromobacter xylosoxidans*, *Gluconobacter diazotrophicus*, *Acinetobacter calcoaceticus*, *Rhizobia*, *Azotobacter* spp., *Bacillus* spp., *Herbaspirillum seropedicae*, and *Azospirillum* spp. (Joo et al. 2005).

### *6.3.2 By Enabling Source Gain (Nitrogen, Phosphorus, and Essential Minerals) or Regulating Plant Hormone Levels*

#### 6.3.2.1 Phosphate Solubilization

Bacteria that solubilize phosphorus are referred to as phosphate solubilizing bacteria (Alori et al. 2017). They supply phosphate in a more acceptable way to the plants and are not deleterious to environment. They transform insoluble organic and inorganic phosphate to a form which can be readily available to plants. Environmental conditions, plant and soil conditions, and bacterial strains all affect actions of phosphate solubilizers (Singh 2015). According to Banerjee et al. (2005), the most powerful phosphate solubilizers are from the genera *Bacillus*, *Rhizobium*, and *Pseudomonas*, as well as non-symbiont nitrogen fixers such as *Azotobacter* and *Azospirillum*. Organic acids produced together with their carboxyl and hydroxyl ions chelate cations or reduce the pH resulting in the release of phosphates (Khot et al. 2012). *Neurospora discreta* survives the stressed environment with high salinity and low precipitation rate and is reported as a powerful phosphate solubilizer (Sharma et al. 2016).

The main sources of organic phosphorus in the soil are organic materials in the form of inositol hexa-phosphate (phytate). Phytate is generally not biologically available to plants because plant roots produce incredibly low amount of phytase enzyme which breaks down phytate. However, many PGPB can solubilize phytate.

Phosphorus mineralization refers to the solubilization of organic phosphorus and degradation of remaining portions of the molecule which is triggered by unavailability of sufficient phosphate in the soil.

#### 6.3.2.2 Nitrogen Fixation

Nitrogen is one of the important nutrients essential for growth of all living organisms including plants and bacteria. The observation of nitrogen shortage in soil has led to the use of large amounts of nitrogenous fertilizers to make up for the necessary plant requirements to accomplish maximum plant yield in most soils. Despite nitrogen's abundance in earth's atmosphere, about 78%, this form of gaseous nitrogen [$N_2$(g)] is not readily accessible to most organisms, i.e., it is unsuitable for plant assimilation until first converted to ammonia (Baas et al. 2014).

Broad range of nitrogen-fixing bacteria have been identified including organisms that fix nitrogen symbiotically with specific plants (mostly legumes). Examples of symbiotic nitrogen fixers are *Rhizobium*, *Sinorhizobium*, *Azorhizobium*, *Allorhizobium*, *Mesorhizobium*, *Bradyrhizobium*, *Frankia*, *Azoarcus*, *Achromobacter*, *Burkholderia*, and *Herbaspirillum* (Babalola 2010; Pérez-Montaño et al. 2014). Rhizobia are Gram-negative bacteria that form symbiotic relationships majorly with leguminous plants (Oldroyd and Dixon 2014). Rhizobial bacteria colonize plant root cells and initiate a complex trend of developmental changes that lead root nodule formation (Gage 2004).

One unsuitable side reaction in the activities of the nitrogenase enzyme in nitrogen fixation is the reduction of $H^+$ to $H_2$ (hydrogen gas) because the hydrogen gas produced is lost to atmosphere leading to a waste of ATP expended in its production. This side reaction significantly lowers the overall efficiency of the nitrogen-fixing process by approximately 30%. On the contrary, some strains of rhizobia have an enzyme called hydrogenase (Sotelo et al. 2021) that can retrieve the lost $H_2$ from the atmosphere and convert it back into H PLUS_SPI to produce ATP used for more nitrogen fixation. These strains assist to save energy while fixing nitrogen at the same time.

### *6.3.3 By Decreasing the Inhibitory Effects of Various Pathogens on Plant Growth and Development in the Forms of Biocontrol Agents*

#### 6.3.3.1 Use of Chemical Fungicide and Their Disadvantages

Synthetic fertilizers and pesticides are used to enhance the crop yield, which are thought to be the most favorable options to protect crops from fungi, pests, and weeds (Kumar et al. 2019). The major constraints in agriculture include depletion of nutrient supply in soils, gap between achievable and actual yields of the crop, and protection of crops from different types of phytopathogens. According to the FAO (Food and Agriculture Organization), 25% of global agricultural production is affected. People are deprived of getting adequate food (Mishra and Arora 2018). Fungal infection in plants is responsible for epidemics like late blight disease, cereal rusts, ergot, brown spot, coffee rust, Sigatoka disease of banana, chestnut blight, downy and powdery mildews, stem rusts, and leaf blight. Chemical control of fungal phytopathogens has immediate effect, easy application, and can control large variety of pests.

Various hazardous chemical families associated with fungicides are dithiocarbamates, nitriles, benzimidazoles, and phenylpyrrole. Possible pathways of environmental contamination due to field application of chemical fungicides are; leaching through soil causing groundwater contamination, evaporation into the surrounding environment which can cause contaminated rainfall, spray drift into the soil, drainage of pesticides from crops with rainfall, presence of pesticides in crop or food products.

Only 0.1% of fungicides reach the target pathogens, while 99.9% leak into the surrounding environment responsible for harmful effects leading to water pollution, soil contamination, increased pathogen resistance, loss of biodiversity, and elimination of useful and beneficial species (e.g., bees). Pesticides affect pollinators like bees interfering with their homing ability, reproductive process, and foraging behavior. Some chemical fungicides reduce the visitation rates of pollinators and reduce the pollen collection, thus, reducing the pollination rate and affecting crop yield. Incorrect applications of pesticides cause health hazards, environmental pollution, and poisoning in humans. Exposure of humans to chemical pesticides can occur directly through oral, dermal route or by inhalation or indirectly due to occupation (Fig. 6.3). Exposure causes various harmful effects like allergic reactions, obesity, cancer, neurological disorders, endocrine problems, etc. High level exposure of chemical fungicide during gestation shows different types of birth defects to a range of 5 to 9% (Kumar et al. 2019).

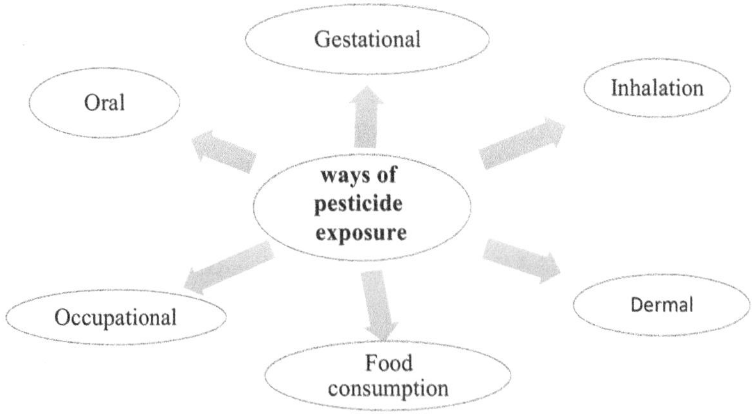


**Fig. 6.3** Different ways of pesticide exposure to humans

Fungal pathogens cause devastating losses of crops and post-harvest fruits. Different chemical fungicides like anilinopyrimidine, benzimidazoles, demethylation inhibitors, dicarboximide, phenylpyrrole, Qo respiration inhibitors, and strobilurin are used to minimize the damage caused due to fungal pathogens. Due to damage caused by chemical pesticides, use of PGPR to prevent fungal diseases has been gaining interest. PGPR are considered as front-line defenses against soilborne pathogens.

#### 6.3.3.2 Biological Control Agents (BCAs) as Alternatives to the Chemical Pesticides for Management of Pest and Diseases in Agriculture

PGPR replace the toxic and inefficient pesticides. PGPR are able to suppress bacterial, fungal, and nematode diseases in plants, and hence they have been used widely in integrated pest management programs (Sivasakthi et al. 2014). The strategies used by PGPR acting as competitors of fungal pathogens are nutrient competition, antibiosis, induced systemic resistance, parasitism, and production of hydrolytic enzymes.

#### 6.3.3.3 PGPR as Biocontrol Agents

PGPR are involved in biocontrol of plant diseases. They colonize rhizosphere and compete with the harmful microorganisms. They also have different mechanisms to prevent plant diseases by controlling the growth of phytopathogens (Sivasakthi et al. 2014). Many PGPR have been used as biocontrol agents. This is an ecofriendly approach; however, they have not replaced the harmful chemicals due to their inconsistency, poor shelf life, and lack of knowledge of actual factors involved in biocontrol (Mishra and Arora 2018).

Properties of PGPR as biocontrol agents include rapid growth in vitro and rapid mass production, utilize root exudates and seed exudates, colonize and multiply faster in rhizosphere and plant tissues, produce different types of bioactive secondary metabolites, compete with phytopathogens, and rapid adaptability to the environmental stress (Sivasakthi et al. 2014). PGPR synthesize several secondary metabolites, which act against phytopathogens, hence can be used in different preparations (Table 6.2). These are phenazines, phloroglucinol, 2,4-diacetylphloroglucinol, pyoluteorin, pyrrolnitrin, and cyclic peptides.

**Table 6.2** PGPRs showing biocontrol activity against phytopathogens

| PGPRs | Crops | Diseases/pathogen | References |
|---|---|---|---|
| *Bacillus amyloliquefaciens* | Tomato | *Tomato mottle virus* | Murphy et al. (2000) |
| | Bell pepper | *Myzus persicae* | Herman et al. (2008) |
| *Pseudomonas fluorescens* | Tobacco | *Tobacco necrosis virus* | Park and Kloepper (2000) |
| *Bacillus* spp. | Cucumber | *Cotton aphids* | Stout et al. (2002) |
| *Pseudomonas aeruginosa* | Mung bean | *Root rot* | Siddiqui et al. (2001) |

### 6.3.4 Bacteria to Bacteria Communication by PGPR in the Promotion of Plant Growth and Colonization

Rhizosphere is rich in microbial community which have beneficial effects on plant growth and plant health. Survival and activities in these niches are under the control of density-dependent mechanism called quorum sensing (Pierson et al. 1994). The communication includes intra as well as inter-species interactions.

#### 6.3.4.1 Quorum Sensing Activity

Quorum sensing is a density-dependent mechanism of cell-to-cell interaction. A property like luminescence, production of certain chemical, is expressed only after achieving specific density. It is the basis of bacterial communication. Quorum sensing depends on the extracellular concentration of the signaling molecule. When an optimum concentration of this molecule is reached, it is detected by the group of bacteria and they respond to it. Quorum sensing is different in Gram-positive and Gram-negative bacteria due to differences in the signal molecule production. In Gram-negative bacteria they have a quorum sensing apparatus that consists of LuxI-type (I) protein, Lux R type (R) regulator that acts as signal receptor, and AHL (N-acyl homoserine lactone) synthase. At a low population density, bacteria produce basal level of AHL. When a threshold amount is reached, signaling molecule interacts with R protein forming R-AHL which interacts with the target promoters that induce expression of target gene. Gram-positive organisms have peptide auto inducers as signaling molecule (Dong and Zhang 2005). The phenomenon is also seen in PGPR which are present in the close vicinity of rhizosphere soil. Microbe-Microbe and Plant-Microbe interactions are highest in region and are responsible for various properties like hormone, antibiotic, siderophore. Etc. production.

### 6.3.5 Role of PGPR ACC Deaminase in Stress Agriculture

As plants are immobile, they are confronted to various kinds of stresses such as drought, flooding, salinity, heat, cold, exposure to heavy metals and nutrient deficiency, and phytopathogen and pests attack (Table 6.3) (Shen et al. 2013). The rise in synthesis of ethylene from its immediate precursor, ACC, is secreted by plants as root exudates. It has been found in almost all plants growing under stress conditions (Liu et al. 2015).

ACC deaminase has been widely reported in numerous microbial species of Gram-negative, Gram-positive bacteria, rhizobia, endophytes, and fungi (Jia et al. 1999). It is extensively studied in numerous species of PGPR like *Agrobacterium genomovars*, *Azospirillum lipoferum* (Blaha et al. 2006), *Alcaligenes*, *Bacillus* (Belimov et al. 2001), *Burkholderia* (Blaha et al. 2006) (Pandey et al. 2005),

**Table 6.3** Role of bacterial ACC deaminase in stress agriculture

| Type of stress | Role of ACC deaminase | References |
|---|---|---|
| **Salinity stress** | The plants inoculated with PGPR containing AC deaminase were able to thrive better through the salinity stress while demonstrating a normal growth pattern as salinity-induced ethylene is successfully reduced and it could decrease the negative impact of salinity onto plant growth | Raut et al. (2017) O'Donnell et al. (1996) Mayak et al. (2004) |
| **Drought stress** | Drought stimulates enhanced ethylene production in plant tissues which leads to unusual growth of a plant. ACC deaminase PGPR reduces the production of ethylene exposed to water stress | Mayak et al. (2004) |
| **Waterlogging stress** | In flooding, ACC, synthesized in roots, is transported to plant shoots where it is converted to ethylene by ACC oxidase. ACC deaminase PGPR showed substantial tolerance to flooding stress implying that bacterial ACC deaminase lowered the effects of stress induced ethylene | Mamoona et al. (2021) |
| **Temperature stress** | A fluctuation in temperature leads to hormonal imbalances in plants, and thus their growth is significantly affected. Accelerated ethylene production under high and chilling temperatures has widely been reported in plant tissues. Plants with ACC deaminase expression can overcome this unfavorable situation by lowering ethylene level | Choi et al. (2021) |
| **Pathogenicity stress** | The broadly recognized mechanisms of biocontrol mediated by PGPR are competition for an ecological niche or a substrate, generating inhibitory allelochemicals, and stimulating systemic resistance (ISR) in host plants to a broad spectrum of pathogens. ACC deaminase rhizobacteria have antagonistic consequences against microbial pathogens | Dobbelaere et al. (2003) Bloemberg et al. (2000) Lugtenberg and Kamilova (2009) Wang et al. (2000) Pandey et al. (2005) Blaha et al. (2006) Belimov et al. (2001) |
| **Heavy metals stress** | When present in excess, heavy metals may act as toxicants and suppress the plants growth. The application of PGPR containing ACC deaminase activity in phytoremediation of heavy metal contaminated soil environment has been proved | Belimov et al. (2001) Arshad et al. (2008) |
| **Organic contaminants stress** | Organic pollutants in the soil environment, if present above permissible limits, hinder plant growth via several mechanisms including abnormal growth of affected plant species. The significance of PGPR containing ACC deaminase activity in improving the growth of plants in the presence of organic contaminants has been recently reviewed | Khalid et al. (2004) |

(continued)

**Table 6.3** (continued)

| Type of stress | Role of ACC deaminase | References |
|---|---|---|
| **Air pollutants stress** | Air pollution, in addition to harming plants, prevents many enzyme systems and metabolic processes of plants. It is likely that PGPR can be utilized as a gene source for genetic modification of plants expressing the enzyme ACC deaminase against plant harm by air pollutants | Vahala et al. (2003) Wang et al. (2000) |
| **Wilting of flowers** | The wilting of ornamental flowers caused by ethylene production is a major impediment in the success of flowering business. Shelf life of flowers could be increased to manifold by treating them with suspensions of PGPR involving ACC deaminase action | Felske et al. (2000) |
| **Nodulation** | It has been well proven that ethylene and its precursor ACC are the destructive regulator of nodulation in numerous plant species. PGPR containing ACC deaminase activity endorses nodulation in legumes through inhibition of ethylene biosynthesis, and consequently, they augment symbiosis and nitrogen fixation in plants | Belimov et al. (2001) Belimov et al. (2001) Jia et al. (1999) |

*Enterobacter* (Penrose and Glick 2001), *Pseudomonas* (Blaha et al. 2006), and *Rhizobium* (Ma et al. 2003; Uchiumi et al. 2004).

#### 6.3.5.1 Mode of Action of Bacterial ACC Deaminase

The bacterial ACC deaminase having a low affinity for ACC can effectively compete with plant enzyme, ACC oxidase, which has high affinity for the same substrate. This results in the reduction of the plant's endogenous ethylene concentration. The biological activity of PGPR relates to the relative amounts of ACC deaminase and ACC oxidase in the system under consideration (Nayani et al. 1998). For PGPR to be able to lower plant ethylene levels, the ACC deaminase level should be at least 100–1000-fold greater than the ACC oxidase level.

PGPR synthesize and secrete indole-3-acetic acid (IAA), which gets adsorbed on the seed or root surface of plants from tryptophan and other small molecules present in seed or root exudates. Some of the newly synthesized IAA is taken up by the plant, and in conjunction with the endogenous plant IAA can further stimulate plant cell proliferation and elongation. IAA stimulates the activity of the enzyme ACC synthetase (Kende 1993).

A significant portion of ACC is exuded from plant roots or seeds and taken up by the soil microbes or hydrolyzed by the ACC deaminase to yield ammonia and α-keto-butyrate. The uptake and subsequent hydrolysis of ACC by microbes decreases the amount of ACC outside the plant (Mayak et al. 2004). The balance between the internal and the external ACC levels is retained through exudation of more ACC into the rhizosphere. Soil microbial communities containing ACC deaminase activity trigger plants to produce more ACC than the plant would otherwise require. They stimulate ACC exudation from plant roots, while supplying microorganisms with a distinctive source of nitrogen (ACC). The growth of microorganism containing ACC deaminase is enhanced in the close vicinities of plant roots as compared to the other soil microbes. By doing so, not only the ACC level is lowered within the plant but also the biosynthesis of the stress hormone ethylene is inhibited. Thus, a plant inoculated with bacteria containing ACC deaminase exhibits more root growth.

#### 6.3.5.2 Recent Advances at Molecular Level

Recently, efforts have been made to introduce specific genes into plants to enable them to cope with complex environmental stresses. Introduction of specific genes responsible for expression of enzymes like ACC deaminase from microbial species directly into crop plants has received great attention in the last few decades. Genetic manipulation of ACC deaminase trait in bacteria has not been much attempted. This is most likely due to (1) ACC deaminase trait is widely found among soil indigenous microbial species and (2) survival and functioning of wild-type microbial species containing ACC deaminase is better than genetically engineered microorganism expressing ACC deaminase genes (Nadeem et al. 2006). However, the genetically modified bacteria could be helpful for improving better understanding of processes responsible for induction of tolerance in plants inoculated with ACC deaminase bacteria against both biotic and abiotic stresses. The use of PGPR containing ACC deaminase activity along with other innovations could prove to be a cost-effective and environment-friendly strategy to ensure sustainable agriculture.

### *6.3.6 Formation of Biofilm by PGPR*

Success of the plant microbe interaction depends on effective root colonization and subsequent biofilm formation. PGPR exhibit regulatory mechanisms like quorum sensing making them stable entities in biofilms (Sanghi et al. 2016). Biofilms are microbial communities associated with a surface which are enclosed in a matrix of polysaccharides allowing growth and survival even in adverse conditions. Biofilms are not washed off easily. Biofilms are formed in a specific manner showing initial surface attachment, microcolony formation, three-dimensional community structure, maturation, and detachment (Fig. 6.4) (Hall-Stoodley et al. 2004). Certain bacteria produce exopolysaccharides which is an important factor in biofilm formation. Essential nutrients are circulated to plants properly by effective root colonization by exopolysaccharides (EPS) producing microbes. Functions performed by these biofilms include protection against abiotic stress and plant defense response.

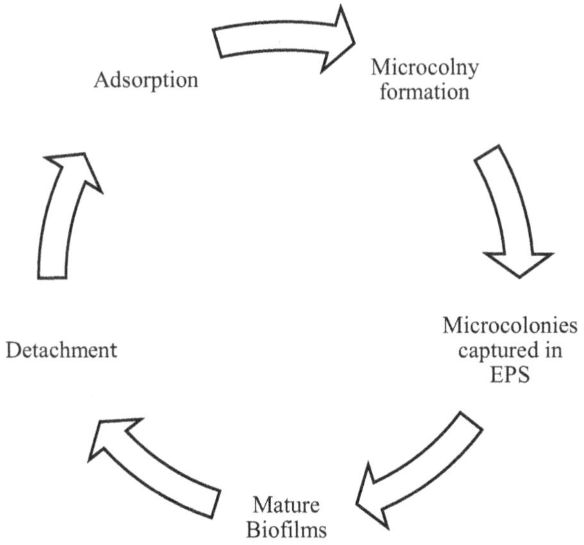


**Fig. 6.4** Steps involved in biofilm formation

#### 6.3.6.1 PGPR Biofilms in Plant Microbe Interaction

Bacteria exhibit different properties when in biofilms than planktonic form. They perform well in inhibiting other competing organisms and outgrow in number, nutrition uptake, and adaptation to changing environments. Cells in biofilm show modulations in their metabolic function. EPS acts as barrier for antimicrobial agents and toxins produced by soil microbiota. Some metal ions are also sequestered in the process. Presence of water in the biofilm allows nutrient uptake, exchange of metabolites, and removal of toxins. The most common groups associated with plant roots and leaves are species of *Pseudomonas*, *P. putida* and *P.* fluorescens, and *Bacillus* which form biofilms on plant surfaces. Various symbiotic (species of *Rhizobium*) and non-symbiotic (Species of *Azospirillum*) nitrogen fixing bacteria have ability to form biofilm. Endospore-forming *Bacillus* produce antimicrobial substances in a biofilm. It was shown in a study that plant polysaccharides such as xylan, pectin, and arabinogalactan were stimulated at initiation stage of biofilm formation (Raut et al. 2017).

#### 6.3.6.2 PGPR Biofilms in Advanced Agriculture

Current concerns regarding hazardous effects of pesticides have led to search for alternative strategies which include use of biofertilizers and biocontrol agents. They consist of plant growth-promoting rhizobacteria having beneficial effects on plant growth as well as protect plant against plant pathogens. However, inability to establish themselves in rhizosphere along with indigenous species limits their commercial use in field applications. This problem can be overcome by application of PGPR in biofilm form giving high output yield, improved soil fertility, and enhanced production.

#### 6.3.6.2.1 PGPR Biofilms as Biofertilizers

Biofertilizers can be used as an alternative to chemical fertilizer for improved soil fertility and crop yield without any pollution and health hazards. Some of the PGPR provide nutrients to plants in a form that can be easily assimilated. Nitrogen fixing bacteria inoculants of *Burkholderia* spp., *Azotobacter* spp., *Bacillus polymyxa*, and *Azospirillum* spp. are commercialized biofertilizers (Vessey 2003). Phosphorus is a limiting nutrient which is not easily accessible to plants. Phosphate solubilizing microorganisms produce organic acids and help in solubilizing this P usually complexed with other minerals. Beneficial biofilms which can be developed in vitro and applied as biofertilizers effectively are called biofilm biofertilizers. Biofilms of *Pseudomonas*, *Trichoderma*, *Penicillium*, and *Bradyrhizobium* showed increased IAA production, phosphate solubilization, siderophore production, nitrogen fixation rates, etc. (Bais et al. 2004). Application of biofilm inocula results in cotton seed germination, soyabean dry weight, increased wheat root, and shoot length. Biofilm formation was observed on the roots of *Arabidopsis thaliana* improving plant productivity and growth (Elshahat et al. 2016).

#### 6.3.6.2.2 PGPR Biofilms as Plant Growth-Promoting Agents

Efficacy of the bioinoculants depends on successful root colonization and environmental factors. Rhizospheric colonization is thus considered as a critical step in the application of microorganisms for valuable purposes. The organisms in the biofilm are responsible for production of IAA and hydrogen cyanide. There is increase in root and shoot dry weight. Some attributes of photosynthesis like transpiration rate, stomatal conductance, SPAD chlorophyll value, and photosynthesis rate improve in a mixed culture biofilm (Ali et al. 2020).

#### 6.3.6.2.3 PGPR Biofilms as Biocontrolling Agents

*B. subtilis* is utilized as a biocontrol agent against *F. oxysporum*, the causative agent of tomato wilt disease. Studies revealed that *B. subtilis* produced lipopeptide and inhibited phytopathogenic fungi by antibiosis mechanism. Usage of consortium of PGPR may often have more impact on biological control and plant growth than a single strain (Krishnamurthy and Gnanamanickam 1998). Antibiotics produced by these PGPR biofilms act as biocontrol agents. Biofilms are usually formed on root tips making plants less sensitive to infection (Bais et al. 2004). *Pseudomonas florescence* coats plant roots by biofilm formation and protects plants against various bacterial and fungal pathogens. *Bacillus amyloliquefaciens* is used as a biocontrol agent against *Rhizoctonia solani*. *Bacillus* spp. inocula on cotton plants showed increased production of jasmonic acid reducing larval feeding. *Bacillus pumilus* biofilm played important role in prevention of pine seedlings damping off disease caused by a *Rhizoctonia solani* (Raut et al. 2017). PGPR biofilms have proved their

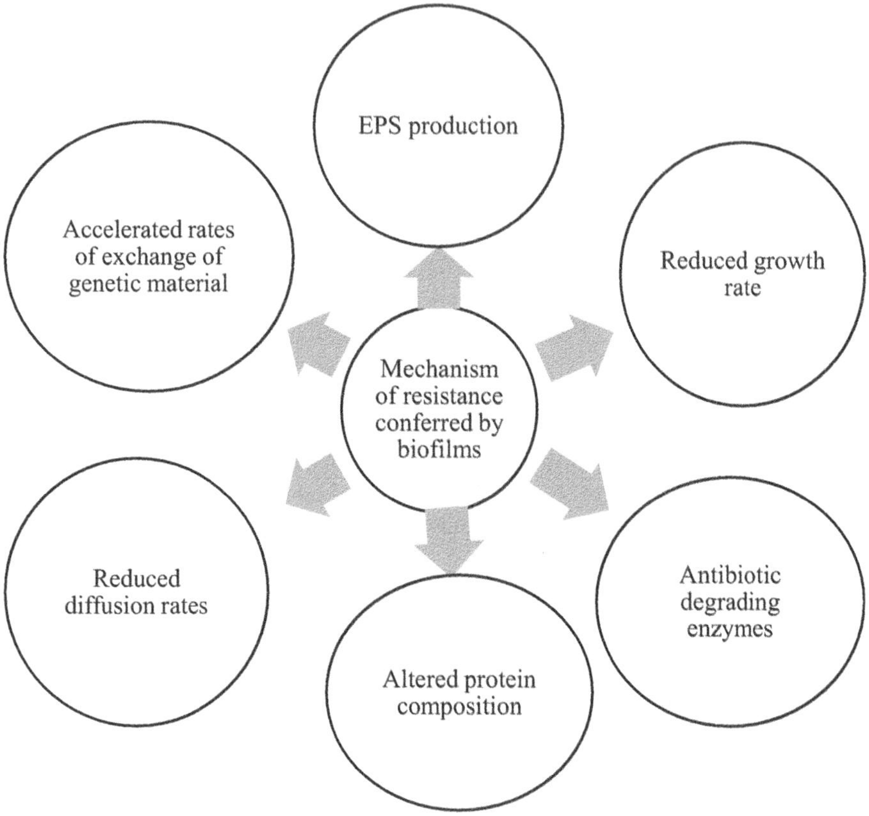


**Fig. 6.5** Mechanism of resistance by biofilms

potential as biocontrol agents with increased microbial activity and hence have better prospects in modern agriculture (Fig. 6.5).

## 6.4 Fungal Diseases Associated with Plants

The phytopathogens include bacteria, fungi, viruses, viroid, and protozoa. Most of the phytopathogenic fungi belong to the Ascomycetes and the Basidiomycetes. The fungicides and other agriculture practices are used to control the fungal diseases. However, new species of fungi often emerge which are resistant to various fungicides. Biotrophic fungal pathogens colonize living plant tissue and get nutrients from living host cells. Necrotrophic fungal pathogens infect and kill host tissue and obtain nutrients from the dead host cells. The different types of fungal infections include the black spot, downy mildew, powdery mildew, blight, rust, wilt, club root, and anthracnose (Table 6.4).

**Table 6.4** Fungal diseases associated with plants

| Name of the disease | Plants affected | Symptoms | Causative agent |
|---|---|---|---|
| Black spot | Rose, oak, maple, sycamore, walnut | Black, gray, or brown spots formed on the leaves of the plant, causing them to drop | *Pseudomonas*, *Asterina*, *Asterinella*, *Diplotheca*, *Glomerella*, *Gnomonia*, *Schizothyrium*, *Placosphaeria*, *Stigmea* |
| Downy mildew | Sunflower, rosemary, primula, osteospermum, *Impatiens walleriana*, coleus, statice, verbena, ornamental cabbage, basil, and cineraria | Leaves and stems turn yellow, prevent flowering | *Plasmopara viticola*, *Bremia*, *Peronospora*, *Phytophthora*, *Plasmopara*, *Sclerospora*, *Pseudoperonospora* |
| Powdery mildew | Rose, grape, grass, wheat, barley, Onion, apple, pear | The powdery fungus grows on the upper surface of the plant leaves, white, yellow, or brown in color. It spreads below the leaves or stems | *Podosphaera xanthii*, *Sphaerotheca fuliginea*, *Erysiphe cichoracearum* |
| Blight | Tomato, potato, Apple, rice, wheat | Brown, discolored leaves tend to dry and curl inwards, in moist conditions shows a white fungal growth. Both tomatoes and potatoes develop brown patches turning into rotten sores | *Phytophthora infestans*, *Erwinia amylovora*, *Xanthomonas oryzae*, *Alternaria triticina* |
| Rust | Wheat, barley, rye, oats, beans, sugarcane | Raised spots below leaves and on the stem, covered with reddish orange spore masses; leaves turn yellow green and black and cause leaves drop | *Puccinia coronata*, *Phragmidium*, *Puccinia graminis*, *Uromyces appendiculatus* |
| Wilt | Chrysanthemum, pepper, tomato, potato, cucumber, muskmelon, pumpkin, gourds | The wilting leaves can turn yellow or brown | *Erwinia tracheiphila*, *Ralstonia solanacearum*, *Solanum tuberosum*, *Zingiber officinale* |
| Clubroot | Cabbage, cauliflower, Brussels sprouts, turnip, swede, wallflowers, *Saxifraga*, broccoli | Roots become swollen and distorted, stunted growth, and purplish, wilting foliage | *Plasmodiophora brassicae* |
| Anthracnose | Tomatoes, cucumbers, beans, melon, sycamore, ash, oak, and maple | Small, sunken spots that appear on fruits and pods, which have pinkish sores in the center | *Colletotrichum* or *Gloeosporium* |

**Table 6.5** Production antifungal metabolites by PGPR against phytopathogens

| PGPR | Antifungal metabolite production by PGPR | Fungicidal activity against the phytopathogen |
|---|---|---|
| *P. fluorescens* BL915 | Pyrrolnitrin | *Rhizoctonia solani* |
| *Pseudomonas* spp. | 2, 4-DAPG | Membrane destruction of *Pythium* spp. |
| *Pseudomonas* spp. | Phenazine | Antagonistic activity against *fusarium oxysporum* |
| *Bacillus* spp. | Circulin, polymyxin, and colistin | Pathogenic fungi as well as gram-negative and gram-positive bacteria |
| *Bacillus subtilis* | Fengycin and iturins | *Podosphaera fusca* |
| *E. coli* | Colicins | Broad range of inhibition of fungal, yeast, gram-positive and gram-negative species |
| *B. Megaterium* | Megacins | |
| *Enterobacter cloacae* | Cloacins | |
| *P. Pyogenes* | Pyocins | |
| *Serratia marcescens* | *P. Pyogenes* | |

## 6.5 Production of Antifungal Metabolites by PGPR

Application of PGPR for the purpose of controlling or lessening the harmful effects of phytopathogens is known as biological control. PGPRs control the destructive effects of pathogenic agents on plants by generating growth inhibitors, i.e., antibiotics, bacteriocins, siderophores, and lytic enzymes or by rising natural resistance of host plant (Table 6.5).

### *6.5.1 PGPR-Mediated Induced Systemic Resistance (ISR) in Plants*

The plant defense mechanisms are triggered by some stimulus prior to infection by a plant pathogen; the possibility of the disease can be reduced. Interactions between plants and pathogens can lead either compatible response to a successful infection or incompatible response to resistance. Induced resistance is a state of enhanced defensive capacity developed by a plant when appropriately stimulated (de Laat and van Loon 1982). Systemic acquired resistance (SAR) and induced systemic resistance (ISR) are two forms of induced resistance. SAR can elicit a quick local reaction, or hypersensitive reaction; the pathogen is restricted to a small area of the site of infection. SAR is triggered by accumulation of pathogenesis-related proteins or salicylic acid (Gundlach et al. 1992).

Selected strains of PGPR repress diseases by antagonism between the bacteria and soilborne pathogens and by inducing a systemic resistance in plant against both root and foliar pathogens. Rhizobacteria mediated ISR look like that of pathogen induced systemic acquired resistance (SAR). Both types of induced resistance make uninfected plant parts more resistant to a broad spectrum of plant pathogens. Several rhizobacteria generate the salicylic acid (SA)-dependent SAR pathway by producing SA at the root surface while other rhizobacteria produce different signaling pathway independent of SA (Reshma et al. 2018).

## 6.6 Conclusion

Plant diseases or phytopathogens alter plants right from the planting stage up to the harvesting and storage of the crop. The plants have developed several different mechanisms by which they defend themselves. PGPR can promote plant growth by both direct and indirect mechanisms (Glick 1995). Direct mechanisms incorporate the production of auxin, ACC deaminase, cytokinin, gibberellin, nitrogen fixation, phosphorus solubilization, and sequestration of iron by bacterial siderophores. Indirect mechanisms include ACC deaminase, antibiotics, cell wall degrading enzymes, competition, hydrogen cyanide, induced systemic resistance, quorum quenching, and siderophores.

These biocontrol bacteria have shown root colonization properties, broad-spectrum antifungal activity, and the ability to promote plant growth. The profiles of antifungal compounds include 2,4-DAPG, phenazine, pyochelin, rhamnolipids, pyoverdines, surfactins, and AHLs which varied among different organisms. The plant inoculation study validates their innate biocontrol and biofertilizer potentials where in vitro antifungal activity shows positive correlation with in vivo disease suppression activity (Ali et al. 2020). PGPR strains launched into the rhizosphere might play vital roles in the transition of soil from a suppressive to favorable condition. Such strains can be used in sustainable agriculture (Wang et al. 2019).

The capacity of bacteria to elicit a defense response in plant, called induced systemic resistance (IRS), includes the induction of synthesis of defense metabolites, but without causing a disease itself, boosting the plant's defensive capacity. PGPR can be used as biocontrol agent and as ecological replacement to the use of agrochemicals (Salomon et al. 2017).

Many fungi play an important role in soil in increasing soil fertility. Some fungi like *Trichoderma*, *Penicillium*, *Aspergillus*, etc. show antagonistic effect against plant pathogens. These beneficial microorganisms are killed due to use of fungicides. Instead of fungicides use of biocontrol agents which show antimicrobial activity should be used.

ACC deaminase activity and quorum sensing play a key role in the expression of several rhizobacterial qualities as well as in plant–bacteria interactions by inducing systemic resistance and enhancing plant tolerance to stress. Several ecological and interdependent crucial characters of bacteria, like antibiotic, siderophore, or enzyme secretion, virulence factors of phytopathogens, as well as plant-microbe communications, are synchronized through quorum sensing (Altaf et al. 2017).

EPS released assists in biofilm formation that enhances plant growth and provides protection from pathogens.

Use of biocontrol agents which can promote growth and also elicit the expression of defense enzymes can be a suitable alternative to environmentally hazardous agrochemicals (Reshma et al. 2018).

The PGPR will require a systematic strategy designed to fully utilize all these beneficial factors, applying combinations of different mechanisms of action allowing crop yields to be maintained or even increased while chemical treatments are reduced (Beneduzi et al. 2012). Nature has a huge biodiversity of PGPR. The knowledge on the regulation of various strategies for bio-fungicidal properties of rhizobacteria for plant growth promotion and disease resistance will lead to development of PGPR with improved reliability and efficacy.

## References

Ahmad F, Ahmad I, Khan MS (2008) Screening of free-living rhizospheric bacteria for their multiple plant growth promoting activities. Microbiol Res 163(2):173–181. https://doi.org/10.1016/j.micres.2006.04.001

Ahmed A, Hasnain S (2014) Auxins as one of the factors of plant growth improvement by plant growth promoting rhizobacteria. Pol J Microbiol 63(3):261–266

Ali S, Hameed S, Shahid M, Iqbal M, Lazarovits G, Imran A (2020) Functional characterization of potential PGPR exhibiting broad-spectrum antifungal activity. Microbiol Res 232. https://doi.org/10.1016/j.micres.2019.126389

Alori ET, Glick BR, Babalola OO (2017) Microbial phosphorus solubilization and its potential for use in sustainable agriculture. Front Microbiol 8:971. https://doi.org/10.3389/fmicb.2017.00971

Altaf MM, Khan MSA, Abulreesh HH, Ahmad I (2017) Quorum sensing in plant growth- promoting rhizobacteria and its impact on plant-microbe interaction. Plant-Microbe Interactions in Agro-Ecological Perspectives, pp 311–331

Anith KN, Momol MT, Kloepper JW, Marois JJ, Olson SM, Jones JB (2004) Efficacy of plant growth-promoting rhizobacteria, acibenzolar-S-methyl, and soil amendment for integrated management of bacterial wilt on tomato. Plant Dis:669–673. https://doi.org/10.1094/PDIS.2004.88.6.669

Arora H, Sharma A, Sharma S, Haron FF, Gafur A, Sayyed RZ, Datta R (2021) Pythium damping-off and root rot of *Capsicum annuum* L.: impacts, diagnosis, and management. Microorganisms 9:823. https://doi.org/10.3390/microorganisms9040823

Arshad M, Shaharoona B, Mahmood T (2008) Inoculation with pseudomonas spp. containing ACC-deaminase partially eliminates the effects of drought stress on growth, yield, and ripening of pea (Pisum sativum L.). Pedosphere 18(5):611–620. https://doi.org/10.1016/s1002-0160(08)60055-7

Baas P, Mohan JE, Markewitz D, Knoepp JD (2014) Assessing heterogeneity in soil nitrogen cycling: a plot-scale approach. Soil Sci Soc Am J 78(S1):S237–S247. https://doi.org/10.2136/sssaj2013.09.0380nafsc

Babalola OO (2010) Beneficial bacteria of agricultural importance. Biotechnol Lett 32:1559–1570

Bais HP, Fall R, Vivanco JM (2004) Biocontrol of Bacillus subtilis against infection of Arabidopsis roots by pseudomonas syringae is facilitated by biofilm formation and Surfactin production. Plant Physiol 134:307–319. https://doi.org/10.1104/pp.103.028712

Banerjee MR, Yesmin L, Vessey JK, Rai M (2005) Plant-growth-promoting rhizobacteria as biofertilizers and biopesticides. Handbook of microbial biofertilizers. New York: Food Products Press, Binghamton, USA, pp. 137–181

Basu A, Prasad P, Das SN, Kalam S, Sayyed RZ, Reddy MS, El Enshasy H (2021) Plant growth promoting Rhizobacteria (PGPR) as green bioinoculants: recent developments, constraints, and prospects. Sustainability 13:1140. https://doi.org/10.3390/su13031140

Beg QK, Gupta R (2003) Purification and characterization of an oxidation-stable, thiol-dependent serine alkaline protease from Bacillus mojavensis. Enzym Microb Technol 32(2):294–304. https://doi.org/10.1016/S0141-0229(02)00293-4

Belimov AA, Safronova VI, Sergeyeva TA, Egorova TN, Matveyeva VA, Tsyganov VE, Borisov AY, Tikhonovich IA, Kluge C, Preisfeld A, Dietz KJ, Stepanok VV (2001) Characterization of plant growth promoting rhizobacteria isolated from polluted soils and containing 1-aminocyc lopropane-1-carboxylate deaminase. Can J Microbiol 47(7):642–652. https://doi.org/10.1139/w01-062

Beneduzi A, Ambrosini A, Passaglia LMP (2012) Plant growth-promoting rhizobacteria (PGPR): their potential as antagonists and biocontrol agents. Genet Mol Biol 35(4 Suppl):1044–1051. https://doi.org/10.1590/s1415-47572012000600020

Blaha D, Prigent-Combaret C, Mirza MS, Moënne-Loccoz Y (2006) Phylogeny of the 1-aminocyc lopropane-1-carboxylic acid deaminase-encoding gene acdS in phytobeneficial and pathogenic Proteobacteria and relation with strain biogeography. FEMS Microbiol Ecol 56(3):455–470. https://doi.org/10.1111/j.1574-6941.2006.00082.x

Bloemberg GV, Wijfjes AHM, Lamers GEM, Stuurman N, Lugtenberg BJJ (2000) Simultaneous imaging of Pseudomonas fluorescens WCS365 populations expressing three different autofluorescent proteins in the rhizosphere: new perspectives for studying microbial communities. Mol Plant-Microbe Interact 13(11):1170–1176. https://doi.org/10.1094/MPMI.2000.13.11.1170

Bostock RM (2005) Signal crosstalk and induced resistance: straddling the line between cost and benefit. Annu. Rev. Phytopathology 43:545–580

Choi, J, Choudhury AR, Park S-Y, Oh M-M, Sa T (2021) Inoculation of ACC Deaminase-Producing Brevibacterium linens RS16 Enhances Tolerance against Combined UV-B Radiation and Heat Stresses in Rice (Oryza sativa L.). Sustainability, 13(18):10013. https://doi.org/10.3390/su131810013

Datta C, Basu PS (2000) Indole acetic acid production by a rhizobium species from root nodules of a leguminous shrub, Cajanus cajan. Microbiol 155(2):123–127. https://doi.org/10.1016/S0944-5013(00)80047-6

de Laat A, van Loon L (1982) Regulation of ethylene biosynthesis in virus-infected tobacco leaves: II. Time course of levels of intermediates and in vivo conversion rates. Plant Physiol 69(1):1–6. https://doi.org/10.1104/pp.69.1.240

Dobbelaere S, Vanderleyden J, Okon Y (2003) Plant growth-promoting effects of diazotrophs in the rhizosphere. CRC Crit Rev Plant Sci 22(2):107–149

Dominguez RB, Hayat A, Alonso GA, Gutiérrez JM, Muñoz R, Marty J-L (2017) Nanomaterial-based biosensors for food contaminant assessment. Elsevier Inc 19:809–839. https://doi.org/10.1016/B978-0-12-804301-1.00019-9

Dong YH, Zhang LH (2005) Quorum sensing and quorum-quenching enzymes. J Microbiol 43(spc1):101–109

Elshahat MR, Ahmed AA, Enas AH, Fekria MS (2016) Plant growth promoting rhizobacteria and their potential for biocontrol of phytopathogens. African J Microbiol Res 10:486–504. https://doi.org/10.5897/ajmr2015.7714

Felske A, Wolterink A, Van Lis R, De Vos WM, Akkermans ADL (2000) Response of a soil bacterial community to grassland succession as monitored by 16S rRNA levels of the predominant ribotypes. Appl Environ Microbiol 66:3998–4003. https://doi.org/10.1128/AEM.66.9.3998-4003.2000

Ferreira MJ, Silva H, Cunha A (2019) Siderophore-producing rhizobacteria as a promising tool for empowering plants to cope with iron limitation in saline soils: a review. Pedosphere, 29(4):409–420. https://doi.org/10.1016/S1002-0160(19)60810-6

Gage DJ (2004) Infection and invasion of roots by symbiotic, nitrogen-fixing rhizobia during nodulation of temperate legumes. Microbiol Mol Biol Rev 68(2):280–300. https://doi.org/10.1128/mmbr.68.2.280-300.2004

Gerze A, Omay D, Guvenilir Y (2005) Partial purification and characterization of protease enzyme from Bacillus subtilis megatherium. In: Davison BH, Evans BR, Finkelstein M, Mcmillan JD (eds) Twenty-sixth symposium on biotechnology for fuels and chemicals. ABAB symposium, vol 121. Humana Press Appl Biochem Biotechnol, pp 1–3. https://doi.org/10.1385/abab:121:1-3:0335

Glick BR (1995) The enhancement of plant growth by free-living bacteria. Can J Microbiol 41(2):109–117

Gundlach H, Müller MJ, Kutchan TM, Zenk MH (1992) Jasmonic acid is a signal transducer in elicitor-induced plant cell cultures. Proc Natl Acad Sci 89(6):2389–2393. https://doi.org/10.1073/pnas.89.6.2389

Gupta G, Snehi SK, Singh V (2017) Role of PGPR in biofilm formations and its importance in plant health. Biofilms in Plant and Soil Health, pp 27–42

Haas D, Défago G (2005) Biological control of soil-borne pathogens by fluorescent pseudomonads. Nat Rev Microbiol 3(4):307–319

Hall-Stoodley L, Costerton JW, Stoodley P (2004) Bacterial biofilms: from the natural environment to infectious diseases. Nat Rev Microbiol 2:95–108

Herman MAB, Nault BA, Smart CD (2008) Effects of plant growth-promoting rhizobacteria on bell pepper production and green peach aphid infestations in New York. Crop Prot 27(6):996–1002. https://doi.org/10.1016/j.cropro.2007.12.004

Hugenholtz P, Goebel BM, Pace NR (1998) Impact of culture-independent studies on the emerging phylogenetic view of bacterial diversity. J Bacteriol 180(18):4765–4774

Hynes RK, Leung GCY, Hirkala DLM, Nelson LM (2008) Isolation, selection, and characterization of beneficial rhizobacteria from pea, lentil, and chickpea grown in western Canada. Can J Microbiol 54(4):248–258. https://doi.org/10.1139/W08-008

Jadhav HP, Shaikh SS, Sayyed RZ (2017) Role of hydrolytic enzymes of Rhizoflora in biocontrol of fungal Phytopathogens: an overview. In: Rhizotrophs: plant growth promotion to bioremediation. Springer, pp 183–203

Jadhav HP, Sayyed RZ, Shaikh SS, Bhamre HM, Sunita S, Enshasy HE (2020a) Statistically designed bioprocess for enhanced production of alkaline protease in *Bacillus cereus* HP_RZ17. J Scient Indus Res 79:491–498

Jadhav HP, Sonawane MS, Khairnar MH, Sayyed RZ (2020b) Production of alkaline protease by rhizospheric *Bacillus cereus* HP_RZ17 & *Paenibacillus xylanilyticus* HP_RZ19. Environ Sustainability 3:5–13

Jia YJ, Kakuta Y, Sugawara M, Igarashi T, Oki N, Kisaki M, Shoji T, Kanetuna Y, Horita T, Matsui H, Honma M (1999) Synthesis and degradation of 1-aminocyclopropane-1-carboxylic acid by *Penicillium citrinum*. Biosci Biotechnol Biochem 63(3):542–549. https://doi.org/10.1271/bbb.63.542

Jones JDG, Dangl JL (2006) The plant immune system. Nature 444:323–329

Joo GJ, Kim YM, Kim JT, Rhee IK, Kim JH, Lee IJ (2005) Gibberellins-producing rhizobacteria increase endogenous gibberellins content and promote growth of red peppers. J Microbiol 43(6):510–515

Kende H (1993) Ethylene biosynthesis and its regulation in higher plants. Annu Rev Plant Physiol Plant Mol Biol 35:155–189. https://doi.org/10.1146/annurev.pp.44.060193.001435

Khalid A, Arshad M, Zahir ZA (2004) Screening plant growth-promoting rhizobacteria for improving growth and yield of wheat. J Appl Microbiol 96(3):473–480. https://doi.org/10.1046/j.1365-2672.2003.02161.x

Khammas K, Ageron E, Grimont PA, Kaiser P (1989) Azospirillum irakense sp. nov., a nitrogen-fixing bacterium associated with rice roots and rhizosphere soil. Res Microbiol 140(9):679–693. https://doi.org/10.1016/0923-2508(89)90199-x

Khot LR, Sankaran S, Maja JM, Ehsani R, Schuster EW (2012) Applications of nanomaterials in agricultural production and crop protection: a review. Crop Prot 35:64–70

Kim KJ, Yang YJ, Kim JG (2003) Purification and characterization of chitinase from Streptomyces sp. M-20. J Biochem Mol Biol 36(2):185–189. https://doi.org/10.5483/bmbrep.2003.36.2.185

Kim YC, Jung H, Kim KY, Park SK (2008) An effective biocontrol bioformulation against Phytophthora blight of pepper using growth mixtures of combined chitinolytic bacteria under different field conditions. Eur J Plant Pathol 120(4):373–382. https://doi.org/10.1007/s10658-007-9227-4

Kirchhof G, Eckert B, Stoffels M, Ivo Baldani J, Reis VM, Hartmann A (2001) Herbaspirillum frisingense sp. nov., a new nitrogan-fixing bacterial species that occurs in C4-fibre plants. Int J Syst Evol Microbiol 51(Pt 1):157–168. https://doi.org/10.1099/00207713-51-1-157

Kour D, Rana KL, Kaur T, Yadav N, Yadav AN, Rastegari AA, Saxena AK (2019) Microbial biofilms: functional annotation and potential applications in agriculture and allied sectors. In: New and future developments in microbial biotechnology and bioengineering: microbial biofilms current research and future trends in microbial biofilms, pp 283–301

Kour D, Rana KL, Yadav AN, Yadav N, Kumar M, Kumar V, Vyas P, Dhaliwal HS, Saxena AK (2020) Microbial biofertilizers: bioresources and eco-friendly technologies for agricultural and environmental sustainability. Biocatal Agric Biotechnol 101487

Kredics L, Antal Z, Szekeres A, Hatvani L, Manczinger L, Vágvölgyi C, Erzsébet N (2005) Extracellular proteases of *Trichoderma* species: A review. Acta Microbiol Immunol Hung 52:169–184

Krishnamurthy K, Gnanamanickam SS (1998) Biological control of rice blast by Pseudomonas fluorescens strain Pf7-14: evaluation of a marker gene and formulations. Biol Control 13(3):158–165. https://doi.org/10.1006/bcon.1998.0654

Kumar S, Nehra M, Dilbaghi N, Marrazza G, Hassan AA, Kim KH (2019) Nano-based smart pesticide formulations: emerging opportunities for agriculture. J Control Release 294:131–153. https://doi.org/10.1016/j.jconrel.2018.12.012

Kusale SP, Attar YC, Sayyed RZ, Malek RA, Ilyas N, Suriani NL, Khan N, Enshasy HA (2021a) Production of plant beneficial and antioxidants metabolites by *Klebsiella variicola* under salinity stress. Molecules 26:1894. https://doi.org/10.3390/molecules26071894

Kusale SP, Attar YC, Sayyed RZ, Enshasy HE, Hanapi Z, Ilyas N, Elgorban AE, Bahkali AH, Marraiki N (2021b) Inoculation of *Klebsiella variicola* alleviated slat stress salinity and improved growth and nutrients in wheat and maize. Agronomy 11:927. https://doi.org/10.3390/agronomy11050927

Lamsal K, Kim SW, Kim YS, Lee YS (2012) Application of rhizobacteria for plant growth promotion effect and biocontrol of anthracnose caused by Colletotrichum acutatum on pepper. Mycobiology 40(4):244–251. https://doi.org/10.5941/MYCO.2012.40.4.244

Liu JL, Xie BM, Shi XH, Ma JM, Guo CH (2015) Effects of two plant growth-promoting rhizobacteria containing 1-aminocyclopropane-1-carboxylate deaminase on oat growth in petroleum-contaminatedsoil. Int J Environ Sci Technol 12(12):3887–3894. https://doi.org/10.1007/s13762-015-0798-x

Liu SH, Zeng GM, Niu QY, Liu Y, Zhou L, Jiang LH, Fei TX, Xu P, Zhang C, Cheng M (2017) Bioremediation mechanisms of combined pollution of PAHs and heavy metals by bacteria and fungi: a mini review. Bioresour Technol 224:25–33. https://doi.org/10.1016/j.biortech.2016.11.095

Lucas García JA, Probanza A, Ramos B, Ruiz Palomino M, Gutiérrez Mañero FJ (2004) Effect of inoculation of bacillus licheniformis on tomato and pepper. Agronomie 24:169–176. https://doi.org/10.1051/agro:2004020

Lugtenberg B, Kamilova F (2009) Plant-growth-promoting rhizobacteria. Annu Rev Microbiol 63:541–556

Ma W, Sebestianova SB, Sebestian J, Burd GI, Guinel FC, Glick BR (2003) Prevalence of 1-aminocyclopropane-1-carboxylate deaminase in Rhizobium spp. Antonie Leeuwenhoek 83:285–291
Mamoona R, Muhammad A, Aziz UD, Kazim A, Humaira G, Muhammad MR, Muhammad H, Muhammad A (2021) Molecular mechanisms of the 1-Aminocyclopropane-1-carboxylic acid (ACC) deaminase producing *Trichoderma asperellum* MAP1 in enhancing wheat tolerance to waterlogging stress. Front Plant Sci 18. https://doi.org/10.3389/fpls.2020.614971
Mayak S, Tirosh T, Glick BR (2004) Plant growth-promoting bacteria confer resistance in tomato plants to salt stress. Plant Physiol Biochem 42(6):565–572. https://doi.org/10.1016/j.plaphy.2004.05.009
McCann HC, Guttman DS (2008) Evolution of the type III secretion system and its effectors in plant-microbe interactions. New Phytol 177(1):33–47
Mishra J, Arora NK (2018) Secondary metabolites of fluorescent pseudomonads in biocontrol of phytopathogens for sustainable agriculture. Appl Soil Ecol 125:35–45. https://doi.org/10.1016/j.apsoil.2017.12.004
Mubeen F, Shiekh MA, Iqbal T, Khan QM, Malik KA, Hafeez FY (2006) In vitro investigations to explore the toxicity of fungicides for plant growth promoting rhizobacteria. Pakistan J Bot 38:1261–1269
Murphy JF, Zehnder GW, Schuster DJ, Sikora EJ, Polston JE, Kloepper JW (2000) Plant growth-promoting rhizobacterial mediated protection in tomato against tomato mottle virus. Plant Dis 84(7):779–784. https://doi.org/10.1094/PDIS.2000.84.7.779
Nadeem S, Hussain I, Naveed M, Asghar H, Zahir Z, Arshad M (2006) Performance of plant growth promoting rhizobacteria containing ACC-deaminase activity for improving growth of maize under salt-stressed conditions. Pak J Agric Sci 43:3–4
Naiman AD, Latrónico A, García de Salamone IE (2009) Inoculation of wheat with Azospirillum brasilense and Pseudomonas fluorescens: impact on the production and culturable rhizosphere microflora. Eur J Soil Biol 45:44–51. https://doi.org/10.1016/j.ejsobi.2008.11.001
Nandi M, Selin C, Brawerman G, Fernando WGD, de Kievit T (2017) Hydrogen cyanide, which contributes to pseudomonas chlororaphis strain PA23 biocontrol, is upregulated in the presence of glycine. Biol Control 108:47–54. https://doi.org/10.1016/j.biocontrol.2017.02.008
Nayani S, Mayak S, Glick BR (1998) Effect of plant growth-promoting rhizobacteria on senescence of flower petals. Indian J Exp Biol 36(8):836–839
Nithyapriya S, Sundaram L, Sayyed RZ, Reddy MS, Dailin DJ, Enshasy HE, Suriani NL, Herlambang S (2021) Production, purification, and characterization of bacillibactin siderophore of *Bacillus subtilis* and its application for improvement in plant growth and oil content in sesame. Sustainability 13:5394. https://doi.org/10.3390/su13105394
O'Donnell PJ, Calvert C, Atzorn R, Wasternack C, Leyser HMO, Bowles DJ (1996) Ethylene as a signal mediating the wound response of tomato plants. Science 274(5294):1914–1917. https://doi.org/10.1126/science.274.5294.1914
Oldroyd GED, Dixon R (2014) Biotechnological solutions to the nitrogen problem. Curr Opin Biotechnol 26:19–24. https://doi.org/10.1016/j.copbio.2013.08.006
Olivares FL, Baldani VLD, Reis VM, Baldani JI, Döbereiner J (1996) Occurrence of the endophytic diazotrophs Herbaspirillum spp. in roots, stems, and leaves, predominantly of Gramineae. Biol Fertil Soils 21:197–200. https://doi.org/10.1007/BF00335935
Pal KK, BMG (2006) Biological control of plant pathogens. The Plant Health Instructor. https://doi.org/10.1094/PHI-A-2006-1117-02
Pandey P, Kang SC, Maheshwari DK (2005) Isolation of endophytic plant growth promoting Burkholderia sp. MSSP from root nodules of Mimosa pudica. Curr Sci 89:1
Park KS, Kloepper JW (2000) Activation of PR-1a promoter by rhizobacteria that induce systemic resistance in tobacco against pseudomonas syringae pv. Tabaci. Biol Control 1:2–9. https://doi.org/10.1006/bcon.2000.0815
Patel S, Sayyed R, Saraf M (2016) Bacterial determinants and plant defense induction : their role as bio-control agent in agriculture plant soil microbes, Plant soil microbes. Springer, Switzerland, pp 187–204

Penrose DM, Glick BR (2001) Levels of ACC and related compounds in exudate and extracts of canola seeds treated with ACC deaminase-containing plant growth-promoting bacteria. Can J Microbiol 47(4):368–372. https://doi.org/10.1139/w01-014

Pérez-Montaño F, Alías-Villegas C, Bellogín RA, Del Cerro P, Espuny MR, Jiménez-Guerrero I, López-Baena FJ, Ollero FJ, Cubo T (2014) Plant growth promotion in cereal and leguminous agricultural important plants: from microorganism capacities to crop production. Microbiol Res 169(5–6):325–336. https://doi.org/10.1016/j.micres.2013.09.011

Pierson LS, Keppenne VD, Wood DW (1994) Phenazine antibiotic biosynthesis in pseudomonas aureofaciens 30-84 is regulated by PhzR in response to cell density. J Bacteriol 176(13):3966–3974. https://doi.org/10.1128/jb.176.13.3966-3974.1994

Raaijmakers JM, Vlami M, de Souza JT (2002) Antibiotic production by bacterial biocontrol agents, Antonie van Leeuwenhoek. Int J Gen Mol Microbiol 81(1–4):537–347. https://doi.org/10.1023/A:1020501420831

Ramette A, Moënne-Loccoz Y, Défago G (2006) Genetic diversity and biocontrol potential of fluorescent pseudomonads producing phloroglucinols and hydrogen cyanide from Swiss soils naturally suppressive or conducive to Thielaviopsis basicola-mediated black root rot of tobacco. FEMS Microbiol Ecol 55(3):369–381. https://doi.org/10.1111/j.1574-6941.2005.00052

Raut V, Shaikh I, Naphade B, Prashar K, Adhapure N (2017). Plant growth promotion using microbial IAA producers in conjunction with Azolla: A novel approach. Chem Biol Technol Agric 4:1. https://doi.org/10.1186/s40538-016-0083-3

Reddy MS, Yellareddygari SKR, Krishna Kumar VK, Sudini H, Kloepper JW, Sairam SS, Wang Q, Liu AS, Sarma YR, Reddy ECS, Vinh NC, Archana G, Naik MK, Soesanto L, XZhou XG, FernandoWGD, Inam-ul-Haq M, Park KS, Egamberdieva D, Sayyed RZ, Zhang S, Du B, Zhi-lin Y, Zhi-lin Y (2019) Commercial potential of biofertilizers and bio-fungicides (PGPR) for sustainableagriculture in Asia and the scope of Asian PGPR Society, Proceedings of the 2nd Asian PGPR Conference, China Academic Press, Beijing, China, 2011, pp 3–4

Reshma P, Naik MK, Aiyaz M, Niranjana SR, Chennappa G, Shaikh SS, Sayyed RZ (2018) Induced systemic resistance by 2,4diacetylphloroglucinol positive fluorescent pseudomonas strains against rice sheath blight. Ind J Exp Biol 56(3):207–212

Ruiu L (2020) Plant-growth-promoting bacteria (PGPB) against insects and other agricultural pests. Agronomy 10(6):861. https://doi.org/10.3390/agronomy10060861

Sagar A, Sayyed RZ, Ramteke PW, Sharma S, Marraiki N, Elgorban AE, Syed A (2020) ACC deaminase and antioxidant enzymes producing halophilic *Enterobacter* sp. PR14 promotes the growth of rice and millets under salinity stress. Physiol Mol Biology of Plants 26:1847–1854

Sagar A, Rai S, Ilyas N, Sayyed RZ, Al-Turki AI, Enshasy HE, Simarmata T (2022) Halotolerant Rhizobacteria for salinity stress mitigation: diversity. Mechanism and Molecular Approaches, Sustainability 14:490. https://doi.org/10.3390/su14010490

Sairam KV, Reddy NS, Reddy MS, Ilao RI, Faylon PS, Dar WD, Sayyed R, Sudini H, Kumar KV, Armanda A (2013) Global success of prathista industries organic Agri-inputs for sustainable agriculture. In: Recent advances in biofertilizers and biofungicides sustainable agriculture proceedings of the 3rd Asian Conference on plant growth-promoting rhizobacteria other microbials, manila, Philipp, pp 21–24

Salomon MV, Funes Pinter I, Piccoli P, Bottini R (2017) Use of plant growth-promoting rhizobacteria as biocontrol agents: induced systemic resistance against biotic stress in plants. Microbial Applicat. https://doi.org/10.1007/978-3-319-52669-0_7

Sanghi R et al (2016) Advances and applications through fungal Nanobiotechnology. Appl Microbiol Biotechnol. https://doi.org/10.1007/978-3-319-42990-8

Sayyed RZ, Chincholkar SB, Reddy MS, Gangurde NS, Patel PR (2013) Siderophore producing PGPR for crop nutrition and Phytopathogen suppression. In: Maheshwari DK (ed) . Bacteria in Agrobiology: Disease Management, pp 449–471

Sayyed RZ, Seifi S, Pate PR, Shaikh SS, Jadhav HP, Enshasy HE (2019) Siderophore production in groundnut rhizosphere isolate, Achromobacter sp. RZS2 influenced by physicochemical factors and metal ions. Environmental Sustainability 1(3):295–301

Shaikh SS, Sayyed RZ (2015) Role of plant growth-promoting rhizobacteria and their formulation in biocontrol of plant diseases. In: Plant microbes Symbiosis: applied facets springer link, pp 337–351

Shaikh SS, Wani SJ, Sayyed RZ, Thakur R, Gulati A (2018) Production, purification and kinetics of chitinase of *Stenotrophomonas maltophilia* isolated from rhizospheric soil. Indian J Exp Biol 56(4):274–278

Singh GG (2015) Plant growth promoting Rhizobacteria (PGPR): current and future prospects for development of sustainable agriculture. J Microb Biochem Technol 7:96–102. https://doi.org/10.4172/1948-5948.1000188

Sharma SB, Sayyed RZ (2013) Mrugesh H Trivedi & Thivakaran Gobi, Phosphate solubilizing microbes: sustainable approach for managing phosphorus deficiency in agricultural soils, vol 2. Springer Plus, p 587

Sharma S, Sayyed R, Sonawane M, Trivedi M, Thivakaran G (2016) Neurospora sp SR8, a novel phosphate solubiliser from rhizosphere of soil of sorghum in Kachh, Gujarat. Indian J Exp Biol 54:644–649

Sharma A, Gupta A, Dalela M, Sharma S, Sayyed RZ, Enshasy HE, Elsayed EA (2020) Linking organic metabolites as produced by *Purpureocillium lilacinum* 6029 cultured on Karanja deoiled cake medium for the sustainable management of root-knot nematodes. Sustainability 12(9):8276. https://doi.org/10.3390/su12198276

Shen X, Hu H, Peng H, Wang W, Zhang X (2013) Comparative genomic analysis of four representative plant growth-promoting rhizobacteria in pseudomonas. BMC Genomics 14:271. https://doi.org/10.1186/1471-2164-14-271

Siddiqui IA, Ehetshamul-Haque S, Shahid Shaukat S (2001) Use of rhizobacteria in the control of root rot-root knot disease complex of mungbean. J Phytopathol 149(6):337–346. https://doi.org/10.1046/j.1439-0434.2001.00630.x

Siddiqui IA, Shaukat SS, Sheikh IH, Khan A (2006) Role of cyanide production by Pseudomonas fluorescens CHA0 in the suppression of root-knot nematode, Meloidogyne javanica in tomato. World J Microbiol Biotechnol 22:641–650. https://doi.org/10.1007/s11274-005-9084-2

Sivasakthi S, Usharani G, Saranraj P (2014) Biocontrol potentiality of plant growth promoting bacteria (PGPR) - *Pseudomonas fluorescens* and *Bacillus subtilis*: a review. African J Agric Res 9(16):1265–1277. https://doi.org/10.5897/AJAR2013.7914

Sokurenko EV, Hasty DL, Dykhuizen DE (1999) Pathoadaptive mutations: gene loss and variation in bacterial pathogens. Trends Microbiol 7:191–195

Sotelo M, Ureta AC, Muñoz S, Sanjuán J, Monza J, Palacios J (2021) Introduction of H2-uptake hydrogenase genes into Rhizobial strains improves symbiotic nitrogen fixation in Vicia sativa and *Lotus corniculatus* forage legumes. Front Agron 3:661534. https://doi.org/10.3389/fagro.2021.661534

Spaepen S, Vanderleyden J, Remans R (2007) Indole-3-acetic acid in microbial and microorganism-plant signaling. FEMS Microbiol Rev 31(4):425–448. https://doi.org/10.1111/j.1574-6976.2007.00072.x

Stout MJ, Zehnder GW, Baur ME (2002) Potential for the use of elicitors of plant resistance in arthropod management programs. Arch Insect Biochem Physiol 51(4):222–235. https://doi.org/10.1002/arch.10066

Sukmawati D, Family N, Hidayat I, Sayyed RZ, Elsayed EA, Dailin DJ, Hanapi SZ, Wadaan MA, Enshasy HE (2021) Biocontrol Activity of *Aureubasidium pullulans* and Candida orthopsilosis isolated from *Tectona grandis* L Phylloplane against *Aspergillus* sp. In: Post-Harvested Citrus Fruit Sustainability, vol 13, p 7479. https://doi.org/10.3390/su13137479

Tsavkelova EA, Klimova SY, Cherdyntseva TA, Netrusov AI (2006) Microbial producers of plant growth stimulators and their practical use: A review. Prikl Biokhim Mikrobiol 42(2):133–143

Uchiumi T, Oowada T, Itakura M, Mitsui H, Nukui N, Dawadi P, Kaneko T, Tabata S, Yokoyama T, Tejima T, Saeki K, Oomori H, Hayashi M, Maekawa T, Sriprang R, Murooka Y, Tajima S, Simomura K,Nomura M, Suzuki A, Shimoda S, Sioya K, Abe M, Minamisawa K (2004)

Expression islands clustered on symbiosis island of Mesorhizobium loti genome. J Bacteriol 186:2439–2448

Vahala J, Ruonala R, Keinänen M, Tuominen H, Kangasjärvi J (2003) Ethylene insensitivity modulates ozone-induced cell death in birch. Plant Physiol 132(1):185–195. https://doi.org/10.1104/pp.102.018887

Vessey JK (2003) Plant growth promoting rhizobacteria as biofertilizers. Plant Soil 255:571–586

Vinay JU, Naik MK, Rangeshwaran R, Chennappa G, Sohel Shaikh SS, Sayyed RZ (2016a) Detection of antimicrobial traits in fluorescent pseudomonads and molecular characterization of an antibiotic pyoluteorin, 3. Biotech 6(227):1–11

Vinay JU, Naik MK, Rangeshwaran R, Chennappa G, Shaikh SS, Sayyed RZ (2016b) Detection of antimicrobial traits in fluorescent pseudomonads and molecular characterization of an antibiotic pyoluteorin, *3*. Biotech 6(227):1–11

Wang C, Knill E, Glick BR, Défago G (2000) Effect of transferring 1-aminocyclopropane-1-carboxylic acid (ACC) deaminase genes into Pseudomonas fluorescens strain CHA0 and its gac a derivative CHA96 on their growth-promoting and disease-suppressive capacities. Can J Microbiol 46(10):898–907. https://doi.org/10.1139/w00-071

Wang X, Li Q, Sui J, Zhang J, Liu Z, Du J, Xu R, Zhou Y, Liu X (2019) Isolation and characterization of antagonistic bacteria paenibacillus jamilae HS-26 and their effects on plant growth. Biomed Res Int 1:1–13. https://doi.org/10.1155/2019/3638926

Wani SJ, Shaikh SS, Sayyed RZ (2016) Statistical-based optimization and scale-up of siderophore production process on laboratory bioreactor, 3. Biotech 6:69

Zachow C, Müller H, Monk J, Berg G (2017) Complete genome sequence of pseudomonas brassicacearum strain L13-6-12, a biological control agent from the rhizosphere of potato. Stand Genomic Sci 12:6. https://doi.org/10.1186/s40793-016-0215-1

Zakaria AK, Sayyed RZ, Enshasy HE, Biosynthesis of Antibiotics by PGPR and Their Roles in Biocontrol of Plant Diseases (2019) Plant growth promoting Rhizobacteria for sustainable stress management Vol II Rhizobacteria in biotic stress management. Springer, Singapore, pp 1–36

Zaman HB, Farooq M, Fatima S, Sayyed RZ, Baba ZA, Sheikh TA, Reddy MS, Enshasy HA, Gafur A, Suriani NL (2021) Bacterial plant biostimulants: a sustainable way towards improving growth, productivity, and health of crops. Sustainability 21(13):2856. https://doi.org/10.3390/su13052856

Zope VP, Enshasy EI, Sayyed RZ (2019) Plant growth promoting Rhizobacteria: an overview in agricultural perspectives. In: Sayyed RZ (ed) Plant growth promoting Rhizobacteria for sustainable stress management Vol II Rhizobacteria in biotic stress management. Springer, Singapore, pp 345–362

# Chapter 7
# Antifungal Compounds of Plant Growth-Promoting *Bacillus* Species

**Sunanda Chakraborty, Tofazzal Islam, and Sunita Mahapatra**

## Contents

7.1 Introduction ........ 135
7.2 Antagonistic Nature of *Bacillus* ........ 137
7.3 Antifungal Compounds Produced by *Bacillus* Spp. ........ 138
7.4 Biofilm Formation Induced by Metabolites of *Bacillus* ........ 143
7.5 Induced Systemic Resistance in Plants ........ 144
7.6 Use of Antimicrobial Metabolites Produced by *Bacillus* in Agriculture ........ 145
7.7 Conclusions and Future Outlook ........ 146
References ........ 148

## 7.1 Introduction

In this era of ever-increasing population and the pressing need for higher yields, pathogenic fungi pose as a major biotic stress and are vital factors of reducing global crop yield (Savary et al. 2012). Outbreaks caused by fungal diseases have increased in frequency and are a recurrent threat to global food security (Islam et al. 2016). The fungal phytopathogens are highly diverse. Their mode of nutrition ranges from biotrophic to necrotrophic. They are able to infect various plant parts including leaves, buds, fruits stem, roots, etc. of the plants. They are able to survive in extreme conditions by the formation of asexual and resting spores. The economic impact, however, is largely dependent on the affected plant part, as well as on the cultivar.

S. Chakraborty · S. Mahapatra (✉)
Department of Plant Pathology, Bidhan Chandra Krishi Viswavidyala, Mohanpur, Nadia, India

T. Islam
Institute of Biotechnology and Genetic Engineering (IBGE), Bangabandhu Sheikh Mujibur Rahman Agricultural University, Gazipur, Bangladesh



R. Z. Sayyed et al. (eds.), *Antifungal Metabolites of Rhizobacteria for Sustainable Agriculture*, Fungal Biology, https://doi.org/10.1007/978-3-031-04805-0_7

Traditionally, farmers mostly rely on the use of chemical fungicides for treating infected crops, which are neither environmentally viable nor an economically feasible option. Therefore, researchers have shifted towards the use of different biocontrol agents as an alternative to the conventional management practices. Among them, the genus *Bacillus* shows high promise as a natural alternative to the synthetic fungicides for the management of fungal diseases in crop plants (Islam et al. 2019; Jamali et al. 2018).

*Bacillus* is a genus of Gram-positive and endospore-forming bacteria (Rai et al. 2015). The members of this genus are cosmopolitan in nature. They are either obligate aerobes or facultative anaerobes, which ensures their survival in diverse and stressed environment (Sharma et al. 2015; Yadav et al. 2016; Kushwaha et al. 2020a). Being an inherent member of the crop rhizospheric and endophytic microbiota, they serve the pivotal roles in maintaining soil fertility and plant growth promotion (Goswami et al. 2019; Shafi et al. 2017; Kushwaha et al. 2021; Blake et al. 2021). Characteristic traits such as thick cell wall, peptide and antibiotic secretion, formation of endospores, and production of antimicrobial (Jadhav et al. 2017) play an important role in their survival under adverse situations.

The interest in *Bacillus* as a potent biocontrol agent has increased manifold, owing to the shift towards sustainable agriculture (Jamali et al. 2018). On one hand, it produces a number of toxic antimicrobial metabolites such as volatiles, lipopeptides, smaller peptides, enzymes, etc. that are capable of restricting growth and reproduction of phytopathogenic fungi. On the other hand, *Bacillus* spp. are able to survive in the extreme environments by the production of endopspores that increase their feasibility as biocontrol agents (Martínez-Álvarez et al. 2016; Fan et al. 2017; Islam et al. 2019). Furthermore, *Bacillus* increases growth and development of the plants through phosphate solubilization, fixation of atmospheric nitrogen, higher nutrient uptake, and enhanced phytohormone production and supplements plant growth (Mahapatra et al. 2022a). It has also been reported to cause a shift in the rhizospheric microbiota and expression of host genes that helped in suppression of diseases (You et al. 2016).

To utilize *Bacillus* spp. as biocontrol agents at full potential, the knowledge of antimicrobial metabolites produced by *Bacillus*, along with their mode of action, is important. This review aimed to develop a better understanding of the antifungal metabolites produced by *Bacillus* spp. against various fungal plant pathogens. This report describes the various classes of antimicrobial metabolites and their mechanism of action employed by the various strains of *Bacillus* species to manage the phytopathogens. Finally, it summarizes the field application of antifungal metabolites produced by *Bacillus* spp., its prospects and concerns, to bridge the knowledge gap and help develop *Bacillus* species as a reliable and effective weapon against the management of the ever-evolving fungal pathogens.

## 7.2 Antagonistic Nature of *Bacillus*

The introduction of *Bacillus* spp. into the host-pathogen system leads to an array of interactions which affect the phytopathogens in different manners. The interactions can either be in a simple direct manner such as competition, antibiosis, etc., or in an indirect manner, such as strengthening the host defense, or shifting the microbiota towards a more unfavorable environment for the pathogen. *Bacillus* strains exhibit a number of traits such as biofilm formation, quorum sensing, secretion of enzymes, and production of diverse antimicrobial substances including lipopeptides (Mondol et al. 2013; Kushwaha et al. 2020b), which work in a synergistic manner in managing the pathogens in the host-pathogen-environment system. Apart from checking the pathogen population, *Bacillus* is also instrumental in all-round growth of the plant, leading to higher crop yield and quality (Shafi et al. 2017; Rahman et al. 2018).

Nutrients are considered to be a limiting factor in the development of microorganisms in the rhizosphere as well as plant tissues. Among those nutrients, iron has a significant role in determining the inoculum load of the pathogen. In such situations, antagonistic microbes, such as *Bacillus*, produce siderophores, which attract rhizospheric iron and facilitate its colonization, while reducing iron availability to the phytopathogens at the same time (Kesaulya et al. 2018). The bacteria produce different types of siderophores, phenol-catecholates, hydroxamates, and carboxylates, which help in iron sequestration, its mineralization, and transport into the plants (Kesaulya et al. 2018). Zhang and co-workers (2011) observed *Bacillus* to rapidly colonize the rhizosphere of banana plants under hydroponic, sand, and soil conditions. They also observed a reduction in *Fusarium* wilt of banana in the plants studied, indicating the antagonistic nature of the *Bacillus* spp. against *Fusarium* spp. Similar results were obtained by Kesaulya and co-workers (2018), when working with the same host-antagonist system. They also recorded the different *Bacillus* strains to produce catechols and salicyclate as siderophores.

*Bacillus* spp. trigger the production of a number of oxidative enzymes like peroxidase (PO) and polyphenol oxidase (PPO) that bring significant structural alterations in the host cell wall, thereby strengthening it against the invading pathogen. The production of oxidative enzymes initiates a cascade of enzymatic reactions leading to the production of quinines and hydrogen peroxide and toxic free radicals which interfere with the growth and development of the pathogen. In many instances, these oxidative enzymes have been found to be highly effective against fugal pathogens (Chakraborty et al. 2021). According to a study by Rais et al. (2017), strains KFP-5, KFP-7, and KFP-17 of *Bacillus* spp. were observed to induce antioxidative enzymes in rice against *Pyricularia oryzae* infection. The enzymes superoxide dismutase, peroxidase, polyphenol oxidase and phenylalanine ammonia-lyase were assessed to increase manifold in rice roots and leaves under soil and hydroponic conditions, respectively.

Many members of the *Bacillus* species have been known to produce chitinolytic enzymes, which are highly effective on the fungal cell wall, as observed by Podile and Prakash (1996). Wang and his co-workers (2013) observed that the chitinases produced by *B. amyloliquefaciens* V656 were effectively able to restrict the growth of *F. oxysporum*. *B. thuringiensis* was known to inhibit the growth of *Sclerotium rolfsii* in soybean plant by a similar mechanism. The chitinase purified from *B. subtilis* TV-125A reportedly inhibited *Fusarium culmoram* (root rot of vegetables) (Senol et al. 2014). The success of chitinolytic enzymes produced by *Bacillus* spp. was further reflected in the works of many researchers against *Rhizoctonia solani* (Solanki et al. 2012a), *Glomerella cingulata* (Islam et al. 2012), *Aspergillus flavus* (Akocak et al. 2015), *Phoma medicaginis* (Slimene et al. 2015), *Alternaria alternata*, and *Fusarium oxysporum* (Brzezinska et al. 2020).

Finally, *Bacillus* spp. have been observed to produce a vast array of antimicrobial compounds such as bacillomycin and zwittermicin A, which provides them with a unique significance in the field of biocontrol of pathogens (Kaspar et al. 2019). Exhaustive research on *Bacillus* has reported that 4–5% of the *Bacillus* genome is devoted to the production of antibiotics and has the capability to produce dozens of structurally diverse antimicrobial secondary metabolites (Mondol et al. 2013). Among antimicrobial metabolites, lipopeptides (LPs) belonging to the fengycin, iturins, and surfactins group are most widely studied (Kushwaha et al. 2020b). The type of LP produced also determines the diversity of *Bacillus* in a particular ecological niche. Fengycin have been obtained from *B. subtilis*, *B. amyloliquefaciens*, *B. cereus*, and *B. thuringiensis*, while surfactins and its variants have been obtained from *B. coagulans*, *B. licheniformis*, and *B. pumilus*. LPs are considered to play a pivotal role in triggering ISR in plants (Rahman et al. 2015). The induction of ISR leads to physical, biochemical, and cytological modifications, such as lignifications, phytoalexin production, and synthesis of defensive enzymes like peroxidases, catalases, and β-1,3-glucanases.

## 7.3 Antifungal Compounds Produced by *Bacillus* Spp.

The potential of different strains of *B. subtilis* group in producing secondary antimicrobial metabolites is well established (Fig. 7.1). These molecules are mostly antimicrobial peptides (AMPs). Their structures are diverse, ranging from cyclic to hydrophobic and containing unusual moieties, viz., D-amino acids (AA) or intramolecular thioether bonds. In addition to AMPs, volatile metabolites are also known to constitute a major share in the family of antimicrobials. Due to the extensively diverse nature of these molecules, their classifications are rather complex and can be based on numerous criteria such as their sources, properties, biosynthetic machinery, biological functions, covalent bonding pattern, three-dimensional structure, or molecular targets (Wang et al. 2015). The antifungal metabolites produced by *Bacillus* spp. against fungi are discussed briefly in the following sub-sections.

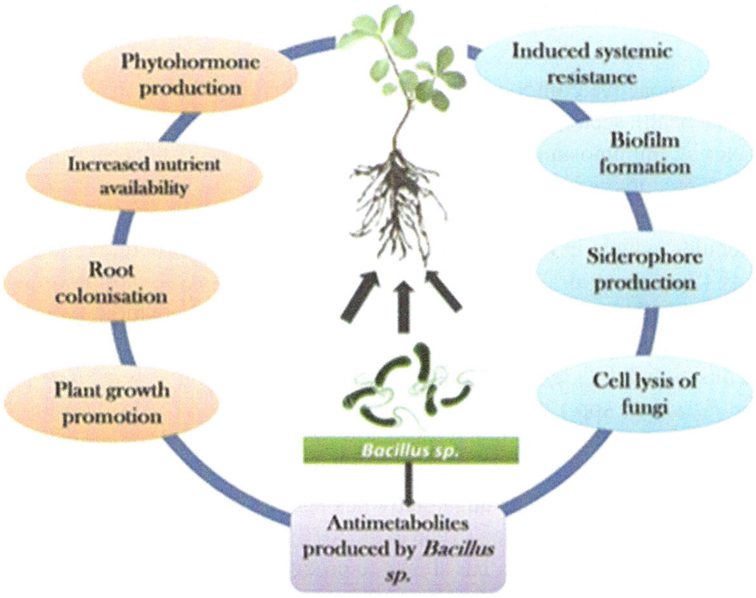


**Fig. 7.1** Direct (right) and indirect (left) impacts of *Bacillus* antimicrobial metabolites on the phytopathogens

### 7.3.1 Ribosomal Peptides (RPs)

Ribosome synthesized peptides (RPs) are produced by *Bacillus* during active cell multiplication. These are usually translated to mature compounds from short precursors (ca. 100 AA) through post-translational modifications (Stoica et al. 2019). These modifications are mediated by a number of enzymes which in turn generate a diverse range of chemical structures. Most of these peptides were originally referred to as "bacteriocins or bacteriocin-like inhibitory substances (BLIS)," which exhibit inhibiting activity against closely related bacteria. Besides bacteriocins, certain *Bacillus* also produces lytic enzymes of biocontrol potential, such as cellulases, proteases, glucanases, and chitinases, which are commonly referred to as cell wall degrading enzymes (CWDE) (Caulier et al. 2018; Sharma et al. 2019; Singh et al. 2013). They are predominantly active against fungi since the fungal cell wall is constituted of chitin and glucan, along with various glycoproteins embedded in it.

### 7.3.2 Polyketides (PKs)

Among the bioactive compounds produced by *Bacillus*, PKs are well known from the human health sector for their broad-spectrum activity encompassing antifungal, antibacterial, immunosuppressive, antitumor, and many more antagonistic abilities. Acyl CoA precursors such as malonate act as the major compounds for synthesis of PKs. They are biosynthesized by the organization of polyfunctional megasynthases into modules, which, in turn, provide them with significant structural and chemical diversity (Zakaria et al. 2019). *Bacillus* mainly produces three antimicrobial PKs: difficidin, bacillaene, and macrolactin. These compounds exhibit antimicrobial activities by inhibiting protein synthesis in a selective manner.

Bacillaene exhibits antifungal activity against various fungi (e.g., *Trichoderma* spp. or *Fusarium* spp.) (Müller et al. 2014); difficidin and oxydifficidin mainly inhibit bacterial pathogens such as *Erwinia amylovora* and *Xanthomonas oryzae* (Aleti et al. 2015). Macrolactins, and their succinyl and malonyl-derivatives, have been observed to exhibit antifungal activities against *Fusarium oxysporum* f. spp. *cubense*, causing wilt of banana (Yuan et al. 2012). Another member of the *Bacillus* group, *B. laterosporus* isolate, has shown to produce a number of novel polyketides such as basiliskamide A and basiliskamide B, which are highly effective against both *C. albicans* and *A. fumigatus* (Mondol et al. 2013).

### 7.3.3 Non-ribosomal Peptides (NRPs)

Non-ribosomal peptides form a diverse family of secondary metabolites with increasing interest in many various fields as antibiotics, immune-suppressors, surfactants, pigments, or antitumor molecules (Zhou et al. 2018). NRPs facilitate pore formation in fungal membranes and inhibition of certain enzymes, which ultimately leads to cell lysis. These include different types of lipopeptides such as iturins, kurstakins, surfactins, and fengycins (Jacques 2011; Sharma et al. 2019; Jamali et al. 2018). A variety of *Bacillus* strains have been reported to produce NRPs against fungal pathogens, which are elaborated in Table 7.1. Iturins and fengycins are mostly studied for their strong antagonistic activity against several plant pathogenic fungi (Fig. 7.2), while surfactins mainly exhibit antibacterial and antiviral activities (Ongena and Jacques 2008).

Several authors confirmed the presence of iturins, bacillomycin, fengycins, and surfactin in antagonistic *Bacillus* strains by performing PCR assays (Kushwaha et al. 2020b; Solanki et al. 2015; Sharma et al. 2019; Singh et al. 2014). Certain surfactins, however, are able to manage numerous plant and human fungal pathogens such as *Botrytis cinerea*, *Candida albicans*, *F. oxysporum*, or *Rhizoctonia solani* (Romero 2013).

*Bacillus* spp. produce thiotemplate NRPs such as bacillibactin and itoic acid, which act as siderophores and chelate iron in the plant rhizosphere effectively.

**Table 7.1** Different lipopeptides produced by *Bacillus* inhibitory against fungal phytopathogens

| Bacillus species | Lipopeptides | Pathogen | Disease | Reference |
|---|---|---|---|---|
| *Bacillus subtilis* | Surfactin/ fengycin | *Ralastonia solanacearum* | Wilt | Fan et al. (2017) |
| | Mycosubtilin | *Fusarium oxysporum*, *Pythium aphanidermatum* | *Fusarium* wilt, damping off | Mihalache et al. (2018) |
| | Iturin/surfactin | *Botrytis cinerea*, *Bremia lactucae* | Gray mold of tomato, downy mildew of lettuce | Hinarejos et al. (2016) |
| | Fengycin | *Rhizoctonia solani* | Cotton damping-off disease | Fan et al. (2017) |
| | Iturin A/ fengycin | *Penicillium digitatum* | Postharvest diseases of mandarin | Waewthongrak et al. (2015) |
| | Bacillomycin ls | *Fusarium oxysporum* | *Fusarium* wilt | Luo et al. (2015) |
| *B. Amyloliquefaciens* | Iturin/fengycin | *Bipolaris sorokiniana*, *fusarium oxysporum f. lycopersici* | Spot blotch of wheat; *fusarium* crown and root rot of tomato | Gong et al. (2015) |
| | Iturin/surfactin | *Alternaria alternata*, *aspergillus flavus*, *Colletotrichum acutatum* | Pre- and postharvest diseases in fruits and vegetables | Dimkić et al. (2015) |
| | Bacillomycin D | *Fusarium oxysporum f.* sp. *spinaciae* | Fusarium wilt on spinach | Zhao et al. (2014) |
| | Bacillomycin L/D | *Rhizoctonia solani* | Damping off of soybean | Chowdhury et al. (2015) |
| *B. Thuringiensis* | Kurstakin | *Colletotrichum gloeosporioides* | Anthracnose of fruits | Mnif and Ghribi (2015) |
| *B. subtilis* | Gageopeptides A–D and gageotetrin B | *Magnaporthe oryzae Triticum* | Wheat blast | Chakraborty et al. (2020) |

These siderophores limit the bioavailability of iron and antagonize the growth of other surrounding microbes, for instance, *F. oxysporum* f. spp. *capsici* (Yu et al. 2011). *Bacillus* group are also able to synthesize other antimicrobial NRPs through non-thiotemplate mechanism. The integration of NRPs into the target microbes facilitates the peptidase-mediated cleavage of the microbes. This, in turn, produces fungitoxic L-APPA moiety which interferes with threonine metabolism in the phytopathogenic fungi (Gao et al. 2018).

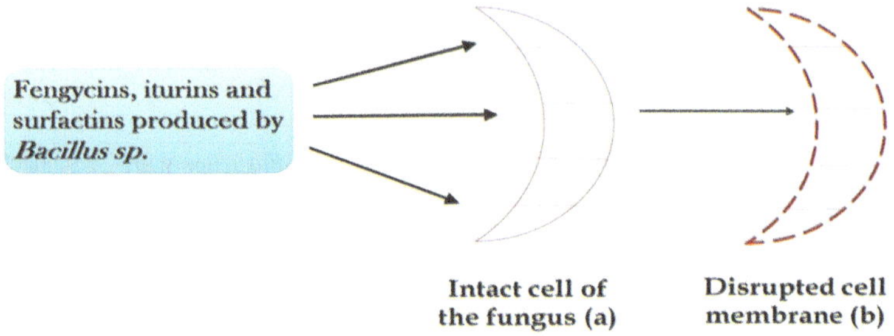


**Fig. 7.2** Figurative depiction of action of LPs on fungal cell

The non-ribosomal peptide synthetases (NRPS) have the ability to synthesize a vast array of peptides by linking unusual amino acids such as d-amino acids, *N*-methylated or hydroxy- amino acids. The multimodular nature of these enzymes facilitates the synthesis of a range of products (Roongsawang et al. 2011). These modules consist of adenylation (A), condensation (C), thiolation (T), and thioesterase (Te) domain, where the specific amino acids are incorporated (Gao et al. 2018).

Based on these features of NRPS, researchers are trying to develop new LPs by engineering of different amino acids into the NRPS. Symmank et al. (2002) genetically engineered the biosynthesis of surfactin to develop a new lipohexapeptide. They deleted a complete amino acid module that created a modified peptide synthetase. The remaining modules were recombined at different highly conserved motifs to develop the new compound. Gao et al. (2018) reported that the deletion of the entire NRPS module, responsible for plipastatin production, inactivated the enzyme, while deletion of individual domains created three new plipastatin derivatives.

### *7.3.4 Volatiles*

Besides RPs, NRPs, and PKs, strains from the *Bacillus* genus are able to produce numerous volatile compounds that play crucial roles in managing surrounding microbiota, especially in the soil. Volatiles are instrumental in the biogeochemical cycles of essential elements, in the bioconversion of the food chain, in many metabolic and physiological reactions (e.g., nitrogen mineralization, electron acceptor, or donor reactions) as well as in communication signals triggering quorum sensing and quorum quenching mechanisms (Effmert et al. 2012). In addition to antifungal activity, the volatile compounds produced by biocontrol strains are able to stimulate plant growth and activate plant defenses against plant pathogens (Gao et al. 2017).

Volatile compounds can be either inorganic (VICs) or organic (VOCs) in nature. Currently, exhaustive research is being conducted on VOCs and their role in management of fungal diseases. Owing to their high penetrability and easy dispersion through soil, they are effective as biofumigants against the target pathogenic microbes (Pourakbar et al. 2021). Since the VOCs are environmentally and ecologically safe to use, they can serve as an effective replacement to conventional management practices in the future.

In a study conducted by Zhang et al. (2020), *Bacillus subtilis* ZD01 demonstrated a strong antifungal activity against *Alternaria solani.* Inhibition rate of *A. solani* growth by VOCs was 50.1 ± 2.1% compared with the control after 6 days, suggesting that the VOCs from ZD01 showed a strong inhibitory activity on fungal mycelia growth. Significant decrease in mycelial penetration was observed after treatment with the VOCs, while penetration ability of the pathogen was completely lost when co-cultured with the VOCs produced by *B. subtilis* ZD01. Effective inhibition of symptom development was also observed in in vivo leaf test. Some important volatile compounds reportedly produced by *Bacillus* spp. are summarized in Table 7.2.

## 7.4 Biofilm Formation Induced by Metabolites of *Bacillus*

Successful colonization in the rhizosphere is a decisive factor in maintaining the desired population of *Bacillus* spp. However, it is largely influenced by environmental factors such as changes in temperature, relative humidity, etc., which make it difficult to establish a successful colonization. Hence, *Bacillus* species have evolved quorum-sensing mediated multicellular aggregations known as biofilm formation for imparting stability, ensuring its survival under unfavorable conditions, as well as for facilitating successful colonization in the rhizosphere (Toyofuku et al. 2016). Biofilm formation in the root facilitates concentration of LPs such as iturins and

**Table 7.2** Volatile compounds produced by *Bacillus* sp. inhibitory against fungal pathogens

| Bacillus strain | Pathogen | Volatile compound | Reference |
|---|---|---|---|
| *B. Velezensis* 5YN8 | *B. cinerea* | Benzothiazole and 2,4-dimethyl-6-tert-butylphenol | Jiang et al. (2018); Li et al. (2020) |
| *B. subtilis* | *Fusarium oxysporum* f. spp. *lactucae* | 6-Methyl-2-heptanone | Liu et al. (2018) |
| *B. Megaterium* strain XTBG34 | *F. Oxysporum* | 2-Pentylfuran | Liu et al. (2018); Zou et al. (2010) |
| *Bacillus* species | *F. Oxysporum* | Benzothiazole | Raza et al. (2015) |
| *B. Amyloliquefaciens* NJN-6 | *F. Oxysporum* f. spp. *cubense* | Benzothiazoles phenol and 2,3,6-trimethylphenol | Yuan et al. (2012) |
| *B. Amyloliquefaciens* CPA-8 | *Penicillium crustosum, Monilinia fructicola* | 1,3 pentadiene, acetoin, thiophene | Gotor-Vila et al. (2017) |
| *B. Amyloliquefaciens* L3 | *F. Oxysporum* f. spp. *niveum* | 2-nonanone, 2-ethyl-1-hexanol | Wu et al. (2018) |

fengycins in plant rhizosphere, which triggers further biofilm formation and antibiosis (Chen et al. 2013).

Biofilm formation involves the attachment and growth of *Bacillus* on different surfaces by the production of extracellular polymers and matrix formation, which ultimately leads to alteration in its phenotypic characteristics in terms of growth rate of *Bacillus* and its gene transcription (Mielich-Süss and Lopez 2015; Mahapatra et al. 2020). The biofilm structure is initiated by a rapid and massive migration of related groups of bacteria, known as swarming. It has been reported that surfactin plays a pivotal role in swarming as it is associated with the mother colony of the aggregation (Mielich-Süss and Lopez 2015). Its role in biofilm formation has been well established by many workers in this field (Zeriouh et al. 2014; Aleti et al. 2016). However, the role of other LPs in biofilm formation is still being researched upon. Xu et al. (2013) observed that bacillomycin D, a member of the iturin family, was responsible for the expression of the genes associated with biofilm formation in *B. amyloliquefaciens*. The culture supernatant of *B. subtilis* and *B. mojavensis* were also reported to produce iturin and surfactin as their main component by Youcef-Ali and co-workers (2014).

It has been observed that the presence of antagonistic fungi triggers the production of antifungal metabolites in *Bacillus* strains such as *B. subtilis*, *B. velezensis*, and *B. amyloliquefaciens*. The production of these antifungal metabolites was not only influenced by the bacterial species producing it, but also by the interacting fungal species (Rayanoothala et al. 2021). For instance, *B. subtilis* 98S produced iturins and fengycins in considerably higher concentration when interacting with *Fusarium oxysporum* and *Pythium aphanidermatum*, as compared to *Botrytis cinerea* (Cawoy et al. 2015). Many instances have proved that the *Bacillus* species are able to sense the fungal antagonists around it, and secrete the antimicrobial metabolites accordingly. Chowdhury et al. (2015) reported that *B. velezensis* FZB42 overproduced bacillomycin and reduced its fengycin production in response to *Rhizoctonia solani* in lettuce rhizosphere. Contrastingly, *B. velezensis* strains FZB42 and S499 increased fengycin production when interacting with *Rhizomucor variabilis* (Zihalirwa Kulimushi et al. 2017).

Yaseen et al. (2018) observed that the null mutation in the gene *srfAA* lead to less efficient biofilm formation of *B. subtilis.* Since, gene *srfAA* encodes for surfactin production, it emphasized the role of surfactins in biofilm formation. Further, surfactins were observed to be profusely secreted by *B. velezensis* SQR9 in response to *Phytophthora parasitica* (Li et al. 2014) as well as by *B. subtilis* B9-5against liquid culture of *Rhizopus stolonifer* (DeFilippi et al. 2018).

## 7.5 Induced Systemic Resistance in Plants

A large body of literature is available on the prospect of induced systemic resistance (ISR) as a potent biocontrol mechanism of *Bacillus* spp. (García-Gutiérrez et al. 2013; Rahman et al. 2015; Akram et al. 2016) The ISR is a signalling mechanism that equips the plant with durable and broad-spectrum resistance from future pathogen infections (Reshma e al., 2018; Patel et al., 2016). The bacterial compounds such as flagellin, LPs, and secondary metabolites such as volatile compounds, iturins, surfactins, fengycins (Vinay et al. 2016; Ongena and Jacques 2008), and siderophores (Patel et al. 2016; Nithyapriya et al. 2021) are involved in triggering ISR. The ISR mechanism can be broadly described as three-step process: first, the plant perceives the elicitors; second, signal transduction takes place using jasmonic acid and ethylene signalling pathways; and finally, morphological modifications in the form of lignification, callose deposition, accumulation of phenols, etc. take place, which inhibits future entry of the pathogen into the host cells (Mariutto and Ongena 2015).

As mentioned earlier, *Bacillus* species has been known to produce a plethora of antimicrobial metabolites which help in the management of the pathogen. Chowdhury and co-workers (2015) observed that the LPs produced by *B. amyloliquefaciens* activated the expression of plant defensive genes in lettuce. The bacteria produced surfactin, which triggered the expression of PDF 1.2 gene, encoding defensin (a host defense peptide). The role of surfactin was established when the gene expression was missing in surfactin-deficient mutants. Farace and collaborators (2015) reported mycosubtillin (an iturin) as an activator of JA and SA signalling pathways, while surfactin triggered SA-regulated pathway. Plipastatin was shown to trigger the early signalling pathways only, in treated grapevine plants against *Botrytis cinerea.* Surfactins and iturins were able to trigger the expression of chitinases, peroxidases, and β-1,3-glucanases in strawberries, which helped in successful inhibition of *Colletotrichum gloeosporioides* (Yamamoto et al. 2015). Rodriguez et al. (2018) observed that surfactin, produced by *Bacillus* spp., enhanced the activity of peroxidase and phenolic compounds in groundnut when exposed to *Sclerotium rolfsii.* Similar results have been observed in a number of crops including rice (Chandler et al. 2015), maize (Gond et al. 2015), and tomato (Abdallah et al. 2017).

The induction of defense responses provides a durable protection of the plant against the invading pathogen. According to Rahman et al. (2015), the signalling is mostly JA dependent, with some lipopeptides triggering SA-dependent signalling pathway as well (Surovy et al. 2019). Taken together, the lipopeptides trigger a defense state which is characterized by acquisition of plant memory, enhanced plant performance, and a robust response to pathogen attack.

## 7.6 Use of Antimicrobial Metabolites Produced by *Bacillus* in Agriculture

There has been growing research over the use of antimicrobial metabolites as an alternative to existing practices of pathogen management. Lipopeptides such as surfactins and iturins are highly effective against a wide range of pathogen including bacteria, oomycetes, fungi, and viruses (Raaijmakers et al. 2010; Kushwaha et al.

2020b; Singh et al. 2014; Solanki et al. 2015; Chakraborty et al. 2020; Hamid et al. 2021). The use of LPs as effective biocontrol compounds is not limited to cereals such as wheat (Mahapatra et al. 2022b), rice, and maize but has also been reported in fruit and vegetables of economic importance. LPs find their use as direct inhibitors of pathogen growth as well as indirect influence to the pathogen population by triggering ISR, plant growth, and facilitating the availability of nutrients such as iron and phosphate to the host plant.

With the discovery of increasing number of *Bacillus* antimicrobial metabolites, there has been a surge in the research works worldwide. Dorighello et al. (2020) observed that *Phakopsora pachyrhizi*, causing Asian soybean rust, is a major threat to soybean production. However, he reported that the use of *B. subtilis* QST-713 (producer of lipopeptides), along with *Bacillus pumilus*, successfully inhibited rust development in soybean, upon exposure to the pathogen. In particular, *B. subtilis* was able to reduce disease severity by 99% in tests with detached leaves and by 23% under field conditions.

In wheat, *Bacillus* LP efficacy was observed against *Zymoseptoria tritici* (Mejri et al. 2018), *Gaeumannomyces graminis* var. *tritici* (Zhang et al. 2017; Yang et al. 2018), and *Fusarium graminearum* (Gong et al. 2015). The latter was also inhibited by *Bacillus* LPs in maize. In potatoes, *Bacillus* biosurfactants were effective against *Fusarium solani* (Mnif and Ghribi 2015). Li et al. (2016) reported that *B. amyloliquefaciens* SYBC H47 produced bacillomycin L, fengycin, and surfactin, which were able to inhibit conidia germination and mycelial growth in B*otryosphaeria dothidea*. 57.5% reduction in disease severity was observed in field trials when peach was exposed to *B. dothidea*, after being inoculated with the biocontrol agent. Some of the other recent works on various *Bacillus* strains against phytopathogenic fungi are summarized in Table 7.3.

## 7.7 Conclusions and Future Outlook

Most of the works done so far have focused on different *Bacillus* species, especially *B. velezensis*, *B. subtilis*, and *B. pumilus*, as they are highly potent as biofungicides. Mostly, *Bacillus* endospores are applied as dried formulations or liquid suspensions, but the CLPs (surfactins, iturins, and fengycins) secreted are unable to achieve the quantity adequate for effective antibiosis (Rayanoothala et al. 2021). A possible way out is the use of spores along with culture supernatants which would ensure the availability of antimicrobial metabolites. However, only a few bioformulations available commercially such as Serenade (R) (from *B. velezensis* QST713) and Double Nickel 55 (from *B. velezensis* D747) contain both spores and bioactive compounds. More research needs to be done to develop stable and effective bioformulations. Further, the shelf life, the concentration of bioactive metabolites, and the required concentration of active ingredient for application need to be mentioned on the label of the bioformulations, to ensure a wider on-field acceptability.

**Table 7.3** Use of *Bacillus* spp. as biocontrol agents against fungal diseases of plants

| Bacillus strain | Pathogen | Disease caused | Host | References |
|---|---|---|---|---|
| *Bacillus velezensis* CC09 | *Blumeria graminis* | Powdery mildew | Wheat | Cai et al. (2017) |
| *B. Velezensis* strains 5yn8 | *Botrytis cinerea* | Gray mold | Pepper | Jiang et al. (2018) |
| *B. atrophaeus* cab-1 | *Sphaerotheca fuliginea* | Powdery mildew | Cucumber | Zhang et al. (2013) |
| *B. amyloliquefaciens* Q-426 | *Fusarium oxysporum* f. spp. *spinaciae* | Wilt | Spinach | Zhao et al. (2013) |
| *B. subtilis Y-1* | *Fusarium* spp. | Rot | Apple | Ju et al. (2014) |
| *B. pumilus* | *Fusarium oxysporum* f. spp. *lycopersici* | Wilt | Tomato | Heidarzadeh and Baghaee-Ravari (2015) |
| *Bacillus amyloliquefaciens* MB101 and *Bacillus subtilis* MB14 | *Rhizoctonia solani* | Damping-off | Tomato | Solanki et al. (2012b); Solanki et al. (2015) |
| *B. amyloliquefaciens* LJ02 | *Sphaerotheca fuliginea* | Powdery mildew | Cucurbits | Li et al. (2015) |
| *B. amyloliquefaciens* strain BUZ-14 | *Botrytis cinerea, Moniliniafructicola, M. laxa, Penicillium digitatum,* | Postharvest diseases | Orange, apple, grape, and stone fruit | Calvo et al. (2017) |
| *B. Atrophaeus* | *Colletotrichum gloeosporioides* | Anthracnose | Avocado | Guardado-Valdivia et al. (2018) |
| *Bacillus* spp. B25 | *Fusarium verticillioides* | Stalk rot and ear rot | Maize | Douriet-Gámez et al. (2018) |
| *B. vallismortis, B. velezensis* BM21 | *Fusarium graminearum* | Corn stalk rot | Maize | Li et al. (2019); Wang et al. (2020) |
| *B. Megaterium* strain AB4 | *Alternaria japonica* | Leaf spot | Broccoli | Acurio Vásconez et al. (2020) |
| *B. Amyloliquefaciens* A3, *B. velezensis* A2 | *Dickeya dadantii* | Stem and root rot | Sweet potato | Hossain et al. (2020) |
| *Bacillus subtilis IBFCBF-4* | *Fusarium oxysporum f.* spp. *niveum* | Fusarium wilt | Watermelon | Zhu et al. (2020) |
| *B. subtilis* (B-CM191, B-CV235, and B-CL-122) | *Fusarium oxysporum* f. spp. *ciceri* race 1 | Wilt | Chickpea | Singh et al. (2014) |

Companies such as Dupont and Bayer have shown huge interest in developing cyclic LPs as effective biocontrol agents. Scientists are also trying to develop genetically engineered *Bacillus* spp. that would produce the desired antifungal metabolites in increased quantities. The patent WO2017125583-A1 claimed that genetic modification of *Bacillus thuringiensis* leads to an overexpression of *krsE* gene, which in turn resulted in increased kurstakin production. The metabolite showed inhibitory action against *B. cinerea* and *Galactomyces geotrichum* as well.

The antimicrobial metabolites produced by different strains of *Bacillus* spp. have remarkable potentials to be utilized in management of fungal phytopathogens. Huge literature has been found depicting the roles of antimicrobial metabolites, especially those belonging to iturin, fengycin, and surfactin families of LPs. These LPs help directly by limiting pathogen growth, killing the infecting pathogen, or indirectly by activating the plant defense systems. These properties render them as an excellent option for biocontrol, which is further necessitated by the growing fungicide resistance and awareness towards sustainable agricultural practices.

Hence, the futuristic approach recommends encouraging research on bioprospecting of *Bacillus* spp. and their isolation from plant rhizosphere. Further genomics and post genomics studies generate essential knowledge for a better understanding of the antimicrobial metabolites and their application in agricultural perspective to secure better productivity and efficient management of the evolving fungal pathogens (Surovy et al. 2019).

## References

Abdallah RAB, Stedel C, Garagounis C, Nefzi A, Jabnoun-Khiareddine H, Papadopoulou KK, Daami-Remadi M (2017) Involvement of lipopeptide antibiotics and chitinase genes and induction of host defense in suppression of fusarium wilt by endophytic *bacillus* spp. in tomato. Crop Prot 99:45–58

Acurio Vásconez RD, Moya EMT, Yépez LAC, Chiluisa-Utreras VP, VacaSuquillo ID (2020) Evaluation of *bacillus megaterium* strain AB4 as a potential biocontrol agent of *Alternaria japonica*, a mycopathogen of *Brassica oleracea* var. italica. Biotechnol Rep 26(e00454)

Akocak PB, Churey JJ, Worobo RW (2015) Antagonistic effect of chitinolytic *pseudomonas* and *bacillus* on growth of fungal hyphae and spores of aflatoxigenic *aspergillus flavus*. Food Biosci 10:48–58

Akram W, Anjum T, Ali B (2016) Phenylacetic acid is ISR determinant produced by *Bacillus fortis* IAGS162, which involves extensive re-modulation in metabolomics of tomato to protect against fusarium wilt. Front Plant Sci 7:498

Aleti G, Sessitsch A, Brader G (2015) Genome mining: prediction of lipopeptides and polyketides from *bacillus* and related Firmicutes. Comput Struct Biotechnol 13:192–203

Aleti G, Lehner S, Bacher M, Compant S, Nikolic B, Plesko M, Schuhmacher R, Sessitsch A, Brader G (2016) Surfactin variants mediate species-specific biofilm formation and root colonization in bacillus. Environ Microbial 18(8):2634–2645

Blake C, Christensen MN, Kovács ÁT (2021) Molecular aspects of plant growth promotion and protection by *Bacillus subtilis*. Mol Plant-Microbe Interact 34(1):15–25

Brzezinska MS, Kalwasińska A, Świątczak J, Żero K, Jankiewicz U (2020) Exploring the properties of chitinolytic bacillus isolates for the pathogens biological control. Microb Pathog 148:104462

Cai XC, Liu CH, Wang BT, Xue YR (2017) Genomic and metabolic traits endow *bacillus velezensis* CC09 with a potential biocontrol agent in control of wheat powdery mildew disease. Microbiol Res 196:89–94

Calvo H, Marco P, Blanco D, Oria R, Venturini ME (2017) Potential of a new strain of *bacillus amyloliquefaciens* BUZ-14 as a biocontrol agent of postharvest fruit diseases. Food Microbial 63:101–110

Caulier S, Gillis A, Colau G, Licciardi F, Liépin M, Desoignies N, Modrie P, Legrève A, Mahillon J, Bragard C (2018) Versatile antagonistic activities of soil-borne *bacillus* spp. and *Pseudomonas* spp. against *Phytophthora infestans* and other potato pathogens. Front Microbiol 9:143

Cawoy H, Debois D, Franzil L, De Pauw E, Thonart P, Ongena M (2015) Lipopeptides as main ingredients for inhibition of fungal phytopathogens by *Bacillus subtilis/amyloliquefaciens*. Microb Biotechnol 8(2):281–295

Chakraborty M, Mahmud NU, Gupta DR, Tareq FS, Shin HJ, Islam T (2020) Inhibitory effects of linear Lipopeptides from a marine *Bacillus subtilis* on the wheat blast fungus *Magnaporthe oryzae Triticum*. Front Microbiol 11:665. https://doi.org/10.3389/fmicb.2020.00665

Chakraborty S, Debnath D, Mahapatra S, Das S (2021) Role of endophytes in plant disease management. In: Singh KP, Jahagirdar S, Sarma BK (eds) Emerging trends in plant pathology. Springer, Singapore, pp 399–424. https://doi.org/10.1007/978-981-15-6275-4_19

Chandler S, Van Hese N, Coutte F, Jacques P, Monica H, De Vleesschauwer D (2015) Role of cyclic lipopeptides produced by *Bacillus subtilis* in mounting induced immunity in rice (*Oryza sativa* L.). Physiol Mol Plant Pathol 91:20–30

Chen Y, Yan F, Chai Y, Liu H, Kolter R, Losick R, Guo JH (2013) Biocontrol of tomato wilt disease by *Bacillus subtilis* isolates from natural environments depends on conserved genes mediating biofilm formation. Environ Microbiol 15(3):848–864

Chowdhury SP, Hartmann A, Gao X, Borriss R (2015) Biocontrol mechanism by root-associated *bacillus amyloliquefaciens* FZB42–a review. Front Microbiol 6:780

DeFilippi S, Groulx E, Megalla M, Mohamed R, Avis TJ (2018) Fungal competitors affect production of antimicrobial Lipopeptides in *Bacillus subtilis* strain B9–5. J Chem Ecol 44(4):374–383

Dimkić I, Berić T, Stević T, Pavlović S, Šavikin K, Fira D, Stanković S (2015) Additive and synergistic effects of *bacillus* spp. isolates and essential oils on the control of phytopathogenic and saprophytic fungi from medicinal plants and marigold seeds. Biol Control 87:6–13

Dorighello DV, Forner C, de Campos RMVB, Bettiol W (2020) Management of Asian soybean rust with *Bacillus subtilis* in sequential and alternating fungicide applications. Australas Plant Pathol 49(1):79–86

Douriet-Gámez NR, Maldonado-Mendoza IE, Ibarra-Laclette E, Blom J, Calderón-Vázquez CL (2018) Genomic analysis of *bacillus sp.* strain B25, a biocontrol agent of maize pathogen *fusarium verticillioides*. Curr Microbiol 75(3):247–255

Effmert U, Kalderas J, Warnke R, Piechulla B (2012) Volatile mediated interactions between bacteria and fungi in the soil. J Chem Ecol 38(6):665–703

Fan HY, Ru JJ, Zhang YY, Wang Q, Li Y (2017) Fengycin produced by *Bacillus subtilis* 9407 plays a major role in the biocontrol of apple ring rot disease. Microbiol Res 199:89–97

Farace G, Fernandez O, Jacquens L, Coutte F, Krier F, Jacques P, Clément C, Barka EA, Jacquard C, Dorey S (2015) Cyclic lipopeptides from *Bacillus subtilis* activate distinct patterns of defence responses in grapevine. Mol Plant Pathol 16(2):177–187

Gao Z, Zhang B, Liu H, Han J, Zhang Y (2017) Identification of endophytic *bacillus velezensis* ZSY-1 strain and antifungal activity of its volatile compounds against *Alternaria solani* and *Botrytis cinerea*. Biol Control 105:27–39

Gao L, Guo J, Fan Y, Ma Z, Lu Z, Zhang C, Zhao H, Bie X (2018) Module and individual domain deletions of NRPS to produce plipastatin derivatives in *Bacillus subtilis*. Microb Cell Factories 17(1):1–13

García-Gutiérrez L, Zeriouh H, Romero D, Cubero J, de Vicente A, Perez-Garcia A (2013) The antagonistic strain *Bacillus subtilis* UMAF6639 also confers protection to melon plants against cucurbit powdery mildew by activation of jasmonate- and salicylic acid-dependent defence responses. Microb Biotechnol 6:264–274

Gond SK, Bergen MS, Torres MS, White JF Jr (2015) Endophytic *bacillus* spp. produce antifungal lipopeptides and induce host defence gene expression in maize. Microbiol Res 172:79–87

Gong AD, Li HP, Yuan QS, Song XS, Yao W, He WJ, Zhang JB, Liao YC (2015) Antagonistic mechanism of iturin a and plipastatin a from *bacillus amyloliquefaciens* S76-3 from wheat spikes against *fusarium graminearum*. PLoS One 10(2):e0116871

Goswami SK, Kashyap PL, Awasthi S (2019) Deciphering rhizosphere microbiome for the development of novel bacterial consortium and its evaluation for salt stress management in solanaceous crops in India. Indian Phytopathol 72:479–488. https://doi.org/10.1007/s42360-019-00174-1

Gotor-Vila A, Teixidó N, Di Francesco A, Usall J, Ugolini L, Torres R, Mari M (2017) Antifungal effect of volatile organic compounds produced by *bacillus amyloliquefaciens* CPA-8 against fruit pathogen decays of cherry. Food Microbiol 64:219–225

Guardado-Valdivia L, Tovar-Pérez E, Chacón-López A, López-García U, Gutiérrez-Martínez P, Stoll A, Aguilera S (2018) Identification and characterization of a new *bacillus atrophaeus* strain B5 as biocontrol agent of postharvest anthracnose disease in soursop (*Annona muricata*) and avocado (*Persea americana*). Microbiol Res 210:26–32

Hamid B, Zaman M, Farooq S, Fatima S, Sayyed RZ, Baba ZA, Sheikh TA, Reddy MS, El Enshasy H, Gafur A, Suriani NL (2021) Bacterial plant biostimulants: a sustainable way towards improving growth, productivity, and health of crops. Sustainability 21(13):2856. https://doi.org/10.3390/su13052856

Heidarzadeh N, Baghaee-Ravari S (2015) Application of *Bacillus pumilus* as a potential biocontrol agent of fusarium wilt of tomato. Arch Phytopathol Plant Protect 48(13–16):841–849

Hinarejos E, Castellano M, Rodrigo I, Bellés JM, Conejero V, López-Gresa MP, Lisón P (2016) *Bacillus subtilis* IAB/BS03 as a potential biological control agent. Eur J Plant Pathol 146(3):597–608

Hossain A, Islam Masum MM, Wu X, Abdallah Y, Ogunyemi SO, Wang Y, Sun G, Li B, An Q (2020) Screening of bacillus strains in biocontrol of pathogen *Dickeya dadantii* causing stem and root rot disease of sweet potato. Biocontrol Sci Tech 30(11):1180–1198

Islam MR, Jeong YT, Lee YS, Song CH (2012) Isolation and identification of antifungal compounds from *Bacillus subtilis* C9 inhibiting the growth of plant pathogenic fungi. Mycobiology 40(1):59–65

Islam MT, Croll D, Gladieux P, Soanes DM, Persoons A, Bhattacharjee P, Hossain MS, Gupta DR, Rahman MM, Mahboob MG, Cook N (2016) Emergence of wheat blast in Bangladesh was caused by a south American lineage of *Magnaporthe oryzae*. BMC Biol 14(1):1–1. https://doi.org/10.1186/s12915-016-0309-7

Islam MT, Rahman MM, Pandey P, Boehme MH, Haesaert G (2019) Bacilli and Agrobiotechnology: Phytostimulation and biocontrol. Springer, Cham. https://doi.org/10.1007/978-3-030-15175-1

Jacques P (2011) Surfactin and other lipopeptides from *bacillus* spp. In: Soberón-Chávez G (ed) Biosurfactants, Microbiology monographs, vol, vol 20. Springer, Berlin, Heidelberg, pp 57–91. https://doi.org/10.1007/978-3-642-14490-5_3

Jadhav HP, Shaikh SS, Sayyed RZ (2017) Role of hydrolytic enzymes of Rhizoflora in biocontrol of fungal Phytopathogens. In: An overview in Rhizotrophs : plant growth promotion to bioremediation. Springer, pp 183–203

Jamali H, Sharma A, Kashyap PL, Roohi SAK (2018) Exploitation of multifarious abiotic stresses, antagonistic activity and plant growth promoting attributes of bacillus amyloliquefaciens AH53 for sustainable agriculture production. Int J Curr Microbiol App Sci 7(10):751–763. https://doi.org/10.20546/ijcmas.2018.710.083

Jiang CH, Liao MJ, Wang HK, Zheng MZ, Xu JJ, Guo JH (2018) *Bacillus velezensis*, a potential and efficient biocontrol agent in control of pepper graymold caused by *botrytis cinerea*. Biol Control 126:147–157

Ju R, Zhao Y, Li J, Jiang H, Liu P, Yang T, Bao Z, Zhou B, Zhou X, Liu X (2014) Identification and evaluation of a potential biocontrol agent, *Bacillus subtilis*, against *fusarium* spp. in apple seedlings. Ann Microbiol 64(1):377–383

Kaspar F, Neubauer P, Gimpel M (2019) Bioactive secondary metabolites from *Bacillus subtilis*: a comprehensive review. J Nat Prod 82(7):2038–2053

Kesaulya H, Hasinu JV, Tuhumury GN (2018) Potential of bacillus spp produces siderophores insuppressing thewilt disease of banana plants. In: *IOP Conference Series: Earth and Environmental Science*, vol 102, No. 1. IOP Publishing, p 012016

Kushwaha P, Kashyap PL, Kuppusamy P, Srivastava AK, Tiwari RK (2020a) Functional characterization of endophytic bacilli frompearl millet (*Pennisetum glaucum*) and their possible role in multiple stress tolerance. Plant Biosyst 154(4):503–514. https://doi.org/10.1080/11263504.2019.1651773

Kushwaha P, Kashyap PL, Srivastava AK, Tiwari RK (2020b) Plant growth promoting and antifungal activity in endophytic bacillus strains from pearl millet (Pennisetum glaucum). Braz J Microbiol 51(1):229–241. https://doi.org/10.1007/s42770-019-00172-5

Kushwaha P, Srivastava R, Pandiyan K, Singh A, Chakdar H, Kashyap PL, Bhardwaj AK, Murugan K, Karthikeyan N, Bagul SY, Srivastava AK, Saxena AK (2021) Enhancement in plant growth and zinc biofortification of chickpea (Cicer arietinum L.) by *Bacillus altitudinis*. J Soil Sci Plant Nutr 21:922–935. https://doi.org/10.1007/s42729-021-00411-5

Li B, Li Q, Xu Z, Zhang N, Shen Q, Zhang R (2014) Responses of beneficial *bacillus amyloliquefaciens* SQR9 to different soilborne fungal pathogens through the alteration of antifungal compounds production. Front Microbiol 5:636

Li Y, Gu Y, Li J, Xu M, Wei Q, Wang Y (2015) Biocontrol agent *bacillus amyloliquefaciens* LJ02 induces systemic resistance against cucurbits powdery mildew. Front Microbiol 6:883

Li X, Zhang Y, Wei Z, Guan Z, Cai Y, Liao X (2016) Antifungal activity of isolated *bacillus amyloliquefaciens* SYBC H47 for the biocontrol of peach gummosis. PLoS One 11(9):e0162125

Li Y, Wang R, Liu J, Xu L, Ji P, Sun L, Pan H, Jiang B, Li L (2019) Identification of a biocontrol agent *Bacillus vallismortis* BV23 and assessment of effects of its metabolites on *fusarium graminearum* causing corn stalk rot. Biocontrol Sci Tech 29(3):263–275

Li X, Gao X, Zhang S, Jiang Z, Yang H, Liu X, Jiang Q, Zhang X (2020) Characterization of a *bacillus velezensis* with antibacterial activity and inhibitory effect on common aquatic pathogens. Aquaculture 523:735165

Liu CH, Wu K, Chu TW, Wu TM (2018) Dietary supplementation of probiotic, *Bacillus subtilis* E20, enhances the growth performance and disease resistance against *vibrio alginolyticus* in parrot fish (*Oplegnathus fasciatus*). Aquac Int 26(1):63–74

Luo C, Liu X, Zhou H, Wang X, Chen Z (2015) Nonribosomal peptide synthase gene clusters for lipopeptide biosynthesis in *Bacillus subtilis* 916 and their phenotypic functions. Appl Environ Microbiol 81(1):422

Mahapatra S, Rayanoothala P, Solanki MK, Das S (2020) Wheat microbiome: present status and future perspective. In: Solanki M, Kashyap P, Kumari B (eds) Phytobiomes: current insights and future vistas. Springer, Singapore, pp 191–223. https://doi.org/10.1007/978-981-15-3151-4_8

Mahapatra S, Chakraborty C, Samanta M, Das S, Islam T (2022a) Current Understanding and Future Directions of Biocontrol of Plant Diseases by Bacillus spp., with Special Reference to Induced Systemic Resistance. In: Islam MT, Rahman M, Panday P (eds) Bacilli in Agrobiotechnology. Bacilli in climate resilient agriculture and bioprospecting. Springer, Cham

Mahapatra S, Chakraborty S, Rayanoothala P, Das S, Bishnoi SK, Kumar S (2022b) Antimicrobial Agents for Wheat Disease Management: Mode of Action and Its Application. In: Kashyap PL, Gupta V, Gupta OP, Sendhil R, Gopalreddy K, Jasrotia P, Singh GP (eds) New Horizons in Wheat and Barley Research. Springer, Singapore. https://doi.org/10.1007/978-981-16-4134-3_6

Mariutto M, Ongena M (2015) Molecular patterns of rhizobacteria involved in plant immunity elicitation. In advances in botanical research. Academic Press 75:21–56

Martínez-Álvarez JC, Castro-Martínez C, Sánchez-Peña P, Gutiérrez-Dorado R, Maldonado-Mendoza IE (2016) Development of a powder formulation based on *Bacillus cereus sensulato* strain B25 spores for biological control of *fusarium verticillioides* in maize plants. World J Microbiol Biotechnol 32:75

Mejri S, Siah A, Coutte F, Magnin-Robert M, Randoux B, Tisserant B, Krier F, Jacques P, Reignault P, Halama P (2018) Biocontrol of the wheat pathogen *Zymoseptoria tritici* using cyclic lipopeptides from *Bacillus subtilis*. Environ Sci Pollut Res 25(30):29822–29833

Mielich-Süss B, Lopez D (2015) Molecular mechanisms involved in *Bacillus subtilis* biofilm formation. Environ Microbiol 17(3):555–565
Mihalache G, Balaes T, Gostin I, Stefan M, Coutte F, Krier F (2018) Lipopeptides produced by Bacillus subtilis as new biocontrol products against fusariosis in ornamental plants. Environ Sci Pollut Res 25(30):29784–29793
Mnif I, Ghribi D (2015) Potential of bacterial derived biopesticides in pest management. Crop Prot 77:52–64
Mondol MA, Shin HJ, Islam MT (2013) Diversity of secondary metabolites from marine *bacillus species*: chemistry and biological activity. Mar Drugs 11(8):2846–2872
Müller S, Strack SN, Hoefler BC, Straight PD, Kearns DB, Kirby JR (2014) Bacillaene and sporulation protect *Bacillus subtilis* from predation by *Myxococcus xanthus*. Appl Environ Microbiol 80(18):5603–5610
Nithyapriya S, Lalitha S, Sayyed RZ, Reddy MS, Dailin DJ, El Enshasy HA, Luh N, Herlambang S (2021) Production, purification, and characterization of bacillibactin siderophore of *Bacillus subtilis* and its application for improvement in plant growth and oil content in sesame. Sustainability 13:5394. https://doi.org/10.3390/su13105394
Ongena M, Jacques P (2008) Bacillus lipopeptides: versatile weapons for plant disease biocontrol. Trends Microbiol 16(3):115–125
Patel PR, Shaikh SS, Sayyed RZ (2016) Dynamism of PGPR in bioremediation and plant growth promotion in heavy metal contaminated soil. Indian J Exp Biol 54:286–290
Podile A, Prakash A (1996) Lysis and biological control of *aspergillus Niger* by *Bacillus subtilis* AF 1. Can J Microbiol 42(6):533–538
Pourakbar L, Moghaddam SS, Enshasy HAE, Sayyed RZ (2021) Antifungal Activity of the Extract of a Macroalgae, Gracilariopsis persica, against Four Plant Pathogenic Fungi. Plants 10(9):1781
Raaijmakers JM, De Bruijn I, Nybroe O, Ongena M (2010) Natural functions of lipopeptides from *bacillus* and *pseudomonas*: more than surfactants and antibiotics. FEMS Microbiol Rev 34(6):1037–1062
Rahman A, Uddin W, Wenner NG (2015) Induced systemic resistance responses in perennial ryegrass against *Magnaporthe oryzae* elicited by semi-purified surfactin lipopeptides and live cells of *bacillus amyloliquefaciens*. Mol Plant Pathol 16:546–558
Rahman M, Sabir AA, Mukta JA, Khan MMA, Mohi-Ud-Din M, Miah MG, Rahman M, Islam MT (2018) Plant probiotic bacteria *bacillus* and *Paraburkholderia* improve growth, yield and content of antioxidants in strawberry fruit. Sci Rep 8(1):1–11
Rai P, Sharma A, Saxena P, Soni AP, Chakdar H, Kashyap PL, Srivastava AK, Sharma AK (2015) Comparison of molecular and phenetic typing methods to assess diversity of selected members of the genus *bacillus*. Microbiology 84:236–246. https://doi.org/10.1134/S0026261715020113
Rais A, Jabeen Z, Shair F, Hafeez FY, Hassan MN (2017) *Bacillus* spp., a bio-control agent enhances the activity of antioxidant defense enzymes in rice against *Pyricularia oryzae*. PLoS One 12(11):e0187412
Rayanoothala P, Divya M, Mahapatra S, Das S (2021) Microbial biofilm: formation, quorum sensing, and its applications in plant disease management. In: Emerging trends in plant pathology. Springer, Singapore, pp 385–398. https://doi.org/10.1007/978-981-15-6275-4_18
Raza W, Yuan J, Ling N, Huang Q, Shen Q (2015) Production of volatile organic compounds by an antagonistic strain *Paenibacillus polymyxa* WR-2 in the presence of root exudates and organic fertilizer and their antifungal activity against *fusarium oxysporum* f. spp. *niveum*. Biol Control 80:89–95
Rodríguez J, Tonelli ML, Figueredo MS, Ibáñez F, Fabra A (2018) The lipopeptide surfactin triggers induced systemic resistance and priming state responses in *Arachis hypogaea* L. Eur J Plant Pathol 152(3):845–851
Romero D (2013) Bacterial determinants of the social behavior of *Bacillus subtilis*. Res Microbiol 164(7):788–798
Roongsawang N, Washio K, Morikawa M (2011) Diversity of nonribosomal peptide synthetases involved in the biosynthesis of lipopeptide biosurfactants. Int J Mol Sci 12(1):141–172

Savary S, Ficke A, Aubertot JN, Hollier C (2012) Crop losses due to diseases and their implications for global food production losses and food security. Food Sec 4:519–537
Senol M, Nadaroglu H, Dikbas N, Kotan R (2014) Purification of Chitinase enzymes from *Bacillus subtilis* bacteria TV-125, investigation of kinetic properties and antifungal activity against *fusarium culmorum*. Ann Clin Microbiol Antimicrobial 13(1):35
Shafi J, Tian H, Ji M (2017) Bacillus species as versatile weapons for plant pathogens: a review. Biotechnol Biotechnol Equip 31(3):446–459
Sharma A, Singh P, Kumar S, Kashyap PL, Srivastava AK, Chakdar H, Singh RN, Kaushik R, Saxena AK, Sharma AK (2015) Deciphering diversity of salt-tolerant bacilli from saline soils of eastern indo-gangetic plains of India. Geomicrobiol J 32:170–180. https://doi.org/10.1080/01490451.2014.938205
Sharma A, Kashyap PL, Srivastava AK, Bansal YK, Kaushik R (2019) Isolation and characterization of halotolerant bacilli from chickpea (*Cicer arietinum* L.) rhizosphere for plant growth promotion and biocontrol traits. Eur J Plant Pathol 153:787–800. https://doi.org/10.1007/s10658-018-1592-7
Singh RK, Kumar PD, Solanki MK, Singh P, Srivastva AK, Kumar S, Kashyap PL, Saxena AK, Singhal PK, Arora DK (2013) Optimization of media components for chitinase production by chickpea rhizosphere associated *Lysinibacillus fusiformis* B-CM18. J Basic Microbiol 53(5):451–460. https://doi.org/10.1002/jobm.201100590
Singh RK, Kumar DP, Singh P, Solanki MK, Srivastava S, Kashyap PL, Kumar S, Srivastava AK, Singhal PK, Arora DK (2014) Multifarious plant growth promoting characteristics of chickpea rhizosphere associated *bacilli* help to suppress soil-borne pathogens. Plant Growth Regul 73(1):91–101. https://doi.org/10.1007/s10725-013-9870-z
Slimene IB, Tabbene O, Gharbi D, Mnasri B, Schmitter JM, Urdaci MC, Limam F (2015) Isolation of a chitinolytic *bacillus licheniformis* S213 strain exerting a biological control against *Phoma medicaginis* infection. Appl Biochem Biotechnol 175(7):3494–3506
Solanki MK, Robert AS, Singh RK, Kumar S, Pandey AK, Srivastava AK, Arora DK (2012a) Characterization of mycolytic enzymes of *bacillus* strains and their bio-protection role against *Rhizoctonia solani* in tomato. Curr Microbiol 65(3):330–336
Solanki MK, Kumar S, Pandey AK, Srivastava S, Singh RK, Kashyap PL, Srivastava AK, Arora DK (2012b) Diversity and antagonistic potential of *bacillus* spp. associated to the rhizosphere of tomato for the management of *Rhizoctonia solani*. Biocontrol Sci Tech 22(2):203–217. https://doi.org/10.1080/09583157.2011.649713
Solanki MK, Singh RK, Srivastava S, Kumar S, Kashyap PL, Srivastava AK (2015) Characterization of antagonistic-potential of two *bacillus* strains and their biocontrol activity against *Rhizoctonia solani* in tomato. J Basic Microbiol 55(1):82–90. https://doi.org/10.1002/jobm.201300528
Stoica RM, Moscovici M, Tomulescu C, Casarica A, Babeanu N, Popa O, Kahraman HA (2019) Antimicrobial compounds of the genus *bacillus*: a review. Rom Biotechnol Lett 24(6):1111–1119. https://doi.org/10.25083/rbl/24.6/1111.1119
Surovy MZ, Gupta DR, Mahmud NU, Dame ZT, Roy PK, Islam MT (2019) Genomics and post-genomics approaches for elucidating molecular mechanisms of plant growth-promoting *bacilli*. In: Islam M, Rahman M, Pandey P, Boehme M, Haesaert G (eds) Bacilli and Agrobiotechnology: Phytostimulation and biocontrol. Bacilli in climate resilient agriculture and bioprospecting. Springer, Cham. https://doi.org/10.1007/978-3-030-15175-1_10
Symmank H, Franke P, Saenger W, Bernhard F (2002) Modification of biologically active peptides: production of a novel lipohexapeptide after engineering of *Bacillus subtilis* surfactin synthetase. Protein Eng 15(11):913–921
Toyofuku M, Inaba T, Kiyokawa T, Obana N, Yawata Y, Nomura N (2016) Environmental factors that shape biofilm formation. Biosci Biotechnol Biochem 80(1):7–12
Vinay JU, Naik MK, Rangeshwaran R, Chennappa G, Shaikh SS, Sayyed RZ (2016) Detection of antimicrobial traits in fluorescent pseudomonads and molecular characterization of an antibiotic pyoluteorin, 3. Biotech 6(227):1–11

Waewthongrak W, Pisuchpen S, Leelasuphakul W (2015) Effect of *Bacillus subtilis* and chitosan applications on green mold (*Penicilium digitatum*Sacc.) decay in citrus fruit. Postharvest Biol Tec 99:44–49

Wang B, Yuan J, Zhang J, Shen Z, Zhang M, Li R, Ruan Y, Shen Q (2013) Effects of novel bioorganic fertilizer produced by *bacillus amyloliquefaciens* W19 on antagonism of *fusarium* wilt of banana. Biol Fertil Soils 49(4):435–446

Wang T, Liang Y, Wu MB, Chen ZJ, Yang LR (2015) Natural products from *Bacillus subtilis* with antimicrobial properties. Chin J Chem Eng 23(4):744–754

Wang S, Sun L, Zhang W, Chi F, Hao X, Bian J, Li Y (2020) *Bacillus velezensis* BM21, a potential and efficient biocontrol agent in control of corn stalk rot caused by *fusarium graminearum*. Egypt J Biol Pest Control 30(1):9

Wu Y, Chen T, Liu Y, Lv X, Li J, Du G, Ledesma-Amaro R, Liu L (2018) CRISPRi allows optimal temporal control of N-acetylglucosamine bioproduction by a dynamic coordination of glucose and xylose metabolism in *Bacillus subtilis*. Metab Eng 49:232–241

Xu Z, Shao J, Li B, Yan X, Shen Q, Zhang R (2013) Contribution of bacillomycin D in *bacillus amyloliquefaciens* SQR9 to antifungal activity and biofilm formation. Appl Environ Microbiol 79(3):808–815

Yadav AK, Manna S, Pandiyan K, Singh A, Kumar M, Chakdar H, Kashyap PL, Srivastava AK (2016) Isolation and characterization of biosurfactant producing *bacillus* sp. from diesel fuel-contaminated site. Microbiology 85:6–62. https://doi.org/10.1134/S0026261716010161

Yamamoto S, Shiraishi S, Suzuki S (2015) Are cyclic lipopeptides produced by *bacillus amyloliquefaciens* S13-3 responsible for the plant defence response in strawberry against *Colletotrichum gloeosporioides*? Lett Appl Microbiol 60(4):379–386

Yang L, Han X, Zhang F, Goodwin PH, Yang Y, Li J, Xia M, Sun R, Jia B, Zhang J, Quan X (2018) Screening *bacillus species* as biological control agents of *Gaeumannomyces graminis* var. *tritici* on wheat. Biol Control 118:1–9

Yaseen Y, Diop A, Gancel F, Bechet M, Jacques P, Drider D (2018) Polynucleotide phosphorylase is involved in the control of lipopeptide fengycin production in Bacillus subtilis. Arch Microbiol 200(5):783–791

You C, Zhang C, Kong F, Feng C, Wang J (2016) Comparison of the effects of biocontrol agent *Bacillus subtilis* and fungicide metalaxyl–mancozeb on bacterial communities in tobacco rhizospheric soil. Ecol Eng 91:119–125

Youcef-Ali M, Chaouche NK, Dehimat L, Bataich EI, Mounira KARA, Cawoy HE, Thonart P (2014) Antifungal activity and bioactive compounds produced by *Bacillus mojavensis* and *Bacillus subtilis*. AfrJ Microbiol Res 8(6):476–484

Yu X, Ai C, Xin L, Zhou G (2011) The siderophore-producing bacterium, *Bacillus subtilis* CAS15, has a biocontrol effect on *fusarium* wilt and promotes the growth of pepper. Eur J Soil Biol 47(2):138–145

Yuan J, Raza W, Shen Q, Huang Q (2012) Antifungal activity of *bacillus amyloliquefaciens* NJN-6 volatile compounds against *fusarium oxysporum* f. spp. *cubense*. Appl Environ Microbiol 78(16):5942–5944

Zakaria AK, Sayyed RZ, Enshasy HE (2019) Biosynthesis of antibiotics by PGPR and their roles in biocontrol of plant diseases. In: Plant growth promoting Rhizobacteria for sustainable stress management Vol II Rhizobacteria in biotic stress management. Springer, Singapore, pp 1–36

Zeriouh H, de Vicente A, Pérez-García A, Romero D (2014) Surfactin triggers biofilm formation of *Bacillus subtilis* in melon phylloplane and contributes to the biocontrol activity. Environ Microbiol 16(7):2196–2211

Zhang N, Wu K, He X, Li SQ, Zhang ZH, Shen B, Yang XM, Zhang RF, Huang QW, Shen QR (2011) A new bioorganic fertilizer can effectively control banana wilt by strong colonization with *Bacillus subtilis* N11. Plant Soil 344(1–2):87–97

Zhang X, Li B, Wang Y, Guo Q, Lu X, Li S, Ma P (2013) Lipopeptides, a novel protein, and volatile compounds contribute to the antifungal activity of the biocontrol agent *bacillus atrophaeus* CAB-1. Appl Microbiol Biotechnol 97(21):9525–9534

Zhang DD, Guo XJ, Wang YJ, Gao TG, Zhu BC (2017) Novel screening strategy reveals a potent bacillus antagonist capable of mitigating wheat take-all disease caused by *Gaeumannomyces graminis* var. *tritic*i. Lett Appl Microbiol 65(6):512–519

Zhang D, Yu S, Yang Y, Zhang J, Zhao D, Pan Y, Fan S, Yang Z, Zhu J (2020) Antifungal effects of volatiles produced by *Bacillus subtilis* against *Alternaria solani* in potato. Front Microbiol 11:1196

Zhao Q, Ran W, Wang H, Li X, Shen Q, Shen S, Xu Y (2013) Biocontrol of Fusarium wilt disease in muskmelon with Bacillus subtilis Y-IVI. BioControl 58(2):283–292

Zhao P, Quan C, Wang Y, Wang J, Fan S (2014) *Bacillus amyloliquefaciens* Q-426 as a potential biocontrol agent against *Fusarium oxysporum* f. spp. *spinaciae*. J Basic Microbiol 54(5):448–456

Zhou M, Liu F, Yang X, Jin J, Dong X, Zeng KW, Liu D, Zhang Y, Ma M, Yang D (2018) Bacillibactin and bacillomycin analogues with cytotoxicities against human cancer cell lines from marine *bacillus* spp. PKU-MA00093 and PKU-MA00092. Mar Drugs 16(1):22

Zhu J, Tan T, Shen A, Yang X, Yu Y, Gao C, Li Z, Cheng Y, Chen J, Guo L, Sun X (2020) Biocontrol potential of *Bacillus subtilis* IBFCBF-4 against *fusarium* wilt of watermelon. J Plant Pathol 102:433–441. https://doi.org/10.1007/s42161-019-00457-6

Zihalirwa Kulimushi P, Argüelles Arias A, Franzil L, Steels S, Ongena M (2017) Stimulation of fengycin-type antifungal lipopeptides in *bacillus amyloliquefaciens* in the presence of the maize fungal pathogen *Rhizomucor variabilis*. Front Microbiol 8:850. https://doi.org/10.3389/fmicb.2017.00850

Zou C, Li Z, Yu D (2010) *Bacillus megaterium* strain XTBG34 promotes plant growth by producing 2-pentylfuran. J Microbiol 48(4):460–466

# Chapter 8
# Bioactive Biomolecules from Plant Growth-Promoting Rhizobacteria (PGPR)

**Manila Bajimaya, Sunita Basnet, Sailesh Malla, and Laxmi Prasad Thapa**

**Contents**


8.1 Introduction .......... 157
8.2 Mechanisms of Fungal Inhibition by PGPR Metabolites .......... 161
8.3 Conclusion .......... 174
References .......... 175


## 8.1 Introduction

The rapid population growth and climatic changes imposed abiotic and biotic stresses on crops and plants which create a big challenge in crop production and yield globally. Sustainable agriculture techniques to combat abiotic stresses, diseases, and pathogens are very essential to fulfill the demand-supply chain of the increased population as well as to conserve the environment. Among the plant diseases, fungal pathogens are the biggest global threat which not only cause huge losses in agriculture and food production but also significantly affect consumers' health (Fisher et al. 2012). After the Second World War, industrialization in


M. Bajimaya
Department of Fungal Biology, Polytechnic Research Institute of Nepal (PORIN), Koteshwor, Nepal

S. Basnet
Department of Natural Product and Bioresources, Polytechnic Research Institute of Nepal (PORIN), Koteshwor, Nepal

S. Malla (✉)
Strain Development Department, Discovery, R&D, Chr. Hansen A/S, Hørshlom, Denmark
e-mail: dksaim@chr-hansen.com

L. P. Thapa (✉)
Department of Molecular Biotechnology, Nepal Academy of Science and Technology, Khumaltar, Lalitpur, Nepal





R. Z. Sayyed et al. (eds.), *Antifungal Metabolites of Rhizobacteria for Sustainable Agriculture*, Fungal Biology, https://doi.org/10.1007/978-3-031-04805-0_8

agriculture introduced extensive use of chemical fertilizers and pesticides to enhance agriculture productivity. However, prolonged use of such chemically synthesized pesticides and fertilizers reduces microbial biodiversity in soil, increases pathogen resistance, as well as degrades the soil quality. Furthermore, direct contact with such toxic chemical fertilizers for a prolonged time causes serious health issues of the farmers, such as skin diseases, respiratory tract problems, lung cancer, etc. (Damalas and Koutroubas 2016). Hence, there is increasing attention to biological control of plant diseases.

Plants and soil microbes interact to help each other for their growth and development. In addition, the plant microbes' interaction maintains the terrestrial ecosystem. Plants can also use these growth-promoting microbes as a weapon against various pathogens, including fungi and insects. These growth-promoting microbes act as natural probiotics for the plants by defending them against various diseases. One might wonder how these tiny microbes are able to protect huge plants. Scientists, driven by curiosity on these microbes, have found solution to these questions by their various experimental models. These growth-promoting microbes not only provide micronutrients and phytohormones to the plants from complex soil system, but also they produce a plethora of secondary metabolites which can act as a bioweapon to the plant pathogens in their vicinity (Backer et al. 2018). Hence, the application of these beneficial microbes is certainly a promising sustainable solution to improve agricultural production, i.e., microbial solution is one of the most eco-friendly alternatives to the extensive use of pesticides and chemical fertilizer to combat fungal crop diseases.

A group of free-living soil microbes that inhabit the rhizosphere and benefit plants are known as plant growth-promoting rhizobacteria (PGPR) (Kloepper and Schroth 1978). Interestingly, the PGPR includes both Gram-positive bacteria such as *Actinomycetes* sp., *Bacillus* sp., as well as Gram-negative bacteria such as *Azospirillum sp.*, *Burkholderia sp.*, *Bradyrhizobium* sp., *Herbaspirillum* sp., fluorescent *Pseudomonas sp.*, *Pantoea sp.*, *Klebsiella*, etc. These PGPR are capable to produce various classes/families of bioactive compounds, for example, polyketides, non-ribosomal polyketide synthases (NRPS), siderophores, surfactants, hydrolytic enzymes (e.g., protease, lipase, etc.), volatile compounds, etc. With the secretion of one or more of these metabolites, the PGPR can inhibit the growth of pathogens. Furthermore, the PGPR can also consume nutrients from the soil by scavenging the essential micronutrients for pathogens to survive. Because of these beneficial properties, PGPR are included in the commercial plant health-care products to enhance agricultural yields and productivity. For instance, *Bacillus* sp. are the most commercially available PGPR as they have dormant endospores which are stable against various factors such as heat, desiccation, UV irradiation, and organic solvents (Brumm et al. 1991; Gates et al. 2010). Some of the commercial plant health-care products derived from *Bacillus* strains are listed in Table 8.1.

Here in this chapter, we summarize and discuss the list of various bioactive compounds produced by PGPR against fungal pathogens for the protection of crops or plants.

**Table 8.1** Various plant health-care products in the market containing *Bacillus* sp. as PGPR and their application (Keswani et al. 2020)

| Bacillus strain | Known properties | Trade name | Company |
|---|---|---|---|
| *Bacillus subtilis* GB03 | EPA-registered (71065–2) biological and seed treatment fungicide | Kodiak™ | Bayer crop science, former Gustafsson LLC |
| | EPA-registered (71065–2) biofungicide that prevents and controls plant diseases. It produces a broad-spectrum iturin antibiotic that disrupts the cell-wall formation of pathogens, and it triggers an advantageous induced systemic resistance (ISR) in plants, whereby a plant's natural immune system is activated to fight plant diseases | Companion | Growth Products Ltd., White Plains, NY 10603 |
| *Bacillus amyloliquefaciens* GB99 + *Bacillus subtilis* GB122 | Combination of strong ISR activity (GB99) with phytostimulation (GB122) | BioYield™ | Bayer crop science, previously Gustafsson |
| *Bacillus subtilis* MBI600 | EPA-registered (71840–8) biofungicide that provides protection against soilborne pathogens, such as *Rhizoctonia solani*, *Pythium* sp., and *fusarium* spp. to help prevent damping-off and other root diseases | Subtilex®, INTEGRAL® | Becker underwood, Saskatoon, Canada acquired by BASF |
| *Bacillus subtilis* MBI600 | Produced by "BioStacked®" technology, enhancing growth of soy beans and peanuts | Vault® | Becker underwood, Saskatoon, Canada |
| *Bacillus subtilis* QST713 | EPA-registered (69592–11) biofungicide, annex 1 listing of the EU agrochemical registration directive (91/414) | SERENADE max | Bayer crop science, previously AgraQuest |
| *Bacillus subtilis* QST713 | EPA-registered (69592-EI, 2012) biofungicide for food crops | SERENADE SOIL® | Bayer crop science, previously AgraQuest |
| *Bacillus subtilis* QST713 | Aqueous suspension biofungicide, recommended for leafy and fruiting vegetables, herbs and spices, and ornamentals | CEASE® | BioWorks, Inc., victor, New York, USA |
| *Bacillus amyloliquefaciens* FZB42 | Biofertilizer, plant growth-promoting activity, provides protection against various soilborne diseases, stimulation of ISR | RhizoVital® | ABiTEP GmbH, Berlin |

(continued)

**Table 8.1** (continued)

| Bacillus strain | Known properties | Trade name | Company |
|---|---|---|---|
| *Bacillus subtilis* FZB24 | EPA-registered biofungicide. FZB24 has been originally isolated by FZB Berlin, the forerunner of ABiTEP GmbH. Registration as a biofungicide for the USA was performed by Taegro Inc. and then sold to Novozymes without agreement with ABiTEP GmbH where the product is still offered | Taegro® | Syngenta, Basel, previously Novozyme, Davis, California and earth biosciences |
| *Bacillus subtilis* CMB26 | Microbial fungicide that controls and has inhibition germination effect on powdery mildew, *Cladosporium fulvum*, and *Botrytis cinerea* | POMEX | NIN co. ltd., |
| *Bacillus subtilis* CX9060 | EPA-registered 71,840-RG,-RE (2012) fungicide, bactericide for food crops, turf, and ornamentals | | Certis Columbia, MD USA |
| *Bacillus subtilis* E4-CDX | Rhizosphere bacterium that competes with harmful pathogens for space around the roots of the grass plant. Once established, this unique strain physically protects the roots and inhibits the advance of soilborne fungi | Easy start® TE-max | COMPO expert GmbH, Münster, Germany |
| *Bacillus amyloliquefaciens* D747 | EPA-registered (70051-RNI, 2011) a broad-spectrum preventive biofungicide for control or suppression of fungal and bacterial plant diseases (powdery mildew, *Sclerotinia*, *Botrytis*, *Alternaria*, bacterial leaf spot, bacterial spot and speck, fire blight, *Xanthomonas*, *Monilinia*) | Double nickel 55™ | Certis Columbia, MD USA |
| *Bacillus amyloliquefaciens* D747 | Annex 1 listing of the EU agrochemical registration directive. Launched to Italy by Intrachem bio Italia SpA for control of *botrytis* and other fungal diseases of grapes, strawberries, and vegetables, and bacterial diseases, such as fire blight in pome fruit and PSA in kiwi fruit | Amylo-X® | Certis Columbia, MD USA/Intrachem bio Italia SpA |
| *Bacillus pumilus* QST2808 | EPA-registered (69592–13) biofungicide, powdery mildew control | Sonata® | Bayer crop science, previously AgraQuest Inc. |
| *Bacillus pumilus* GB34 (=INR7) | EPA-registered biofungicide (264–985), suppression of root diseases | YieldShield | Bayer crop science, previously Gustafsson |

(continued)

**Table 8.1** (continued)

| Bacillus strain | Known properties | Trade name | Company |
|---|---|---|---|
| *Bacillus pumilus* BU F-33 | EPA-registered (71840-RG, –RE, 2013) plant growth stimulator, induced systemic resistance | | Becker underwood, Saskatoon, Canada |
| *Bacillus pumilus* GHA 181 | EPA-registered fungicide (2012), food crops, seeds, ground cover, and ornamentals | | Premier horticulture |

## 8.2 Mechanisms of Fungal Inhibition by PGPR Metabolites

Most commonly, rhizobacteria are referred to as PGPR, and these microbes also play major roles in the inhibition of plant pathogens. These PGPR are categorized according to their beneficial effects for the plants such as nitrogen fixer, enzyme/hormones producers, biocontrol agents, etc. (Bloemberg and Lugtenberg 2001). For biocontrol properties, i.e., to act against plant pathogens, various types of metabolites with antimicrobial or antifungal or nematicide properties are produced by PGPR. Hence, the soil microbes that are able to produce fungicides can be used as biocontrol agent to protect crops and plants against abroad range of soilborne pathogenic fungi. The most common widespread soilborne plant pathogenic fungi include *Aspergillus* sp., *Botrytis* sp., *Claviceps* sp., *Candida* sp., *Fusarium* sp., *Pythium* sp., *Gaeumannomyces* sp., *Trichoderma* sp., etc. In general, the bioactive secondary metabolites from PGPR can either inhibit the fungal spore germination, stop the mycelial growth of fungi, or kill the fungi to protect the plant from fungal diseases (Olanrewaju et al. 2017). In addition, the rhizobacteria can also induce the plant resistant mechanisms and degrade the pathogenicity factor (such as micro-toxin) of the pathogen. Furthermore, the rhizobacteria can produce hydrolytic enzymes such as chitinase to degrade cell wall of pathogens (Jadhav et al. 2017). In the Table 8.2, the biocontrol mechanism exhibited by various Gram-positive *Bacillus* sp. against various fungi causing plant diseases are listed.

### *8.2.1 Pathogenic Fungal Inhibition by Antibiotics from PGPR*

The plant roots are inhabited by microbes that can uptake root extrudates and trade off vital secondary metabolites to kill plant pathogens including fungi, bacteria, and nematodes. Hence, this phenomenon of production of antibiotics by PGPR is one of the major natural defense mechanisms against plant pathogens to remain plant healthy in natural environment.

PGPR can be either Gram-negative or Gram-positive bacteria. Both Gram-negative and Gram-positive rhizobacteria produce varieties of potent secondary metabolites that are used against warfare under the soil to protect plant. For instance, some of the antifungal secondary metabolites produced by Gram-negative

**Table 8.2** Biocontrol mechanisms exhibited by *Bacillus* species (Miljaković et al. 2020)

| Bacillus species | Mechanism (s) | Target pathogen (s) | Plant disease |
|---|---|---|---|
| *Bacillus amyloliquefaciens* | Bacteriocins, surfactin, fengycin | Various pathogenic bacteria, *fusarium oxysporum*, *fusarium avenaceum*, *Mucor* sp. | Several diseases of field and vegetable crops |
| *Bacillus amyloliquefaciens* | Iturin | *Verticillium dahliae* | Wilt of cotton |
| *Bacillus amyloliquefaciens/ bacillus subtilis* | Iturin, surfactin/ surfactin, fengycin | *Fusarium graminearum* | Head blight of wheat |
| *Bacillus subtilis* | Surfactin, lytic enzymes | *Fusarium* sp. | Clove rot of garlic |
| *Bacillus subtilis* | Iturin, fengycin | *Pythium ultimum* | Damping-off of bean |
| *Bacillus* sp., *bacillus amyloliquefaciens* | Surfactin, iturin, fengycin, siderophore | *Sclerotinia sclerotiorum* | White mold of common bean |
| *Bacillus pumilus*, *bacillus amyloliquefaciens* | Lipopeptides | *Pseudomonas syringae* pv. *Aptata* | Leaf spot disease of sugar beet |
| *Bacillus subtilis* | Iturin, surfactin, plipastatin, bacillomycin, difficidin | *Gaeumannomyces graminis* var. *tritici* | Take-all of wheat |
| *Bacillus licheniformis* | Surfactin | *Magnaporthe grisea* | Blast disease of rice |
| *Bacillus subtilis* | Chitinase | *Rhizoctonia solani* | Stem canker and black scurf of potato |
| *Bacillus amyloliquefaciens* | Protease | *Fusarium oxysporum* f. spp. *lycopersici* | Wilt disease of tomato |
| *Bacillus amyloliquefaciens* | Siderophores, cellulase, lipase, protease, chitinase | *Clavibacter michiganensis* | Bacterial canker of tomato |
| *Bacillus* sp. | Protease, glucanase, chitinase, siderophores | *Fusarium verticillioides* | Stalk and ear rot of maize |
| *Bacillus velezensis* | Protease, chitinase, cellulase, glucanase | *Botrytis cinerea* | Gray mold disease of pepper |
| *Bacillus subtilis* | Siderophores | *Fusarium oxysporum* f. spp. *capsici* | Wilt of pepper |
| *Bacillus subtilis* | Siderophores, lytic enzymes | *Bipolaris sorokiniana* | Spot blotch of wheat |

*Pseudomonas* spp. are oomycin A, cepaciamide A, ecomycins, DDR, viscosinamide, butyrolactones, *N*-butylbenzene, sulfonamide, pyocyanin, phenazines, phenazine-1-carboxylic acid, phenazine-1-carboxamide, pyrrolnitrin, pyoluteorin,

2,4-diacetylphloroglucinol, and rhamnolipids (Fernando et al. 2005a, b).The Gram-positive PGPR are also abundantly present in soil. One of the examples is spore-forming *Bacillus* sp. which inhibit growth, induced systemic resistance in plant, and competing for ecological niches with plant pathogens. Interestingly, *Bacillus* sp. inhibit the growth of *S. sclerotiorum* fungus, a white mold that destroys seeds of plants. Some of the commonly found *Bacillus* sp. in soil are *B. amyloliquefaciens*, *B. licheniformis*, *B. pumilus* and *B. subtilis*, *B. velezensis*, etc. (Miljaković et al. 2020). These *Bacillus* strains are sources of bioactive secondary metabolites, and most of the antifungal compounds produced by these strains belong to Lipopeptide family (Penha et al. 2020). Chemical structures of some lipopeptides, produced by *Bacillus* sp. with antifungal activities, are given in Fig. 8.1.

#### 8.2.1.1 Lipopeptides

Lipopeptides are made up of lipid and amino acids which consist of hydrophobic head and hydrophilic tail that act as biosurfactants having fungicidal property. They are non-ribosomally synthesized by peptide synthetases (NRPS). *B. licheniformis*, *B. pumilus*, *B. cereus*, *B. amyloliquefaciens*, *B. megaterium*, *B. globigii*, and *B. thuringiensis* produce wide variety of lipopeptides. Based on amino acid sequences, the lipopeptides have been classified into three major groups (or families): surfactins, iturins, and fengycin (Meena and Kanwar 2015). The lipopeptides are categorized into different groups according to their length, branching of the fatty acid side chains, and amino acid substitutions in the peptide rings (Kowall et al. 1998).

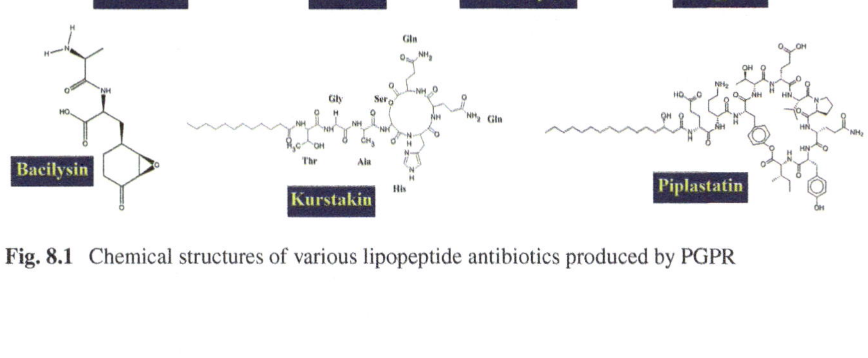


**Fig. 8.1** Chemical structures of various lipopeptide antibiotics produced by PGPR

#### 8.2.1.1.1 Iturins

Iturins have small molecular mass of ~1.1. kDa and are composed up of alpha and beta amino acid. All strains of *B. subtilis* are capable to produce iturin family of lipopeptides (Meena and Kanwar 2015). They are effective against pathogenic fungi, *P. ultimum*, *M. phaseolina*, *F. oxysporum*, *S. sclerotiorum*, and *R. solani*, that degrade plant health. Iturin acts mainly by making the fungal spores more permeable, thus preventing the spore germination. Iturin D is a secondary metabolite of *B. subtilis* which inhibits *C. trifolii*. It causes lysis of conidia forming inflated germ tubes on germinating conidia which inhibit germination of conidia and hence directly control the *C. trifolii* by suppressing its life cycle.

#### 8.2.1.1.2 Bacillomycin

Bacillomycins, belonging to iturin family of lipopeptides, are majorly produced by *B. subtilis*, and they have very strong antifungal activity. There are various types of Bacillomycin such as Bacillomycin D, Bacillomycin F, BacillomycinLc, Bacillomycin L, etc. BacillomycinLc was originated from Bacillomycin L but consist of asparagine-1 and glutamate-5 instead of aspartate-1 and glutamine-5 (Eshita et al. 1995).

#### 8.2.1.1.3 Surfactins

Surfactins are natural lipopeptides that are amphiphilic in nature. They have a polar amino acid head and hydrocarbon tail making them natural biosurfactant owing antifungal activity. It consists of β-hydoxy heptacyclic depsipeptides with possible alanine, valine, leucine, or isoleucine amino acid variations at positions 2, 4, and 7 in the cyclic depsipeptide moiety and C-13 to C-16 variation in the β-hydroxy fatty acid chains. They initiate selective cationic channels in the bilayer phospholipid membrane (Sheppard et al. 1991). There are various types of surfactin, for example, a new surfactin WH1 fungin, produced by *B. amyloliquefaciens*, inhibits glucan synthase and decreases callose synthesis on fungal cell wall which causes apoptosis in fungal cells. Similarly, Leu7-surfactin synthesized by *B. mojavensis* controls *Fusarium verticillioides* (Snook et al. 2009). Surfactin can inhibit various *Fusarium* sp., *F. oxysporum*, *F. moniliforme*, and *F. graminearum*, as well as *Colletotrichum gloeosporiodes* (Kim et al. 2010; Ntushelo et al. 2019).

#### 8.2.1.1.4 Plipastatin and Fengycin

Plipastatin and fengycin are very related molecules, except the most important difference between these two molecules lies in the peptide moiety at the position of the D-tyrosine, which is encoded by the fourth gene *ppsD* of plipastatin operon and the second gene *fenB* of fengycin operon (Hussein 2019). Both of these compounds are very effective against filamentous fungi.

Plipastatins A and B are antifungal antibiotics belonging to a family of cyclic lipopeptides. Plipastatin in combination with iturins can completely kill conidial spores of *F. graminearum*. Plipastatin A causes vacuolation and conglobation in young hyphae and branch tips. Plipastatin A treated *Fusarium* had severe damage in their plasma membranes, and the membrane is separated from cell walls thus significantly reducing the fungal survivability.

Fengycin is made from a cyclic lipodecapeptide containing a β-hydroxy fatty acid with a side chain consisting of 16–19 carbon atoms. It consists of isoforms, having difference in length, branching of the β-hydroxy fatty acid moiety, and amino-acid composition of the peptide ring. For instance, fengycin A has D-alanine at position 6. Similarly, fengycin B has D-valine at position 6. They protect plants from broad range of fungi: *Magnaporthe grisea*, *Plasmodiophora brassicae*, *Botryosphaeria dothidea*, *C. gloeosporiodes*, *F. verticillioides*, *F. solani*, *F. graminearum*, and *F. oxysporum*.

Fengycin usually inhibits filamentous fungi by suppressing the functions of the enzymes phospholipase A2 and aromatase. However, fengycin has no action against yeast and bacteria. Fengycin majorly affects the integrity of biological membranes in a molar-ratio-concentration dependent manner. Therefore, the higher the concentration, the greater the disruption membranes causing fungal death. Fengycin also causes structural deformations of the hyphae and suppresses in mycotoxin production as well as inhibits fungal proliferation by making hyphae more permeable and preventing from fungal rot. Hence, fengycin affects cell membrane structure making it more porous, creating primary lesions in the fungal cell membrane disrupting cell integrity and finally causing the death of fungus.

#### 8.2.1.1.5 Zwittermicin A

Zwittermicin A is a highly water-soluble antibiotic which is majorly produced by *B. cereus* group and *B. thuringiensis*. It is an amino polyol having structure similar to polyketide antibiotics exhibiting wide antimicrobial potential. Zwittermicin A is mainly used to treat pathogenic fungi oomycete in agriculture. Hence, this antifungal compound has huge agricultural demand (Sansinenea et al. 2012).

### *8.2.2 Pathogenic Fungal Inhibition by Siderophores from PGPR*

For the growth and development of most of the living organism, iron is one of the essential elements. Despite being the fourth most abundant element in the soil, iron is not easily accessible to the microorganisms as it gets oxidized to form insoluble oxyhydroxide polymers at biological pH and aerobic conditions (Khan et al. 2018).

However, microbes have evolved a number of ways for iron scavenging. Among those, one of the ways includes siderophore-mediated acquisition of iron through a specific receptor and transport systems. Siderophores are low molecular weight (500–1500 daltons range), iron-chelating organic compounds, which are mainly produced by bacteria and fungi under iron-limiting conditions. Because of the very high affinity for iron, siderophores are the strongest soluble $Fe^{3+}$ binding agents known (Fereira et al. 2019). The major function of siderophore is to sequester iron from the surroundings and provide it to microbes. Iron is an essential constituent of cytochrome, heme, as well as non-heme protein and cofactors of various enzymes. So, siderophores can dramatically reduce the amount of iron available to phytopathogens in the rhizosphere which ultimately inhibits the fungal growth (Maindad et al. 2014). Hence, siderophores are broadly used in agriculture to control various plant diseases and to promote plant growth.

In general, siderophores form stable hexadentate or octahedral complex with $Fe^{3+}$. Depending upon the nature of the functional group(s) used for $Fe^{3+}$ coordination (i.e., chemical nature), siderophores are mainly classified into three groups: i) catecholate and phenolate type, ii) hydroxamate type, and iii) carboxylate type (i.e., derivatives of citric acid) (Fig. 8.2). Catecholates and phenolates are also known as "aryl caps" siderophores. However, a mixed type of siderophore comprises of a combination of any of the above three types to chelate iron, i.e., a mixed type contains at least two different iron-binding moieties (Fig. 8.3).

Catecholate type of siderophore consists of catecholate and hydroxyl groups to bind $Fe^{3+}$ with adjacent hydroxyls or catechol ends and forms a hexadentate octahedral complex. These siderophores are only produced by bacteria. It has dihydroxybenzoic acid (DHBA) coupled to an amino acid. Hydroxylate type of siderophores are found in both bacteria and fungi (Paul and Dubey 2015; Saha et al. 2016).

Hydroxamate siderophores are mainly produced by ectorhizosphere and rhizoplane bacteria. These siderophores consist of N5-acyl-N5-Hydroxyornithine or N6-acyl-N6-Hydroxylysine (Winkelmann 2002). A bidentate ligand forms between

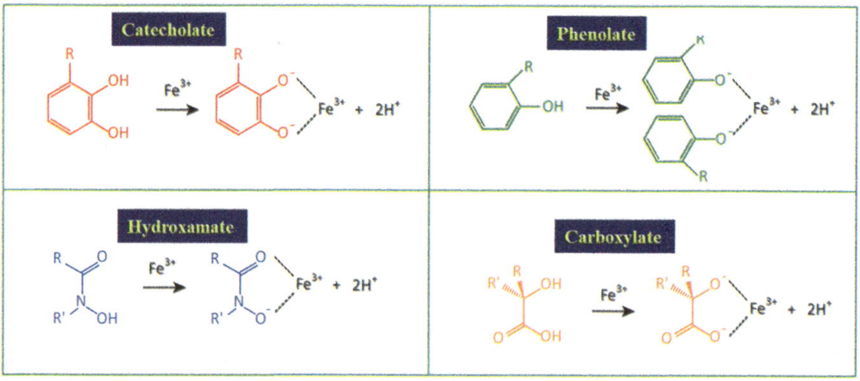


**Fig. 8.2** Four moieties confer iron-binding capacity to siderophores: catecholate, phenolate, hydroxamate, and carboxylate. (Figure adapted from Wilson et al. (2016))

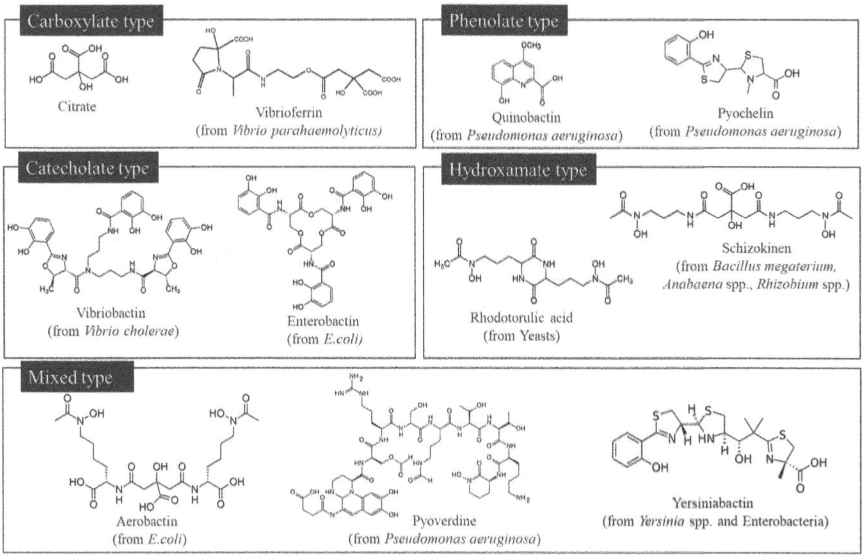


**Fig. 8.3** Chemical structures of various types of siderophores produced by microbes

two oxygen molecules coming from each hydroxamate group and iron. Each hydroxamate is capable of forming a hexadentate octahedral complex with $Fe^{+3}$ ion (Saha et al. 2016).

Carboxylate type siderophores consist of citric acid or β-hydroxyaspartic acid that binds with iron, and they contain both carboxyl and hydroxyl groups. These siderophores are produced in only a few species of bacteria (e.g., *Rhizobium meliloti*, *Staphylococcus hyicus*, etc.) and fungi like members of Mucorales belonging to Zygomycota (Paul and Dubey 2015; Saha et al. 2016;Winkelmann 2002).

The PGPR such as *Rhizobium* sp., *Pseudomonas* sp., *Bacillus* sp., etc. have great potential to produce various types of siderophores under iron limiting conditions. In general, at least one type of siderophore is produced by the rhizobial species. The trihydroxamate vicibactin (Dilworth et al. 1998) and dihydroxymate. Schizokinen siderophores are isolated from *R. leguminosarum* (Storey et al. 2006). The *Rhizobium* spp. isolated from root nodules of groundnut produced dihydroxymate type siderophore which showed antifungal activity on charcoal rot disease caused by *Aspergillus* spp. (Srinivasan 2017). Similarly, several fluorescent *Pseudomonas* strains, particularly *P. aeruginosa*, *P. putida*, and *P. fluorescens*, isolated from plant rhizosphere produce siderophores. *Pseudomonas fluorescens* ATCC 17400 produces two siderophores, pyoverdine and quinolobactin, which display strong in vitro antagonism against the oomycete *Pythium* (Mossialos et al. 2000). Pyoverdine is produced by various *Pseudomonas* sp. including *P. aeruginosa* and *P. putida*. Likewise, *P. aeruginosa* can also produce pyochelin (Cornelis and Dingemans 2013). Similarly, bacillibactin siderophore is produced by *Bacillus* sp. to antagonize the growth of pathogenic *Fusarium* sp. in the soil (Yu et al. 2011).

Based on the mode of biosynthesis, siderophores are classified into two groups: i) non-ribosomal peptide synthetase (NRPS)-dependent route and ii) NRPS-independent route (Carroll and Moore 2018). Many of the siderophores are peptides that are biosynthesized by NRPS modular multienzymes. Biosynthesis of enterobactin siderophore in *E. coli*, *vibriobactin* production in *Vibrio cholerae*, and *yersiniabactin* synthesis in *Yersinia pestis* are mediated by NRPS route (Crosa and Walsh 2002). On the other hand, there has been increasing interest in the study on NRPS independent siderophore (NIS) produced by various microbes. Example of NIS includes aerobactin, desferrioxamine, etc. A hybrid type of siderophore, i.e., NRPS-NIS, is also found in nature. A unique example of the hybrid NRPS-NIS is petrobactin which is produced by various *Bacillus* sp. (Pfleger et al. 2007).

### *8.2.3 Volatile Organic Compounds (VOCs) by PGPR for Fungal Inhibition*

The co-existence of plant and PGPR is known to exist since the beginning of human civilization. The beneficial synergy between the volatile organic compounds (VOCs) produced by PGPR and plant species plays an important role in the sustainable agriculture industry. The global population growth demands sustainable and sufficient crop production with less or no usage of insecticides, pesticides, or any form of synthetic chemical fertilizers. This challenge has been faced by crop scientists worldwide and is even more challenging due to the customers' preference for organic plant products. VOCs on the other hand are low molecular weight, gaseous compounds produced by bacterial cells under normal conditions, and they are active even at their low concentrations (Raza et al. 2016; Wheatley 2002). Some of the working principles of PGPR are inducing plant resistance against phyto-pathogens, regulating balance in hormonal and nutritional level, and easing nutrients uptake by solubilizing the nutrients thereby boosting the plant growth rate (Tahir et al. 2017a; Ryu et al. 2003). In addition, the PGPR-based VOCs have shown an extra degree of advantage over other biocontrol and growth-regulating mechanisms because VOCs do not need any direct physical contact with the pathogen or plant parts, while most of the other processes require direct physical contact (Yunus et al. 2016; Xie et al. 2014). Numerous bacterial species are reported to have an influence on plant growth promotion and induction in systemic resistance (Bailly and Weisskopf 2012; Kai and Piechulla 2014). For instance, 2, 3-butanediol and acetoin are the two biochemicals displaying plant growth-promoting property and are produced by *B. subtilis* GB03 and *B. amyloliquefaciens* IN937a (Ryu et al. 2003). Likewise, 2-pentylfuran secreted by *B. megaterium* XTBG34 stimulates *Arabidopsis thaliana* growth (Zou et al. 2010). In addition to their growth-promoting activity, the VOCs elicit plant tolerance against both biotic and abiotic elements by inducing systemic resistance. The 2, 3-butanediol produced by *B.* species significantly induced resistance in the *Arabidopsis* plant against the pathogen *Erwinia carotovora* subsp. *Carotovora* (Ryu et al. 2004). On the other hand, 3-pentanol and 2-butanone control the pathogen *P. syringae* pv. *lachrymans* on the leaf spot of cucumber (Song and Ryu 2013).

#### 8.2.3.1 Collection and Analysis of Volatiles Organic Compounds

It is quite tricky to collect, analyze, and quantify the amount of VOC produced by PGPR. Unlike the analysis of liquid or solid compounds, special techniques are required for VOC analysis and quantification. The major approved procedures for the collection and assay of VOCs produced by PGPRs are Filter Headspace Collection (FHC) and Headspace-solid phase microextraction methods. The FHC technique mainly requires adsorbent materials such as super Q, Tenax, charcoal, etc. that favor the binding of compounds with specific chemical properties. In this method, organic solvents such as dichloromethane, pentane, hexane, etc. are employed to extract compounds, which result in an inadequate resolution of highly volatile and early-eluting compounds (Ryu et al. 2004).

In the Headspace-solid phase microextraction method, the VOCs present in the culture broth are adsorbed on an extracting phase. The superiority of this method is explained by the two fold increase in the analyzing compounds than the latter mentioned method. Another advantage of this method over the solvent extraction technique is the rapidity and requires only sample volumes. Therefore, this method is suitable for the analysis of VOCs present in fermented broths with low analyte concentration. The use of a thin polymeric film immobilized on a fused silica core allows evaluation of samples under varying experimental conditions and can be used to resolve small inorganic or organic gaseous molecules such as HCN, $NH_3$, and ethylene (Kai and Piechulla 2009).

Gas chromatography-mass spectrometry (GC-MS) analysis is the best technique for the analysis of VOCs as this analyzing instrument allows the determination of the number of different compounds and their relative proportions in the VOCs mixture. The identification of a particular chemical component is made by a comparison of its retention time with its standard one. The GC-MS compound library incorporated within the software makes it easier to determine the compounds if it already exists. Apart from the known compounds, a significant number of unknown compounds may be present in the mixture, and for that Nuclear Magnetic Resonance analysis (NMR) technique is applied (Farag et al. 2006).

#### 8.2.3.2 Different Types of Volatile Organic Compounds and Their Applications

The VOCs produced by PGPR are basically biosynthesized through catabolic pathways such as glycolysis, protein, and lipid degradation pathway. These various chemical classes of VOCs are mainly fatty acid derivatives, sulfur, acids, terpenes, and nitrogen-containing compounds (Penuelas et al. 2014). The most possible role of microbial VOCs is to act as communication signals during inter-and intra-organisms communication as well as a cell-to-cell communication. Beside, VOCs also work as growth promoter through better nutrient uptake in the host plants, or by promoting growth inhibition of disease-causing agents. Likewise, the population, development, establishment, and survival of microbes of different bacterial communities in the defined ecological region crucially depend on VOCs produced by PGPR. Numerous PGPR-based bioactive VOCs were identified and also tested their applications against different plant diseases and phytopathogens. Some of the VOCs and their applications are listed in Table 8.3.

An array of soilborne microbes such as *Bacillus*, *Actinomycetes*, *Pseudomonas*, and *Enterobacter* species produce various types of VOCs. The key mechanism of VOCs is based on suppressing plant pathogens, thereby stimulating plant growth and inducing systemic resistance against plant diseases. VOCs like 2, 3-butanediol, acetonin, and 2-pentylfuran stimulate the growth of plant species in their vicinity by resentencing various types of plant pathogens. The VOCs produced from soilborne microbes are bio-based; hence they are environmentally sound, and they can help to increase crop production and pathogen control without accumulating the chemical pollutants in soil.

## 8.2.4 Hydrolytic Enzymes from PGPR for Pathogen Inhibition

Another important biocontrol mechanism of PGPR includes the synthesis of various hydrolytic enzymes, which catalyze the hydrolysis of chemical bonds. Thus, hydrolytic enzymes inhibit the growth of phytopathogens through hydrolysis of their cell wall, proteins, and DNA materials. Polymeric compounds such as cellulose, chitin, protein, and glucose can be degraded by hydrolytic enzymes from PGPR. For example, 1,4-β-D glycosidic linkages present in the cell wall polysaccharides of plant pathogens are degraded by cellulase enzyme, causing the ultimate death of plant pathogens. Many of the rhizobacterial sp. such as *S. marcescens*, *B. subtilis*, *B. cereus*, *B. thuringiensis*, etc. have a potential to produce hydrolytic enzymes for the biocontrol of plant pathogens (Felse and Panda 1999; Someya et al. 2000).

### 8.2.4.1 Types of Hydrolytic Enzymes

Different types of hydrolytic enzymes from PGPR are found in soil, and each type of enzyme has a specific function, and they are named depending upon the compound they fix, break, or degrade. Presence of hydrocarbons, chemical fertilizers, as well as heavy metals in the soil could have a profound effect on plant growth and yield. In addition to the pathogen inhibition properties, hydrolytic enzymes also contribute to soil remediation by removing toxic chemicals and heavy metals that are deposited into the soil through chemically polluted water and other environmental factors (Metin et al. 2016). Various hydrolytic enzymes from PGPR are described below:

**Table 8.3** Various types of bioactive VOCs produced by soilborne microbes and their applications

| Producers | Bioactive compounds | Applications | References |
|---|---|---|---|
| *Bacillus subtilis* GB03, *B. amyloliquefaciens* IN937a | 2, 3-Butanediol | Induce resistance in the *Arabidopsis* against plant pathogen *Erwinia carotovora* | Ryu et al. (2003) |
| *Bacillus megaterium* XTBG34 | 2-Pentylfuran | Stimulates *Arabidopsis thaliana* growth | Zou et al. (2010) |
| *Pseudomonas fluorescens* SS101 | 13-Tetradecadien-1-ol, 2-butanone and 2-methyl-n-1-tridecene | Promote *Nicotiana tabacum* growth | Park et al. (2015) |
| *Paenibacillus polymyxa* E681 | Tridecane | Significantly induced systemic resistance in plants | Lee et al. (2012) |
| *Bacillus amyloliquefaciens* L3 | 2-Nonanone and 2-heptanone | Antifungal activities | Wu et al. (2019) |
| *Pseudomonas chlororaphis* | Cyclohexanol | Inhibition of mycelial plug/ sclerotia | Fernando et al. (2005a, b) |
| *Flavobacterium* sp. GSE09 | 2,4-Di-tertbutylphenol | Stimulated pepper fruit ripening | Sang et al. (2010) |
| *Paenibacillus polymyxa* E681 | Hexadecane | Induced resistance systemic against both *Pectobacterium carotovorum* and *pseudomonas syringae* | Park et al. (2013) |
| *Pseudomonas aurantiaca* | Benzothiazole | Inhibition of mycelial plug/ sclerotia | Fernando et al. (2005a, b) |
| *Bacillus subtilis* FB17 | Acetoin | ISR mediated by salicylic acid and ethylene pathway effective against *pseudomonas syringae* DC3000 | Radruppa et al. (2010) |
| *Bacillus subtilis* SYST2 | Albuterol and 1,3-propanediol | Growth promotion and induced systemic resistance against the bacterial wilt pathogen | Tahir et al. (2017b) |

##### 8.2.4.1.1 Chitinase

Enzyme chitinase produced by PGPR lyses the fungal cell wall through degradation of chitin polymer present in the cell wall of plant pathogens. The degradation of chitin polymer into monomers damages the cell wall of the plant pathogens causing the ultimate death of the cell. The hydrolytic enzyme can either be used directly in the biocontrol of microorganisms or indirectly by using purified chitinase enzymes (Kim et al. 2003). Soilborne microbes such as *S. marcescens* and *B. subtilis* NPU001 produce enzyme chitinase that is capable of controlling the plant pathogens like *Rhizoctonia solani* and *F. oxysporum* (Someya et al. 2000; Chang et al. 2010).

The hydrolytic enzyme chitinase produced by *Bacillus* species also plays an important role to control the plant-parasitic nematodes which are the most economically and ecologically damaging pests, causing severe loss of crop production worldwide (Frankland 1887). The use of hydrolytic enzyme is superior to chemical-based nematocides due to its adverse effects on human health and the environment.

#### 8.2.4.1.2 Phytase

Phosphorus is another vital component for plants growth found in soil which, however, is found in insoluble form for the plant to utilize directly. For the availability of the phosphorus, phytase enzyme is required, which converts insoluble phosphate to organic acids like phytic acids. Hence, phytase enzyme catalyzes the degradation reaction of insoluble phosphorus into soluble form so that it can be available for plants. The phytase is produced by a group of PGPR (Richardson 2001).

#### 8.2.4.1.3 Glucanase

Glucanase is produced by a number of soilborne microbes such as bacteria, fungi, and also by higher plants. Glucanases are of two types, exo- or endo-β-1, 3-glucanases. In filamentous ascomycetes, β-1,3-glucan constitutes a prominent cell wall component that is responsible for the rigidity of the cell wall structure. These enzymes cause the degradation of the pathogen's cell wall and further penetrate into the host mycelium (Fridlender et al. 1993). Endo-β-1,3-glucanase activities cleave inside a glucan chain in a more or less random fashion (Fig. 8.4), while the exo-β-1,3-glucanase activities release glucose residues from the non-reducing end.

Glucanase can hydrolyze the cell wall components by hydrolysis of the substrate, which sequentially cleaves glucose residues from the non-reducing end and cleaving linkages random sites along the polysaccharide chain, releasing smaller oligosaccharides (Noronha and Ulhoa 1996). The hydrolytic enzyme β-1,3-glucanase produced by *Streptomyces* sp. strain 385 controls the plant pathogen *F. oxysporum* (Singh et al. 1999).

#### 8.2.4.1.4 Proteases

Proteases play a significant role in the lysis of the cell wall of plant pathogenic fungi because chitin and/or fibrils of β-glucan are embedded into the protein matrix. The protease breaks down major proteins of plant pathogens into peptide chains and/or their constituent amino acids and thereby abolishing the capacity of the pathogen's protein to act against plant cells. The enzyme proteinase produced by *Bacillus* species plays a significant role to control the plant-parasitic nematodes (Frankland 1887). *Pseudomonas fluorescens* LRB3W1 and *Serratia marcescens* B2-F species

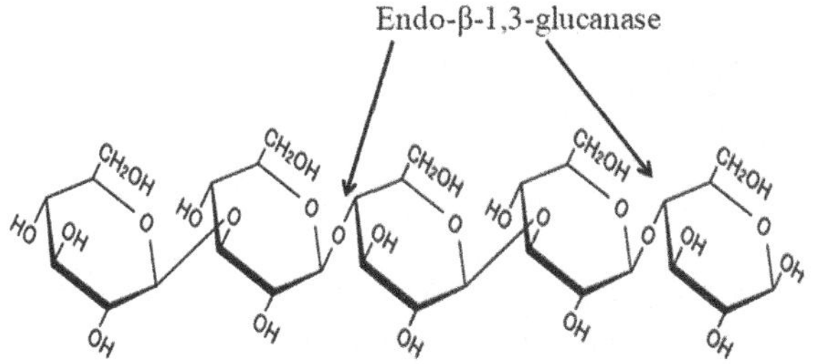


**Fig. 8.4** Mechanism of degradation of β-1,3-glucan by enzyme β-1,3-glucanase

produce the enzyme proteinase that kills the plant pathogen *F. oxysporum* (Someya et al. 2007). Similarly, various *Bacillus* proteases can act against the pathogenic fungi (Elad and Kapat 1999).

8.2.4.1.5 Cellulase

Enzyme cellulase hydrolyzes the 1,4-β-D-glucosidic linkages in cellulose and plays a significant role in nature by recycling this polysaccharide. Cellulose chains form numerous intra- and intermolecular hydrogen bonds, which account for the formation of rigid, insoluble, crystalline microfibrils. The complete degradation of cellulose involves a complex interaction between three different cellulolytic enzymes, cellulose/endoglucanases, exo-cellobiohydrolase/exo-glucanases, and β-glucosidases, which act synergistically to convert cellulose into β-glucose (Lynd et al. 2002). *B. subtilis* YJ1 produces the enzyme cellulase and that controls plant pathogen *F. oxysporum* (Li-Jung et al. 2010).

### 8.2.4.2 Other Enzymes

In addition to the above hydrolytic enzymes, other enzymes such as nitrogenase and aminocyclopropane-1-carboxylate (ACC) deaminase also play major roles to improve growth and maintain good health of the plants and crops.

8.2.4.2.1 Nitrogenase

Nitrogen is an important element for the plant because it is a vital element of chlorophyll; therefore, nitrogen has a direct impact on photosynthesis. The free nitrogen from the atmosphere is fixed and made available to the plants by PGPR. Nitrogen is also a major component of amino acids which is the building block of protein. Therefore, obtaining the metabolically consumable form of nitrogen, i.e., reduction into ammonia, is an important process. The group of PGPR named diazotrophs is a major group of organisms that generates their own nitrogenous monomers and is used for the synthesis of nucleic acids, proteins, etc. through different biochemical pathways (Burns and Hardy 1978). Hence, diazotrophs mediate nitrogen fixation through the nitrogenase enzyme.

#### 8.2.4.2.2 Aminocyclopropane-1-Carboxylate (ACC) Deaminase

The enzyme ACC deaminase reduces the hydrocarbon contamination like ethylene level in the soil. The mode of action of ACC deaminase is initiated by catalyzing the reaction where ACC breaks down to $\alpha$-ketobutyrate and ammonia (Honma and Shimomura 1978). This special enzyme produced by PGPR plays a vital role in plant growth and nutrient uptake through nitrogen fixation. ACC enables plants to uptake metals like iron and soluble phosphate, hence promoting plant growth.

## 8.3 Conclusion

Plant growth-promoting rhizobacteria play a vital role in a plant's health in various ways. In one hand, these PGPR scavenge micronutrients from soil that are essential for the growth and development of pathogens. On the other hand, colonizing in the root of plants by consuming root extrudates, these PGPR biosynthesize and secrete wide ranges of secondary metabolites with antagonistic activity to kill the plant pathogens. The PGPR are naturally soilborne microbes, and they produce the various bioactive metabolites depending upon their need to survive in the environment. Among those bioactive metabolites lipopeptides are the main secondary metabolites that have significant effect against filamentous fungi. In addition, siderophores also display major role to inhibit pathogenic fungi to help plants against fungal diseases. Apart from these secondary metabolites hydrolytic enzymes such as chitinase, gluconase, proteinase, phytase, and cellulose produce and other enzymes like nitrogenase, and ACC deaminase by different plant growth-promoting rhizobacteria play significant role to control the plant pathogens and plant growth promotion. These enzymes destroy the plant pathogens by lysing their cell wall and ultimately killing pathogens. The uses of PGPR-based hydrolytic enzymes are environment-friendly as compared to chemical-based pesticides, nematocides, and fungicides. In addition, without direct contact, volatile compounds produced by these PGPR have both plant growth promotion and diseases control effect. Hence, PGPR helps plant in various ways from growth promotion activities by providing nutrients as well as secreting bioactive metabolites that combat various pathogens causing diseases or induced resistant system in plants. The detailed insights on the biosynthesis of these growth promotion factors and secondary metabolites from PGPR along with the mode of action for fungi inhibition will be beneficial to develop sustainable and next-generation biofungicides to sophisticate sustainable agriculture. Furthermore, effect of the combination of these metabolites against various plant pathogens should be studied in depth to develop the effective biocontrol system replacing the chemical fertilizers and pesticides.

## References

Backer R, Rokem JS, Ilangumaran G, Lamont J, Praslickova D, Ricci E, Subramanian S, Smith DL (2018) Plant growth-promoting Rhizobacteria: context, mechanisms of action, and roadmap to commercialization of biostimulants for sustainable agriculture. Front Plant Sci 23(9):1473

Bailly A, Weisskopf L (2012) Current knowledge and future challenges. Plant Signal Behav 7(1):1–7

Bloemberg GV, Lugtenberg BJ (2001) Molecular basis of plant growth promotion and biocontrol by rhizobacteria. Curr Opin Plant Biol 4(4):343–350

Brumm PJ, Hebeda RE, Teague WM (1991) Purification and characterization of the commercialized, cloned *bacillus megaterium* α-amylase. Part I: purification and hydrolytic properties. Starch 43:315–319

Burns RC, Hardy RWF (1978) Nitrogen fixation in bacteria and higher plants. Berlin: Springer-Verlag Mol Biol BiochemBiophys 21

Carroll CS, Moore MM (2018) Ironing out siderophore biosynthesis: a review of non-ribosomal peptide synthetase (NRPS)-independent siderophore synthetases. Crit Rev Biochem Mol Biol 53(4):356–381

Chang WT, Chen M, Wang SL (2010) An antifungal chitinase produced by *Bacillus subtilis* using chitin waste as a carbon source. World J Microbiol Biotechnol 26(5):945–950

Cornelis P, Dingemans J (2013) *Pseudomonas aeruginosa* adapts its iron uptake strategies in function of the type of infections. Front Cell Infect Microbiol 3:75

Crosa JH, Walsh CT (2002) Genetics and assembly line enzymology of siderophore biosynthesis in bacteria. Microbiol Mol Biol Rev 66(2):223–249

Damalas CA, Koutroubas SD (2016) Farmers' exposure to pesticides: toxicity types and ways of prevention. Toxics 4(1):1

Dilworth MJ, Carson KC, Giles RGF, Byrne LT, Glenn AR (1998) Rhizobium leguminosarumbv. viciae produces a novel cyclic trihydroxamate siderophore, vicibactin. Microbiology 144:781–791

Elad Y, Kapat A (1999) The role of *Trichoderma harzianum* protease in the biocontrol of *Botrytis cinerea*. Eur J Plant Pathol 105:177–189

Eshita SM, Roberto NH, Beale JM, Mamiya BM, Workman RF (1995) BacillomycinLc, a new antibiotic of the iturin group: isolations, structures, and antifungal activities of the congeners. J Antibiot 48(11):1240–1247

Farag MA, Ryu CM, Sumner LW, Paré PW (2006) GC–MS SPME profiling of rhizobacterial volatiles reveals prospective inducers of growth promotion and induced systemic resistance in plants. Phytochemistry 67(20):2262–2268

Felse AP, Panda T (1999) Production of microbial chitinases. Bioprocess Eng 23:127–134

Fereira MJ, Silva H, Cunha A (2019) Siderophore-producing *Rhizobacteria* as a promising tool for empowering plants to cope with iron limitation in saline soils: A Review. Pedosphere 29(4):409–420

Fernando DW, Ramarathnam R, Krishnamoorthy A, Savchuk S (2005a) Identifi cation and use of potential bacterial organic antifungal volatiles in biocontrol. Soil Biol Biochem 37(5):955–964

Fernando WGD, Nakkeeran S, Zhang Y (2005b) Biosynthesis of antibiotics by PGPR and its relation in biocontrol of plant diseases. In: Siddiqui ZA (ed) PGPR: biocontrol and biofertilization. Springer, Dordrecht

Fisher MC, Henk DA, Briggs CJ, Brownstein JS, Madoff LC, McCraw SL (2012) Emerging fungal threats to animal, plant and ecosystem health. Nature 484:186–194

Frankland SG, Frankland RF (1887) Studies on some new microorganisms obtained from air. Royal Soc Lon-don Phil Trans Ser B Biol Sci 178:257–287

Fridlender M, Inbar J, Chet I (1993) Biological control of soil borne plant pathogens by a ß–1 ,3–glucanase producing *Pseudomonas cepacia*. Soil Biol Biochem 25(9):1211–1221

Gates SD, McCartt AD, Lappas P, Jeffries JB, Hanson RK, Hokama LA, Mortelmans KE (2010) *Bacillus* endospore resistance to gas dynamic heating. J Appl Microbiol 109:1591–1598

Honma M, Shimomura T (1978) Metabolism of 1-Aminocyclopropane-1-carboxylic acid. Agric Biol Chem 42(10):1825–1831

Hussein W (2019) Fengycin or plipastatin? A confusing question in bacilli. BioTechnologia 100(1):47–55

Jadhav HP, Shaikh SS, Sayyed RZ (2017) Role of hydrolytic enzymes of *Rhizoflora*in biocontrol of fungal phytopathogens: an overview. Rhizotrophs: Plant Growth Promotion Bioremediation:183–203

Kai M, Piechulla B (2009) Plant growth promotion due to rhizobacterial volatiles-an effect of CO 2. FEBS Lett 583(21):3473–3477

Kai M, Piechulla B (2014) Impact of volatiles of the *rhizobacteria Serratia odorifera* on the moss Physcomitrella patens. Plant Signal Behav 5(4):444–446

Keswani C, Singh HB, García-Estrada C (2020) Antimicrobial secondary metabolites from agriculturally important bacteria as next-generation pesticides. Appl Microbiol Biotechnol 104:1013–1034

Khan A, Singh P, Srivastava A (2018) Synthesis, nature and utility of universal iron chelator-Siderophore: a review. Microbiol Res 212-213:103–111

Kim KJ, Yang YJ, Kim JG (2003) Purification and characterization of Chitinase from *Streptomyces* sp. M–20. J Biochem Mol Biol 36(2):185–189

Kim PI, Ryu J, Kim YH, Chi YT (2010) Production of biosurfactant lipopeptides Iturin a, fengycin and surfactin a from*Bacillus subtilis* CMB32 for control of colletotrichumgloeosporioides. J Microbiol Biotechnol 20(1):138–145

Kloepper JW, Schroth MN (1978) Plant growth promoting rhizobacteria on radishes 1978. Proc 4th Int Conf Plant Path Bact Angers 2:879–882

Kowall M, Vater J, Kluge B, Stein T, Franke P, Ziessow D (1998) Separation and characterization of surfactin isoforms produced by *Bacillus subtilis* OKB 105. J Colloid Interf Sci 204:1–8

Lee B, Farag MA, Park HB, Kloepper JW, Lee SH, Ryu CM (2012) Induced resistance by a long-chain bacterial volatile: elicitation of plant systemic defense by a C13 volatile produced by *Paenibacilluspolymyxa*. PLoS One 7(11):48744

Li-Jung Y, Hsin-Hung L, Zheng-Rong X (2010) Purification and characterization of a cellulase from *bacillus subtilis* YJ1. J Mar Sci Technol 18(3):466–471

Lynd LR, Weimer PJ, van Zyl WH (2002) Microbial cellulose utilization: fundamentals and biotechnology. Microbiol Mol Biol Rev 66(3):506–577

Maindad DV, Kasture VM, Chaudhari H, Dhavale DD, ChopadeBA SDP (2014) Characterization and fungal inhibition activity of siderophore from wheat rhizosphere associated acinetobacter-calcoaceticus strain HIRFA32. Indian J Microbiol 54(3):315–322

Meena KR, Kanwar SS (2015) Lipopeptides as the antifungal and antibacterial agents: applications in food safety and therapeutics. Bio Med Research Inter 2:9

Metin T, Bulent T, Nurgul K, Ulker A, Filiz E, Emrah N (2016) Adem G, Plant growth promoting rhizobacteria's (PGPRS) enzyme dynamics in soil remediation. 11:210–230

Miljaković D, Marinković J, Balešević-Tubić S (2020) The significance of*Bacillus* spp. in disease suppression and growth promotion of field and vegetable crops. Microorganisms 8:1037

Mossialos D, Meyer JM, Budzikiewicz H, Wolff U, Koedam N, Baysse C, Anjaiah V, Cornelis P (2000) Quinolobactin, a new siderophore of *Pseudomonas fluorescens* ATCC 17400, the production of which is repressed by the cognate pyoverdine. Appl Environ Microbiol 66(2):487–492

Noronha EF, Ulhoa CJ (1996) Purification and characterization of an endo–β–1,3–glucanase from *Trichoderma harzianum*. Can J Microbiol 42(10):1039–1044

Ntushelo K, Ledwaba LK, Rauwane ME, Adebo OA, Njobeh PB (2019) The mode of action of *bacillus* species against *fusarium graminearum*, tools for investigation, and future prospects. Toxins (Basel) 11(10):606

Olanrewaju OS, Glick BR, Babalola OO (2017) Mechanisms of action of plant growth promoting bacteria. World J Microbiol Biotechnol 33(11):197

Park HB, Lee B, Kloepper JW, Ryu CM (2013) One shot-two pathogens blocked. Exposure of *Arabidopsis* to hexadecane, a long chain volatile organic compound, confers induced resistance against both *Pectobacteriumcarotovorum* and *pseudomonas syringae*. Plant Signal Behav 8(7):e24619-1–e24619-3

Park YS, Dutta S, Ann M, Raaijmakers JM, Park K (2015) Promotion of plant growth by *pseudomonas Fluorescens* strain SS101 via novel volatile organic compounds. Biochem Biophys Res Commun 461(2):361–365

Paul A, Dubey R (2015) Characterization of protein involved in nitrogen fixation and estimation of co-factor. Appl J Curr Res Biosci Plant Boil 2(1):89–97

Penha RO, Vandenberghe LPS, Faulds C (2020) *Bacillus* lipopeptides as powerful pest control agents for a more sustainable and healthy agriculture: recent studies and innovations. Planta 251:70

Penuelas J, Asensio D, Tholl D (2014) Biogenic volatile emissions from the soil. Plant Cell Environ 37(8):1866–1891

Pfleger BF, Lee JY, Somu RV, Aldrich CC, Hanna PC, Sherman DH (2007) Characterization and analysis of early enzymes for petrobactin biosynthesis in *bacillus anthracis*. Biochemist 46(13):4147–4157

Radruppa T, Biedrzycki ML, Kunjeti SG, Donofrio NM, Czymmek KJ, Paré PW, Bais HP (2010) The rhizobacterial elicitor acetoin induces systemic resistance in *Arabidopsis thaliana*. Commun Integr Biol 3(2):130–138

Raza W, Yousaf S, Rajer FU (2016) Plant growth promoting activity of volatile organic compounds produced by biocontrol strains. Sci Lett 4(1):40–43

Richardson AE (2001) Prospects for using soil microorganisms to improve the acquisition of phosphorus by plants. Aust J Plant Physiol 28:897–906

Ryu C, Farag MA, Hu C, Reddy MS, Wei H, Pare PW (2003) Bacterial volatiles promote growth in *Arabidopsis*. PNAS 100(8):4927–4932

Ryu C, Farag MA, Hu C, Reddy MS, Kloepper JW, Pare PW (2004) Bacterial volatiles induce systemic resistance in *Arabidopsis*. Plant Physiol 134(3):1017–1026

Saha M, Sarkar S, Sarkar B, Sharma B, Bhattacharjee S, Tribedi P (2016) Microbial siderophores and their potential applications: a review. Environ Sci Pollut Res 23:3984–3999

Sang Y, Ryu KC, Lee S, Bee H, Soo K, Jung H (2010) Proteome analysis of *Arabidopsis* seedlings exposed to bacterial volatiles. Planta 232(6):1355–1370

Sansinenea E, Ortiz A, Zwittermicin A (2012) A promising aminopolyol antibiotic from biocontrol bacteria. Current Org Chem 16:978–987

Sheppard JD, Jumarie C, Cooper DG, Laprade R (1991) Ionic channels induced by surfactin in planar lipid bilayer membranes. Biochim Biophys 1064:13–23

Singh PP, Shin YC, Park CS, Chung YR (1999) Biological control of *fusarium* wilt of cucumber by chitinolytic bacteria. Phytopathology 89(1):92–99

Snook ME, Mitchell T, Hinton DM, Bacon CW (2009) Isolation and characterization of Leu7-surfactin from the endophytic bacterium *Bacillus mojavensis* RRC 101, a biocontrol agent for *fusarium verticillioides*. J Agric Food Chem 57:4287–4292

Someya N, Kataoka N, Komagata T (2000) Biological control of cyclamen soil borne diseases by *Serratia marcescens* strain B2. Plant Dis 84(3):334–340

Someya N, Tsuchiya K, Yoshida T, Noguchi MT, Akutsu K, Sawada H (2007) Fungal cell wall degrading enzyme-producing bacterium enhances the biocontrol efficacy of antibiotic producing bacterium against cabbage yellows. J Plant Dis Protec 114(3):108–112

Song GC, Ryu CM (2013) Two volatile organic compounds trigger plant self defense against a bacterial pathogen and a sucking insect in cucumber under open field conditions. Int J Mol Sci 14(5):9803–9819

Srinivasan T (2017) Studies on antifungal activity of siderophores produced by *Rrhizobium* spp. isolated from groundnut (Arachis hypogaea). J Agric Food Sci Res 8:4

Storey EP, Boghozian R, James L, Douglas L, Chakraborty R (2006) Characterization of 'Schizokinen'; a dihydroxamate-type siderophore produced by *rhizobium leguminosarum* IARI 917. Biometals 19:637–649

Tahir HAS, Gu Q, Wu H, Niu Y, Huo R, Gao X (2017a) *Bacillus* volatiles adversely affect the physiology and ultra-structure of *Ralstonia solanacearum* and induce systemic resistance in tobacco against bacterial wilt. Sci Rep 7(1):40481

Tahirl HAS, Gu Q, Wu H, Raza W, Safdar A, Huang Z, Rajer FU, Gao X (2017b) Effect of volatile compounds produced by *Ralstonia solanacearum* on plant growth promoting and systemic resistance inducing potential of *bacillus* volatiles. BMC Plant Biol 17(1):33

Wheatley RE (2002) The consequences of volatile organic compound mediated bacterial and fungal interactions. Antonie Van Leeuwenhoek 81:357–364

Wilson BR, Bogdan AR, Miyazawa M, Hashimoto K, Tsuji Y (2016) Siderophores in iron metabolism: from mechanism to therapy potential. Trends Mol Med 22:1077–1090

Winkelmann G (2002) Microbial siderophore-mediated transport. Biochem Soc Trans 30(4):691–696

Wu Y, Zhou J, Li C, Ma Y (2019) Antifungal and plant growth promotion activity of volatile organic compounds produced by *Bacillus amyloliquefaciens*. Microbiol Open:1–14

Xie S, Wu H, Zang H, Wu L, Zhu Q, Gao X (2014) Plant growth promotion by spermidine-producing *Bacillus subtilis* OKB105. MPMI 27(7):655–663

Yu X, Ai C, Xin L, Zhou G (2011) The siderophore-producing bacterium, *Bacillus subtilis* CAS15, has a biocontrol effect on fusarium wilt and promotes the growth of pepper. Eur J Soil Biol 47(2):138–145

Yunus FN, Iqbal M, Jabeen K, Kanwal Z, Rashid F (2016) Antagonistic activity of *Pseudomonas fluorescens* against fungal plant pathogen *aspergillus Niger*. Sci Lett 4(1):66–70

Zou C, Li Z, Yu D (2010) *Bacillus megaterium* strain XTBG34 promotes plant growth by producing 2-pentylfuran. J Microbiol 48(4):460–466

# Part II
# Volatile Compounds and Hydrolytic Enzymes of PGPR

# Chapter 9
# Metabolomic Profiling of Plant Growth-Promoting Rhizobacteria for Biological Control of Phytopathogens

**Zakira Naureen, Syed Abdullah Gilani, Binta Kondoor Benny, Hajra Sadia, Fauzia Y. Hafeez, and Azra Khanum**

## Contents

9.1 Introduction .......... 181
9.2 Microbial Metabolites and Their Role in Biotic Interactions .......... 183
9.3 Secondary Metabolites Involved in Antagonism .......... 184
9.4 Factors Affecting Secondary Metabolite Production in Soil Bacteria .......... 193
9.5 Metabolomic Profiling of PGPR .......... 194
9.6 Techniques Used to Study Metabolomic Profiles .......... 196
9.7 Genome Mining .......... 199
9.8 Conclusion .......... 200
References .......... 201

## 9.1 Introduction

Metabolomics is the study of small metabolites such as amino acids, nucleic acids, lipids, or carbohydrates and complex secondary metabolites present in biological systems inside the cells or extracellular fluids. Metabolomics utilizes the analytical techniques such as gas or liquid chromatography, mass spectrometry (MS), and

Z. Naureen (✉) · S. A. Gilani · B. K. Benny
Department of Biological Sciences and Chemistry, College of Arts and Science, University of Nizwa, Nizwa, Sultanate of Oman
e-mail: zakira@unizwa.edu.om

H. Sadia
National University of Science and Technology, Islamabad, Pakistan

F. Y. Hafeez
COMSATS University, Islamabad, Pakistan

A. Khanum
University institute of Biochemistry and Biotechnology, PMAS University of Arid Agriculture Rawalpindi, Rawalpindi, Pakistan

 
R. Z. Sayyed et al. (eds.), *Antifungal Metabolites of Rhizobacteria for Sustainable Agriculture*, Fungal Biology, https://doi.org/10.1007/978-3-031-04805-0_9

nuclear magnetic resonance (NMR) spectroscopy to create metabolomic profiles that enable identification and relative quantification of metabolites at a given time. Metabolomic profiles are basically the secondary metabolites produced under extracellular and intracellular environmental condition. Metabolomic study of plant-microbe interactions in the rhizosphere is an emerging field and encompasses analysis of entire metabolome and metagenome to understand the complex inter- and intra-species interactions in the root zone (Reuben et al. 2008; Wen et al. 2020; Ilyas et al. 2020; Tantray et al. 2021).

Rhizosphere being a complex and dynamic microenvironment is the house of numerous interactions which can be broadly categorized into plant root-microbe interactions and microbe-microbe interactions (Mhlongo et al. 2018; Reuben et al. 2008). These can be in turn subdivided into plant-root-pathogen interaction; plant root-PGPR interaction; PGPR-PGPR interaction; and PGPR-pathogen interaction. As a result, there is a complex metabolomic interaction in this microenvironment which is governed by several factors such as plant root exudates that help shape a microbial community within the vicinity, extracellular secretions of rhizobacterial metabolites, and presence of other soil microorganisms (Naureen et al. 2005; Wen et al. 2020). Collectively, the metabolites secreted from plant root and the surrounding microbial community determines the overall metabolome in the rhizosphere. These metabolites secreted from plant and microbes determine which of the microbial communities will survive and which will be out competed from the rhizosphere (Naureen et al. 2005). Under natural environmental conditions, the plant health is largely dependent on presence of plant growth-promoting bacteria and absence of pathogenic bacteria besides the other biotic and abiotic factors. This highlights the importance of favorable interactions between plant and associated PGPR that are characterized, governed, and manifested by the metabolites released by the two parties (Wen et al. 2020).

The most sought-after secondary metabolites of PGPR are the antimicrobial compounds that can kill, eradicate, or suppress phytopathogens in the root zone or prime plants to combat the attacking pathogens. One of the basic methods to get these antimicrobials is to maintain a healthy population of the PGPR bioinoculant in the rhizosphere; however, research has shown that this is a challenging task, and therefore it is envisaged that these secondary microbial metabolites could possibly be deployed to control phytopathogens and improve plant growth (Basu et al. 2021). Besides that, although indigenous microbial communities in the rhizosphere are abundant, only a few members of these communities can benefit the plants. This implicates selection and screening of useful bacterial strains that can produce a variety of secondary metabolites manifesting plant growth-promoting or biocontrol activities.

Hence analysis of these metabolites to generate a profile will not only enable us to understand the complex interactions in the rhizosphere but will help identify novel secondary metabolites that can serve as plant growth enhancers or biological control agents. In addition, these metabolomic profiles will help in selection of multifaceted microbial strains for integrated pest management.

Conventional techniques based on microbiological assays, biochemical analysis, and enzymatic studies have been used to study rhizobacterial secondary metabolites; however, these methods fail to give an insight into effect of these metabolites on the system in context. In addition, most of these techniques can target only a limited number of metabolites at a time and require several assays or experiments to get a complete profile. On the other hand, metabolomic profiles generated via high-throughput data analysis tools help to overcome these drawbacks and provide a comprehensive biochemical status report (Pham et al. 2005; Vignoli et al. 2019).

Consequently, this chapter will focus on PGPR metabolome with respect to two major categories of secondary metabolites involved in biological control of phytopathogens. In addition to that, the factors effecting the secondary metabolite production and techniques used to study these metabolites will be discussed.

## 9.2 Microbial Metabolites and Their Role in Biotic Interactions

Secondary metabolites of PGPR were speculated to be metabolic spin offs from primary metabolism, with coincidental activity. However, the diversity in structure and function of these metabolites highlights their importance in the rhizosphere. Several studies have reported that the soil bacteria produce these secondary metabolites to acquire a niche close to plant roots so that they can get a continuous supply of root exudates. In doing so they have to compete the neighboring communities, associate with symbiotic relatives, communicate via signaling, and sequester nutrients. To accomplish all these tasks, the soil bacteria synthesize a variety of soluble and volatile secondary metabolites that mediate microbe-microbe and plant-microbe interactions. Some of these metabolites act as weapons in microbial warfare, providing a competitive advantage to producer strains in niche exclusion (Sharma et al. 2020; Jauri and Kinkel 2014; Schulz-Bohm et al. 2015), while others such as homoserine lactone act as chemical signals that mediate quorum sensing; still others behave as metal chelating compounds such as siderophores that sequester iron and antimicrobial compounds that suppress or kill the soil-borne plant pathogens, the ultimate competitors of soil bacteria for food and shelter (Berg et al. 2014; Naureen et al. 2017). Secondary metabolites with antimicrobial activity play a significant role in shaping microbial communities and mediating predation or protection against predators (Korp et al. 2016). For instance, knocking out genes responsible for antibiotic production in the predatory soil bacterium *Myxococcus xanthus* results in a diminished growth rate on prey cells (Xiao et al. 2011). Besides that, these antimicrobial secondary metabolites might act as signaling molecules in inter- and intra-species interactions, affecting various cellular functions such as biofilm formation, virulence, motility, and nutrient use (Linares et al. 2006).

PGPR have a mutualistic relationship with plants, and to maintain this relationship, they must produce secondary metabolites that are beneficial for both parties. For instance, PGPR can drastically alter plant's root system and enhance plant biomass production by emitting complex volatiles metabolites that act without actual physical contact (Ryu et al. 2003). These metabolites can not only directly antagonize phytopathogenic fungi but also induce systemic resistance in plants against several phytopathogens (Van Dam et al. 2016). Conclusively it can be stated that PGPR have a million reasons to produce these secondary metabolites based on the ecological functions they serve for the bacteria and the host plant; however, the production of these metabolites is under strict regulatory controls that allow production of only particular metabolites depending upon the environmental stimulus received. In the next section, we will have a detailed overview of microbial metabolites produced by PGPR strains that act as antimicrobials and play a significant role in the rhizosphere.

## 9.3 Secondary Metabolites Involved in Antagonism

There is a huge repertoire of research studies reporting production of plant growth-promoting properties of PGPR strains such as nitrogen fixation, phytohormone production, P-solubilization, zinc mobilization, etc. (Naureen et al. 2017). Similarly, there are several reports of biocontrol activities such as production of antagonistic metabolites and hydrolytic enzymes. In this section we will describe only those categories of metabolites that have antagonistic activity and can be used to provide biological control of plant pathogens.

PGPR produce a myriad of metabolites such as soluble and volatile organic compounds having diverse physiochemical and biological properties. These metabolites play an essential role in shaping the overall plant-microbe and microbe-microbe interactions in the root zone. For instance, the volatile metabolites are small molecules (<300 Da) belonging to various chemical classes and can diffuse through air- and water-filled pores and even without water in soil and act as chemical signals for promotion of cooperation or competition among soil inhabitants that are not living in direct vicinity (Schulz and Dickschat 2007; Schmidt et al. 2015; Hossain et al. 2015; Ledger et al. 2016, Kannojia et al. 2019). However, the soluble secondary metabolites are highly soluble in water owing to their higher polarity and act for shorter distances exhibiting strong biological activities in promoting competition and niche exclusion because of their high degree of functionalization.

The main classes of secondary metabolites involved in biological control (soluble and volatile compounds) produced by soil bacteria are shown in Figs. 9.1 and 9.2.

### *9.3.1 Soluble Compounds*

#### 9.3.1.1 Bacteriocins

Competition for nutrients and acquisition of a niche close to plant roots to get a continuous supply of root exudates is one of the fierce battles among microbial inhabitants of rhizosphere. Consequently, the rhizobacteria secrete several soluble

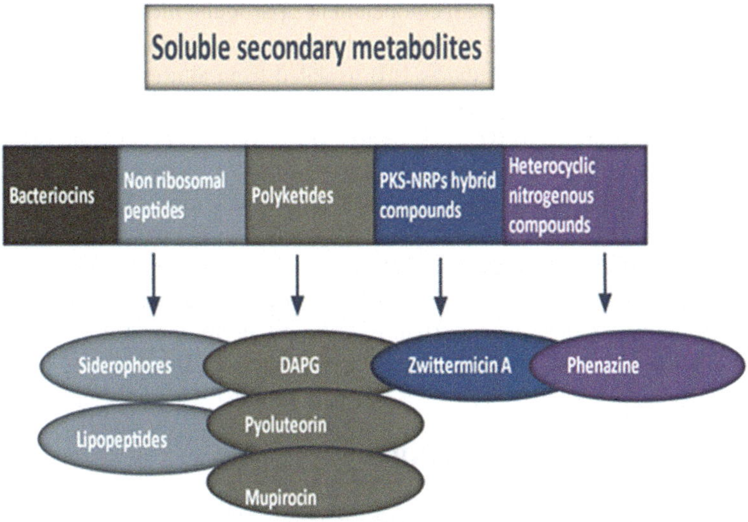


**Fig. 9.1** Soluble secondary metabolites involved in biological control of phytopathogens

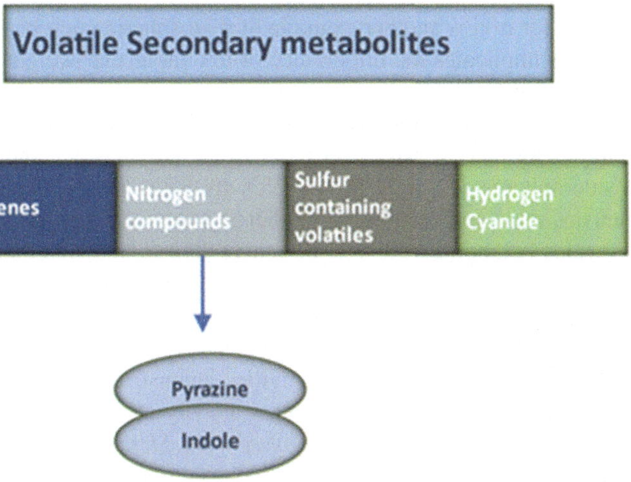


**Fig. 9.2** Volatile secondary metabolites involved in biological control of phytopathogens

compounds having strong direct or indirect antagonistic activity against their enemy, for example, bacteriocins. Bacteriocins are antimicrobial peptides (ribosomal or non-ribosomal) that exhibit bactericidal and bacteriostatic activities (Mojgani 2017). They exhibit antagonistic activity against same species (narrow spectrum) or against a diverse range of species belonging to other genera (broad spectrum) (Cotter et al. 2005; Parret et al. 2003). Their importance in shaping the soil microbial population dynamics can be estimated by the fact that 99% of bacterial inhabitants of rhizosphere produce at least one bacteriocin out of 500 bacteriocins identified and characterized to date to remain competent in a particular niche (Riley and Wortz 2002). *Bacillus* spp. are major producers of bacteriocins having strong antimicrobial activity; for example, *B. subtilis* 14B, isolated from almond rhizosphere, produces bacteriocins (Bac14B) active against *A. tumefaciens*, thus controlling the crown gall (Hammami et al. 2009). Similarly, *Bacillus clausii* GM17 produces Bac GM17 exhibiting both antibacterial and antifungal activities. In addition, *Pseudomonas putida* BW11M1, isolated from banana roots, inhibits the plant pathogenic strain of *P. putida* GR12-2R3 by production of putidacin (Parret et al. 2003). In addition to bacteria, halophilic Archaea belonging to family *Halobacteriaceae* have been reported for bacteriocin production (Li et al. 2003). The bacteriocins can play an important role in agriculture industry in minimizing the use of chemical fertilizers and pesticides (Mojgani 2017).

#### 9.3.1.2 Non-ribosomal Peptides (NRPs)

There are two important classes of NRPs: siderophores and lipopeptides that are synthesized by large non-ribosomal peptide synthetases (NRPSs).

(a) **Siderophore**

Competition for iron is another example of rhizospheric microbe-microbe interactions, and this implicates the importance of low molecular weight, high affinity, iron-chelating compounds produced by microorganisms under iron-limited conditions termed as siderophores. These siderophores are used in sequestering, solubilizing, transporting, and storing iron (Chu et al. 2010; Hider and Kong 2010). Bacterial siderophores scavenge limited iron from the vicinity and thus deprive other organisms of iron, thereby limiting their growth (Naureen et al. 2005). Many bacterial species belonging to bacterial genera *Rhizobium*, *Aeromonas*, *Bacillus*, *Pseudomonas*, and *Serratia* can produce various kinds of siderophores having antimicrobial properties. For instance, pyoverdines produced by *Pseudomonas*; ornibactin from *Burkholderia*; bacillibactin from *Bacillus*, and desferrioxamine from *Streptomyces* have shown antifungal activities against many fungal species. For example, pyoverdins demonstrate antifungal activity against the phytopathogenic *Aspergillus niger*, *A. oryzae*, *A. flavus*, *F. oxysporum*, *M. phaseolina*, and *Sclerotium rolfsii* (Taguchi et al. 2010). Similarly, siderophore-producing *Bacillus subtilis* and *Serratia marcescens* have been reported to antagonize *M. grisea* eventually suppressing rice blast in *Super Basmati* rice (Naureen et al. 2009). This indicates the importance of these metabolites in manifesting biological control of phytopathogenic fungi and their applications in integrated pest management (IPM).

(b) **Lipopeptides (LPs)**

In addition to siderophores, NRPs also synthesize linear or cyclic oligopeptides linked with a tail of lipids that exhibit surfactant, antimicrobial, antipredation, and cytotoxic properties. These lipopeptides have diverse structures and functions owing to the length and composition of lipid tail together with the number, type, and configuration of the amino acids in the peptide moiety.

Several bacterial species belonging to *Streptomyces*, *Bacillus*, and *Pseudomonas* produce lipopeptides and thus play an important role in shaping of microbial communities in the vicinity of plant roots (Raaijmakers et al. 2006).

Cyclic lipopeptides are mostly produced by fluorescent *Pseudomonas* species and are considered effective biocontrol agents due to their ability to act on biological membranes (Raaijmakers et al. 2006). For instance, visconamide produced by florescent *Pseudomonas* DR54 exhibits tremendous antifungal and biosurfactant properties and suppresses *R. solani* in rice (Nielsen et al. 2000; Thrane et al. 2000). Similarly, *P. fluorescens* 96.578 produces tensin that inhibits *R. solani* in sugar beet (Nielsen et al. 2000). Another cyclic lipopeptide, massetolides A, produced by *P. fluorescence strain SS101* suppresses late blight in tomato (Nielsen et al. 2000). In addition, these also act as biosurfactants and antimicrobials against a vast majority of pathogens (Raaijmakers et al. 2006; Raju et al. 2016). Some of these cyclic lipopeptides also act as toxins; examples include viscosin, amphisin, tolaasin, and syringomycin (Nybroe and Sørensen 2004).

Moreover, cyclic lipopeptides produced by *Bacillus* spp., such as surfactins, iturins, and fengycins, have been successfully deployed against a wide range of phytopathogens including bacteria, fungi, and oomycetes (Ongena and Jacques 2008).

#### 9.3.1.3 Polyketides

One of the most plausible properties of PGPR is antibiosis and that makes them attractive from agriculture point of view. PGPR manifest antibiosis by producing a myriad of secondary metabolites such as antibiotics, antifungal compounds, and hydrolytic enzymes that help them to get established into a niche and in turn help plants by eradicating phytopathogens. There are numerous studies reporting selection and screening of PGPR strains based on antibiotic production (Shilev 2013). PGPR mainly produce diffusible or volatile antibiotics in addition to cell wall lysing enzymes such as chitinases, cellulases, and glucanases. Most of these antibiotic and antifungal compounds belong to yet another class of secondary metabolites synthesized by multidomain enzymes termed as polyketide synthases (PKSs). These are collectively termed as polyketides and are produced by bacterial species belonging to the genera *Pseudomonas* (El-Sayed et al. 2003), Actinobacteria (Stinear et al. 2004), *Bacillus* (Chen et al. 2006), *Myxococcus* (Frank et al. 2007), and *Burkholderia* (Partida-Martinez and Hertweck 2005).

PKSs can be classified into three types based on their structure and function and are involved in biosynthesis of a variety of secondary metabolites effective in treatment of infectious diseases and cancer (Hertweck et al. 2007). Type III PKSs are important from agriculture point of view as they synthesize important metabolites such as polyhydroxy phenols, 2,4-diacetylphloroglucinol (2,4-DAPG), and antibiotics (Bangera and Thomashow 1999).

(a) **2,4-Diacetylphloroglucinol (DAPG)**

DAPG is a polyketide antibiotic produced by *P. fluorescence* and has received considerable attention in the past few years because of its tremendous potential in combating many phytopathogens, thus providing an effective biocontrol of many crop diseases (Duffy et al. 2004; Lanteigne et al. 2012). Indigenous DAPG-producing bacteria are the reason behind monoculture soils becoming suppressive that restrain the growth of many phytopathogens (Weller et al. 2002). The DAPG biosynthesis is governed by a set of genes phlACBDEF which are highly conserved in antagonistic pseudomonads (Almario et al. 2017). DAPG-producing biocontrol pseudomonads use DAPG as a signal to synergistically stimulate themselves and other pseudomonads for the expression of their DAPG biosynthetic genes. In addition to that, DAPG also exerts influence on synthesis of other antifungal metabolites in *Pseudomonas* species such as repression of pyoluteorin (Haas and Keel 2003).

Strain consortiums based on different genotypes of DAPG-producing pseudomonads have proven successful in enhanced and consistent DAPG production in various soil environments as well as improved growth, activity, and competitiveness in the rhizosphere (Maurhofer et al. 2004)

(b) **Pyoluteorin**

A majority of *P. fluorescence* and *Streptomyces* spp. have been reported to produce another phenolic polyketide antibiotic pyoluteorin that inhibits oomycete fungi such as *Pythium ultimum*, thereby suppressing several crop diseases (Howell and Stipanovic 1980). Pyoluteorin contains a resorcinol ring linked to a bichlorinated pyrrole moiety. The resorcinol ring is derived through polyketide biosynthesis, while biosynthesis of pyrrole moiety remains uncharacterized. Biosynthesis of pyoluteorin by rhizobacteria depends upon various factors such as cell density, nutrient source, temperature, and presence or absence of DAPG (Cuppels et al. 1986). Pyoluteorin has bactericidal, herbicidal, and fungicidal properties (Howell and Stipanovic 1980).

(c) **Mupirocin**

Mupirocin also known as pseudomonic acid is yet another polyketide antibiotic having a unique chemical structure and is produced by *P. fluorescence*. It contains 9-hydroxynonanoic acid linked to monic acid A by an ester linkage. Mupirocin is derived from acetate units incorporated into monic acid A and 9-hydroxynonanoic acid through polyketide synthesis. Mupirocin exhibits bactericidal activity via inhibiting isoleucyl-Trna synthetase, thus preventing the incorporation of isoleucine into newly synthesized proteins (Gurney and Thomas 2011). *P. fluorescens* avoids the inhibitory effects of mupirocin by altering the target sites, isoleucyl-tRNA synthetase. Although mupirocin and its derivatives exhibit strong antibacterial activity against *Mycoplasma*, *Haemophilus influenzae*, *Neisseria gonorrhoeae*, *Staphylococci* spp., and *Streptococci* spp., it is less antagonistic toward Gram-positive *Bacilli* and anaerobes (Sutherland et al. 1985). Several other PGPR strains also produce mupirocin that helps them to afford biocontrol of phytopathogenic fungi (Fernando et al. 2005a, b).

#### 9.3.1.4 PKS-NRPS Hybrid Compounds

Perhaps the best possible explanation of diversity of bacterial secondary metabolites is the presence of hybrid NRPS-PKS clusters that allow production of a wide range of secondary metabolites (Du et al. 2001; Garcia et al. 2012). Hybrid NRPS-PKS metabolites isolated from numerous Gram-positive and Gram-negative soil bacteria have shown promising results against both phyto and human pathogenic microorganisms, for example, the immunosuppressant rapamycin from *Streptomyces hygroscopicus* (Schwecke et al. 1995), *rhizoxin* from the fungal endosymbiont *Burkholderia rhizoxinica* (Partida-Martinez and Hertweck 2007) that also occurs in other soil microorganisms (Brendel et al. 2007; Loper et al. 2008), epothilone from the *myxobacterium Sorangium cellulosum* used in cancer therapy (Julien et al. 2000), and zwittermicin A produced by *Bacillus* spp. that acts against oomycetes and other phytopathogenic fungi.

(a) **Zwittermicin A**

Zwittermicin A is an aminopolyol antibiotic produced by *Bacillus* spp. with a promising broad spectrum antagonistic activity against many phytopathogens (Emmert et al. 2004). Zwittermicin A is synthesized by a mixed NRPS/PKS pathway owing to the existence of NRPS and PKS homologs (Emmert et al. 2004; Kevany et al. 2009). Zwittermicin A produced by *B. cereus* and *B. thuringiensis* significantly inhibits oomycetes, phytopathogenic fungi, as well as insects owing to its fungicidal and insecticidal activities (Emmert et al. 2004). It has been reported to suppress damping-off disease caused by *Phytophthora medicaginis* in alfalfa (Pathma et al. 2019), fruit rot of cucumber (Smith et al. 1993), and *Phytophthora parasitica* var. nicotianae infection in tobacco (Stabb et al. 1994). Besides that, it has shown strong antagonistic activity against several phytopathogenic fungi, bacteria, and protists.

#### 9.3.1.5 Heterocyclic Nitrogenous Compounds

Rhizobacteria secrete several extracellular heterocyclic nitrogenous compounds with antimicrobial activity to eradicate their competitors. These are synthesized through shikimic acid pathway, and among these compounds, phenazine is a powerful green pigment having a strong antimicrobial activity (Chin-A-Woeng et al. 2003)

(a) **Phenazine**

Several members of *Pseudomonas*, *Brevibacterium*, *Burkholderia*, and Streptomyces produce nitrogen-containing low molecular weight heterocyclic compound having brightly colored pigment termed as phenazine (Tambong and Hofte 2001). Several derivatives of this compound have also been isolated from *Pseudomonas* spp. such as pyocyanin, 2-hydroxy phenazines, and hydroxyphenazine-1-carboxylic acid (PCA; Turner and Messenger 1986). The antimicrobial activity of phenazine is largely dependent on redox transformation of the compound that results in accumulation of toxic superoxide radicals in the target cells (Wang and Newman 2008). Phenazine has been found effective in combating seed-borne pathogens of barley and oats (Khan et al. 2016). Phenazine and its derivatives confer rhizosphere competence and long-term survival to the strains that produce it; however, their antagonistic activity is dependent upon chemotaxis and mobility of the producing strains. Strains lacking motility fail to exhibit antagonism as they are unable to colonize in the rhizosphere (Zheng and Sinclair 2000).

### *9.3.2 Volatile Compounds*

Tremendous attention has been paid to the volatile compounds produced by soil bacteria in the past few decades. Certain PGPR strains can produce microbial volatile organic compounds (mVOCs) including alcohols, alkanes, alkenes, esters, ketones, sulfur, and terpenoids (Ryu et al. 2004; Bacnchio et al. 2009). Under normal conditions these compounds have low molecular weight and high vapor pressure, which results in their vaporization and release in the atmosphere. Structural and function characterization of these mVOCs is a promising research arena, and the repertoire of these volatile compounds is on an ongoing increase. Not only these are attractive from agricultural and medical point of view but are bringing new insights into inter- and intra-species communications (Santoro et al. 2011; Tahir et al. 2017). These compounds have potentials to improve plant growth and induce systemic resistance (ISR) against pathogenic organisms, affording enhanced crop productivity (Ryu et al. 2004; Tahir et al. 2017; Naseem et al. 2018; Lee et al. 2012; Khan and Bano 2019). Some VOCs from *Paraburkholderia phytofirmans* have been reported to enhance plant growth and salinity (Khan and Bano 2019).

Bacterial volatiles perform a variety of functions that shape the overall plant-microbe interactions. One of the main functions of these VOCs is biological control of plant pathogens such as fungi and insects. VOCs have been reported to cause morphological changes in fungi and inhibit spore germination and mycelial growth, thus eradicating the fungi (Effmert et al. 2012). For instance, 14 bacterial species including four *Pseudomonas* spp., isolated from canola and soybean plants, produced 23 VOCs such as benzothiazole, cyclohexanol, n-decanal, dimethyl trisulfide, 2-ethyl-1-hexanol, and nonanal that prevented sclerotia and ascospores germination and inhibited mycelial growth of *Sclerotinia sclerotiorum*, both in vitro and in soil experiments (Fernando et al. 2005a, b). In addition to this, a comprehensive survey of 1018 bacterial isolates belonging to families *Alcaligenaceae*, *Bacillales*, *Micrococcaceae*, *Rhizobiaceae*, and *Xanthomonadaceae* showed that 328 isolates produced VOCs which inhibited spore germination and mycelial growth of two nematicidal fungi (Zou et al. 2007). Different bacterial strains were found to produce different VOCs, and some of these such as phenylacetaldehyde, acetamide, benzothiazole, 1-butanamine, benzaldehyde, methanamine, and 1-decene, were fungistatic. Several VOCs have also been reported to alter gene expression and enzymatic activity in fungi; however, the mechanism of such activity is not well understood (Effmert et al. 2012). Similarly, VOC produced by two *Bacillus* endophytes significantly reduced the weight and number of the sclerotia of *S. sclerotiorum* (Massawe et al. 2018) while those of *Burkholderia ambifaria* (Groenhagen et al. 2013) and a range of other rhizobacterial isolates (Velivelli et al. 2015) inhibit the growth of the pervasive soil-borne pathogen *Rhizoctonia solani*. In addition to fungi and oomycetes, the mVOCs produced by *Bacillus* species possess inhibitory activity against bacterial pathogens such as *Clavibacter michiganensis* and *Xanthomonas oryzae* that cause ring rot of potato and bacterial blight of rice, respectively (Rajer et al. 2017).

Although the bacterial VOCs have received considerable attention in past 30 years, the studies regarding their mechanism of affording biocontrol are scarce and deserve a thorough research. The capability of mVOCs to suppress the incidence of phytopathogens demonstrates their potential as effective eco-friendly ingredients of Integrated nutrient and pest management systems.

#### 9.3.2.1 Terpenes

Terpenes were formally considered as plant-derived metabolites only; however, the recent studies have shown that these are produced by bacteria, fungi, and amoeba (Yamada et al. 2015; Chen et al. 2016). Terpenes are metabolites derived from dimethylallyl pyrophosphate and isopentenyl pyrophosphate, either via the mevalonate pathway or from the deoxyxylulose phosphate pathway (Dickschat 2011). Bacteria produce many terpenes such as geosmin and novel sesquiterpenes, yet the biological and ecological functions of most of them are still unknown (Gurtler et al. 1994).

Perhaps the best explored bacterial species for terpene synthesis are streptomyces species which produce novel sesquiterpenes and monoterpenes. For instance, an odoriferous *Streptomyces albidoflavus* isolated from corn seeds produces a novel antibacterial sesquiterpene, named albaflavenone (Gurtler et al. 1994). Other species of streptomyces and some fungi also produce albaflavenone (Reidys et al. 2011; Takamatsu et al. 2011; Moody et al. 2012). A comparative genomic analysis of six *Collimonas* strains resulted in identification of terpene synthase genes in two *C. pratensis* strains which after heterologous expression and biochemical characterization in *Escherichia coli* were found to be responsible for the synthesizing a mixture of sesquiterpenes, the major sesquiterpene being Germacrene D-4-ol (Song et al. 2015). In addition to that, four antimicrobial monoterpenes, i.e., g-terpinene, /–pinene, b-pinene, and b-myrcene, were detected in the headspace of *C. pratensis* strain Ter91. Moreover b-pinene was observed to inhibit *Staphylococcus aureus* and *Rhizoctonia solani*, whereas the mixture of all four monoterpenes inhibited *E. coli* growth (Song et al. 2015).

#### 9.3.2.2 Nitrogen Compounds

A variety of nitrogen-containing compounds are produced by rhizobacteria such as pyrazines, pyrroles, thiazoles, pyridines, indole, and aniline derivatives (Groenhagen et al. 2014; Dickshat et al. 2010). Among these the most widespread nitrogen compound occurring in many soil bacteria are pyrazines, whereas indole serves as an important signaling molecule.

(a) **Pyrazines (1,4-Diazabenzenes)**

Pyrazines are volatile organic compounds having strong antimicrobial activities (Rajini et al. 2011). Mostly these are produced in plants; however, some *Bacillus*, *Pseudomonas*, *Streptomyces*, and *Chondromyces* (Dickshat et al. 2005, 2010; Rajini et al. 2011) also produce pyrazines by two different biosynthetic pathways. For instance, in *Corynebacterium glutamicum* the acetolactate synthase enzyme is responsible for pyrazine production via acetolactate and its higher homologs (Dickshat et al. 2010). Conversely, in myxobacteria, pyrazines are formed via reduction of branched amino acids such as valine to valinal followed by dimerization (Nawrath et al. 2010a, b). Their antimicrobial properties make them attractive from biocontrol perspective.

(b) **Indole**

Indole and its derivatives are known for their antagonistic activity against antibiotic-resistant bacterial pathogens. Indole can inhibit quorum sensing and production of virulence factors (Lee et al. 2015). Indole and its derivatives are synthesized from tryptophan by the enzyme tryptophanase in both Gram-positive and Gram-negative bacteria (Lee and Lee 2010; Pandey et al. 2013). The biological functions of indole include antibiotic tolerance, cellular signaling, plasmid stability, controlling spore formation, cell division, and biofilm formation (Wang et al. 2001; Di Martino et al. 2003). In plants they control plant defense systems, growth, and root development (Lee et al. 2015). Although indole is a stable compound within the bacteria producing it, several non-indole-producing bacteria can modify or degrade indole by diverse oxygenases, such as monooxygenases, dioxygenases, and P450 family members (Wikof et al. 2009). Although, indole derivatives are abundantly synthesized in soil microbial communities, their biological roles and mechanisms of action is not clear.

#### 9.3.2.3 Sulfur-Containing Volatiles

Many rhizobacterial species synthesize structurally diverse volatile sulfur compounds and alkyl sulfides ranging from small compounds, such as dimethyl sulfide (DMS), dimethyl disulfide (DMDS), and dimethyl trisulfide (DMTS), to large complex compounds such as 2-methyltetrahydrothiophen-3-one that formed from homocysteine in bacteria (Nawrath et al. 2010a, b). The volatile sulfur compounds mediate plant-microbe and interspecific microbe-microbe interactions (Tyc et al.

2015). For instance, DMDS inhibits quorum sensing (Tyc et al. 2015) and fungal growth while stimulates bacterial growth (Chernin et al. 2011).

#### 9.3.2.4 Hydrogen Cyanide (HCN)

HCN is one of the most well-known volatile antagonistic secondary metabolites synthesized by a variety of PGPR strains such as *Pseudomonas* spp. *Chromobacterium* spp., and *Bacillus* spp. (Naureen et al. 2009; Ahemad and Kibret 2014; Naureen et al. 2017). Glycine is the precursor of HCN, and the enzyme responsible to catalyze the reaction is HCN synthase (Castric 1994). PGPR strains producing HCN have been reported to suppress the growth of several plant pathogenic fungi in soil (Fischer et al. 2013). For instance, Lanteigne et al. (2012) reported that simultaneous production of DAPG and HCN by *Pseudomonas* spp. controls the bacterial canker of tomato. Moreover, HCN production by *P. fluorescens* suppresses several soilborne pathogens such as black root rot of tobacco, caused by *Thielaviopsis basicola* (Voisard et al. 1989). On the other hand, mutant *P. fluorescens* defective in the antibiotic enzymes and HCN synthesis were unable to suppress tobacco black root rot (Voisard et al. 1989). However, transformation of the mutants with plasmid harboring functional hcnABC genes restored HCN production as well as biocontrol capability confirming the antifungal behavior of HCN.

## 9.4 Factors Affecting Secondary Metabolite Production in Soil Bacteria

The production of secondary metabolites in bacteria is under strict regulatory control and is influenced by a variety of biotic and abiotic factors such as availability of nutrients, plant root exudates, temperature, pH, moisture, light, microbial communities in the vicinity, and secondary metabolites produced by other microbial community members.

Bacterial genes are organized in operons which are mostly turned off and are induced by several inducers, for example, under limited availability or absence of glucose and presence of lactose, lactose act as an inducer of Lac operon. Similarly, unavailability of nutrients such as carbon, nitrogen, phosphate, or other key nutrient sources triggers secondary metabolite synthesis (Bibb 2005; Sanchez et al. 2010; van Wezel and McDowell 2011) Conversely high concentrations of glucose, phosphate, or ammonium generally repress secondary metabolite production (Masuma et al. 1986). However, this varies from species to species, for example, high phosphate concentration might trigger production of secondary metabolites in some bacterial species (Gotoh et al. 1992) but not in others. Similarly, under iron-limiting conditions, the bacterial species produce siderophores to sequester iron. This not only helps the microbial strains to get iron but also serves the plants in two ways.

First is depriving the pathogens of iron and secondly facilitating iron uptake by plants. Another example is production of organic acids to solubilize phosphate under phosphate limiting conditions which facilitates plants to acquire phosphate. In addition to that, bacteria produce many antagonistic compounds as mentioned above depending upon the composition of microbial community in the root zone. Although the selecting pressure comes from root exudates, the survival of fittest bacterial strain depends upon many other factors such as production of bacteriocins, antibiotics, HCN, and other secondary metabolites. Conclusively it can be stated that in an attempt to acquire the root zone niche and to be able to get root exudates, PGPR species produce several secondary metabolites that not only help them colonize and dominate other microbial inhabitants but also improve plant growth and eradicate phytopathogens. In addition, several of these metabolites induce systemic resistance in plants and are attractive from seed priming perspective.

## 9.5 Metabolomic Profiling of PGPR

The fact that a multitude of various environmental factors control the expression of genes involved in secondary metabolite production by rhizobacteria makes identification and characterization of these bacteria very important from the perspective of selection of potent strains for potential application in agriculture. Several approaches have been used to explore bacterial strains for production of secondary metabolites ranging from simple microbiological assays to sophisticated modern techniques coupled with high throughput data analysis that has given the field of PGPR metabolomics a new dimension. However, isolating PGPR strain with multifaceted secondary metabolite production is a challenging task and so does the stimulation of the PGPR strains to produce secondary metabolites. The reason is the management of a variety of factors that govern the regulation of various genes involved in secondary metabolite production. There are several approaches to activate the bacterial strains to produce secondary metabolites; however, to discuss them here is beyond the scope of this chapter. Therefore, in next section we will discuss the two main approaches that are used to induce the secondary metabolite production in culturable PGPRs.

### *9.5.1 The OSMAC (One Strain-Many Compounds) Approach*

This approach aims at culturing a single bacterial strain under specific environmental conditions. The term OSMAC was first used by Zeeck and coworkers, and it simply meant that a single strain could produce different metabolites under different growth conditions (Bode et al. 2002). OSMAC involves alteration of growth parameters such as composition of media, for example, using different carbon sources, concentration of phosphate and nitrogen, changes in temperature and pH, aeration rate, addition of growth enhancers or inhibitors, etc. This implicates that variations in abiotic factors would ultimately result in expression of various genes corresponding to production of varying metabolites. For example, adding low concentrations of scandium and/or lanthanum in streptomyces fermentation medium resulted in induction of several secondary metabolite-biosynthetic gene clusters and activated the silent biosynthetic gene clusters such as actinorhodin biosynthesis in *Streptomyces lividans* (Kawai et al. 2007).

Several reports have mentioned that even a minute change in the prevailing growth environment results in an intense effect on the biosynthesis of microbial metabolites. For example, variation in phosphate concentration may result in overproduction of a specific metabolite or may result in synthesis of an entirely novel type of secondary metabolite (Bode et al. 2002; Martin 2004; Romano et al. 2014). The OSMAC approach has been deployed to induce the cryptic coelichelin biosynthetic gene cluster in *S. coelicolor* (Ikeda et al. 1999) and to identify the cryptic fungal biosynthetic gene clusters (Ancheeva et al. 2017).

### *9.5.2 The OSMAC with Co-cultivation Approach*

In addition to altering the abiotic conditions, changes in microbial community also influence production of secondary metabolites by the PGPR. This is because complex microbe-microbe interactions result in activation of some gene clusters and suppression of others. For instance, soil bacteria growing under limited carbon can be induced to produce various antibiotics upon challenging with other bacterial species (Garbeva and de Boer 2009).

Transcriptomic or metabolomic analysis coupled with OSMAC analyses revealed that soil bacteria such as *Pseudomonas fluorescens* Pf0–1 can fine-tune their competitive strategies when combated with divergent bacterial competitors under specific environmental conditions (Garbeva et al. 2011). There are several cryptic gene clusters in soil bacteria that remain silent until the bacteria are co-cultivated with a symbiont or a competitor indicating the influence of microbe-microbe interaction on expression of genes involved in biosynthesis of secondary metabolites (Traxler et al. 2013; Tyc et al. 2014).

In this context Traxler et al. (2013) have analyzed the induction of secondary metabolite production by using the newly developed techniques of nanoDESI and MALDI-TOF imaging mass spectrometry. They have reported that when co-cultivated with other actinomycetes, *S. coelicolor* A3 produces many secondary metabolites that were not produced while it was cultured alone. In addition to that, most of the compounds were specific for the five pairwise interactions. Similarly, *Streptomyces lividans*, a remarkably close relative of *S. coelicolor*, produces a mixture of red pigments prodiginines and actinorhodins while interacting with mycolic acid-producing bacteria, such as *Rhodococcus erythropolis*, *Tsukamurella pulmonis*, and *Corynebacterium glutamicum* (Onaka et al. 2011). *S. coelicolor* also produces actinorhodin, while interacting with *Bacillus*, *Myxococcus*, and *Serratia* species (Yang et al. 2009; Traxler et al. 2013; Perez et al. 2011). In a recent study, *Lysinibacillus sphaericus* ZA9 strain was observed to produce a novel secondary metabolite 2-penthyl-4-quinolinecarboxylic acid in response to competition with various fungal isolates (Naureen et al. 2017).

Co-cultivation of bacterial strains followed by analysis of their interactions that wake up the sleeping cryptic gene clusters helps isolating the PGPR strains for production of novel secondary metabolites that can be utilized in agriculture and medicine.

## 9.6 Techniques Used to Study Metabolomic Profiles

The advancement in the field of metabolomics owes a great deal to sophisticated techniques like nuclear magnetic resonance (NMR) spectroscopy and MS coupled with LC or GC analysis that enables metabolomic profiling of bacterial isolates (Table 9.1). 1H NMR being a non-destructive and highly reproducible method enables high-throughput characterization of metabolites in intricate biological mixtures (Pham et al. 2005; Vignoli et al. 2019; Tantray et al. 2021). Based on the chemical nature of metabolites, NMR resolves number of peaks generated by a metabolite, their location, and ratio of heights that help in elucidation of their chemical structures. NMR however requires careful sample preparation so that the peak resolution can be achieved. On the other hand, MS-based techniques are highly sensitive and selective and can be easily combined with diverse separation methods such as GC or LC, and most metabolic studies are performed on these hybrid instruments (Svatos 2011). These techniques give detailed information regarding detected mass and fragmentation spectra that can be utilized to determine the molecular formula of a metabolite from the accurate mass. Gas chromatography-mass spectrometry (GC-MS) is used to analyze volatile compounds and volatile derivatives of polar primary metabolites. Liquid chromatography-mass spectrometry (LC-MS) is a fast highly accurate technique that uses a column for effective quantification of a wide range of known and unanticipated secondary metabolites simultaneously (Fang and Dorrestein 2014). For instance, Yasmin et al. (2017) have reported metabolomic profiling of rhizospheric *Pseudomonas aeruginosa* BRP3 strain in controlling bacterial blight of rice using LC-MS. LC-MS is mostly used to analyze polar and semipolar, non-volatile compounds. These techniques have enabled detection of volatile molecules from bacteria which used to remain unidentified through conventional techniques such as microbiological assays (Fernie et al. 2004). Extremely sensitive techniques such as GC-Q-TOF–MS have made the analysis of complex volatile mixtures in ultra-low concentrations possible.

In addition to NMR and MS, mass spectrometry imaging (IMS) has emerged as an advanced tool to study the metabolome at cellular and subcellular levels (Luzzatto-Knaan et al. 2015). Several modifications of IMS methods allow analysis of samples in their native state. For example, ambient ionization IMS utilizes ionization at atmospheric pressure to analyze samples in their native state, with

**Table 9.1** Techniques for metabolomic profiling of PGPR metabolites

| Techniques | Metabolites studied | References |
|---|---|---|
| Thin layer chromatography (TLC) | Study of nod metabolites; antifungal metabolites of pseudomonas fluorescence | Spaink et al. (1992) and Koche et al. (2012) |
| Mass spectrometry (MS) | Characterization of microbial siderophores by mass spectrometry | Pluháček et al. (2016) |
| Proton transfer reaction mass spectrometry | Study of rhizosphere volatile organic compounds and their induction by biotic stresses | Steeghs et al. (2004) |
| Tandem analysis methods Liquid chromatography/ electrospray ionization mass spectrometry | Identification of isoflavone conjugates from roots of lupine species | Kachlicki et al. (2005) |
| Gas chromatography/ mass spectrometry | Identification of signaling molecules during ectomycorrhizae formation Analysis of bacterial secondary metabolites of pseudomonas species | Menotta et al. (2004), Fang and Dorrestein (2014) and Anantha et al. (2016) |
| Gas chromatography/ combustion/isotope-ratio mass spectrometry | To study the nature and dynamics of plant sugars in the rhizosphere | Derrien et al. (2003) |
| Nuclear magnetic resonance (NMR) | Secondary metabolites such as 2,3-butanediol, acetoin, siderophore Characterization of a phenazine-producing strain *pseudomonas chlororaphis* GP72 with broad-spectrum antifungal activity from green pepper rhizosphere | Vignoli et al. (2019), Pham et al. (2005), Liu et al. (2007) and Tantray et al. (2021). |
| Liquid chromatography-mass spectrometry (LC-MS) | Bacterial volatile compounds; Identification of secondary metabolites produced by *Pseudomonas chlororaphis* and *P. aurantiaca* strains isolated from cactus, cotton, and Para grass | Fernie et al. (2004) and Shahid et al. (2017) |
| Fourier transformed-infrared spectroscopy (FT-IR) | Identification of secondary metabolites of seawater bacterial population; Characterization of streptomyces isolates | Rajarajan et al. (2019) and Maleki and Mashinchian, (2011) |
| Raman spectroscopy | Understanding the metabolomics of plant-microbe interactions | Polisetti et al. (2016) |
| Near infrared spectroscopy (NIR) | Bacterial identification, secondary metabolite profiling | Kamnev, (2008), Verma et al. (2018) and Reuben et al. (2008) |
| Proton transfer reaction mass spectrometry | VOCs in the rhizosphere during plant-microbe interactions | Blake et al. (2009) |
| Liquid extraction surface analysis (LESA) | Thiazolyl peptide antibiotics from different actinobacteria | Kai et al. (2012). |

(continued)

**Table 9.1** (continued)

| Techniques | Metabolites studied | References |
|---|---|---|
| Laser ablation electrospray ionization-mass spectrometry (LAESI-MS) | Colonization of *P. aeruginosa* and *S. aureus* in a polymicrobial biofilm | Dean et al. (2015) |
| Matrix-assisted laser desorption/ionization method (MALDI) | Extracellular metabolites produced by five *Lysobacte*r strains; secondary metabolites of intact cyanobacteria<br>Study of aconitum alkaloids from aconite roots; rapid typing of *Bacillus subtilis* by their secondary metabolites | Yang et al. (2009), de Bruijn et al. (2015), Erhard et al. (1997), Sun et al. (1998) and Leenders et al. (1999) |
| Desorption electrospray ionization-mass spectrometry (DESI-MS) | Characterization and mapping of secondary metabolites of *Streptomyces* sp. from caatinga | Rodrigues et al. (2018) |

minimum or no pre-preparation. The ionization of sample takes place before entering the mass spectrometer, thus allowing only ions to enter the mass spectrometer. More than 30 different ambient ionization methods have been developed in the past decade, and the most widespread of these methods is the matrix-assisted laser desorption/ionization method (MALDI) (Cornett et al. 2007; Kaltenpoth et al. 2016). MALDI-IMS can be applied to bacterial metabolites produced during monoculture or co-culturing with other species. MALDI-IMS was used to identify many interaction-specific metabolites produced by *Streptomyces coelicolor* while co-cultivated with other actinomycetes and which were not produced in its monoculture (Tyc et al. 2017). Similarly, in vitro antagonistic interactions between *Streptomyces* and *Bacillus* strains have been studied using MALDI-IMS (Yang et al. 2009). Similarly, MALDI-IMS was used to analyze the spatial distribution of extracellular metabolites produced by five *Lysobacter* strains. The results showed that metabolomic data obtained corresponded well with the consequent gene clusters in the genomes of these five strains (de Bruijn et al. 2015). Combining MALDI-IMS with fluorescence in situ hybridization facilitated simultaneous detection of antibiotic production and taxonomic identification of model bacterium *Streptomyces* (Kaltenpoth et al. 2016). MALDI-IMS coupled with live colony nanoDESI mass spectrometry was applied to study the spatial distribution of lipopeptide massetolide A during interactions of *Pseudomonas fluorescens* SS101 and the protozoan *Naegleria americana* (Song et al. 2015).

Another advancement in MS is the laser ablation electrospray ionization mass spectrometry (LAESI-MS). LAESI-MS was applied to visualize the colonization of *P. aeruginosa* and *S. aureus* in a polymicrobial biofilm as well as to analyze ions following LL-37 antimicrobial peptide treatment of the biofilm. This shows the potential of this ambient ionization method in studying the biofilms and interspecific interactions (Dean et al. 2015).

Another combination of liquid extraction surface analysis (LESA) automated chip-based nano-electrospray ionization and high-resolution mass spectrometry was used for extraction and detection of thiazolyl peptide antibiotics from different actinobacteria (Kai et al. 2012). LESA proved to be a powerful technique in early detection of natural products produced by bacterial colonies on culture plates as early as within the first 2 min. This allowed extraction and detection of antibiotics at a low cost but high mass accuracy. This method can be applied as a high-throughput screening procedure for identification and discovery of novel antibiotics. Another versatile method for detecting molecular mass range of 50–1200 Da solid, liquid, or gaseous molecules is direct analysis in real-time high-resolution mass spectrometry (DART-HRMS). This method permits rapid qualitative and quantitative analysis of bacterial metabolites (Gross 2014).

Proton transfer reaction mass spectrometry has also been used in studying the VOCs in the rhizosphere during plant-microbe interactions by a process known as soft ionization that prevents extensive fragmentation of volatile compounds, thus enabling real-time analysis (Blake et al. 2009). Spectroscopic analyses such as near infrared spectroscopy (NIR), Fourier transform infrared spectroscopy, and Raman spectroscopy have also been used in bacterial identification as well as secondary metabolite profiling (Kamnev 2008; Reuben et al. 2008; Verma et al. 2018). Raman spectroscopy has been deployed successfully in understanding the metabolomics of plant-microbe interactions. For example, Polisetti et al. (2016) have studied the chemical profile of *Pantoea* spp. YR343 during co-cultivation with *Arabadopsis thaliana*. These advance techniques have enabled researchers to study, analyze, and exploit the microbial metabolome for identification of novel secondary metabolites that not only play an important role in rhizosphere but have potentials for applications in agriculture and medicine (Tantray et al. 2021).

## 9.7 Genome Mining

"Genome mining" has emerged as a powerful in silico tool to detect the biosynthetic pathways of secondary metabolite synthesis but also their possible functional and chemical interactions (Ziemert et al. 2016). Genome mining requires genome sequences of sequenced organisms to identify uncharacterized biosynthetic gene clusters encoding enzymes responsible for synthesis of these secondary metabolites followed by the experimental identification of their products. In fact the main purpose of genome mining is to identify new biosynthetic gene clusters (BGCs) that encode for polyketide synthases (PKS) and non-ribosomal peptide synthases (NRPS) that catalyze the two most important biosynthetic routes of natural product synthesis (Timmermans et al. 2017). It also enables the comparison between the target and known gene clusters to predict their structure and function using different associated web database (Boddy 2014). These web databases and tools are added to improve the performance of genome mining (Ziemert et al. 2016). Three important webtools in this regard are:

(i) "Antibiotics and Secondary Metabolite Analysis Shell" (antiSMASH). Launched in 2011, this is a web server that associates the gene clusters with a series of specific algorithms, thus enabling compounds analysis (Blin et al. 2013). In addition, this allows the prediction of sequences and detailed analysis of identified gene clusters, thus providing predicted image of amino acid stereochemistry structure (Boddy 2014).

(ii) "Prediction Informatics for Secondary Metabolomes" (PRISM) is an open-web tool, comprising of genomic prediction of secondary metabolomes by using different algorithms to compare the new genetic information with 57 virtual enzymatic reactions (Skinnider et al. 2015).

(iii) "Integrated Microbial Genomes Atlas of Biosynthetic gene Clusters" (IMG/ABC) (Machado et al. 2017) was first launched in 2015. It is a huge open-web database of known predicted microbial BGCs, capable of associating BGCs with secondary metabolites (SMs) and analyzing both BGCs and SMs. Consequently, it enables in predicting the function of BGCs by comparing its function of similar known BGCs (Hadjithomas et al. 2017).

Genome mining coupled with other analytical tools has been successfully used in metabolic profiling of bacterial strains. For instance, Vater et al. (2018) have used genome mining in combination with mass spectrometry to search for lipopeptide biosynthetic genes in *Paenibacillus polymyxa* E681 that led to the discovery of a novel lipoheptapeptide paenilipoheptin. In a recent study, Nelkner et al. (2019) have unraveled the genetic potential of the biocontrol agent *Pseudomonas brassicacearum* using genome mining coupled with comparative genomics and transcriptomics. Although genome mining holds significant promise in the field of PGPR metabolomics, there are scarce reports regarding use of genome mining for PGPR metabolomic profiling. Future studies in this regard would definitely unravel the hidden treasure of novel PGPR metabolites involved in biological control of phytopathogens.

## 9.8 Conclusion

Plant growth-promoting rhizobacteria contain a repertoire of biosynthetic gene clusters encoding a variety of secondary metabolites that can not only mediate the biotic interactions in the rhizosphere but can be successfully utilized as alternatives to chemical fungicides, pesticides, and insecticides. Although countless research reports have documented the isolation and characterization of the PGPR from soil and then their successive screening for the presence or absence of plant growth-promoting and biocontrol potential, however, these basic techniques cannot encompass most of the secondary metabolites having novel properties. Advance analytical techniques coupled with high throughput data analysis therefore provide an excellent tool to generate microbial metabolomic profiles that can help selection and propagation of PGPRs with maximum beneficial traits both for agricultural industry and environment. In silico techniques like genome mining have also emerged as outstanding approaches to fish for the biosynthetic gene clusters that encode enzymes for PKS or NPRS or hybrid systems and can help select even unculturable bacterial genes for heterologous expression and subsequent secondary metabolite production.

Although metabolomics is not in its infancy, there are scarce reports focusing on metabolomic profiling of PGPR strains themselves and not the metabolome of plants hosting them. Future studies in this regard will help to unravel microorganism harboring novel genes encoding novel metabolites that can reshape our environmental, agricultural, and medicinal practices.

## References

Ahemad M, Kibret M (2014) Mechanisms and applications of plant growth promoting rhizobacteria: current perspective. J King Saud Univ Sci 26:120

Almario J, Bruto M, Vacheron J, Prigent-Combaret C, Moënne-Loccoz Y, Muller D (2017) Distribution of 2, 4-diacetylphloroglucinol biosynthetic genes among the Pseudomonas spp. reveals unexpected polyphyletism. Front Microbiol 8:1218

Anantha PS, Deventhiran M, Saravanan P, Anand D, & Rajarajan S (2016) A comparative GC-MS analysis of bacterial secondary metabolites of Pseudomonas species. The Pharma Innovation, 5(4, Part B):84

Ancheeva E, Küppers L, Akone SH, Ebrahim W, Liu Z, Mándi A et al (2017) Expanding the metabolic profile of the fungus Chaetomium spp. through co-culture with autoclaved Pseudomonas aeruginosa. Eur J Org Chem 2017(22):3256–3264

Banchio E, Xie X, Zhang H, Paré PW (2009) Soil bacteria elevate essential oil accumulation and emissions in sweet basil. J Agric Food Chem 57:653–657

Bangera MG, Thomashow LS (1999) Identification and characterization of a gene cluster for synthesis of the polyketide antibiotic 2,4-diacetylphloroglucinol from Pseudomonas fluorescens Q2–87. J Bacteriol 181:3155–3163

Basu A, Prasad P, Das SN, Kalam S, Sayyed RZ, Reddy MS, El Enshasy H (2021) Plant growth promoting rhizobacteria (PGPR) as green bioinoculants: recent developments, constraints, and prospects. Sustain For 13(3):1140

Berg G, Mahnert A, Moissl-Eichinger C (2014) Beneficial effects of plant-associated microbes on indoor microbiomes and human health? Front Microbiol 5:15

Bibb MJ (2005) Regulation of secondary metabolism in Streptomycetes. Curr Opin Microbiol 8:208–215

Blake RS, Monks PS, Ellis AM (2009) Proton-transfer reaction mass spectrometry. Chem Rev 109(3):861–896

Blin K, Medema MH, Kazempour D, Fischbach MA, Breitling R, Takano E, Weber T (2013) antiSMASH 2.0—a versatile platform for genome mining of secondary metabolite producers. Nucleic Acids Res 41(W1):W204–W212

Boddy CN (2014) Bioinformatics tools for genome mining of polyketide and non-ribosomal peptides. J Ind Microbiol Biotechnol 41(2):443–450

Bode HB, Bethe B, Höfs R, Zeeck A (2002) Big effects from small changes: possible ways to explore nature's chemical diversity. Chembiochem 3(7):619–627

Brendel N et al (2007) A cryptic PKS-NRPS gene locus in the plant commensal Pseudomonas fluorescens Pf-5 codes for the biosynthesis of an antimitotic rhizoxin complex. Org Biomol Chem 5:2211–2213

Castric P (1994) Influence of oxygen on the Pseudomonas aeruginosa hydrogen cyanide synthase. Curr Microbiol 29(1):19–21

Chen XH et al (2006) Structural and functional characterization of three polyketide synthase gene clusters in Bacillus amyloliquefaciens FZB 42. J Bacteriol 188:40244036
Chen X et al (2016) Terpene synthase genes in eukaryotes beyond plants and fungi: occurrence in social amoebae. Proc Natl Acad Sci U S A 113:12132–12137
Chernin L et al (2011) Quorum-sensing quenching by rhizo- bacterial volatiles. Environ Microbiol Rep 3:698–704
Chin-A-Woeng TF, Bloemberg GV, Lugtenberg BJ (2003) Phenazines and their role in biocontrol by Pseudomonas bacteria. New Phytol 157(3):503–523
Chu BC et al (2010) Siderophore uptake in bacteria and the battle for iron with the host; a bird's eye view. Biometals 23:601–611
Cornett DS et al (2007) MALDI imaging mass spectrometry: molecular snapshots of biochemical systems. Nat Methods 4:828–833
Cotter PD et al (2005) Bacteriocins: developing innate immunity for food. Nat Rev Microbiol 3:777–788
Cuppels DA, Howell CR, Stipanovic RD, Stoessl A, Stothers JB (1986) Biosynthesis of pyoluteorin: a mixed polyketide-tricarboxylic acid cycle origin demonstrated by [1, 2-13C2] acetate incorporation. Z Naturforsch C 41(5–6):532–536
de Bruijn I et al (2015) Comparative genomics and metabolic profiling of the genus Lysobacter. BMC Genomics 16:1–16
Dean SN et al (2015) Analysis of mixed biofilm (Staphylococcus aureus and Pseudomonas aeruginosa) by laser ablation electrospray ionization mass spectrometry. Biofouling 31:151–161
Derrien D, Marol C, & Balesdent J (2003) The dynamics of neutral sugars in the rhizosphere of wheat. An approach by 13 C pulse-labelling and GC/C/IRMS. Plant and Soil 267(1):243–253
Di Martino P et al (2003) Indole can act as an extracellular signal to regulate biofilm formation of Escherichia coli and other indole- producing bacteria. Can J Microbiol 49:443–449
Dickschat JS (2011) Isoprenoids in three-dimensional space: the stereochemistry of terpene biosynthesis. Nat Prod Rep 28:1917–1936
Dickschat JS et al (2005) Novel pyrazines from the myxobacterium Chondromyces crocatus and marine bacteria. Eur J Org Chem 2005:4141–4153
Dickschat JS et al (2010) Pyrazine biosynthesis in Corynebacterium glutamicum. Eur J Org Chem 2010:2687–2695
Du L et al (2001) Hybrid peptide–polyketide natural products: biosynthesis and prospects toward engineering novel molecules. Metab Eng 3:78–95
Duffy B, Keel C, Défago G (2004) Potential role of pathogen signaling in multitrophic plant-microbe interactions involved in disease protection. Appl Environ Microbiol 70(3):1836–1842
Effmert U, Kalderás J, Warnke R, Piechulla B (2012) Volatile mediated interactions between bacteria and fungi in the soil. J Chem Ecol 38(6):665–703
El-Sayed AK et al (2003) Characterization of the mupirocin biosynthesis gene cluster from Pseudomonas fluorescens NCIMB 10586. Chem Biol 10:419–430
Emmert EA, Klimowicz AK, Thomas MG, Handelsman J (2004) Genetics of zwittermicin A production by Bacillus cereus. Appl Environ Microbiol 70(1):104–113
Erhard M, von Döhren H, & Jungblut P (1997) Rapid typing and elucidation of new secondary metabolites of intact cyanobacteria using MALDI-TOF mass spectrometry. Nature biotechnology 15(9):906–909
Fang J, Dorrestein PC (2014) Emerging mass spectrometry techniques for the direct analysis of microbial colonies. Curr Opin Microbiol 19:120–129
Fernando WD, Nakkeeran S, Zhang Y (2005a) Biosynthesis of antibiotics by PGPR and its relation in biocontrol of plant diseases. In: PGPR: biocontrol and biofertilization. Springer, Dordrecht, pp 67–109
Fernando WD, Ramarathnam R, Krishnamoorthy AS, Savchuk SC (2005b) Identification and use of potential bacterial organic antifungal volatiles in biocontrol. Soil Biol Biochem 37(5):955–964
Fernie AR et al (2004) Metabolite profiling: from diagnostics to systems biology. Nat Rev Mol Cell Biol 5:763–769

Fischer S, Príncipe A, Alvarez F, Cordero P, Castro M, Godino A et al (2013) Fighting plant diseases through the application of bacillus and pseudomonas strains. In: Symbiotic endophytes. Springer, Berlin, Heidelberg, pp 165–193

Frank B et al (2007) Spiroketal polyketide formation in sorangium: identification and analysis of the biosynthetic gene cluster for the highly cytotoxic spirangienes. Chem Biol 14:221–233

Garbeva P, de Boer W (2009) Inter-specific interactions between carbon-limited soil bacteria affect behavior and gene expression. Microb Ecol 58:36–46

Garbeva P et al (2011) Transcriptional and antagonistic responses of Pseudomonas fluorescens Pf0–1 to phylogenetically different bacterial competitors. ISME J 5:973–985

Garcia I et al (2012) Elucidating the biosynthetic pathway for the polyketide-nonribosomal peptide collismycin A: mechanism for formation of the 2,20-bipyridyl ring. Chem Biol 19:399–413

Gotoh T et al (1992) Farnesyl diphosphate synthase and solanesyl diphosphate synthase reactions of diphosphate-modified allylic analogs – the significance of the diphosphate linkage involved in the allylic substrates for prenyltransferase. J Biol Chem 112:20–27

Groenhagen U, Baumgartner R, Bailly A, Gardiner A, Eberl L, Schulz S, Weisskopf L (2013) Production of bioactive volatiles by different Burkholderia ambifaria strains. J Chem Ecol 39(7):892–906

Groenhagen U et al (2014) Streptopyridines, volatile pyridine alkaloids produced by Streptomyces spp. FORM5. Beilstein J Org Chem 10:1421–1432

Gross JH (2014) Direct analysis in real time – a critical review on DART-MS. Anal Bioanal Chem 406:63–80

Gurney R, Thomas CM (2011) Mupirocin: biosynthesis, special features and applications of an antibiotic from a gram-negative bacterium. Appl Microbiol Biotechnol 90(1):11–21

Gurtler H et al (1994) Albaflavenone, a sesquiterpene ketone with a zizaene skeleton produced by a Streptomycete with a new rope morphology. J Antibiot 47:434–439

Haas D, Keel C (2003) Regulation of antibiotic production in root-colonizing Pseudomonas spp. and relevance for biological control of plant disease. Annu Rev Phytopathol 41(1):117–153

Hadjithomas M, Chen IMA, Chu K, Huang J, Ratner A, Palaniappan K et al (2017) IMG-ABC: new features for bacterial secondary metabolism analysis and targeted biosynthetic gene cluster discovery in thousands of microbial genomes. Nucleic Acids Res 45(D1):D560–D565

Hammami I et al (2009) Optimization and biochemical characterization of a bacteriocin from a newly isolated Bacillus subtilis strain 14B for biocontrol of Agrobacterium spp. strains. Lett Appl Microbiol 48:253–260

Hertweck C et al (2007) Type II polyketide synthases: gaining a deeper insight into enzymatic teamwork. Nat Prod Rep 24:162–190

Hider RC, Kong X (2010) Chemistry and biology of siderophores. Nat Prod Rep 27:637–657

Hossain MJ, Ran C, Liu K, Ryu CM, Rasmussen-Ivey CR, Williams MA, Hassan MK, Choi S-K, Jeong H, Newman M et al (2015) Deciphering the conserved genetic loci implicated in plant disease control through comparative genomics of bacillus amyloliquefaciens subsp. plantarum. Front Plant Sci 6:631

Howell CR, Stipanovic RD (1980) Suppression of Pythium ultimum-induced damping-off of cotton seedlings by Pseudomonas fluorescens and its antibiotic, pyoluteorin. Phytopathology 70(8):712–715

Ikeda H, Nonomiya T, Usami M, Ohta T, Ōmura S (1999) Organization of the biosynthetic gene cluster for the polyketide anthelmintic macrolide avermectin in Streptomyces avermitilis. Proc Nat Acad Sci 96(17):9509–9514. ISME J 9, 2329–2335

Ilyas N, Mumtaz K, Akhtar N, Yasmin H, Sayyed RZ, Khan W, Enshasy HAE, Dailin DJ, Elsayed EA, Ali Z (2020) Exopolysaccharides producing bacteria for the amelioration of drought stress in wheat. Sustain For 12(21):8876

Jauri PV, Kinkel LL (2014) Nutrient overlap, genetic relatedness and spatial origin influence interaction-mediated shifts in inhibitory phenotype among Streptomyces spp. FEMS Microbiol Ecol 90:264–275

Julien B et al (2000) Isolation and characterization of the epothilone biosynthetic gene cluster from Sorangium cellulosum. Gene 249:153–160

Kachlicki P, Marczak Ł, Kerhoas L, Einhorn J, & Stobiecki M (2005) Profiling isoflavone conjugates in root extracts of lupine species with LC/ESI/MSn systems. Journal of mass spectrometry 40(8):1088–1103

Kai M et al (2012) Direct mass spectrometric screening of antibiotics from bacterial surfaces using liquid extraction surface analysis. Rapid Commun Mass Spectrom 26:2477–2482

Kaltenpoth M et al (2016) Linking metabolite production to taxonomic identity in environmental samples by (MA)LDI-FISH. ISME J 10:527–531

Kamnev AA (2008) FTIR spectroscopic studies of bacterial cellular responses to environmental factors, plant-bacterial interactions and signalling. Spectroscopy 22(2–3):83–95

Kannojia P, Choudhary KK, Srivastava AK, Singh AK (2019) PGPR bioelicitors: induced systemic resistance (ISR) and proteomic perspective on biocontrol. In: PGPR Amelioration in sustainable agriculture. Woodhead Publishing, pp 67–84

Kawai K et al (2007) The rare earth, scandium, causes antibiotic overproduction in Streptomyces spp. FEMS Microbiol Lett 274:311–315

Kevany BM, Rasko DA, Thomas MG (2009) Characterization of the complete zwittermicin A biosynthesis gene cluster from Bacillus cereus. Appl Environ Microbiol 75(4):1144–1155

Khan N, Bano A (2019) Rhizobacteria and abiotic stress management. In: Plant growth promoting Rhizobacteria for sustainable stress management. Springer, Singapore, pp 65–80

Khan H, Parmar N, Kahlon RS (2016) Pseudomonas-plant interactions I: plant growth promotion and defense-mediated mechanisms. In: Pseudomonas: molecular and applied biology. Springer, Cham, pp 419–468

Koche MD, Gade RM, & Deshmukh AG (2012) Antifungal activity of secondary metabolites produced by Pseudomonas fluorescens. Pesticide Research Journal, 24(1):21–24

Korp J et al (2016) Antibiotics from predatory bacteria. Beilstein J Org Chem 12:594–607

Lanteigne C, Gadkar VJ, Wallon T, Novinscak A, Filion M (2012) Production of DAPG and HCN by Pseudomonas spp. LBUM300 contributes to the biological control of bacterial canker of tomato. Phytopathology 102(10):967–973

Ledger T, Rojas S, Timmermann T, Pinedo I, Poupin MJ, Garrido T, Richter P, Tamayo J, Donoso R (2016) Volatile-mediated effects predominate in Paraburkholderia phytofirmans growth promotion and salt stress tolerance of Arabidopsis thaliana. Front Microbiol 7:1838

Lee JH, Lee J (2010) Indole as an intercellular signal in microbial communities. FEMS Microbiol Rev 34:426–444

Lee B, Farag MA, Park HB, Kloepper JW, Lee SH, Ryu CM (2012) Induced resistance by a long-chain bacterial volatile: elicitation of plant systemic defense by a CVolatile produced by Paenibacillus polymyxa. PLoS One 7:e48744

Lee JH et al (2015) Roles of indole as an interspecies and interkingdom signaling molecule. Trends Microbiol 23:707–718

Leenders F, Stein TH, Kablitz B, Franke P, & Vater J (1999) Rapid typing of Bacillus subtilis strains by their secondary metabolites using matrix-assisted laser desorption/ionization mass spectrometry of intact cells. Rapid communications in mass spectrometry, 13(10):943–949

Li Y, Xiang H, Liu J, Zhou M, Tan H (2003) Purification and biological characterization of halocin C8, a novel peptide antibiotic from Halobacterium strain AS7092. Extremophiles 7(5):401–407

Liu H, He Y, Jiang H, Peng H, Huang X, Zhang X, ... & Xu Y (2007) Characterization of a phenazine-producing strain Pseudomonas chlororaphis GP72 with broad-spectrum antifungal activity from green pepper rhizosphere. Current microbiology 54(4):302–306

Linares JF et al (2006) Antibiotics as intermicrobial signaling agents instead of weapons. Proc Natl Acad Sci U S A 103:19484–19489

Loper JE et al (2008) Isolation and identification of rhizoxin analogs from Pseudomonas fluorescens Pf-5 by using a genomic mining strategy. Appl Environ Microbiol 74:3085–3093

Luzzatto-Knaan T et al (2015) Mass spectrometry tools and workflows for revealing microbial chemistry. Analyst 140:4949–4966

Machado H, Tuttle RN, Jensen PR (2017) Omics-based natural product discovery and the lexicon of genome mining. Curr Opin Microbiol 39:136–142

Maleki H, & Mashinchian O (2011) Characterization of Streptomyces isolates with UV, FTIR spectroscopy and HPLC analyses. BioImpacts: BI, 1(1), 47
Martín JF (2004) Phosphate control of the biosynthesis of antibiotics and other secondary metabolites is mediated by the PhoR-PhoP system: an unfinished story. J Bacteriol 186(16):5197–5201
Massawe VC, Hanif A, Farzand A, Mburu DK, Ochola SO, Wu L et al (2018) Volatile compounds of endophytic bacillus spp. have biocontrol activity against Sclerotinia sclerotiorum. Phytopathology 108(12):1373–1385
Masuma R et al (1986) Production of nanaomycin and other antibiotics by phosphate-depressed fermentation using phosphate-trapping agents. J Antibiot 39:1557–1564
Maurhofer M, Baehler E, Notz R, Martinez V, Keel C (2004) Cross talk between 2, 4-diacetylp hloroglucinol-producing biocontrol pseudomonads on wheat roots. Appl Environ Microbiol 70(4):1990–1998
Menotta M, Gioacchini AM, Amicucci A, Buffalini M, Sisti D, & Stocchi V (2004) Headspace solid-phase microextraction with gas chromatography and mass spectrometry in the investigation of volatile organic compounds in an ectomycorrhizae synthesis system. Rapid communications in mass spectrometry 18(2):206–210
Mhlongo MI, Piater LA, Madala NE, Labuschagne N, Dubery IA (2018) The chemistry of plant–microbe interactions in the rhizosphere and the potential for metabolomics to reveal signaling related to defense priming and induced systemic resistance. Front Plant Sci 9:112
Mojgani N (2017) Bacteriocin-producing rhizosphere bacteria and their potential as a biocontrol agent. In: Rhizotrophs: plant growth promotion to bioremediation. Springer, Singapore, pp 165–181
Moody SC et al (2012) Investigating conservation of the albaflavenone biosynthetic pathway and CYP170 bifunctionality in Streptomycetes. FEBS J 279:1640–1649
Naseem H, Ahsan M, Shahid MA, Khan N (2018) Exopolysaccharides producing rhizobacteria and their role in plant growth and drought tolerance. J Basic Microbiol 58:1009–1022
Naureen Z, Yasmin S, Hameed S, Malik KA, Hafeez FY (2005) Characterization and screening of bacteria from rhizosphere of maize grown in Indonesian and Pakistani soils. J Basic Microbiol Int J Biochem Physiol Genet Morphol Ecol Microorganisms 45(6):447–459
Naureen Z, Price AH, Hafeez FY, Roberts MR (2009) Identification of rice blast disease-suppressing bacterial strains from the rhizosphere of rice grown in Pakistan. Crop Prot 28(12):1052–1060
Naureen Z, Rehman NU, Hussain H, Hussain J, Gilani SA, Al Housni SK et al (2017) Exploring the potentials of Lysinibacillus sphaericus ZA9 for plant growth promotion and biocontrol activities against phytopathogenic fungi. Front Microbiol 8:1477
Nawrath T et al (2010a) The biosynthesis of branched dialkylpyrazines in myxobacteria. Chem Biodivers 7:2129–2144
Nawrath T et al (2010b) The biosynthesis of the aroma volatile 2-methyltetrahydrothiophen-3-one in the bacterium Chitinophaga Fx7914. Chembiochem 11:1914–1919
Nelkner J, Torres Tejerizo G, Hassa J, Lin TW, Witte J, Verwaaijen B et al (2019) Genetic potential of the biocontrol agent pseudomonas brassicacearum (formerly P. trivialis) 3Re2-7 unraveled by genome sequencing and mining, comparative genomics and transcriptomics. Gene 10(8):601
Nielsen TH, Thrane C, Christophersen C, Anthoni U, Sørensen J (2000) Structure, production characteristics and fungal antagonism of tensin–a new antifungal cyclic lipopeptide from Pseudomonas fluorescens strain 96.578. J Appl Microbiol 89(6):992–1001
Nybroe O, Sørensen J (2004) Production of cyclic lipopeptides by fluorescent pseudomonads. In: Pseudomonas. Springer, Boston, MA, pp 147–172
Onaka H et al (2011) Mycolic acid-containing bacteria induce natural-product biosynthesis in Streptomyces species. Appl Environ Microbiol 77:400–406
Ongena M, Jacques P (2008) Bacillus lipopeptides: versatile weapons for plant disease biocontrol. Trends Microbiol 16(3):115–125
Pandey R et al (2013) Indole: a novel signaling molecule and its applications. Indian J Biotechnol 12:297–310

Parret AH et al (2003) Plant lectin-like bacteriocin from a rhizosphere-colonizing Pseudomonas isolate. J Bacteriol 185:897–908
Partida-Martinez LP, Hertweck C (2005) Pathogenic fungus harbours endosymbiotic bacteria for toxin production. Nature 437:884–888
Partida-Martinez LP, Hertweck C (2007) A gene cluster encoding rhizoxin biosynthesis in "Burkholderia rhizoxina", the bacterial endosymbiont of the fungus Rhizopus microsporus. Chembiochem 8:41–45
Pathma J, Raman G, Sakthivel N (2019) Microbiome of rhizospheric soil and vermicompost and their applications in soil fertility, pest and pathogen management for sustainable agriculture. In: Soil fertility Management for Sustainable Development. Springer, Singapore, pp 189–210
Perez J et al (2011) Myxococcus xanthus induces actinorhodin overproduction and aerial mycelium formation by Streptomyces coelicolor. Microb Biotechnol 4:175–183
Pham LH, Vater J, Rotard W, Mügge C (2005) Identification of secondary metabolites from Streptomyces violaceoruber TÜ22 by means of on-flow LC–NMR and LC–DAD–MS. Magn Reson Chem 43(9):710–723
Pluháček T, Lemr K, Ghosh D, Milde D, Novák J, & Havlíček V (2016) Characterization of microbial siderophores by mass spectrometry. Mass spectrometry reviews 35(1):35-47
Polisetti S, Bible AN, Morrell-Falvey JL, Bohn PW (2016) Raman chemical imaging of the rhizosphere bacterium Pantoea spp. YR343 and its co-culture with Arabidopsis thaliana. Analyst 141(7):2175–2182
Raaijmakers JM et al (2006) Cyclic lipopeptide production by plant-associated pseudomonas spp.: diversity, activity, biosynthesis, and regulation. Mol Plant-Microbe Interact 19(7):699–710
Rajarajan PN, Jeganathan P, Rajeswari K, Sumathy N, & Devi AU (2019) GC-MS, FTIR and H, C NMR structural analysis and identification of secondary metabolites from seawater bacterial population. Asian Journal of Pharmacy and Pharmacology, 5(6):1138–1143
Rajer FU, Wu H, Xie Y, Xie S, Raza W, Tahir HAS, Gao X (2017) Volatile organic compounds produced by a soil-isolate, Bacillus subtilis FA26 induce adverse ultra-structural changes to the cells of Clavibacter michiganensis ssp. sepedonicus, the causal agent of bacterial ring rot of potato. Microbiology 163(4):523–530
Rajini KS et al (2011) Microbial metabolism of pyrazines. Crit Rev Microbiol 37:99–112
Raju R, Kandhasamy S, Nalliappan GK, Natarajan KV, Gandhi K, Chandrasekaran B (2016) Cyclic depsipeptide producing fluorescent pseudomonads exerts antifungal activity against fungal pathogens of maize (Zea mays). Afr J Microbiol Res 10(42):1767–1774
Reidys CM et al (2011) Topology and prediction of RNA pseudoknots. Bioinformatics 27:1076–1085
Reuben S, Bhinu VS, Swarup S (2008) Rhizosphere metabolomics: methods and applications. In: Secondary metabolites in soil ecology. Springer, Berlin, Heidelberg, pp 37–68
Riley MA, Wertz JE (2002) Bacteriocins: evolution, ecology, and application. Annu Rev Microbiol 56:117–137
Rodrigues JP, Prova SS, Moraes LAB, & Ifa DR (2018) Characterization and mapping of secondary metabolites of Streptomyces sp. from caatinga by desorption electrospray ionization mass spectrometry (DESI–MS). Analytical and bioanalytical chemistry, 410(27):7135–7144
Romano S, Dittmar T, Bondarev V, Weber RJ, Viant MR, Schulz-Vogt HN (2014) Exo-metabolome of Pseudovibrio spp. FO-BEG1 analyzed by ultra-high resolution mass spectrometry and the effect of phosphate limitation. PLoS One 9(5):e96038
Ryu C-M et al (2003) Bacterial volatiles promote growth in Arabidopsis. Proc Natl Acad Sci U S A 100:4927–4932
Ryu CM, Farag MA, Hu CH, Reddy MS, Kloepper JW, Pare PW (2004) Bacterial volatiles induce systemic resistance in Arabidopsis. Plant Physiol 134:1017–1026
Sanchez S et al (2010) Carbon source regulation of antibiotic production. J Antibiot 63:442–459
Santoro M, Zygadlo J, Giordano W, Banchio E (2011) Volatile organic compounds from rhizobacteria increase biosynthesis of essential oils and growth parameters in peppermint (Mentha piperita). Plant Physiol Biochem 49:1077–1082
Schmidt R et al (2015) Volatile affairs in microbial interactions

Schulz S, Dickschat JS (2007) Bacterial volatiles: the smell of small organisms. Nat Prod Rep 24:814–842

Schulz-Bohm K et al (2015) A fragrant neighborhood: volatile mediated bacterial interactions in soil. Front Microbiol 6:1212

Schwecke T et al (1995) The biosynthetic gene cluster for the polyketide immunosuppressant rapamycin. Proc Natl Acad Sci U S A 92:7839–7843

Shahid I, Rizwan M, Baig DN, Saleem RS, Malik KA, & Mehnaz S (2017) Secondary metabolites production and plant growth promotion by Pseudomonas chlororaphis and P. aurantiaca strains isolated from cactus, cotton, and para grass. Journal of microbiology and biotechnology 27(3):480–491

Sharma A, Gupta A, Dalela M, Sharma S, Sayyed RZ, Enshasy HAE, Elsayed EA (2020) Linking organic metabolites as produced by Purpureocillium Lilacinum 6029 cultured on Karanja deoiled cake medium for the sustainable management of root-knot nematodes. Sustain For 12(19):8276

Shilev S (2013) Soil rhizobacteria regulating the uptake of nutrients and undesirable elements by plants. In: Plant microbe symbiosis: fundamentals and advances. Springer, New Delhi, pp 147–167

Skinnider MA, Dejong CA, Rees PN, Johnston CW, Li H, Webster AL et al (2015) Genomes to natural products prediction informatics for secondary metabolomes (PRISM). Nucleic Acids Res 43(20):9645–9662

Smith KP, Havey MJ, Handelsman J (1993) Suppression of cottony leak of cucumber with Bacillus cereus strain UW85. Plant Dis 77(2):139–142

Song CX et al. (2015) Exploring the genomic traits of fungus- feeding bacterial genus Collimonas. BMC Genom. Published online December 24, 2015 https://doi.org/10.1186/s12864-015-2289-3

Spaink HP, Aarts A, Stacey G, Bloemberg GV, Lugtenberg BJ & Kennedy EP (1992) Detection and separation of Rhizobium and Bradyrhizobium Nod metabolites using thin-layer chromatography. Mol Plant Microbe Interact 5(1):72–80

Stabb EV, Jacobson LM, Handelsman JO (1994) Zwittermicin A-producing strains of Bacillus cereus from diverse soils. Appl Environ Microbiol 60(12):4404–4412

Steeghs M, Bais HP, de Gouw J, Goldan P, Kuster W, Northway M, ... & Vivanco JM (2004) Proton-transfer-reaction mass spectrometry as a new tool for real time analysis of root-secreted volatile organic compounds in Arabidopsis. Plant Physiology 135(1):47–58

Stinear TP et al (2004) Giant plasmid-encoded polyketide synthases produce the macrolide toxin of mycobacterium ulcer- ans. Proc Natl Acad Sci U S A 101:1345–1349

Sun W, Liu S, Liu Z, Song F, & Fang S (1998) A study of aconitum alkaloids from aconite roots in Aconitum carmichaeli Debx using matrix-assisted laser desorption/ionization mass spectrometry. Rapid Communications in Mass Spectrometry 12(13):821–824

Sutherland R, Boon RJ, Griffin KE, Masters PJ, Slocombe B, White AR (1985) Antibacterial activity of mupirocin (pseudomonic acid), a new antibiotic for topical use. Antimicrob Agents Chemother 27(4):495–498

Svatos A (2011) Single-cell metabolomics comes of age: new developments in mass spectrometry profiling and imaging. Ana- lyt Chem 83:5037–5044

Taguchi F, Suzuki T, Inagaki Y, Toyoda K, Shiraishi T, Ichinose Y (2010) The siderophore pyoverdine of Pseudomonas syringae pv. tabaci 6605 is an intrinsic virulence factor in host tobacco infection. J Bacteriol 192(1):117–126

Tahir HA, Gu Q, Wu H, Raza W, Hanif A, Wu L, Colman MV, Gao X (2017) Plant growth promotion by volatile organic compounds produced by Bacillus subtilis SYST2. Front Microbiol 8:171

Takamatsu S, Lin X, Nara A, Komatsu M, Cane DE, Ikeda H. (2011). Characterization of a silent sesquiterpenoid biosynthetic pathway in Streptomyces avermitilis controlling epi-isozizaene albaflavenone biosynthesis and isolation of a new oxidized epi-isozizaene metabolite. Microb Biotechnol. 4(2):184–91

Tambong JT, Höfte M (2001) Phenazines are involved in biocontrol of Pythium myriotylum on cocoyam by Pseudomonas aeruginosa PNA1. Eur J Plant Pathol 107(5):511–521

Tantray YR, Wani N, Malik NA, Wani MS, Jan I (2021) Role of metabolomics in deciphering the plant-microbe interaction. Plant-Microbe Dynam Recent Adv Sustain Agric 71

Thrane C, Nielsen TH, Nielsen MN, Sørensen J, Olsson S (2000) Viscosinamide-producing Pseudomonas fluorescens DR54 exerts a biocontrol effect on Pythium ultimum in sugar beet rhizosphere. FEMS Microbiol Ecol 33(2):139–146

Timmermans ML, Paudel YP, Ross AC (2017) Investigating the biosynthesis of natural products from marine proteobacteria: a survey of molecules and strategies. Mar Drugs 15(8):235

Traxler MF et al (2013) Interspecies interactions stimulate diversification of the Streptomyces coelicolor secreted metabolome. MBio 4:e00459–e0e513

Turner JM, Messenger AJ (1986) Occurrence, biochemistry and physiology of phenazine pigment production. Adv Microb Physiol 27:211–275

Tyc O et al (2014) Impact of interspecific interactions on anti- microbial activity among soil bacteria. Front Microbiol 5:567

Tyc O et al (2015) Volatiles in inter-specific bacterial interactions. Front Microbiol 6:1412

Tyc O, Song C, Dickschat JS, Vos M, Garbeva P (2017) The ecological role of volatile and soluble secondary metabolites produced by soil bacteria. Trends Microbiol 25(4):280–292

van Dam NM et al (2016) Calling in the dark: the role of volatiles for communication in the rhizosphere. In: Blande DJ, Glinwood R (eds) Deciphering chemical language of plant communication. Springer, pp 175–210

van Wezel GP, McDowall KJ (2011) The regulation of the secondary metabolism of Streptomyces: new links and experimental advances. Nat Prod Rep 28:1311–1333

Vater J, Herfort S, Doellinger J, Weydmann M, Borriss R, Lasch P (2018) Genome mining of the lipopeptide biosynthesis of Paenibacillus polymyxa E681 in combination with mass spectrometry: discovery of the lipoheptapeptide paenilipoheptin. Chembiochem 19(7):744–753

Velivelli SL, Kromann P, Lojan P, Rojas M, Franco J, Suarez JP, Prestwich BD (2015) Identification of mVOCs from Andean rhizobacteria and field evaluation of bacterial and mycorrhizal inoculants on growth of potato in its center of origin. Microb Ecol 69(3):652–667

Verma A, Kumar S, Kumar G, Saini JK, Agrawal R, Satlewal A, Ansari MW (2018) Rhizosphere metabolite profiling: an opportunity to understand plant-microbe interactions for crop improvement. In crop improvement through microbial biotechnology. Elsevier, pp 343–361

Vignoli A, Ghini V, Meoni G, Licari C, Takis PG, Tenori L et al (2019) High-throughput metabolomics by 1D NMR. Angew Chem Int Ed 58(4):968–994

Voisard C, Keel C, Haas D, Dèfago G (1989) Cyanide production by Pseudomonas fluorescens helps suppress black root rot of tobacco under gnotobiotic conditions. EMBO J 8(2):351–358

Wang Y, Newman DK (2008) Redox reactions of phenazine antibiotics with ferric (hydr) oxides and molecular oxygen. Environ Sci Technol 42(7):2380–2386

Wang DD et al (2001) Indole can act as an extracellular signal in Escherichia coli. J Bacteriol 183(14):4210–4216

Weller DM, Raaijmakers JM, Gardener BBM, Thomashow LS (2002) Microbial populations responsible for specific soil suppressiveness to plant pathogens. Annu Rev Phytopathol 40(1):309–348

Wen T, Zhao M, Liu T, Huang Q, Yuan J, Shen Q (2020) High abundance of Ralstonia solanacearum changed tomato rhizosphere microbiome and metabolome. BMC Plant Biol 20:1–11

Wikoff WR et al (2009) Metabolomics analysis reveals large effects of gut microflora on mammalian blood metabolites. Proc Natl Acad Sci U S A 106:3698–3703

Xiao Y et al (2011) Antibiotic production by Myxobacteria plays a role in predation. J Bacteriol 193:4626–4633

Yamada Y et al (2015) Terpene synthases are widely distributed in bacteria. Proc Natl Acad Sci U S A 112:857–862

Yang YL et al (2009) Translating metabolic exchange with imaging mass spectrometry. Nat Chem Biol 5:885–887

Yasmin S, Hafeez FY, Mirza MS, Rasul M, Arshad HM, Zubair M, Iqbal M (2017) Biocontrol of bacterial leaf blight of rice and profiling of secondary metabolites produced by rhizospheric Pseudomonas aeruginosa BRp3. Front Microbiol 8:1895

Zheng XY, Sinclair JB (2000) The effects of traits of Bacillus megaterium on seed and root colonization and their correlation with the suppression of Rhizoctonia root rot of soybean. BioControl 45(2):223–243

Ziemert N, Alanjary M, Weber T (2016) The evolution of genome mining in microbes–a review. Nat Prod Rep 33(8):988–1005

Zou CS, Mo MH, Gu YQ, Zhou JP, Zhang KQ (2007) Possible contributions of volatile-producing bacteria to soil fungistasis. Soil Biol Biochem 39(9):2371–2379

# Chapter 10
# Hydrolytic Enzymes from PGPR Against Plant Fungal Pathogens

**Suneeta Panicker and R. Z. Sayyed**

## Contents

10.1 Introduction .......... 211
10.2 Plant Growth-Promoting Rhizobacteria (PGPR) .......... 212
10.3 Plant Diseases Caused By Fungi .......... 214
10.4 Fungal Cell Wall Composition .......... 215
10.5 Synergism of Hydrolytic Enzymes .......... 228
10.6 Commercially Available Bio-Fungicides .......... 229
10.7 Conclusion and Future Perceptive .......... 230
References .......... 231

## 10.1 Introduction

The term rhizosphere was first coined by Hiltner (1904) with an intention to describe the soil microorganisms around and inside roots. With time it has segregated into rhizoplane which include the microbes living on the root surface and endophytes those living inside the root (Zhang et al. 2017).

The rhizosphere is the narrow zone of soil specifically influenced by the root system (Dobbelaere et al. 2003). This zone is rich in nutrients when compared with the bulk soil due to the accumulation of a variety of plant exudates, such as amino acids and sugars, providing a rich source of energy and nutrients for bacteria (Gray and Smith 2005). Due to this reason, the bacterial number found around the roots is usually 10–100 times greater than the bulk soil. The rhizosphere is a complex

S. Panicker (✉)
Department of Microbiology, Dr. D.Y. Patil Arts, Commerce and Science College, Pune, India

R. Z. Sayyed
Department of Microbiology, PSGVP Mandal's S I Patil Arts, G B Patel Science and SKTV Sangh Commerce College, Shahada, India



R. Z. Sayyed et al. (eds.), *Antifungal Metabolites of Rhizobacteria for Sustainable Agriculture*, Fungal Biology, https://doi.org/10.1007/978-3-031-04805-0_10

habitat containing living plant roots with associated soil where root exudates attract and promote microbial growth and metabolism and hence contains a wide variety of microorganisms. The bacteria that colonize in the rhizosphere are called rhizobacteria (Schroth and Hancock 1982). The activities of the diverse rhizosphere microbial community greatly influence many aspects of plant physiology and growth, including protecting the host plant from deleterious pathogenic microorganisms. These bacteria, in the soil near plant roots, on the surface of plant root systems, in spaces between root cells, or inside specialized cells of root nodules that are able to stimulate plant growth through a wide array of mechanisms, are known as plant growth-promoting rhizobacteria (PGPR) (Kloepper and Schroth 1978). PGPR employ a variety of mechanisms to promote plant growth and development. The more widely recognized mechanisms include (1) biofertilization (enhanced nutrient availability), (2) suppression of diseases through biocontrol, (3) induction of disease resistance in plants, (4) production of phytohormones, and (5) production of volatile signal compounds (Whipps 2001). On the broad scale, PGPR include the legume-nodulating rhizobia, and as such can be separated into those residing outside plant cells (extracellular or free-living ePGPR) and the rhizobia that live inside plant cells (intracellular or symbiotic iPGPR) (Gray and Smith 2005; Freitas et al. 2007).

## 10.2 Plant Growth-Promoting Rhizobacteria (PGPR)

The plants are invariably accompanied with highly organized and balanced set of microorganisms (Lebeis 2014; Bulgarelli et al. 2015). This set of microorganisms is known as the phytomicrobiome (Smith et al. 2017), and the plant with the phytomicrobiome is named as holobiont (Berg et al. 2016; Smith et al. 2017). The group of microorganisms specially associated with the roots are called rhizomicrobiome, and the nitrogen-fixing rhizobia associated with legumes are the well-studied and identified examples of it (Gray and Smith 2005).

PGPR occupy the roots of both monocotyledons and dicotyledons. They are capable of amplifying the growth of the plants directly or indirectly. They modify the frame work of the host plant root system by using various mechanisms like production of phytohormones and other signals which help the plants in increasing the lateral root branching and root hair development. They are also known to improve functioning of roots, boost the nutrition of plants, and affect the overall physiology of the plant. Few PGPR genera showing plant growth-promoting activity are *Pseudomonas*, *Azospirillum*, *Azotobacter*, *Bacillus*, *Burkholderia*, *Enterobacter*, *Rhizobium*, *Erwinia*, *Mycobacterium*, *Mesorhizobium*, *Flavobacterium*, etc. (Bhattacharyya and Jha 2012).

The causative agent of stem blight is shown to be inhibited by *Bacillus subtilis*, *Pseudomonas chlororaphis*, and *endophytic P. fluorescens*. Similarly *Macrophomina phaseolina*, the pathogen causing root rot of black gram, was reduced by the treatment of seed and application to the soil of *P. fluorescens*. *P. fluorescens* could also decrease the sheath blight of rice after it was applied to the seeds and foliar. *Burkholderia cepacia* strains are known to be used as biocontrol agents against *Fusarium* spp. *Aspergillus niger* and *Fusarium udum* were shown to be controlled in a better way by *B. subtilis* in combination with the chitin or chitin-containing materials.

PGPR are potentially useful in stimulating plant growth and increasing crop yields (Sayyed et al. 2010; Hamid et al. 2021; Basu et al. 2021; Khan et al. 2021). Thus, the rhizosphere of crop plants is a promising source of PGPR (Lucas et al. 2001; Barriuso et al. 2005). It is clearly indicated through the worldwide literature that the use of PGPR in agriculture is increasing speedily, and as a result, a significant increase in growth and yield of agronomically important crops has been observed (Figueiredo et al. 2008; Araujo 2008). The PGPR are highly specific in their plant growth-promoting ability, and hence some bacteria are specific to certain plant species, cultivar, and genotype (Gupta et al. 2000; Lucy et al. 2004). PGPR are known to produce many compounds that are released in the soil. These compounds can improve the health of soil in the long term and thus can significantly reduce the use of chemical fertilizers. They are also involved in producing antimicrobials (biocontrol of phytopathogen) (Sukmawati et al. 2021; Sharma et al. 2020) and hence can potentially minimize the use of chemical pesticides (Lugtenberg and Kamilova 2009). However, for appropriate usage of PGPR in agriculture, the mechanisms of plant growth promotion and the biocontrol by them need to be determined in a more extensive way. This is possible by studying the rhizosphere microbial consortium, its diversity and structure with respect to natural selection; interpopulational relations like symbiosis, parasitism, mutualism, or competence; and succession (Barriuso et al. 2008).

PGPR has become an environment-friendly, outstanding option for the plant disease management (Liu et al. 2017). They are preferred because they are nontoxic and naturally occurring, easy to apply, stimulate the development of the plant, and improve the health of the soil, and most importantly they assist the plant in stress tolerance (abiotic and biotic) (Sagar et al. 2021). They are also favored due to various methods they employ to carry out these functions (Patel et al. 2016; Khan and Sayyed 2019).

The plant growth is affected in two different ways by the PGPR: indirectly or directly. In the direct way, the plants are benefited by compounds like phytohormones, synthesized by these bacteria, or they facilitate the absorption of nutrients from the soil by changing the solubility of the nutrients (Kusale et al. 2021a; Glick 1995). By the indirect way for plant growth promotion, PGPR reduces or prevents the harmful effects of plant pathogens by synthesizing antagonistic substances or by inducing resistance to pathogens (Nasab and Sayyed; 2019; Zope et al. 2019; Glick 1995). Thus, either through one or a combination of these methods, a particular PGPR may influence the growth and development of the plant. These mechanisms include production of auxin phytohormone (Patten and Glick 2002) or nitrogen fixing associated with roots (Döbereiner 1992) or decrease of plant ethylene levels (Kusale et al. 2021b; Sagar et al. 2020; Glick et al. 2007); synthesis of hydrolytic enzymes, such as chitinases, glucanases, proteases, and lipases, that can lyse pathogenic fungal cells (Jadhav et al. 2020a, b; Shaikh et al. 2018; Neeraja et al. 2010; Maksimov et al. 2011); competition for nutrients and suitable colonization of niches at the root surface (Shaikh and Sayyed 2015; Kamilova et al. 2005); regulation of plant ethylene levels through the ACC-deaminase enzyme, which can act to modulate the level of ethylene in a plant in response to stress imposed by the infection (Glick and Bashan 1997; Van Loon 2007); production of siderophores (Nithyapriya et al. 2021) and antibiotics; etc. (Vinay et al. 2016; Reshma et al. 2018). These interactions between the plants and the PGPR are currently in commercial use (Podile and Kishore 2006) as promising agents for sustainable agriculture. PGPR as biocontrol agents have been tried in many plants like wheat, barley, maize, tomatoes, canola, soy, lentils, peas, potatoes, etc. (Gray and Smith 2005).

## 10.3 Plant Diseases Caused By Fungi

Every year, millions of revenues are lost due to the destruction of the crops by the fungal and oomycete pathogens. Both these organisms are phylogenetically faraway, but their pathogenesis is almost same which involves spreading of pathogenic particles, attachment to the host receptor, identification and invasion, and encroaching growth and development of abrasions (Fig. 10.1). Both these pathogens have distinctive capacity to violate cuticles in order to manifest the infection briskly (Soanes et al. 2007). Thus, these two are the significant causes of many plant diseases leading to major annual economic loss.

Plant pathogens also referred to as phytopathogens cause a wide variety of diseases by infecting either the whole plant or a particular part of the plant. The effect of such infections may vary from simple mild symptoms to disastrous in which large plantations of crops are demolished, causing heavy loss of the food crop yield. This results in today's shortfall of food supply due to which many millions of lives are insufficiently fed. Various factors like different genotypes and pathogen count in

**Fig. 10.1** Fungal and oomycota pathogenesis

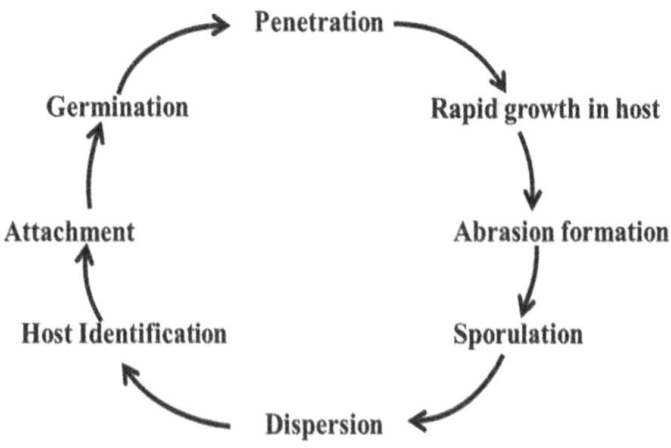

a given space and time influence the power of these pathogens and thus become a challenging task to manage them (Strange and Scott 2005). One way to combat the loss is use of fungicides, but repeated use of fungicides resulted in evolved resistance in these pathogens leading to severe loss of yield. Hence, biological methods are proving to be more beneficial as compared to the chemical ones. Globally, it is reported that most of the fungus species and/their products are pathogenic to the plants.

Oomycota belongs to a different class of eukaryotic microbes which are like fungus but not true fungus. Most of them are deadly pathogenic to animals or plants. As opposite to the fungi that have chitin as major cell wall component, the oomycete cell wall has cellulose as a main constituent (Fawke et al. 2015). These oomycotic pathogens follow varied forms of lifestyles like necrotrophic, biotrophic, and hemibiotrophic (a combination of both). Many are true obligate parasites which completely depend on their hosts, viz., *Hyaloperonospora parasitica*, *Hyaloperonospora arabidopsidis*, *Albugo candida*, and *Plasmopara viticola*. *Phytophthora* spp. are hemibiotrophs, which can live in axenic culture (facultative), while *Pythium ultimum* are necrotrophs.

## 10.4 Fungal Cell Wall Composition

Fungus contains a layered cell wall. They have an outer layer and an inner layer. The inner layer is the one which surrounds the plasma membrane immediately and is highly conserved. It gives the structural skeleton to the cell. The outer layer is diverse among the species. It is potently modified according to the developmental and maturation requirements of the organism. It also has an influence of its reaction with the environment. Their cell wall hardly contains proteins more than 20 percent, and most of them are in the glycoprotein form. The protein portion somewhat contributes in the structure but mostly plays many different roles. Proteins in the outer layer that are close to the outer surface confer the surface properties like adhesive/antigenic/hydrophilic/hydrophobic to the cell wall. Fungal cell wall also contains some amount of lipids and wax which generally control the water movement mainly to avoid dehydration.

The basic function of any cell wall is to protect its internal organelles and constituents from the external or environmental factors. The components of the cell wall function in transportation of nutrients, extracellular degradation of non-permeable substrates, communication, signaling, etc. Therefore, it can be a perfect target for making the agents to destroy the cells. Fungal cell wall has a unique structure and can be targeted for developing antifungal agents. The fungal cell walls mainly consist of chitins, mannans, glucans, and glycoproteins (Bowman and Free 2006). It also contains inert fibrillar materials bound to the proteins, sugars, lipids, and various polysaccharides (Fig. 10.2). The fungal cell wall is mainly composed of polysaccharides, nearly 80%, about 20% of proteins (glycoproteins) and small amount of lipids. These lipids and proteins facilitate the movement of water and

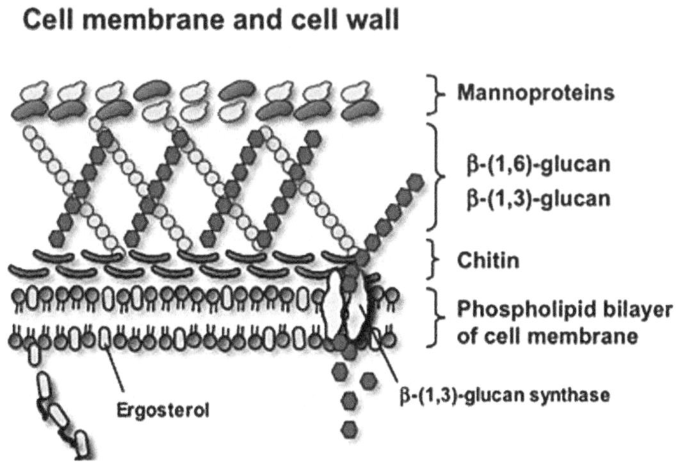


**Fig. 10.2** Diagrammatic representation of fungal cell wall components. The main constituent chitin is shown immediately above the cell membrane. β-1,3- and β-1,6-glucan reside above the chitin layer and the mannoproteins form the outermost layer of the cell wall

protect the cell from dehydration (Cox and Hooley 2009). Its fibrillar structure is built on chitin, chitosan, ß-glucans, and a variety of heteropolysaccharides. Many antifungal compounds produced by different organisms have been extensively studied with respect to their structure and biosynthesis. Unlike true fungi, members of the phylum *Oomycota* lack chitin in their cell walls and instead contain cellulose as a major part.

The transmembrane chitin synthase enzyme passes the N-acetylglucosamine residues from uridine diphosphate-N-acetylglucosamine (UDPGlcNAc) to the growing chitin fiber. In the same manner, β-1,3-glucan synthase transfers glucose from uridine diphosphate-N-glucose extruded β-1,3-glucan fiber (Fesel and Zuccaro 2015).

### *10.4.1 Hydrolytic Enzymes Produced By Rhizospheric Microbes*

Many rhizospheric microbes are known to produce extracellular hydrolytic enzymes that act on various polymers like proteins, chitin, hemicellulose, cellulose, and DNA of the plant pathogens and thus are capable to inhibit the phytopathogens (Pal and Gardener 2006). Recently vast studies referring to the isolation and characterization of these hydrolytic enzymes have determined their ability to control plant pathogens (Shaikh and Sayyed 2015; Shaikh et al. 2016). Enzymes like protease, chitinase, cellulose, and glucanase are known to act on the fungal cell wall leading to its degradation and finally the lysis of the fungal cell (Kobayashi et al. 2002). Rhizospheric microbes producing such hydrolytic enzymes can be used as environment-friendly biocontrol agents (Mabood et al. 2014). Many bacterial species like *B. subtilis*, *B. cereus*, *B. subtilis*, *B. thuringiensis*, *S. marcescens*, *R. solani*, *F. oxysporum*, *S. rolfsii*, *P. ultimum*, etc. are shown to synthesize such enzymes that can affect the cell wall integrity of the pathogens and inhibit them (Felse and Panda 1999; Someya et al. 2000).

Some actinomycete isolates were shown to produce ß-1,3-, ß-1,4-, and ß-1,6-glucanases that hydrolyze glucans from *Phytophthora* cell walls, leading to the lysis of these cells. These isolates were used as biocontrol agents against a pathogen *Phytophthora fragariae* var. rubi which causes raspberry root rot (Valois et al. 1996). The prevalence of disease symptoms caused by *Rhizoctonia solani* in cotton and cucumber was shown to be reduced by chitinase-producing bacteria such as *Enterobacter agglomerans* (Chernin et al. 1995), *Bacillus cereus* (Pleban et al. 1997), and *Paenibacillus illinoisensis* (Jung and Kim 2003). Extracellular proteases produced by *Stenotrophomonas maltophilia W81* were reported to be used to control *Pythium ultimum* in sugar beet (Dunne and Crowley 1997). *Micromonospora carbonacea* was shown to produce cellulase to inhibit *Phytophthora cinnamomi*, a causative agent of *Banksia grandis* root rot (El-Tarabily et al. 1996). Inhibition of *Fusarium oxysporum f.* sp. cucumerinum causing Fusarium wilt of cucumber (*Cucumis sativus*) by *Paenibacillus* sp. 300 and *Streptomyces* sp. 385 by production of chitinase and b-1,3-glucanase was also reported (Singh et al. 1999). Similarly, *Pseudomonas cepacia* produced b-1,3 glucanase which lysed the cell wall of *Rhizoctonia solani*, *Sclerotium rolfsii*, and *Pythium ultimum* and decreased the occurrence of the disease caused by these pathogens (Fridlender et al. 1993). These hydrolytic enzymes produced by PGPR play an important role in the control of various plant pathogens and thereby improve the plant growth, making PGPR an efficacious biocontrol agent (Garbeva et al. 2004; Ran et al. 2005).

#### 10.4.1.1 Lysis of Cell Wall By Hydrolytic Enzymes

In most fungi, their cell wall is made up of chitin ($C_8\ H_{13}O_5N$)n. Chitin is a long-chain, unbranched polymer of glucose derivative like N-acetyl-D-glucosamine (NAG) which is linked by β-1,4 linkage. The hydrolytic enzymes target these glycosidic bonds in chitin to degrade the cell wall of the phytopathogens. *Paenibacillus* spp. strain B2, isolated from the mycorrhizosphere of *Sorghum bicolor* and having an antagonistic activity toward soil-borne fungal pathogens, possessed extracellular and enzymatic activities. Extracellular enzymatic activity such as chitinolytic, cellulolytic, pectinolytic, and proteolytic was determined in the *Paenibacillus* spp. strain B2 which was isolated from the mycorrhizosphere of *Sorghum bicolor*. Using electron microscopy and molecular cytology, the mode of action of these lytic enzymes during the cellular interplay between host and pathogen was studied. Images showed disorganization of the contents of the cell wall of the pathogenic *Phytophthora parasitica* and *Fusarium oxysporum* (Budi et al. 2001). The inhibition of *Sclerotium rolfsii* and *F. oxysporum* through the degradation of cell wall on beans was also reported by Felse and Panda (1999).

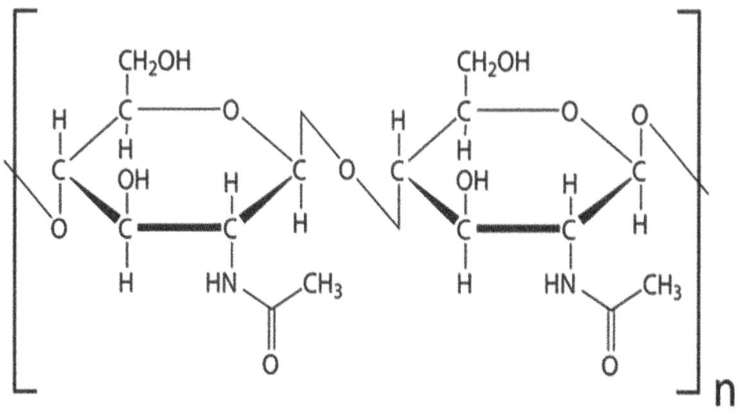


**Fig. 10.3** Covalent structures of *N*-acetylglucosamine and its linear homopolymer, chitin

10.4.1.1.1 Chitinases in Biocontrol of Phytopathogenic Fungi

Chitin is a homopolymer of N-acetyl-D-glucosamine (GlcNAc), linked by 1,4-β linkage and is unbranched (Fig. 10.3). Chitin polymers present in the cell wall of fungal phytopathogens are lysed by chitinase enzyme causing degradation of the fungal cell wall. Investigative studies on lytic activity among biocontrol agents have focused largely on the characterization of enzyme systems capable of degrading fungal cell wall components, of which chitinases are among the most intensively studied (Sitrit et al. 1995; Zhang and Yuen 2001). The microorganisms synthesizing chitinase can be directly used as the biocontrol agents. In another indirect way, the chitinase enzyme can be purified from these organisms and used as BCAs, or through genetic engineering the genes for chitinase enzyme can be manipulated, and those organisms or the modified chitinase can be used as BCA (Kim et al. 2002).

Removal of chitin by the enzymatic action or disruption of its synthesis leads to the osmotic instability of the fungal cell wall. Chitinase enzyme cleaves the glycosidic linkages present in the chitin polymer converting them into monomers (Fig. 10.4). Also the fibrillar structures present in the chitin are randomly cleaved by chitinase at the internal sites causing the release of diacetylchitobiose without releasing monosaccharide or oligosaccharides (Harman et al. 1993; Manocha and Sahai 1993). Molecular and biochemical characterizations have revealed that chitinases, similar to other glycosyl hydrolases, are modular in nature and can differ according to their structural organization. Enzymes can vary both within and between microbes, depending on the numbers, types, and positions of discrete binding and catalytic domains (Gilkes et al. 1991; Warren 1996). Domains that have been described within bacterial chitinases include chitin-binding domains (Svitil and Kirchman 1998; Tanaka et al. 1999), cadherin-like domains (Morimoto et al.1997), and fibronectin type III domains (Watanabe et al. 1990). *Stenotrophomonas maltophilia* represents a rhizosphere bacterial species of potential agronomic importance (Kobayashi et al. 1995; Mazzola et al. 1995). Traits of

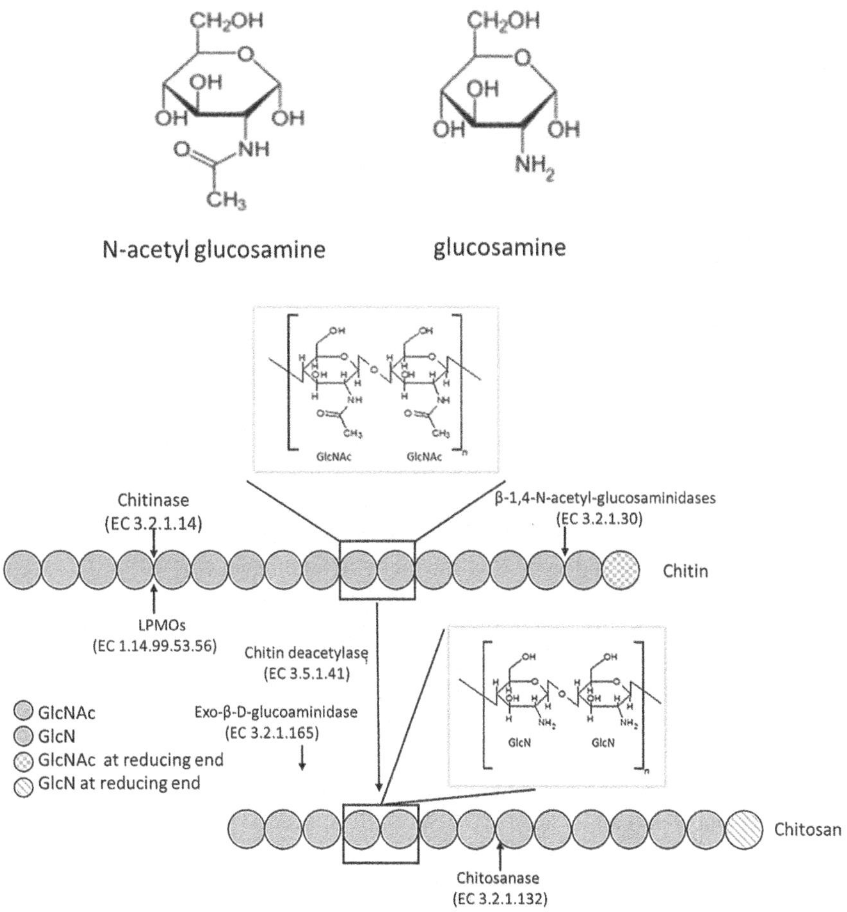


**Fig. 10.4** Enzymatic degradation of fungal cell wall chitin
(**a**) β-1,4-N-acetyl-glucosaminidases (EC 3.2.1.30) separate the chitin polymer into GlcNAc monomers in an exo-type pattern.

*S. maltophilia* associated with biocontrol mechanisms include antibiotic production (Jakobi and Winkelmann 1996), extracellular enzyme activities such as protease and chitinase (Kobayashi et al. 1995; Zhang and Yuen 2001), and rhizosphere colonization potential (Lambert et al. 1990; Kobayashi et al. 1995; Dunne and Crowley 1997). Thus, chitinase [EC 3.2.1.14] plays a crucial role in controlling the fungal diseases in plants through degradation of chitin polymer.

List of microorganisms showing hydrolytic activity:

- Chitinase produced by *S. marcescens* against *R. solani* and *F. oxysporum* (Someya et al. 2000).
- Chitinase produced by *B. subtilis* NPU 001 against *F. oxysporum* (Chang et al. 2010).

- Chitinase produced by *S. plymuthica* C48 against *Botrytis cinerea* (Frankowski et al. 2001).
- β −1,3-glucanase produced by *Paenibacillus* sp. strain 300 and *Streptomyces* sp. strain 385 against *F. oxysporum* (Singh et al. 1999),
- Cellulase produced by *Bacillus subtilis* YJ1 (Li-Jung et al. 2010).
- Cellulase produced by *Cellulomonas* spp. ASN2 (Muhammad et al. 2012).
- Carboxymethyl cellulase and polygalacturonase produced by *Bacillus coagulans* (Odeniyi et al. 2009).
- Cellulase produced by *Bacillus cereus*, *Bacillus subtilis*, and *Bacillus thuringiensis* (Basavaraj et al. 2014).
- Protease produced by *Pseudomonas aeruginosa* PGPR2 against *Macrophomina* spp., *Rhizoctonia* spp., and *Fusarium* spp. (Illakkiam et al. 2013).
- Alkaline protease produced by *Bacillus subtilis* PE-11 (Adinarayana et al. 2003).

There are various reports on isolation of chitinase enzyme from higher plants, insects, crustaceans, fungi, and bacteria (Kramer et al. 1997). Chitinase from different rhizobacteria have been reported and shown to possess antifungal properties. Long back, Skujins et al. (1965) demonstrated fungal growth inhibition by chitinases of *Streptomycetes*, and same was shown by Schlumbaum et al. (1986). Loss of biocontrol activity by *Serratia marcescens* chitinase mutants (inactivated chiA gene) further confirmed the importance of chitinase as an antifungal agent (Jones et al. 1986). Later it was also determined that chitinases from different rhizobacteria exhibit fungal antagonism in vitro (Gay et al. 1992; Fridlender et al. 1993). Valuing the applications of this enzyme, it was isolated, purified, and characterized in 1992 (De La Cruz et al. 1992). Using recombination technology chiA gene of *S. marcescens* was cloned in *E. coli* and used to control the disease caused by *S. rolfsii* and *R. solani* (Oppenheim and Chet 1992). Similarly, chitinase genes from *S. marcescens* were expressed in *Pseudomonas* to control the pathogens *F. oxysporum* var. redolens and *Gaeumannomyces graminis* var. tritici (Sundheim 1992). Transgenic rhizobium showed antifungal activity during symbiosis on alfalfa roots. This was confirmed by the destruction of *R. solani* hyphal tips when they were treated with cell-free nodule extracts (Sitrit et al. 1993).

#### 10.4.1.1.2 Mode of Action of Chitinase

The second most abundant polymer in nature is chitin; hence, chitin-degrading enzymes, chitinases, are considered as the most important biocontrol agents. They work very well against insects and fungal plant pathogens, as ample chitin is present in their cell wall (Havukkala 1991). Chitinase enzymes are of three types, based on their mode of action:

(**b**) Endochitinases (EC 3.2.1.14) arbitrarily cleave at internal sites over the full length of the chitin microfibril.

(**c**) Exochitinases (EC 3.2.1.14) catalyze the continuous release of diacetylchitobiose bit by bit so that no monosaccharides or oligosaccharides are produced (Fig. 10.4) (Harman et al. 1993; Manocha and Sahai, 1993).

*Myrothecium verrucaria* produces extracellular chitinase that is shown to inhibit germination and germ tube elongation of the groundnut rust fungus *Puccinia arachidis*. Also chitinase secreted by *Acremonium obclavatum* were demonstrated to inhibit the germination of uredospores of the peanut rust (Manocha and Balasubramanian 1994).

#### 10.4.1.1.3 Proteases in Biocontrol of Phytopathogenic Fungi

The cell wall is a vital component that preserves the sustainability of fungal cells, giving the typical morphology and guarding the cell against outer injuries. Different classes of proteins are present in the cell wall of fungi. These proteins have different functions and sometimes these functions are species-specific. The various functions of these proteins in the cell wall are shape maintenance, adherence, protection against environmental stress, taking in of the molecules, and intracellular signal transmission and reception of external stimuli, along with production and rebuilding of cell wall components and immune escape (Klis et al. 2006). Glycosyl phosphatidyl inositol (GPI) anchored proteins, transglycosidases, hydrolases, deacetylase, yapsins, adhesins, hydrophobins, and mannoproteins. In spite of the fact that various cell wall proteins are investigated to be immunogenic, they are possibly to be expressed in vivo, and to take part in adaptive immunity, their importance in growth and pathogenesis is not known (Pitarch et al. 2006).

The fungal cell wall is largely made up of mannoproteins, which are also referred as cell wall proteins (CWPs). The exterior outer layer is made of mannoproteins and looks much lighter. It is mainly accountable for regulating the porosity of the cell wall. These proteins are generally heavily glycosylated (either O- or N-glycosylated) with mannose-containing polysaccharides. The carbohydrate part makes up for up to 90% of their molecular weight. In fungi the O-glycosylation happens at serine and/or threonine residues and does the addition of mannose sugar through α-1,2- or α-1,3-type linkages (as occurs in *S. cerevisiae*). β-1,3-glucan is dispersed throughout the cell surface. Here it is covalently attached to some chitin polymers, giving a frame to which mannoproteins are also covalently bound. There is a cross-linking between the branched glucans to each other and to the chitin; furthermore, the glucans are additionally linked to the mannoproteins. Thus, the proteins are also the component of the fungal cell wall that can be considered for the destruction of the pathogen (Klis et al. 2006). Yeast like *S. cerevisiae* and *C. albicans* contain nearly 30–50% protein by dry weight in their cell wall, whereas *N. crassa* hyphal cell wall shows the presence of just 15% protein and 20–30% protein by dry weight in filamentous fungi cell walls. The cell wall proteins are mostly glycoproteins that have N-linked and O-linked oligosaccharides. The amino acid sequence motifs, viz., asparagine-any amino acid-serine and asparagine-any amino acid-threonine, present in the proteins are the points where the N-linked oligosaccharides are attached. There are proteins called PIR (proteins with internal repeats) which are linked to the β-1,3-glucan chains by ester linkages that lead to the formation of polysaccharide-protein complex like β-1,3-glucan–PIR-β1,3-glucan. This complex is found all over the inner layer of fungal cell wall (Yin et al. 2007).

Proteases [E.C. 3.4.24] are the enzymes known to cleave the proteins or peptides. Hence, proteases also play a crucial role in the cell wall lysis of fungal plant pathogens as the main structural components of the cell wall, chitin, and/or fibrils of β-glucan are logged into the protein matrix. By cleaving this protein matrix into peptide chains or amino acids, proteases like chitinases also can destroy the pathogen's cell wall integrity. Few proteases are also known to inactivate the pathogenic extracellular enzymes of the pathogens. Thus, proteolytic activity is equally essential for complete lysis of the fungal cells (Elad and Kapat 1999). Proteases are very common in nature produced by almost all species. Microbial sources are favored for the production of these enzymes due to many reasons: fast growing, easy to grow, easy to perform genetic manipulations to improve the enzymatic properties, etc. (Anwar and Saleemuddin 1998; Beg and Gupta 2003). The enzymes from microbial sources possess several favorable factors like mostly extracellular, produced in bulk quantity, broad substrate range, optimum at various environmental conditions, etc. Many *Bacillus* species like *Bacillus cereus*, *Bacillus stearothermophilus*, *Bacillus mojavensis*, *Bacillus megaterium*, and *Bacillus subtilis* are shown to synthesize protease (Sookkheo et al. 2000; Banik and Prakash 2004; Gerze et al. 2005).

Mannoproteins from the fungal cell wall have important immunostimulatory activities (Pietrella et al. 2006). Generally, the outer layer of the fungal hyphae possesses a film of these proteins which are called as moonlighting proteins (Jeffery 1999). They are named so because these proteins have many functions that are usually unrelated and independent of each other like moonlighting people who have several jobs. Such moonlighting proteins are present in various organisms.

Alkaline proteinase gene (prbl) was shown to exert a crucial inhibitory effect of mycoparasitism (Geremia et al. 1993). It was also demonstrated that this gene was induced in the host cells (*R. solmi*), immediately 4 hours after exposure to the mycoparasites (Flores et al. 1996).

###### *10.4.1.1.3.1 Mode of Action of Protease*

Proteins are polymers of amino acids linked by peptide bonds. Degradation of proteins is possible by hydrolysis of these peptide bonds which result in the release of smaller peptides or individual amino acids from the proteins. Proteins are hydrolyzed by enzymes known as proteases or proteinases. The chemical structures of specific amino acids can be recognized by particular protease in order to cleave a certain peptide bond (Fig. 10.5).

The bacteria secrete proteases in the presence of high protein concentration in order to metabolize those proteins, but these enzymes are also shown to be liberated when grown on chitin or glucose as the only carbon source. This indicates that some additional complex mechanisms occur in these situations. Probably many proteases are produced to search for different proteins, in response to various carbon sources. This will luckily allow the lysis of fungal cell wall. Thus, these bacteria can be highly used as biocontrol agents. *B. subtilis* natto strain BEST195 genome is shown

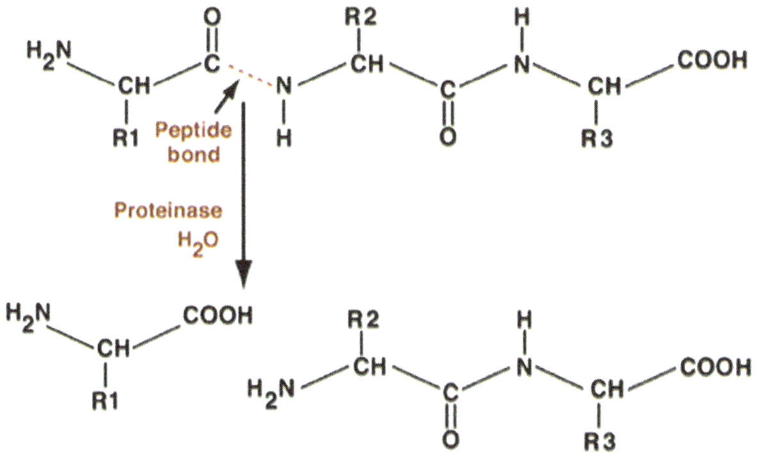


**Fig. 10.5** Proteinase action on peptide bond to release amino acids or peptides

to encode about 615 known or putative peptidases (Kamada et al. 2014). Though these enzymes are expressed even when grown on glucose or chitin, the exact mechanism to trigger their expression is yet unclear.

Treatment of fungus with proteolytic enzymes increases their susceptibility to lysis by chitinase and 1,3-ß-glucanase. This concept was proven by Sivan and Chet (1989). They treated the hyphal walls of *F. oxysporum* with NaOH or proteolytic enzymes and showed that this treatment indeed increased their sensitivity to chitinase and 1,3-ß-glucanase activity. Similarly, Elad et al. (1985) determined that trypsin treatment of the *F. oxysporum f.* spp. cucumerinum cell wall made it more susceptible to the lytic enzymes produced by *Pythium nunn*.

#### 10.4.1.2 Cellulase in Biocontrol of Phytopathogenic Fungi

Both fungi and oomycetes are equally pathogenic to plants as compared to the rest of the pathogens. They are capable of rupturing the cuticles of their host plants and initiate the infection swiftly (Soanes et al. 2007). These infections cause many severe diseases in plants which lead to a significant loss of crop yield and thus a huge economic loss per year. As fungi contain chitin in their cell wall, oomycetes contain cellulose. Thus, to control these pathogens, hydrolytic enzymes like chitinase and cellulose can be used as biocontrol agents.

Cellulose is a linear polysaccharide. It is made up of many hundreds to thousands of d-glucose subunits with one non-reducing end and another reducing end with aldehyde structure. These subunits are linked to each other by β-1,4-glycosidic linkages (Fig. 10.6a) (Holtzapple 2003). Cellulose has a molecular formula of $(C_6H1_2O_6)_n$. Repeating units of two glucose molecules together are called cellobiose (Fig. 10.6b). Both amorphous and crystalline regions are present in varying amounts in the structure of the cellulose (Ciolacu et al. 2011). These regions are interwoven

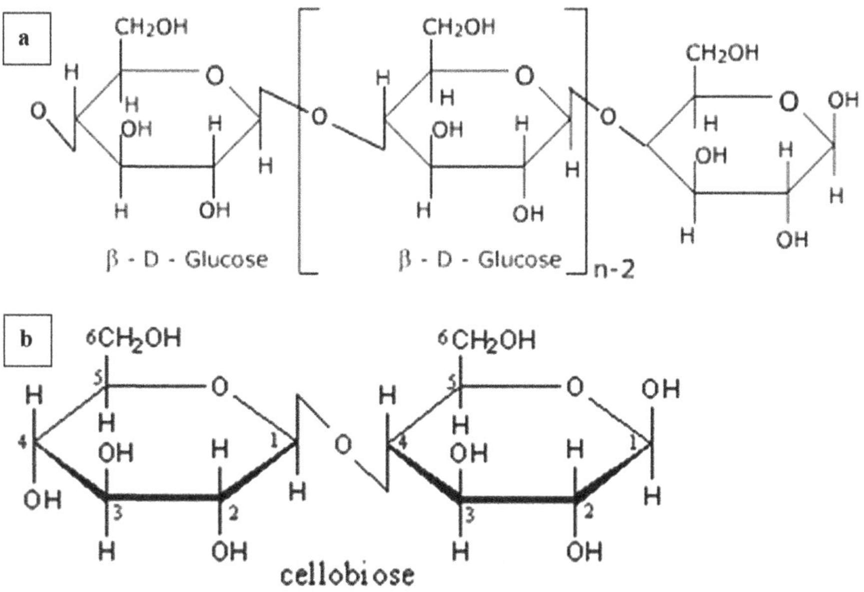


**Fig. 10.6** (**a**) Glucose monomers linked with beta 1,4 linkage to form cellulose polymer. (**b**) Repeating units of two glucose molecules in cellobiose

to form the proper structure of cellulose. Intramolecular and intermolecular hydrogen bonding between glucose monomers builds a big network that gives rise to the crystalline structure of the polymer. Iα, Iβ, II, and III are the four major crystalline forms in cellulose. These hydrogen bonds also lead to the poor solubility of cellulose. The degree of crystallinity in the cellulose structure defines its resistance to degradation by the microbial enzymes.

Many soil microorganisms account for the production of enzyme cellulase which are group enzymes responsible for production of simple sugars by reducing the complex polysaccharide of cellulose (Lynd et al. 2002; Baldrian and Valášková 2008). Cellulases [EC 3.2.1.4] are the members of a class of enzyme that accelerate the hydrolysis of cellulose. They are mainly produced by bacteria, fungi, and protozoa. Plants and animals are also known to synthesize cellulases (Lee and Koo 2001). Cellulase predominantly consists of endoglucanase and exoglucanases including cellobiohydrolases and β-glucosidase, which act together to hydrolyze cellulose.

The amorphous part of the cellulose is acted upon by the endoglucanase enzyme, which randomly cleaves the internal bonds resulting in the formation of new chain ends. These newly created ends are then readily attacked by different kinds of enzymes. Endoglucanases show maximum activity against the soluble form of cellulose. On the other hand, the exoglucanase has the ability to produce cellobiose or glucose from the cellulose by cleaving either the reducing or non-reducing end of it. Exoglucanase is active against the crystalline structure of cellulose. The third enzyme in this group is β-glucosidase that acts on the cellobiose to release the glucose units from the non-reducing end. This enzyme is not active against either form (amorphous/crystalline) of the cellulose (Vocadlo and Davies 2008). Thus, for the complete degradation of the cellulose, a complex interplay between all these cellulolytic enzymes that act synergistically to release the glucose molecules from cellulose is necessary. Lytic enzymes like protease and cellulase produced by PGPR *Bacillus cereus* and endophyte *Pseudomonas aeruginosa* were reported to be significantly lysing the cell wall of *Phytophthora* (Valois et al. 1996) and *Pythium* spp. (El-Tarabily et al. 2009), the oomycete pathogen. Thus, these biological controls were able to reduce the incidence of the rhizome rot and leaf blight disease significantly and helped to increase yield profoundly.

Very recently 17 rhizobacterial strains were isolated from the *Hippophae rhamnoides* L. plant rhizospheric soil and root nodules and were identified by 16SrRNA sequencing as belonging to the genera *Azorhizobium*, *Paenibacillus*, *Frankia*, *Bacillus*, *Pseudomonas*, and *Brevibacillus*. Few of this bacterial population were shown to produce cellulase (Bhadrecha et al. 2020).

##### 10.4.1.2.1 Mode of Action of Cellulase

Interaction between different cellulolytic enzymes is required for the complete degradation of cellulose. There are mainly three enzymes acting synergistically to break the cellulose into glucose. They are endoglucanases [EC 3.2.1.4], exoglucanases [EC 3.2.1.91], and β-glucosidases [EC 3.2.1.21] (Lynd et al. 2002). Endo-1,4-β glucanase cleaves the internal bonds, while exoglucanases and the β-glucanase cut from the ends of cellulose to release two to four units, and finally the cellobiase cleaves the disaccharide cellobiose into two glucose molecules (Fig. 10.7).

#### 10.4.1.3 Glucanases in Biocontrol of Phytopathogenic Fungi

The cell walls of fungi consist of large amounts of polysaccharides called glucans. Glucans are carbohydrates made up of glucose units. It mainly consists of β-1,3-linked backbone having some branches via β-1,6 linkages with the side chains branching from the six-position of the backbone (Kapteyn et al. 2000; Kim et al. 2000). Their structures are extremely varying. Glucose molecules in glucan may be linked by alpha (α) or beta (β) or both, either branched or linear and microfibrillar or amorphous. Largely found form of glucan in the fungal cell wall is Alpha 1,3 glucans sensu stricto (pseudonigerans). The microfibrillar structural existence of the glucans in the cell wall imparts resistance. Beta glucans are more composite with their branched or linear microfibrillar structure containing β-1,3 and β-1,6 linkages. Along with chitin they are the highly essential fungal cell wall constituents present in the cell wall of all fungal phyla except *Microsporidia.* Fungal cell wall contains two groups of glucans, α-glucans and β-glucans. The α-glucans in the cell wall consists of glucose units bound by α-1,3 linkages and they are insoluble in water. α-1,4-joined glucose units are also present in some fungal cell wall in order to inter link the unbranched α-1,3-glucan chains or in an interspersed way with α-1,3-bound

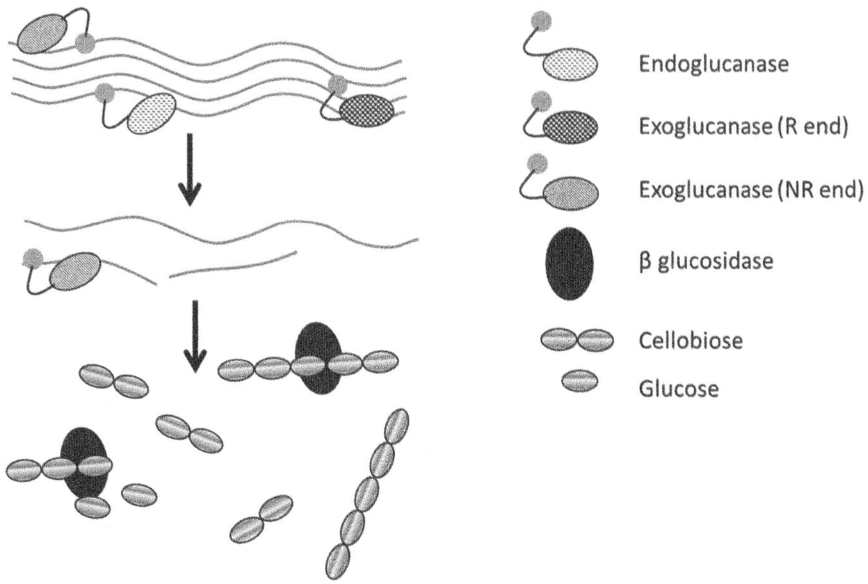


**Fig. 10.7** Mechanism of action of cellulase

glucose units (Ruiz-Herrera 1991). β-glucan is another glucan in the fungal cell wall and it is made of β-1,3-glucan. This is the most ample glucan in the cell wall and accounts for nearly 65–90% of the entire β-glucan component (Bowman and Free 2006). A major portion of β-1,3-glucan is covalently linked with β-1,6-glucan, forming a branched poly-polysaccharides (Shahinian and Bussey 2000). These branched β-glucans are either strongly or roughly linked to the cell wall. They get collected around the fungus in the form of a slime layer or like a gelatinous material (Ruel and Joseleau 1991; Ruiz-Herrera 2012) depending on the molecular weight of the polymer or the degree of branching of the different glucan polysaccharide. The α-glucans and β-1,3-glucan are also present in the plant cell wall, but the β-1,6-glucan is only found in the fungal cell wall and in the cell wall of representatives of Chromista phylum, such as the oomycetes belonging to the *Phytophthora* and *Pythium* genera (Bartnicki-Garcia 1968; Sietsma et al. 1969).

Many studies have shown their significance in the structure of the cell wall by indicating that drastic changes occur in the cell structure, even cell lysis, if the synthesis of β-1,3-glucans is inhibited or is enzymatically digested. One of such study in *S. pombe* showed that when β-1,3-glucan was digested by β-1,3-glucanase or its synthesis was inhibited by aculeacin caused either the self-destruction of the cells or the cells lost their original structure (Miyata et al. 1985).

β-1,3-Glucanases [EC 3.1.1.6] are broadly found in fungi, bacteria, and higher plants (Simmons 1994). β-1,3-Glucanases can be grouped as either exo- or endo-β-1,3-glucanases (β-1,3-glucan: glucanohydrolase). The action of this enzyme leads to degradation of cell wall and then entry into the mycelium of the pathogen (Fridlender et al. 1993). This enzyme has practical implementation the cell wall lysis of yeast, molds, and higher plants (Pang et al. 2004). β-1,3-glucanases of *Bacillus cepacia* were shown to inhibit *Rhizoctonia solani*, *Sclerotium rolfsii*, and *Pythium ultimum* through their hydrolytic activity (Fridlender et al. 1993). Similarly, *F. oxysporum* was inhibited by the β-1,3-glucanases produced by *Paenibacillus* and *Streptomyces* sp. (Singh et al. 1999). One more study reported seven β-1,3-glucanases synthesized from *T. harzianum* under various growth conditions (Vazquez et al. 1998).

#### 10.4.1.3.1 Mode of Action of Glucanase

Glucanase enzymes cleave the β-1,6 linkages in glucans causing degradation of cell wall and later penetration into the mycelium of pathogen (Fridlender et al. 1993). There are two possible ways for their action (Noronha and Ulhoa 1996):

1. Sequential hydrolysis of glucose residues from the non-reducing end by the ex-1,3-glucanases (EC 3.2.1.58).
2. Random cleavage of the linkages within the glucan polymer chain producing tiny oligosaccharides by endo-1,3-glucanases (EC 3.2.1.39).

Long back it was determined that 1,3-β-glucanase can be synthesized upon induction by inducers like starch, laminarin, mannitol, xylose, and glycerol (Reese and Mandels 1959) of which laminarin is shown to increase the secretion of this enzyme maximally (Elad et al. 1982). Thus, we can say that 1,3-β-glucanase is a semiconstitutive enzyme (Bull and Chesters 1966). In another study it was also reproduced that 1,3-β-glucanase when incubated with NaOH-treated cell walls of fusaria was able to liberate more glucose (up to 135-fold) as compared to the untreated cell walls (Sivan and Chet 1989).

The oligosaccharides that are released upon the enzymatic action of chitinase or β-1,3-glucanase on the fungal cell wall are shown to have a very significant role in signal transduction which is required for the defense mechanisms in the plants. Thus, these oligosaccharides have been proven to induce hypersensitive reactions, lignification, and phytoalexin production (Shibuya and Minami 2001).

Cell wall-degrading enzymes (CWDEs) like chitinases, β-1,3-glucanases, and proteases are composite molecules that catalyze the breakdown of a particular substrate into specific products. They are made up of nearly 100–1000 amino acids. The specificity and sensitivity of these CWDEs is determined not only by their amino acid sequence but even by their complicated tertiary structures (Iyer and Ananthanarayan 2008). CWDEs lose their activity or are denatured, if unfolded or if the polypeptides are disordered. They are also sensitive to denaturation by physical agent like temperature, chemical agents like acids, and chelating agents or to the microbial proteases. This sensitive issue of these enzymes is the major hurdle in their use as biocontrol agents. This can be overcome by stabilizing the CWDEs during their synthesis and usage. Direct resistance to these types of biological control agents by the plant pathogens has not yet observed. However, the plant pathogens can develop structures like endospores, chlamydospores, and melanized sclerotia which can easily resist the action of these enzymes (Bardin et al. 2015). Some pathogens have a counter mechanism to suppress the synthesis of these enzymes. In France, the *Coniothyrium minitans* is commercially applied to control *S. sclerotiorum*, but a substantial difference in the susceptibility of this pathogen has been seen in various parts of the country (Nicot et al. 2016). Many strains of *S. sclerotiorum* are known to produce sclerotia, which may vary in thickness, and thus the susceptibility of this pathogen to the *C. minitans* may differ. The hydrolytic enzymes produced by *Trichoderma* spp. can permeabilize and degrade the fungal cell wall (Karlsson et al. 2017). This increased permeability allows different antimicrobial metabolites to enter and reach different targets like genome or other macromolecules.

*Serratia marcescens* is a known biological control for gray mold which is an airborne pathogen of cyclamen plants (Someya et al. 2000). It produces chitinases to degrade the fungal cell wall and induce a defense reaction in plants. Additionally it also produces low molecular weight molecules which have antifungal activity as well. Thus, this organism can be a potent and determined biological control for cyclamen plant diseases.

There are various reports on production of hydrolytic enzymes able to degrade cell walls (Chernin and Chet 2002), iron-chelating siderophores, and several cyclic lipodepsipeptides (LDP) (Dalla Serra et al. 2003) by the species of *Bacillus* and *Pseudomonas* and hence could be good candidates for biocontrol of many plant diseases.

In addition to hydrolytic enzyme (proteolytic, chitinolytic, and cellulolytic) secretion, *Bacillus amyloliquefaciens* BNM340 also secretes surfactin and some iturin-like lipodepsipeptides, such as iturin A. All these properties make this bacterium a potent antifungal agent (Dalla Serra et al. 2003; Pimentel et al. 1991).

Similarly, *Pseudomonas fluorescens* (Mercado-Blanco et al. 2004) possess various characteristics which include production of siderophore, volatile compounds, and antibiotics such as PRN and PLT in addition to the synthesis of CWDEs like chitinase and protease (Dalla Serra et al. 2003; de Souza and Raaijmakers, 2003). Thus, *Pseudomonas fluorescens* can also be classified as a biocontrol agent for plant diseases.

## 10.5 Synergism of Hydrolytic Enzymes

There are several reports on antifungal activity of cell wall-degrading enzymes. The antifungal activity of purified endochitinase and exochitinase (chitobiosidase) was tested by Lorito et al. (1993). They showed that there was inhibition of germ tube elongation and spore germination. The chitin levels in the cell wall of the target fungi determine the degree of inhibition by these enzymes. The chitinolytic enzymes from bacterial sources are found to be more active than those from different sources. Also, they are effective against broad spectra of pathogenic fungi. Moreover, it has also been observed that when these enzymes are used in combinations, it leads to an increase in their antifungal activity. Authors of many studies interpret that when hydrolytic enzymes with complementary modes of action are mixed, they synergistically maximize their efficacy of antifungal activity. It was observed that when all the four CWDEs were used together in a solution, the highest level of antifungal activity was achieved (Lorito et al. 1993; Schirmbiick et al. 1994). Thus, the concept of using synergism of these hydrolytic enzymes can be used as a potent weapon to fight the plant diseases. This synergist activity can be between different enzymes or their combinations with other known antifungal compounds to maximize their action and to reduce the usage of risky, dangerous chemical fungicides. Other antifungal compounds like inhibitors of synthesis of cell wall constituents, sphingolipids, proteins, etc. (Vicente et al. 2003). These combined actions of different antifungal agents will totally inactivate the pathogen.

## 10.6 Commercially Available Bio-Fungicides

Omega Bacigold is a bio-fungicide, commercially available in India. It is a bacterial preparation of *Bacillus subtilis* that is used as a fungicide. This bacterium colonizes on root system in the developing stage of the plant to compete with the fungal pathogens for the space and nutrients. In this way it prevents various diseases caused by fungal pathogens like *Alternaria*, *Aspergillus*, *Fusarium*, *Pythium*, and *Rhizoctonia* which are very well known to cause deadly diseases such as damping-off, powdery mildew, sour rot, downy mildew, early leaf spot, early blight, late blight, bacterial spot, and walnut blight. Though this formulation has not described the production of hydrolytic enzymes by *Bacillus subtilis*, there are many reports of cell wall-degrading enzymes from this bacteria which we have cited throughout this chapter. The manufacturer only claims of production of cyclic lipopeptides antibiotics called Iturins. This antibiotic can either kill or reduce the growth of pathogenic fungi.

Another commercially available bio-fungicide in India is ABTEC Pseudo which has a dual action of plant growth and antifungal activity. It consists of *Pseudomonas fluorescens*. This preparation is active against many root pathogenic fungi, bacteria, nematodes, and several foliar fungal and bacterial pathogens. It is a versatile formulation with four major actions, viz., antibiotic production, induced systemic resistance (ISR), competition, and mycoparasitism. Mycoparasitism refers to the parasitization and disintegration of pathogen hyphae by secreting lytic enzymes like chitinase, proteases, lipases, and ß-1,3-glucanases. Bio Don is one more commercially available biological fungicide made up of a mixture of bacteria like *Bacillus subtilis*, *Bacillus megaterium*, and *Bacillus licheniformis*. This is an antagonistic biological product that colonizes the root and leaf system of the plant and competes and thereby suppresses the growth of the plant disease-causing organisms. It creates defensive mechanism in plants toward pathogens. It releases enzymes which promote growth in plants. All the *Bacillus* species are shown to produce hydrolytic enzymes that can take care of the fungal pathogens. Few more commercially

**Table 10.1** List of few commercially available PGPR strains

| Organism name | Marketed as | Marketed by | Used for |
|---|---|---|---|
| *Delftia acidovorans* | BioBoost | BrettYoung seeds ltd., Manitoba | Seed treatment in canola |
| *Azotobacter* spp. | Bioplin | Kumar Krishi Mitra bio products Pvt. ltd., Pune, India | Soil drenching for sunflower, tomato, and other vegetable crops |
| *Bacillus* spp. | Bioyield | Gustafson, LLC, Plano, Tx | Seed treatment in tomato, tobacco, cucumber, and pepper |
| *Bacillus*, *pseudomonas*, and *Streptomyces* spp. | Compete | Plant health care BV, CA Vught | Soil drenching for turfgrass, nursery, and greenhouse plantations |
| *B. subtilis GB03* | Kodiak | Gustafson, LLC, Dallas, Tx | Seed treatment in fruits and vegetables |

available bio-fungicides are Astratop, Astrazeb, Funginol, Venus, Astraphil, Flick, Sharp, Orgam – F, Bio-Cure-F, Super Rakshak Bio Fungicide, bio-fungicide, RimZim, Shan-101, etc.

There are few PGPR preparations in the market (Table 10.1) that are primarily used for direct plant growth-promoting mechanisms. However, the bacterial species used in these formulations are shown to be responsible for many extra activities, for example, Bioplin which contains *Azotobacter* spp.; Bioyield containing *Bacillus* spp.; Compete that contains a mixture of *Bacillus*, *Pseudomonas*, and *Streptomyces* spp.; Kodiak made of *Bacillus subtilis* GB03. All these bacteria not only promote the plant growth but also produce different metabolites that protect plants from various diseases. We have already mentioned that many species of bacillus and pseudomonas produce antibiotics and CWDEs in order to inhibit the growth of pathogen so that the yield of the crop is improved.

## 10.7 Conclusion and Future Perceptive

The idea of bio-fungicides or the microbial sources for the control of plant fungal pathogens are getting immense attention. The biocontrol agents capable of producing hydrolytic enzymes are proving to be a significant option for the harmful chemical fungicides. These are considered more useful, safe, and long-lasting because they exclusively act on the fungal cell wall and do not affect the plant tissue. They are eco-friendly biocontrol agents as opposed to the other routinely used fungicides. In the existing global situation, the extensive usage of chemical fertilizers and germicides has created an alarming distress. Hence a great deal of focus and attention are paid toward the practice of microbes or microbial products for this purpose. Recent researchers have come up with promising formulated products of rhizobacteria, whose application in the field of agriculture is proving to be an eco-friendly, greener method for long-lasting agriculture to fight the plant pathogens. These rhizobacteria are very well studied for their multipotential capacity not only in plant protection but also in promoting their growth. The plant protection by rhizobacteria is due to their potential of producing different protecting antifungal products like hydrolytic enzymes, antibiotics, and many other metabolites. Growth promotion of plants by rhizobacteria is due to production of metabolites like siderophores, indole-acetic acid, etc. Thus, these truly called PGPRs help in reducing the broad usage and dosage of agrochemicals. These PGPR release their antifungal metabolites in a sustained manner due to which it is difficult for the target organism to develop resistance against them as it is experienced with the chemical fungicides. PGPRs can be used individually or in combinations in sector of plant growth and protection. They are considered as the major determinants in plant growth, health, and yield with sustainability. Substantial research of the cell wall-degrading hydrolytic enzymes can enable the researchers to manipulate the bacterial community for their large-scale production and better performance.

## References

Adinarayana K, Ellaiah P, Prasad DS (2003) Purification and partial characterization of thermostable serine alkaline protease from a newly isolated Bacillus subtilis PE-11. AAPS Pharm Sci Tech 4(4):440–448

Anwar A, Saleemuddin M (1998) Alkaline proteases. A review. Bioresour Technol 6:175–183

Araujo FF (2008) Inoculation of seeds with Bacillus subtilis, formulated with oyster flour and development of corn, soy and cotton. Agrotec Sci 32:456–462

Baldrian P, Valášková V (2008) Degradation of cellulose by basidiomycetous fungi. FEMS Microbiol Rev 32(3):501–521

Banik RM, Prakash M (2004) Laundry detergent compatibility of the alkaline protease from Bacillus cereus. Microbiol Res 159(2):135–140

Bardin M, Ajouz S, Comby M, Lopez-Ferber M, Graillot B, Siegwart M, Nicot PC (2015) Is the efficacy of biological control against plant diseases likely to be more durable than that of chemical pesticides? Front Plant Sci 6:566

Barriuso J, Pereyra MT, Lucas García JA, Megías M, Gutierrez Mañero FJ, Ramos B (2005) Screening for putative PGPR to improve establishment of the symbiosis Lactarius deliciosus-Pinus sp. Microbial Ecol 50:82–89

Barriuso J, Ramos B, Lucas JA, Lobo AP, García-Villaraco A, Gutiérrez Mañero FJ (2008) Ecology, genetic diversity and screening strategies of plant growth promoting rhizobacteria (PGPR). In: Ahmad I, Pichtel J, Hayat S (eds) Plant-bacteria interactions: strategies and techniques to promote plant growth, Wiley-VCH Verlag GmbH & co. KGaA, Weinheim ISBN: 978-3-527-31901-5

Bartnicki-Garcia S (1968) Cell Wall chemistry, morphogenesis, and taxonomy of fungi. Annu Rev Microbiol 22:87–108

Basavaraj I, Patagundi CT, Shivasharan KBB (2014) Isolation and characterization of cellulose producing bacteria from soil. Int J Curr Micro Appl Sci 3(5):59–69

Basu A, Prasad P, Das SN, Kalam S, Sayyed RZ, Reddy MS, El Enshasy H (2021) Plant growth promoting Rhizobacteria (PGPR) as green bioinoculants: recent developments, constraints, and prospects. Sustainability 13:1140. https://doi.org/10.3390/su13031140

Beg QK, Gupta R (2003) Purification and characterization of an oxidation-stable, thiol-dependent serine alkaline protease from Bacillus mojavensis. Enz Microbial Techn 32:294–304

Berg G, Rybakova D, Grube M, Koberl M (2016) The plant microbiome explored: implications for experimental botany. J Exp Bot 67:995–1002
Bhadrecha P, Bala M, Khasa YP, Arshi A, Singh J, Kumar M (2020) Hippophae rhamnoides L. rhizobacteria exhibit diversified cellulase and pectinase activities. Physiol Mol Biol Plants 26(5):1075–1085
Bhattacharyya PN, Jha DK (2012) Plant growth-promoting rhizobacteria (PGPR): emergence in agriculture. World J Microbial Biotechnol 28:1327–1350
Bowman SM, Free SJ (2006) The structure and synthesis of the fungal cell wall. Bioassays 28:799–808
Budi SW, Tuinen D, Arnould C, Dumas-Gaudot E, Gianinazzi-Pearson V, Gianinazzi Erratum S (2001) Hydrolytic enzyme activity of Paenibacillus spp. strain B2 and effects of the antagonistic bacterium on cell integrity of two soil-borne pathogenic fungi. Appl Soil Ecol 16(3):291–292
Bulgarelli D, Garrido-Oter R, Munch PC, Weiman A, Droge J, Pan Y, McHardy AC, Schulze-Lefert P (2015) Structure and function of the bacterial root microbiota in wild and domesticated barley. Cell Host Microbe 17:392–403
Bull AT, Chesters CGC (1966) The biochemistry of laminarin and the nature of laminarinase. Adv Enzymol 28:325–364
Chang WT, Chen M, Wang SL (2010) An antifungal chitinase produced by Bacillus subtilis using chitin waste as a carbon source. World J Microbiol Biotechnol 26:945–950
Chernin L, Chet I (2002) Microbial enzymes in biocontrol of plant pathogens and pests. In: Enzymes in the environment: activity, ecology, and applications. NYRG, New York, pp 171–225
Chernin LZ, Haran IS, Chet I (1995) Chitinolytic Enterobacter agglomerans antagonistic to fungal plant pathogens. Appl Environ Microbiol 61:1720–1726
Ciolacu D, Ciolacu F, Popa VI (2011) Amorphous cellulose—structure and characterization. Cellul Chem Technol 45(1–2):13–21
Cox PW, Hooley P (2009) Hydrophobins: new prospects for biotechnology. Fungal Biol Rev 23:40–47
Dalla Serra M, Menestrina G, Carpaneto A, Gambale F, Fogliano V, Ballio A (2003) Molecular mechanisms of action of syringopeptins, antifungal peptides from pseudomonas syringae pv. Syringae, in Pore-Forming Peptides and Proteins Toxins. Menestrina G, Dalla M, Serra, and Lazarovici P. Taylor & Francis, London, pp. 272–295
De La Cruz J, Rey M, Lorca JM, Hidalgo- Gallego A, Dominguez F, Pintor-Toro JA, Uobell A, Bentiez T (1992) Isolation and characterization of three chitinases from Trichoderma harzianum. European J Biochem 206:859–867
de Souza JT, Raaijmakers JM (2003) Polymorphisms within the prnD and pltC genes from pyrrolnitrin and pyoluteorin-producing pseudomonas and Burkholderia spp. FEMS Microbiol Ecol 43(1):21–34
Dobbelaere S, Vanderleyden J, Okon Y (2003) Plant growth promoting effects of diazotrophs in the rhizosphere. CRC Crit Rev Plant Sci 22:107–149
Döbereiner J (1992) History and new perspectives of diazotrophs in association with non-leguminous plants. Symbiosis 13:1–13
Dunne C, Crowley JJ, Moenne-Loccoz Y, Dowling DN, de Bruijn FJ, O'Gara F (1997) Biological control of Pythium ultimum by Stenotrophomonas maltophilia W81 is mediated by an extracellular proteolytic activity. Microbiol 143:3921–3931
Elad Y, Kapat A (1999) The role of Trichoderma harzianum protease in the biocontrol of Botrytis cinerea. Eur J Plant Pathol 105(2):177–189
Elad Y, Kalfon A, Chet I (1982) Control of Rhizoctonia solani in cotton by seed-coating with Trichoderma spp. spores. Plant Soil 66:279–281
Elad Y, Lifshitz R, Baker R (1985) Enzymatic activity of the mycoparasite Pythium nunn during interaction with host and non-host fungi. Physiol Plant Pathol 27:131–148
El-Tarabily KA, Sykes ML, Kurtboke ID, Hardy GE, Barbosa AM, RFH D (1996) Synergistic effects of a cellulase-producing Micromonospora carbonacea and an antibiotic-producing Streptomyces violascens on the suppression of Phytophthora cinnamomi root rot of Banksia grandis. Can J Bot 74:618–624

El-Tarabily KA, Nassar AH, Hardy GE, Sivasithamparam K (2009) Plant growth promotion and biological control of Pythium aphanidermatum, a pathogen of cucumber, by endophytic actinomycetes. J Appl Microbiol 106(1):13–26
Fawke S, Doumane M, Schornack S (2015) Oomycete interactions with plants: infection strategies and resistance principles. Microbiol Mol Biol Rev 79(3):263–280
Felse AP, Panda T (1999) Production of microbial chitinases. Bioprocess Eng 23(2):127–134
Fesel P, Zuccaro A (2015) β-glucan: crucial component of the fungal cell wall and elusive MAMP in plants. Fungal Genet Biol 90:53–60. https://doi.org/10.1016/j.fgb.2015.12.004
Figueiredo MVB, Burity HA, Martinez CR, Chanway CP (2008) Alleviation of water stress effects in common bean (Phaseolus vulgaris L.) by co-inoculation Paenibacillus x Rhizobium tropici. Appl Soil Ecol 40:182–188
Flores A, Chet I, Herrera-Estrella A (1996) Improved biocontrol activity of Trichoderma harzianum by over-expression of the proteinase-encoding gene prb1. Curr Genet 31(1):30–37
Frankowski J, Lorito M, Scala F, Schmidt R, Berg G, Bahl H (2001) Purification and properties of two chitinolytic enzymes of Serratia plymuthica HRO-C48. Arch Microbiol 176:421–426
Freitas ADS, Vieira CL, Santos CERS, Stamford NP, Lyra MCCP (2007) Characterization of isolated rhizobia from Pachyrhyzus erosus L. cultivated in saline soil of the State of Pernambuco, Brazil. Bragantia 66(3):497–504
Fridlender M, Inbar J, Chet I (1993) Biological control of soil borne plant pathogens by a ß-1,3-glucanase-producing Pseudomonas cepacia. Soil Biol Biochem 25:1211–1221
Garbeva P, van Veen JA, van Elas JD (2004) Assessment of the diversity, and antagonism towards Rhizoctonia solani AG3, of Pseudomonas species in soil from different agricultural regimes. FEMS Microbiol Ecol 47:51–64
Gay PA, Saikurnar KV, Cleveland TE, Tuzun S (1992) Antagonistic effect of chitinolytic bacteria against toxin producing fungi. Phytopathology 82:1074
Geremia RA, Goldman GH, Jacobs D, Ardiles W, Vila SB, Van Montagu M, Herrera-Estrella A (1993) prb1, a gene specifically expressed during simulated mycoparasitism by Trichoderma harzianum. Mol Microbiol 8:603–613
Gerze A, Omay D, Guvenilir Y (2005) Partial purification and characterization of protease enzyme from Bacillus subtilis and bacillus megatherium. Appl Biochem Biotechnol 121-124:335–345
Gilkes NR, Henrissat DG, Kilburn RC, Miller WRAJ Jr (1991) Domains in microbial glycanases: sequence conservation, function, and enzyme families. Microbiol Rev 55:303–315
Glick BR (1995) The enhancement of plant growth by free-living bacteria. Can J Microbiol 41:109–117
Glick BR, Bashan Y (1997) Genetic manipulation of plant growth-promoting bacteria to enhance biocontrol of phytopathogens. Biotechnol Adv 15:353–378
Glick BR, Cheng Z, Czarny J, Duan J (2007) Promotion of plant growth by ACC deaminase-producing soil bacteria. Eur J Plant Pathol 119:329–339
Gray EJ, Smith DL (2005) Intracellular and extracellular PGPR: commonalities and distinctions in the plant-bacterium signaling processes. Soil Biol Biochem 37:395–412
Gupta A, Gopal M, Tilak KV (2000) Mechanism of plant growth promotion by rhizobacteria. Indian Indian J Exp Biol 38:856–862
Hamid B, Zaman M, Farooq S, Fatima S, Sayyed RZ, Baba ZA, Sheikh TA, Reddy MS, El Enshasy H, Gafur A, Suriani NL (2021) Bacterial plant biostimulants: a sustainable way towards improving growth, productivity, and health of crops. Sustainability 13:2856. https://doi.org/10.3390/su13052856
Harman GE, Hayes CK, Lorito M, Broadway RM, Pietro A, Peterbauer C, Tronsmo A (1993) Chitinolytic enzymes of Trichoderma harzianum: purification of chitobiosidase and endochitinast. Phytopathology 83:313–318
Havukkala I (1991) Chitinolytic enzymes and plant pests. In: Ilag LL, Raymundo AK (eds) Biotechnology in the Philippines towards the year 2000, proceedings of the second Asia-Pacific biotechnology congress. SEARCA, University of the Philippines, Los Baños, pp 14–127
Hiltner L (1904) Über neuere erfahrungen und probleme auf dem debiete der bo denbakteriologie und unter besonderer berucksichtigung der grundund und brache. Zbl Bakteriol 2:14–25

Holtzapple MT (2003) Cellulose. In: Caballero B (ed) Encyclopedia of food sciences and nutrition, 2nd edn. Academic Press, pp 998–1007

Illakkiam D, Anuj NL, Ponraj P, Shankar M, Rajendhran J, Gunasekaran P (2013) Proteolytic enzyme mediated antagonistic potential of Pseudomonas aeruginosa against Macrophomina phaseolina. Indian J Exp Biol 51:1024–1031

Iyer PV, Ananthanarayan L (2008) Enzyme stability and stabilization—Aqueous and non-aqueous environment. Process Biochem 43:1019–1132

Jadhav HP, Sayyed RZ, Shaikh SS, Bhamre HM, Sunita K, Enshasy HE (2020a) Statistically designed bioprocess for enhanced production of alkaline protease in *Bacillus cereus* HP_RZ17. J Sci Ind Res 79:491–498

Jadhav HP, Sonawane MS, Khairnar MH, Sayyed RZ (2020b) Production of alkaline protease by rhizospheric *Bacillus cereus* HP_RZ17 & *Paenibacillus xylanilyticus* HP_RZ19. Environ Sustain 3:5–13

Jakobi M, Winkelmann G (1996) Maltophilin: a new antifungal compound produced by Stenotrophomonas maltophilia R3089. J Antibiot 49:1101–1104

Jeffery CJ (1999) Moonlighting proteins. Trends Biochem Sci 24:8–11

Jones JDG, Grady KL, Suslow TV, Bedbrook JR (1986) Isolation and characterization of genes encoding two chitinase enzymes from Serratia marcescens. EMBO J 5:467–473

Jung HK, Kim SD (2003) Purification and characterization of an antifungal antibiotic from Bacillus megaterium KL 39, a biocontrol agent of red-pepper phytophthora blight disease. Kor J Microbiol Biotechnol 31:235–241

Kamada M, Hase S, Sato K, Toyoda A, Fujiyama A, Sakakibara Y (2014) Whole genome complete resequencing of Bacillus subtilis Natto by combining long reads with high-quality short reads. PLoS One 9(10):e109999

Kamilova F, Validov S, Azarova T, Mulders I, Lugtenberg B (2005) Enrichment for enhanced competitive plant root tip colonizers selects for a new class of biocontrol bacteria. Environ Microbiol 7:1809–1817

Kapteyn JC, Hoyer LL, Hecht JE, Muller WH, Andel A, Verkleij AJ, Makarow M, Van Den Ende H, Klis FM (2000) The cell wall architecture of Candida albicans wild-type cells and cell wall-defective mutants. Mol Microbiol 35:601–611

Karlsson M, Atanasova L, Jensen DF, Zeilinger S (2017) Necrotrophic mycoparasites and their genomes. Microbiol Spectrum 5:FUNK-0016-2016

Khan A, Sayyed RZ (2019) Sonia Seifi Rhizobacteria : legendary soil guards in abiotic stress management. In: Sayyed, Arora, Reddy (eds) Plant growth promoting Rhizobacteria for sustainable stress management Vol 1 abiotic stress management. Springer, Singapore, pp 27–342

Khan N, Ali S, Shahi MA, Mustafa A, Sayyed RZ, Curaá JA (2021) Insights into the interactions among roots, rhizosphere and Rhizobacteria for improving plant growth and tolerance to abiotic stresses: a review. Cell 10(6):1551. https://doi.org/10.3390/cells10061551

Kim YT, Kim EH, Cheong C, Williams DL, Kim C-W, Lim S-TJCR (2000) Structural characterization of β-D-(1 →3, 1 →6)-linked glucans using NMR spectroscopy. Carbohydr Res 328:331–341

Kim DJ, Baek JM, Uribe P, Kenerley CM, Cook DR (2002) Cloning and characterization of multiple glycosyl hydrolase genes from Trichoderma virens. Curr Genet 40:374–384

Klis FM, Boorsma A, De Groot PWJ (2006) Cell wall construction in Saccharomyces cerevisiae. Yeast 23:185–202

Kloepper JW, Schroth MN (1978) Plant growth-promoting rhizobacteria on radishes. Proceedings of the fourth international conference on plant pathogen bacteria, 2 INRA pp. 879–882

Kobayashi DY, Guglielmoni M, Clarke BB (1995) Isolation of the chitinolytic bacteria Xanthomonas maltophilia and Serratia marcescens as biological control agents for summer patch disease of turfgrass. Soil Biol Biochem 27:1479–1487

Kobayashi DY, Reedy RM, Bick JA, Oudemans PV (2002) Characterization of chitinase gene from Stenotrophomonas maltophilia strain 34S1 and its involvement in biological control. Appl Environ Microbiol 68(3):1047–1054

Kramer KJ, Muthukrishnan S, Johnson L, White F (1997) Chitinases for insect control. In: Carozzi N, Koziel M (eds) Advances in insect control: the role of transgenic plants. Taylor & Francis, Bristol, pp 185–193

Kusale SP, Attar YC, Sayyed RZ, Malek RA, Ilyas N, Suriani NL, Khan N, Enshasy HE (2021a) Production of plant beneficial and antioxidants metabolites by *Klebsiella variicola* under salinity stress. Molecules 26:1894. https://doi.org/10.3390/molecules26071894

Kusale SP, Attar YC, Sayyed RZ, Enshasy HE, Hanapi Z, Ilyas N, Elgorban AM, Bahkali AH, Marraiki N (2021b) Inoculation of *Klebsiella variicola* alleviated slat stress salinity and improved growth and nutrients in wheat and maize. Agronomy 11:927. https://doi.org/10.3390/agronomy11050927

Lambert B, Meire P, Joos H, Lens P, Swings J (1990) Fast-growing, heterotrophic bacteria from the rhizosphere of young sugar beet plants. Appl Environ Microbiol 56:3375–3381

Lebeis SL (2014) The potential for give and take in plant-microbiome relationships. Front Plant Sci 5:287. https://doi.org/10.3389/fpls.2014.00287

Lee SM, Koo YM (2001) Pilot-scale production of cellulase using T. reesei rut C-30 in fed-batch mode. Microb Biotechnol 11:229–233

Li-Jung Y, Hsin-Hung L, Zheng-Rong X (2010) Purification and characterization of a cellulose from bacillus subtilis YJ1. J Mar Sci Technol 18:466–471

Liu K, Newman M, McInroy JA, Hu CH, Kloepper JW (2017) Selection and assessment of plant growth promoting rhizobacteria (PGPR) for biological control of multiple plant diseases. Phytopathology 107:928–936

Lorito M, Harman GE, Hayes CK, Broadway RM, Woo SL, Di Piettro A (1993) Chitinolytic enzymes produced by Trichoderma harzianum 11 antifungal activity of purified endochitinase and chitobiosidase. Plytopathology 83:302–307

Lucas García J, Probanza A, Ramos B, Gutiérrez Manero FJ (2001) Genetic variability of rhizobacteria from wild populations of Lupinus species based on PCR-RAPDs. J Plant Nutr Soil Sci 164:1–7

Lucy ME, Reed E, Glick B (2004) Applications of free living plant growth-promoting rhizobacteria. Antonie Van Leeuwenhoek 86:1–25

Lugtenberg B, Kamilova F (2009) Plant-growth-promoting Rhizobacteria. Annu Rev Microbiol 63:541–556

Lynd LR, Weimer PJ, van Zyl WH, Pretorious IS (2002) Microbial cellulose utilization: fundamentals and biotechnology. Microbiol Mol Biol Rev 66:506–577

Mabood F, Zhou X, Smith DL (2014) Microbial signalling and plant growth promotion. Can J Plant Sci 94(6):1051–1063

Maksimov IV, Abizgildina RR, Pusenkova LI (2011) Plant growth promoting rhizobacteria as alternative to chemical crop protectors from pathogens (Review). Appl Biochem Microbiol 47:333–345

Manocha MS, Balasubramanian R (1994) Fungal chitinases: their properties and roles in morphogenesis, mycoparasitism and control of pathogenic fungi. In: Petrini O, Ouel-lette GB (eds) Host wall alterations by parasitic fungi. APS Press, St. Paul, pp 81–90

Manocha MS, Sahai AS (1993) Mechanisms of recognition in necrotrophic and biotrophic mycoparasites. Can J Microbiol 39:269–275

Mazzola M, Stahlman PW, Leach JE (1995) Application method affects the distribution and efficacy of rhizobacteria suppressive of downy brome (Bromus tectorum). Soil Biol Biochem 27:1271–2178

Mercado-Blanco J, Rodríguez-Jurado D, Hervás A, Jiménez-Díaz RM (2004) Suppression of Verticillium wilt in olive planting stocks by root-associated fluorescent Pseudomonas spp. Biol Control 30(2):474–486

Miyata M, Kanbe T, Tanaka K (1985) Morphological alterations of the fission yeast Schizosaccharomyces pombe in the presence of aculeacin A: spherical wall formation. J Gen Microbiol 131:611–621

Morimoto K, Karita S, Kimura T, Sakka K, Ohmiya K (1997) Cloning, sequencing and expression of the gene encoding Clostridium paraputrificum chitinase ChiB and analysis of the functions of novel cadherin-like domains and a chitin binding domain. J Bacteriol 179:7306–7314

Muhammad I, Safdar A, Syed Q, Nadeem M (2012) Isolation and screening of cellulolytic bacteria from soil and optimization of cellulase production and activity. Turk J Biochem 37:287–293

Nasab BF and Sayyed RZ (2019) Plant Growth Promoting Rhizobacteria and Salinity Stress : A Journey into the Soil in : Plant Growth Promoting Rhizobacteria for sustainable stress Management Vol 1 Abiotic Stress Management Sayyed, Arora & Reddy Eds, Springer, Singapore, pp. 21–34

Neeraja C, Anil K, Purushotham P, Suma K, Sarma P, Moerschbacher BM, Podile AR (2010) Biotechnological approaches to develop bacterial chitinases as a bioshield against fungal diseases of plants. Crit Rev Biotechnol 30:231–241

Nicot PC, Avril F, Duffaud M, Leyronas C, Troulet C, Villeneuve F, Bardin M (2016) Are there regional differences in the susceptibility of Sclerotinia sclerotiorum strains to Coniothyrium minitans? IOBC/WPRS Bull 117:83–87

Nithyapriya S, Lalitha S, Sayyed RZ, Reddy MS, Dailin DJ, Enshasy HE, Suriani NL, Herlambang S (2021) Production, purification, and characterization of bacillibactin siderophore of *Bacillus subtilis* and its application for improvement in plant growth and oil content in sesame. Sustainability 13:5394. https://doi.org/10.3390/su13105394

Noronha EF, Ulhoa CJ (1996) Purification and characterization of an endo-β-1,3-glucanase from Trichoderma harzianum. Can J Microbiol 42:1039–1044

Odeniyi OA, Onilude AA, Ayodele MA (2009) Production characteristics and properties of cellulase/polygalacturonase by a Bacillus coagulans strain from a fermenting palm fruit industrial residue. Afr J Microbiol Res 3:407–417

Oppenheim AB, Chet I (1992) Cloned chitinases in fungal plant pathogen control strategies. Trends Biotechnol 10:392–394

Pal KK, Gardener BM (2006) Biological control of plant pathogens. Plant Health Instr 25(2). https://doi.org/10.1094/PHI-A-2006-1117-02

Pang Z, Otaka K, Suzuki Y, Goto K, Ohnishi M (2004) Purification and characterization of an endo-1,3-β-glucanase from Arthrobacter spp. J Biol Macromol 4:57–66

Patel S, Sayyed R, Saraf M (2016) Bacterial determinants and plant defense induction : their role as bio-control agent in agriculture plant soil microbes, plant soil microbes. Springer, Cham, pp 187–204

Patten CL, Glick BR (2002) Role of Pseudomonas putida indoleacetic acid in development of the host plant root system. Appl Environ Microbiol 68:3795–3801

Pietrella D, Bistoni G, Corbucci C, Perito S, Vecchiarelli A (2006) Candida albicans mannoprotein influences the biological function of dendritic cells. Cell Microbiol 8:602–612

Pimentel JP, Olivares F, Pitard RM, Urquiaga S, Akiba F, Döbereiner J (1991) Dinitrogen fixation and infection of grass leaves by Pseudomonas rubrisubalbicans and Herbaspirillum seropedicae. Plant Soil 137(1):61–65

Pitarch A, Jiménez A, Nombela C, Gil C (2006) Decoding serological response to Candida cell wall immunome into novel diagnostic, prognostic, and therapeutic candidates for systemic candidiasis by proteomic and bioinformatic analyses. Mol Cell Proteomics 5:79–96

Pleban S, Chernin L, Chet I (1997) Chitinolytic activity of an endophytic strain of Bacillus cereus. Lett Appl Microbiol 25:284–288

Podile AR, Kishore GK (2006) Plant growth-promoting rhizobacteria. In: Gnanamanickam SS (ed) Plant-associated bacteria. Springer, Dordrecht, pp 195–230

Ran LX, Liu CI, Wu GJ, van Loon LC, Bakker PAHM (2005) Suppression of bacterial wilt in Eucalyptus urophylla by fluorescent Pseudomonas spp. in China. Biol Control 32:111–120

Reese ET, Mandels M (1959) β -D-1,3-glucanases in fungi. Can J Microbiol 5:173–185

Reshma P, Naik MK, Aiyaz M, Niranjana SR, Chennappa G, Shaikh SS, Sayyed RZ (2018) Induced systemic resistance by 2,4diacetylphloroglucinol positive fluorescent Pseudomonas strains against rice sheath blight, Indian. J Exp Biol 56(3):207–212

Ruel K, Joseleau JP (1991) Involvement of an extracellular glucan sheath during degradation of Populus wood by Phanerochaete chrysosporium. Appl Environ Microbiol 57(2):374–384

Ruiz-Herrera J (1991) Fungal Cell Wall: structure, synthesis, and Assembly. CRC Press

Ruiz-Herrera J (2012) Fungal Cell Wall: structure, synthesis, and assembly (Chapter 6), 2nd edn. CRC Press, pp 73–95

Sagar A, Sayyed RZ, Ramteke PW, Sharma S, Marraiki N, Elgorban AE, Syed A (2020) ACC deaminase and antioxidant enzymes producing halophilic *Enterobacter* sp. PR14 promotes the growth of rice and millets under salinity stress. Physiol Mol Biol Plants 26:1847–1854

Sagar A, Rai S, Ilyas N, Sayyed RZ, Al-Turki AI, Enshasy HA, Simarmata T (2021) Halotolerant rhizobacteria for salinity stress mitigation: diversity, mechanism and molecular approaches. Sustainability 14:490. https://doi.org/10.3390/su14010490

Sayyed RZ, Gangurde NS, Patel PR, Josh SA, Chincholkar SB (2010) Siderophore production by *Alcaligenes faecalis* and its application for plant growth promotion by Arachis hypogaea. Indian J Biotechnol 9:302–307

Schirmbiick M, Lorito M, Wang YL, Hayes CK, Arsian-Atac I, Scala F, Harman GE, Kubicek CP (1994) Parallel formation and synergism of hydrolytic enzymes and peptaibol antibiotics, molecular mechanisms involved in the antagonistic action of Trichoderma harzianum against phytopathogenic fungi. Appl Environ Microbiol 60:4364–4370

Schlumbaum A, Mauch F, Vogeli V, Boller T (1986) Plant chitinases are potent inhibitors of fungal growth. Nature London 324:365–367

Schroth MN, Hancock JG (1982) Disease-suppressive soil and root-colonizing bacteria. Science 216:1376–1381

Shahinian S, Bussey H (2000) β-1,6-glucan synthesis in Saccharomyces cerevisiae. Mol Microbiol 35(3):477–489

Shaikh SS, Sayyed RZ (2015) Role of plant growth–promoting Rhizobacteria and their formulation in biocontrol of plant diseases. In: Plant microbes symbiosis: applied facets. Springer, pp 337–351

Shaikh SS, Sayyed RZ, Reddy MS (2016) Plant growth promoting rhizobacteria: a sustainable approach to agro–ecosystem. In: Hakeem KR, Akhtar M, Abdullah S (eds) Plant, soil and microbes–interactions and implications in crop science. Springer, Cham, pp 181–201

Shaikh SS, Wani SJ, Sayyed RZ, Thakur R, Gulati A (2018) Production, purification and kinetics of chitinase of *Stenotrophomonas maltophilia* isolated from rhizospheric soil. Indian J of Exp Biol 56(4):274–278

Sharma A, Gupta A, Dalela M, Sharma S, Sayyed RZ, Enshasy HE, Elsayed AE (2020) Linking organic metabolites as produced by *Purpureocillium lilacinum* 6029 cultured on Karanja deoiled cake medium for the sustainable management of root-knot nematodes. Sustainability 12(9):8276. https://doi.org/10.3390/su12198276

Shibuya N, Minami E (2001) Oligosaccharide signalling for defence responses in plant. Physiol Mol Plant Pathol 59:223–233

Sietsma JH, Eveleigh DE, Haskins RH (1969) Cell wall composition and protoplast formation of some oomycete species Biochim. Biophys Acta (BBA) – General Subjects 184(2):306–317

Simmons CR (1994) The physiology and molecular biology of plant 1-3-β-D-glucanase and 1,3;1-4-β-D-glucanase. Crit Rev Plant Sci 13:325–387

Singh PP, Shin YC, Park CS, Chung YR (1999) Biological control of fusarium wilt of cucumber by chitinolytic bacteria. Phytopathology 89:92–99

Sitrit Y, Barak Z, Kapulnik Y, Oppenheim AB, Chet I (1993) Expression of a Serratia marcescens chitinase gene in rhizobium meliloti during symbiosis on alfalfa roots. Mol Plant-Microbe Interact 6:293–298

Sitrit Y, Vorgias CE, Chet I, Oppenheim AB (1995) Cloning and primary structure of the chiA gene from Aeromonas caviae. J Bacteriol 177:4187–4189

Sivan A, Chet I (1989) Degradation of fungal cell walls by lytic enzymes of Trichoderma Harzianum. Microbiology 135:675–682

Skujins JJ, Pogieter HJ, Alexander M (1965) Dissolution of fungal walls by a streptomycete-chitinase and glucanase. Arch Biochem Biophys 111:358–364

Smith DL, Gravel V, Yergeau E (2017) Editorial: signalling in the phytomicrobiome. Front Plant Sci 8:611

Soanes DM, Richards TA, Talbot NJ (2007) Insights from sequencing fungal and oomycete genomes: what can we learn about plant disease and the evolution of pathogenicity? Plant Cell 19:3318–3326

Someya N, Kataoka N, Komagata T, Hirayae K, Hibi T, Akutsu K (2000) Biological control of cyclamen soil borne diseases by Serratia marcescens strain B2. Plant Dis 84:334–340

Sookkheo B, Sinchaikul S, Phutrakul S, Chen ST (2000) Purification and characterization of the highly thermostable proteases from Bacillus stearothermophilus TLS33. Prot Exp Pur 20:142–151

Strange RN, Scott PR (2005) Plant disease: a threat to global food security. Annu Rev Phytopathol 43:1–660

Sukmawati D, Family N, Hidayat I, Sayyed RZ, Elsayed EA, Dailin DJ, Hanapi SZ, Wadaan MA, Enshasy HE (2021) Biocontrol activity of *Aureubasidium pullulans* and Candida orthopsilosis isolated from *Tectona grandis* L. Phylloplane against *Aspergillus* sp. In post-harvested citrus fruit. Sustainability 13:7479. https://doi.org/10.3390/su13137479

Sundheim L (1992) Effect of chitinase encoding genes in biocontrol pseudomonas spp. In: Tjamos EC, Papavizas GC, Cook RJ (eds) Biological control of plant diseases: progress and challenges for the future. Plenum, New York, pp 331–333

Svitil AL, Kirchman DL (1998) A chitin-binding domain in a marine bacterial chitinase and other microbial chitinases: implications for the ecology and evolution of 1,4-ß-glycanases. Microbiology 144:1299–1308

Tanaka T, Fujiwara S, Nishikori S, Fukui T, Takagi M, Imanaka T (1999) A unique chitinase with dual active sites and triple substrate binding sites from the hyperthermophilic archaeon Pyrococcus kodadaraensis KOD1. Appl Environ Microbiol 65:5338–5344

Valois D, Fayad K, Barasubiye T, Garon M, Dery C, Brzezinski R, Beaulieu C (1996) Glucanolytic Actinomycetes antagonistic to Phytophthora fragariae var. rubi, the causal agent of raspberry root rot. Appl Environ Microbiol 62(5):1630–1635

van Loon LC (2007) Plant responses to plant growth-promoting rhizobacteria. Eur J Plant Pathol 119:243–254

Vazquez-Garcidueiias S, Leal-Morales CA, Herrera-Estrella A (1998) Analysis of the ß-1,3-glucanolytic system of the biocontrol agent Trichoderma harzianum. Appl Environ Microbiol 64:1442–1446

Vicente MF, Basilio A, Cabello A, Pelaez F (2003) Microbial natural products as a source of anti-fungals. Clin Microbiol Infect 9:15–32

Vinay JU, Naik MK, Rangeshwaran R, Chennappa G, Shaikh SS, Sayyed RZ (2016) Detection of antimicrobial traits in fluorescent pseudomonads and molecular characterization of an antibiotic pyoluteorin, *3*. Biotech 6(227):1–11

Vocadlo DJ, Davies GJ (2008) Mechanistic insights into glycosidase chemistry. Curr Opin Chem Biol 12:539–555

Warren RAJ (1996) Microbial hydrolysis of polysaccharides. Annu Rev Microbiol 50:183–212

Watanabe T, Suzuki K, Oyanagi W, Ohnishi K, Tanaka H (1990) Gene cloning of chitinase A1 from Bacillus circulans WL-12 revealed its evolutionary relationship to Serratia chitinase and to the type III homology units of fibronectin. J Biol Chem 265:15659–15665

Whipps JM (2001) Microbial interactions and biocontrol in the rhizosphere. J Exp Bot 52:487–511

Yin QY, de Groot PWJ, de Koster CG, Klis FM (2007) Mass spectrometry-based proteomics of fungal wall glycoproteins. Trends Microbiol 16:20–26

Zhang R, Vivanco JM, Shen Q (2017) The unseen rhizosphere root soil-microbe interactions for crop production. Curr Opin Microbiol 37:8–14

Zhang Z, Yuen GY (2001) Chitinases from the plant disease biocontrol agent, Stenotrophomonas maltophilia C3. Phytopathology 91:204–211

Zope VP, Enshasy HE, Sayyed RZ (2019) Plant growth promoting rhizobacteria: an oOverview in agricultural perspectives. In: Plant growth promoting rhizobacteria for sustainable stress management Vol II rhizobacteria in biotic stress management. Sayyed RZ (Eds), Springer-Singapore, 2019, pp 345–362

# Chapter 11
# Induction of Hydrolytic Enzymes: A Criterion for Biological Control Candidates against Fungal Pathogen

**Nusaibah Syd Ali, Tuan Muhammad Syafiq, and Muniroh Md Saad**

**Contents**

11.1 Introduction ... 239
11.2 Potential of Bacteria as a Biological Control Agent (BCA) ... 240
11.3 Potential of *Trichoderma* Fungi as a Biological Control Agent (BCA) ... 242
11.4 Hydrolytic Cell Wall-Degrading Enzymes ... 243
11.5 Conclusion ... 246
References ... 247

## 11.1 Introduction

Biological control of plant diseases has demonstrated a sustainable eco-friendly atmosphere in the agricultural industry. This practice has been adapted in most developed countries such as the United States, United Kingdom, and Japan. The use of chemical control such as fungicides and pesticides has long contaminated the waterways, soil, and agricultural products and affected farmers' and consumers' health. In addition, biological control has been reported as an alternative strategy to control crop disease and reduce damage caused by plant pathogens (Wang et al.

---

N. S. Ali (✉) · T. M. Syafiq · M. Md Saad
Department of Plant Protection, Faculty of Agriculture, Universiti Putra Malaysia, Serdang, Selangor, Malaysia
e-mail: nusaibah@upm.edu.my

 
R. Z. Sayyed et al. (eds.), *Antifungal Metabolites of Rhizobacteria for Sustainable Agriculture*, Fungal Biology, https://doi.org/10.1007/978-3-031-04805-0_11

2000; Gerhardson 2002). Microbes such as bacteria, fungi, actinomycetes, and mycorrhiza have been reported as biological control agents of various diseases in specific or in general. These microbes inhibit or parasitize pathogens by excreting defense metabolites in combination with extracellular cell wall-degrading enzymes, competition for nutrients (i.e., iron, nitrogen, or carbon) in colonization sites, and stimulation of plant resistance mechanisms and development (Howell 2003). Thus, the current chapter presents the evaluation of potential *Trichoderma* spp., *Bacillus* sp., and *Pseudomonas* sp. as bio-inoculants in producing fungal cell wall-hydrolyzing enzymes. These enzymes may facilitate the inhibition growth of fungal pathogens.

## 11.2 Potential of Bacteria as a Biological Control Agent (BCA)

According to Droby et al. (2009), the use of natural antagonistic microorganisms to combat pests or suppress plant diseases is called biological control. In accordance with this, some spore-forming bacteria such as *Bacillus* and *Pseudomonas* spp. can be used as biological control agents against plant pathogens due to its beneficial properties (Pérez-García 2011).

*Bacillus cereus* is a Gram-positive, aerobic-to-facultative, spore-forming rod widely distributed environmentally and bearing close phenotypic and genetic (16S rRNA) relationships to several other *Bacillus* species, specifically *Bacillus anthracis* (Ash et al. 1991). *Bacillus cereus* has a saprophytic life cycle in which spores germinated in the soil, with the production of a vegetative cells that could sporulate, maintaining the life cycle (Arnesen et al. 2008). *Bacillus cereus* is the most common aerobic spore bearer in many types of soil and in sediments, dust, and plants (Ribeiro et al. 2010).

In contrary, *Pseudomonas* is a genus of Gram-negative, non-spore-forming, rod-shaped bacteria. *Pseudomonas* spp. are frequently found in the soil, water, decaying matter, and in some species that are plant and animal pathogens (Azadeh and Meon 2009). *Pseudomonas* species typically give a positive result to oxidase and catalase tests. The genus exhibits a great deal of metabolic diversity which enables it to colonize a wide range of habitats (Azadeh and Meon 2009).

Endophytic bacteria such as *Bacillus* spp. and *Pseudomonas* spp. have the ability to increase plant's immune system against pathogenic fungi attack (Alström 2001; Ramli et al. 2016). Some of the methods to strengthen this immune system are by promoting plant growth, solubilizing mineral nutrient, and inducing systemic resistance (Francis et al. 2010). Most plant growth-promoting (PGP) bacteria reside in the plant rhizosphere, and it is the best site to isolate beneficial endophytic bacteria (Kishore et al. 2005). A data obtained by Nusaibah et al. (2017) proved that *B. cereus* is an endophytic bacterium that could significantly improve vegetative growth parameters of oil palm seedlings such as plant height, number of fronds, stem diameter, bole diameter, and root and top weight. Those findings were in line with previous study by Amir et al. (2005) where *Bacillus* spp. demonstrated the ability to enhance development of roots and tops of the oil palm seedlings.

Some essential mineral nutrients such as nitrogen, phosphorus, and iron are fixed in insoluble forms; they are mostly not accessible in the soil to plants (Johri et al. 2003; Verma et al. 2010). However, microorganisms may increase the availability of essential compounds by secreting certain enzymes or chelators (Francis et al. 2010). For example, in the soil, organic phosphorus is stored mainly as insoluble phytate. Many rhizosphere bacteria can solubilize phosphorus from phytate by secreting active phytases (Francis et al. 2010). The most efficient Gram-positive strains that promote plant growth in this manner belong to the genus *Bacillus* (Jorquera et al. 2008). According to Zakry et al. (2012), *Bacillus* sp. could provide nitrogen and reduce nitrogen fertilizer dependence for young palms by encouraging biological nitrogen fixation. In addition, Amir et al. (2005) also found out that *Bacillus* spp. may help increase the availability of essential nutrients especially nitrogen, phosphorous, and potassium for oil palm seedlings.

In *Pseudomonas* species, *Pseudomonas aeruginosa* is well known in producing rhamnolipid biosurfactant with potential surface-active properties when grown on different carbon substrates. This particular bacterium produces two types of glycolipids both containing rhamnose as the carbohydrate moiety (Desai and Banat 1997). *Pseudomonas aeruginosa* is a root colonizer and IAA, siderophore, and antibiotic producer coupled with a broad spectrum of antimicrobial activity against agriculturally important phytopathogens (Hariprasad et al. 2014). Even though *P. aeruginosa* was reported as an opportunistic human pathogen (De Bentzmann and Plésiat 2011), its occurrence in rhizosphere of agriculture's important crops and their ability to promote plant growth and suppress plant diseases were frequently reported (Lim et al. 2019; Ramli et al. 2016; Bivi et al. 2010; Zaiton et al. 2008). Besides that, pseudomonades have been reported to produce a wide spectrum of compounds with antimicrobial activities against phytopathogenic fungi and bacteria (Cazorla et al. 2006). Strains of pseudomonades often produce more than one antimicrobial compound such as phenazines, phenazine-1-carboxylic acid and 2,4-diacetylphloroglucinol, pyoluteorin and pyrrolnitrin (Paul and Sarma 2006), monoacetylphloroglucinol, and 2,4-diacetylphloroglucinol (Guihen et al. 2004). Secondary metabolites of *P. aeruginosa*, extracted using ethyl acetate, demonstrated the highest inhibition in vitro toward *G. boninense* with 3-demethylubiquinone-9 as potential antimicrobial compounds detected in the crude extract which may have contributed to the inhibition of *G. boninense* growth (Lim et al. 2019).

## 11.3 Potential of *Trichoderma* Fungi as a Biological Control Agent (BCA)

Fungal species belonging to the genus *Trichoderma* are largely spread in the soil and rhizosphere. According to Harman et al. (2004), genus *Trichoderma* has been known as "an opportunistic avirulent plant symbionts" because they can benefit from and provide benefit in direct interactions with plants. Some of the benefits are improved nutrient uptake, increased growth and yields, increased percentage and rate of seed germination, and strengthened plant defenses against various diseases (Harman et al. 2004). Many studies have demonstrated *Trichoderma* spp. as effective biological control agents against various crop diseases (Harman et al. 2004; Kim and Knudsen 2013; Abo-Elyousr et al. 2014). *Trichoderma asperellum*, a less well-studied species, could potentially suppress several plant diseases including *Fusarium* wilt pathogen of tomato (Cotxarrera et al. 2002), fungal pathogens attack on cacao (Bailey et al. 2008), and BSR disease of oil palm (Musa et al. 2018).

The capability of *Trichoderma* spp. to fight plant diseases is mostly due to their direct antagonistic effects on the pathogenic fungi, specifically due to their ability to excrete lytic enzymes such as chitinases and β-1,3-glucanases (Viterbo et al. 2002; Benítez et al. 2004). These enzymes hydrolyze the pathogen's cell wall, thus controlling the growth of fungal pathogens. In addition, *Trichoderma* spp. are effective soil inhabitants and root colonizers especially if the soil is rich with inorganic soil substrates (Yang et al. 2011). *Trichoderma* isolates employ several mechanisms in providing biocontrol of disease, including antibiosis, mycoparasitism, and induced resistance (Gajera et al. 2012; Chaverri and Samuels 2013; Shoresh et al. 2010). *Trichoderma* spp. parasitic mechanism is clearly demonstrated by the direct penetration and parasitism of the pathogens (Harman et al. 2004).

Many studies have been conducted to prove the ability of *Trichoderma* spp. as a BCA for oil palm against BSR disease caused by *Ganoderma* spp. A research conducted by Alizadeh et al. (2011) demonstrated that induced plant defense response of oil palm seedlings by *Trichoderma* sp. contributed significantly toward the resistance of the seedlings against *G. boninense* infection. Besides that, Alizadeh et al. (2011) also found that *Trichoderma* sp. is a good plant growth promoter of oil palm seedlings. Another study suggested that the oil palm treated with *Trichoderma* sp. could produce high level of chitinase that may act as defense mechanism against *Ganoderma* sp. infection (Naher et al. 2012). Musa et al. (2018) showed that the production of hydrolytic enzymes by *T. asperellum*, *Trichoderma brevicompatum*, *Trichoderma harzianum*, and *Trichoderma virens* could minimize *G. boninense* spread. Results obtained by Musa et al. (2018) also suggested that *Trichoderma* spp. have potential as good BCAs, and these findings were supported by previous studies to control BSR disease on oil palm (Sariah et al. 2005).

Therefore, the use of BCA as an alternative way to combat fungal diseases is an ideal option, apart from chemical and cultural control methods.

## 11.4 Hydrolytic Cell Wall-Degrading Enzymes

Plants' defense against fungal infection or other biotic and abiotic stresses is through physically strengthening the cell walls through lignification and suberization and producing various pathogenesis-related (PR) proteins (Ebrahim et al. 2011). Defense enzymes such as peroxidase, polyphenol oxidase, phenylalanine ammonia lyase, chitinase, and β-1, 3-glucanase are associated with resistance stimulus in plants (Prasannath et al. 2014; Gajanayaka et al. 2014; Seneviratne et al. 2014). Pathogenesis-related proteins, namely, chitinase and β-1, 3-glucanase, are two important hydrolytic enzymes found in abundance in many plant species after infection by fungal pathogens. These enzymes significantly increase and play a key role in defense reactions against fungal pathogens by degrading fungal cell walls, where chitin and β-1,3-glucan are the major structural components of the cell walls in many pathogenic fungi. However, when the plants fail to produce sufficient volume of these hydrolyzing enzymes in encountering the pathogenic fungi invasion, the disease may spread throughout the plant effortlessly. This is where the planters require the aid of biological control agents (BCAs) with the ability to produce these hydrolytic enzymes. Application of potential BCAs could help to inhibit the pathogenic fungal pathogens via degradation of its cell wall content which is mainly chitin and glucan.

*Trichoderma* spp. were reported to produce numerous secondary metabolites such as non-ribosomal peptides, terpenoids, pyrones, and indolic-derived compounds (Contreras-Cornejo and Macías-Rodríguez 2016). Some strains of *Trichoderma* are antagonistic to other fungi, thus displaying promising potentials as BCAs of soil-borne diseases (Kacprzak et al. 2014). The biological control potential of *Trichoderma* spp. has been demonstrated on a wide spectrum of plant diseases both in vitro and in vivo (Woo et al. 2014; Kotze et al. 2011). The production of extracellular cell wall-degrading enzymes such as β-1,3-glucanase, cellulase, chitinase, and protease marks an important role of *Trichoderma* spp. in the inhibition of the fungal pathogens (Gajera et al. 2012). The cell wall-degrading enzymes could hydrolyze or degrade the cell wall of fungal pathogens and inhibit their growth (Woo et al. 2014). Musa et al. (2017) demonstrated that *Trichoderma asperellum*, *Trichoderma harzianum*, and *Trichoderma virens* have the ability to produce hydrolytic enzymes such as chitinase, cellulose, β-1,3-glucanase, and protease. Zaiton et al. (2008) and Idris et al. (2010) also reported that endophytic microbes, namely, *T. asperellum* and *Pseudomonas aeruginosa*, are known to induce resistance in oil palm by secreting hydrolytic enzymes such as chitinases and glucanases. These characteristics have been reported to be the major key factors in the degradation of fungal cell walls (McNeil et al. 1984). The production of these cell wall-degrading enzymes could be responsible for the inhibition of *Ganoderma boninense* growth. Similar findings were also reported by Naher et al. (2014) and Yeoh et al. (2012).

### *11.4.1 Hydrolytic Enzymes: The Key Characteristic in Biocontrol Selection*

The ability to secrete hydrolytic enzymes such as chitinase, glucanase, cellulase, and protease would be an important criterion of BCAs utilized to inhibit pathogenic fungal diseases (Ajit et al. 2006; Hoster et al. 2005). In the case studies reported above, all the BCAs studied were able to produce chitinase and glucanase. These cell wall-degrading enzymes could be responsible for the inhibition of pathogenic *G. boninense* growth in vitro (Musa et al. 2017; Naher et al. 2014; Yeoh et al. 2012). Glucanase and chitinase are classified as pathogenesis-related proteins (Sela-Buurlage et al. 1993; Stintzi et al. 1993) and also known as endotype enzymes (Sikorski et al. 2006; Watanabe et al. 1992) that could degrade the structural component of the fungi cell wall directly. 1,3-β-Glucans are polysaccharides, present as a cell wall component of plant pathogenic fungi. Besides that, β-1,3-glucanase produced by bacteria was able to inhibit fungal growth (Leelasuphakul et al. 2006; Manjula and Podile 2005; Aono et al. 1995). Aside from bacteria, the most popular biological control fungi, namely, *Trichoderma* spp., were reported to be able to produce β-1,3-glucanases, chitinase, cellulose, proteases, and volatile and non-volatile antibiotics (Gajera et al. 2012; Elad et al. 1982). In addition, a study by Naher et al. (2012) also demonstrated that *T. harzianum* has the ability to induce higher enzyme activity in oil palm roots and leaves in the plant defense response against basal stem rot disease caused by *Ganoderma* spp. The mechanism of fungal growth inhibition by *Trichoderma* is not through hyphal penetration; thus, it is possible that extracellular metabolites are responsible for fungal degradation (Chérif and Benhamou 1990).

Microbial chitinase has been reported to deteriorate and destroy the cell walls of many pests and pathogens, thus exhibiting antibacterial, antifungal, insecticidal, or nematicidal activity (Edreva 2005). Nawani et al. (2002) documented that bacterial biocontrol agents such as *B. cereus* and *Bacillus alvei* produced the highest chitinase effect after 9 days of incubation period. Therefore, to select a good BCA of plant pathogenic fungi, traits in ability to produce β-glucanase enzymes are crucial. In the case study discussed above, barley was used as the glucan substrate to assess β-glucanase enzyme assay of the BCAs. According to a study by Celestino et al. (2006), glucan from barley was best degraded by endophytic microbes compared to other glucan substrates such as laminarin, xylan, and mannan.

#### 11.4.1.1 Chitinase

Chitinase enzyme has received increased attention due to its wide range of biotechnological applications, specifically as a major characteristic in selecting BCAs of phytopathogenic fungi and harmful insects. This is because chitin is the main component of fungal cell wall and cuticle of insect-pests (Saima et al. 2013). Elicitation of chitinase enzyme by BCAs may degrade the chitin walls of the fungal pathogens

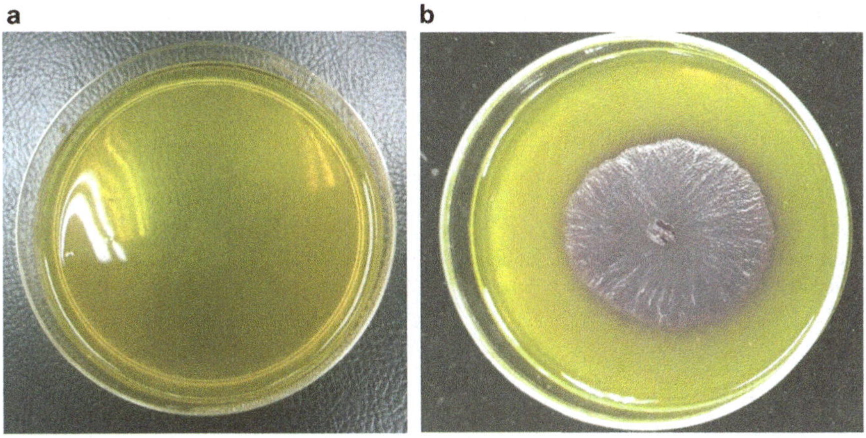


**Fig. 11.1** Negative chitinase activity in a control plate (**a**), positive chitinase activity demonstrating purple zone (**b**)

**Table 11.1** Chitinase production assay

| Microbes | Chitinase production (cm) |
|---|---|
| *Trichoderma asperellum* (UPM16) | 5.35[a] |
| *Trichoderma harzianum* (UPM29) | 5.21[a] |
| *Pseudomonas aeruginosa* (UPMP3) | 3.47[b] |
| *Bacillus cereus* (UPM15) | 2.26[c] |

Means within columns with similar letters are not significantly different by Tukey's Studentized Range (HSD) Test at $P \leq 0.05$ $n = 6$

and simultaneously slows down the vigorousness of the infection. This feature will aid in defense mechanisms against fungal pathogens. Besides that, chitinase was also found to be involved in the development and growth mechanisms of trees (Santos et al. 2008). Chitinase has an acidic isoelectric point and may accumulate in intercellular spaces of a plant (Silipo et al. 2010). Hence, the total chitinase activity not only related to the host-pathogen interaction but also demonstrates potential as a detection tool of pathogenic fungi invasion or as a biomarker for early disease detection (Naher et al. 2011; Yeoh et al. 2012).

Chitinase activity was determined using the growing medium and a method developed by Agrawal and Kotasthane (2012). Observations were recorded based on the diameter of the purple-colored zone (Fig. 11.1) on the tenth day after incubation. In the case study presented in Table 11.1, *T. asperellum* exhibited highest chitinase activity (5.35 cm) followed by *T. harzianum* (5.21 cm), *Pseudomonas aeruginosa* (3.47 cm), and *Bacillus cereus* (2.26 cm). Overall, all the studied endophytic microbes were able to excrete this important enzyme required in the degradation of fungal cell wall. The ability to secrete chitinase enzyme is one of the most important criteria to inhibit fungal pathogens, thus to be selected as a potent BCA.

**Table 11.2** Hydrolysis of β-glucan substrate by candidate biocontrol microbes. Δ Abs 550 nm represents the net absorbance of the reaction mixture after incubation for 0.5 h with the enzyme at 50 °C

| Microbes | Δ $Abs_{550nm}$ |
|---|---|
| *Trichoderma asperellum* (UPM16) | $1.43^a$ |
| *Trichoderma harzianum* (UPM29) | $0.60^b$ |
| *Bacillus cereus* (UPM15) | $0.57^b$ |
| *Pseudomonas aeruginosa* (UPMP3) | $0.37^c$ |

Means within columns with similar letters are not significantly different by Tukey's Studentized Range (HSD) Test at $P \leq 0.05$ $n = 6$

### 11.4.1.2 β-1,3-Glucanase

β-1,3-glucan is a structural component in the cell wall of many yeasts, bacteria, and filamentous fungi (Carsolio et al. 1994). In order to counter most pathogenic fungal infections, plants induce β-1,3-glucanases which are one of the pathogenesis-related (PR) proteins and belong to the PR-2 family and are believed to play an important role in plant defense responses to pathogen infection (Ebrahim et al. 2011). Apart from that, β-1, 3-glucanases were expressed at a low concentration (basal level) in plants; however, when plants are infected by fungal, bacterial, or viral pathogens, β-1,3-glucanase concentration increases drastically. In relation to that, induction of β-1, 3-glucanases and other PR proteins in the plant may also be triggered by some components or degraded components of pathogens. Nevertheless, plant β-1, 3-glucanases are not only induced due to pathogen infection but also by other stress factors such as wounding, drought, exposure to heavy metals, air pollutant ozone, and ultraviolet radiation (Akiyama et al. 2004).

Based on the results of a case study tabulated in Table 11.2, BCA candidates, namely, *T. asperellum T. harzianum P. aeruginosa* and *B. cereus*, were able to hydrolyze barley β-glucan substrate by β-1,3-glucanase activity which was assayed by the reducing sugar method (Miller et al. 1999). Table 11.2 demonstrates the absorbance reading of each treatment mixture whereby the higher the absorbance reading, the more glucan has been hydrolyzed by β-glucanase enzyme produced by the microbes. *T. asperellum* was significantly more efficient compared to *T. harzianum, P. aeruginosa* and *B. cereus* in hydrolyzing β-glucan substrate.

## 11.5 Conclusion

Consequently, it may be concluded here that biocontrol microbes that have the ability to secrete chitinase and β-1,3-glucanase enzymes could efficiently degrade the type of chitin and glucan substrate present in the pathogenic fungal cell walls.

**Acknowledgments** This project was supported financially by the Fundamental Research Grant Scheme (FRGS), administered through the Ministry of Higher Education, Malaysia (Grant No: FRGS/1/2018/5540093).

## References

Abo-Elyousr KA, Abdel-Hafez SI, Abdel-Rahim IR (2014) Isolation of *Trichoderma* and evaluation of their antagonistic potential against *Alternaria porri*. J Phytopathol 162(9):567–574

Agrawal T, Kotasthane AS (2012) Chitinolytic assay of indigenous *Trichoderma* isolates collected from different geographical locations of Chhattisgarh in Central India. Springerplus 1:73

Ajit NS, Verma R, Shanmugam V (2006) Extracellular chitinases of fluorescent pseudomonads antifungal to *fusarium oxysporum* f. spp. *dianthi* causing carnation wilt. Current Microbiol 52:310–316

Akiyama T, Pillai MA, Sentoku N (2004) Cloning, characterization and expression of OsGLN2, a rice endo-1, 3-β-glucanase gene regulated developmentally in flowers and hormonally in germinating seeds. Planta 220:129–139

Alizadeh F, Abdullah SNA, Khodavandi A, Abdullah F, Yusuf UK, Chong PP (2011) Differential expression of oil palm pathology genes during interactions with *Ganoderma boninense* and *Trichoderma harzianum*. J Plant Physiol 168(10):1106–1113

Alström S (2001) Characteristics of bacteria from oilseed rape in relation to their biocontrol activity against *Verticillium dahliae*. J Phytopathol 149(2):57–64

Amir HG, Shamsuddin ZH, Halimi MS, Marziah M, Ramlan MF (2005) Enhancement in nutrient accumulation and growth of oil palm seedlings caused by PGPR under field nursery conditions. Commun Soil Sci Plant Anal 36(15–16):2059–2066

Aono R, Hammura M, Yamamoto M, Asano T (1995) Isolation of extracellular 28-and 42-kilodalton beta-1, 3-glucanases and comparison of three beta-1, 3-glucanases produced by *Bacillus circulans* IAM1165. Appl Environ Microbiol 61:122–129

Arnesen LP, Fagerlund A, Granum PE (2008) From soil to gut: Bacillus cereus and its food poisoning toxins. FEMS Microbiol Rev 32(4):579–606

Ash C, Farrow JAE, Wallbanks S, Collins MD (1991) Phylogenetic heterogeneity of the genus *bacillus* revealed by comparative analysis of small-subunit-ribosomal RNA sequences. Lett Appl Microbiol 13(4):202–206

Azadeh BF, Meon S (2009) Molecular characterization of Pseudomonas aeruginosa UPM P3 from oil palm rhizosphere. Am J Appl Sci 6(11):1915

Bailey BA, Bae H, Strem MD, Crozier J, Thomas SE, Samuels GJ, Holmes KA (2008) Antibiosis, mycoparasitism, and colonization success for endophytic *Trichoderma* isolates with biological control potential in *Theobroma cacao*. Biol Control 46(1):24–35

Benítez T, Rincón AM, Limón MC, Codon AC (2004) Biocontrol mechanisms of *Trichoderma* strains. Int Microbiol 7(4):249–260

Bivi MR, Farhana MSN, Khairulmazmi A, Idris A (2010) Control of Ganoderma boninense: a causal agent of basal stem rot disease in oil palm with endophyte bacteria in vitro. Int J Agric Biol 12(6):833–839

Carsolio C, Gutiérrez A, Jiménez B, Van Montagu M, Herrera-Estrella A (1994) Characterization of ech-42, a *Trichoderma harzianum* endochitinase gene expressed during mycoparasitism. InProc Nat Acad Sc 91:10903–10907

Cazorla FM, Duckett SB, Bergstroem ET, Noreen S, Odijk R, Lugtenberg BJ, Thomas-Oates JE, Bloemberg GV (2006) Biocontrol of avocado dematophora root rot by antagonistic Pseudomonas fluorescens PCL1606 correlates with the production of 2-hexyl 5-propyl resorcinol. Mol Plant-Microbe Interact 19:418–428

Celestino SM, De Freitas SM, Medrano FJ, De Sousa MV, Ferreira Filho EX (2006) Purification and characterization of a novel pectinase from *Acrophialophora nainiana* with emphasis on its physicochemical properties. J Biotech 123:33–42

Chaverri P, Samuels GJ (2013) Evolution of host affiliation and substrate preference in the cosmopolitan fungal genus *Trichoderma* with evidence of interkingdom host jumps. Evolution 67:2823–2837

Chérif M, Benhamou N (1990) Cytochemical aspects of chitin breakdown during the parasitic action of a *Trichoderma* sp. on *fusarium oxysporum* f. sp. *radicis-lycopersici*. Phytopathology 80:1406–1414

Contreras-Cornejo HA, Macías-Rodríguez L, del Val E, Larsen J (2016) Ecological functions of *Trichoderma* spp. and their secondary metabolites in the rhizosphere: interactions with plants. FEMS Microbiol Eco 92:16–21

Cotxarrera L, Trillas-Gay MI, Steinberg C, Alabouvette C (2002) Use of sewage sludge compost and *Trichoderma asperellum* isolates to suppress *fusarium* wilt of tomato. Soil Biol Biochem 34(4):467–476

De Bentzmann S, Plésiat P (2011) The Pseudomonas aeruginosa opportunistic pathogen and human infections. Environ Microbiol 13(7):1655–1665

Desai JD, Banat IM (1997) Microbial production of surfactants and their commercial potential. Microbiol Mol Biol Rev 61:414764

Droby S, Wisniewski M, Macarisin D, Wilson C (2009) Twenty years of postharvest biocontrol research: is it time for a new paradigm? Postharvest Biol Technol 52(2):137–145

Ebrahim S, Usha K, Singh B (2011) Pathogenesis related (PR) proteins in plant defense mechanism. Sci Against Microb Pathog 2:1043–1054

Edreva A (2005) Pathogenesis-related proteins: research progress in the last 15 years. Gen Appl Plant Physiol 31:105–124

Elad Y, Chet I, Henis Y (1982) Degradation of plant pathogenic fungi by *T. harzianum*. Can J Microbiol 28:719–725

Francis I, Holsters M, Vereecke D (2010) The gram-positive side of plant–microbe interactions. Environ Microbiol 12(1):1–12

Gajanayaka GM, Prasannath K, De Costa DM (2014) Variation of chitinase and β-1, 3-glucanase activities in tomato and chilli tissues grown under different crop management practices and agroecological regions. Proceed Peradeniya Univer Inter Res Sess 18:519

Gajera HP, Bamharolia RP, Patel SV, Khatrani TJ, Goalkiya BA (2012) Antagonism of *Trichoderma* spp. against *Macrophomina phaseolina*: evaluation of coiling and cell wall degrading enzymatic activities. J Plant Pathol Microbiol 3:149

Gerhardson B (2002) Biological substitutes for pesticides. Trends Biotech 20:338–343

Guihen E, Glennon JD, Cullinane M, O'Gara F (2004) Rapid analysis of antimicrobial metabolites monoacetylphloroglucinol and 2,4-diacetylphloroglucinol using capillary zone electrophoresis. Electrophoresis 25:1536–1542

Hariprasad P, Chandrashekar S, Singh SB, Niranjana SR (2014) Mechanisms of plant growth promotion and disease suppression by Pseudomonas aeruginosa strain 2apa. J Basic Microbiol 54(8):792–801

Harman GE, Howell CR, Viterbo A, Chet I, Lorito M (2004) *Trichoderma* species—opportunistic, avirulent plant symbionts. Nat Rev Microbiol 2(1):43

Hoster F, Schmitz JE, Daniel R (2005) Enrichment of chitinolytic microorganisms: isolation and characterization of a chitinase exhibiting antifungal activity against phytopathogenic fungi from a novel Streptomyces strain. Appl Microbiol Biotech 66:434–442

Howell CR (2003) Mechanisms employed by *Trichoderma* species in the biological control of plant diseases: the history and evolution of current concepts. Plant Dis 87:4–10

Idris AS, Nasyarudin MN, Maizatul SM, Zaiton S (2010) Gano EB1-A fungal biocontrol agent for Ganoderma in oil palm. Bangi, MPOB Infor Ser, p 443

Johri BN, Sharma A, Virdi JS (2003) Rhizobacterial diversity in India and its influence on soil and plant health. Adv Biochem Eng Biotechnol 84:49–89

Jorquera M, Martínez O, Maruyama F, Marschner P, de la Luz Mora M (2008) Current and future biotechnological applications of bacterial phytases and phytase producing bacteria. Microbes Environ 23:182–191

Kacprzak MJ, Rosikon K, Fijalkowski K, Grobelak A (2014) The effect of *Trichoderma* on heavy metal mobility and uptake by *Miscanthus giganteus*, *Salix* sp., *Phalaris arundinacea*, and *Panicum virgatum*. Appl Env Soil Sci:74–83

Kim TG, Knudsen GR (2013) Relationship between the biocontrol fungus *Trichoderma harzianum* and the phytopathogenic fungus *fusarium solani* f. sp. *pisi*. Appl Soil Ecol 68:57–60

Kishore GK, Pande S, Podile AR (2005) Phylloplane bacteria increase seedling emergence, growth and yield of field-grown groundnut (*Arachis hypogaea L.*). Lett Appl Microbiol 40(4):260–268

Kotze C, Van Niekerk J, Mostert L, Halleen F, Fourie P (2011) Evaluation of biocontrol agents for grapevine pruning wound protection against trunk pathogen infection. Phytopathol Mediterr 50:S247–S263

Leelasuphakul W, Sivanunsakul P, Phongpaichit S (2006) Purification, characterization and synergistic activity of β-1, 3-glucanase and antibiotic extract from an antagonistic Bacillus subtilis NSRS 89-24 against rice blast and sheath blight. Enzyme Microb Technol 38:990–997

Lim P, Gansau JA, Chong KP (2019) Biocontrol of basal stem rot pathogen, *Ganoderma boninense* by *Pseudomonas aeruginosa*. Bangladesh J Botany 48(2):209–215

Manjula K, Podile AR (2005) Production of fungal cell wall degrading enzymes by a biocontrol strain of *Bacillus subtilis* AF 1. Ind J Exp Biol 43:892–896

McNeil M, Darvill AG, Fry SC, Albersheim P (1984) Structure and function of the primary cell walls of plants. Annu Rev Biochem 53:625–663

Miller RN, Holderness M, Bridge PD, Chung GF, Zakaria MH (1999) Genetic diversity of *Ganoderma* in oil palm plantings. Plant Pathol 48:595–603

Musa H, Hassan MA, Isyaku MS, Halidu J, Suleiman AS (2017) Antagonistic potential of *Trichoderma* species against *Ganoderma* disease of oil palm. Nig J Agr Food Env 13:60–66

Musa H, Nusaibah SA, Khairulmazmi A (2018) Assessment on *Trichoderma spp.* mixture as a potential biocontrol agent of *Ganoderma boninense* infected oil palm seedlings. J Oil Palm Res 30(3):403–415

Naher L, Ho CL, Tan SG, Yusuf UK, Abdullah F (2011) Cloning of transcripts encoding chitinases from *Elaeis guineensis* Jacq. And their expression profiles in response to fungal infections. Physiol Mol Plant Pathol 76:96–103

Naher L, Yusuf UK, Siddiquee S, Ferdous J, Rahman MA (2012) Effect of media on growth and antagonistic activity of selected *Trichoderma* strains against *Ganoderma*. Afric J Microbiol Res 6:7449–7453

Naher L, Yusuf UK, Ismail A, Hossain K (2014) *Trichoderma* spp.: a biocontrol agent for sustainable management of plant diseases. Pak J Bot 46:1489–1493

Nawani NN, Kapadnis BP, Das AD, Rao AS, Mahajan SK (2002) Purification and characterization of a thermophilic and acidophilic chitinase from *Microbispora* sp. V2. J Appl Microbiol 93:965–975

Nusaibah SA, Saad G, Hun TG (2017) Antagonistic efficacy of *Trichoderma harzianum* and *Bacillus cereus* against *Ganoderma* disease of oil palm via dip, place and drench (DPD) artificial inoculation technique. Inter J Agricul Biol 19(2)

Paul D, Sarma YR (2006) Plant growth promoting rhizhobacteria (PGPR) mediated root proliferation in black pepper (Piper nigrum L.) as evidenced through GS root software. Arch Phyto Plant Protect 39:311–314

Pérez-García A, Romero D, de Vicente A (2011) Plant protection and growth stimulation by microorganisms: biotechnological applications of *bacilli* in agriculture. Curr Opin Biotechnol 22:187–193

Prasannath K, De Costa DM, Hemachandra KS (2014) Quantification of peroxidase activity in chili tissues grown under two crop protection systems across a temperature gradient. InProc HETC Symp, p 102

Ramli NR, Mohamed MS, Seman IA, Zairun MA, Mohamad M (2016) The potential of endophytic bacteria as a biological control agent for *Ganoderma* disease in oil palm. Sains Malaysiana 45(3):401–409

Ribeiro NF, Heath CH, Kierath J, Rea S, Duncan-Smith M, Wood FM (2010) Burn wounds infected by contaminated water: case reports, review of the literature and recommendations for treatment. Burns 36(1):9–22

Saima M, Kuddus M, Roohi I, Ahmad IZ (2013) Isolation of novel chitinolytic bacteria and production optimization of extracellular chitinase. J Gen Eng Biotech 11:39–46

Santos P, Fortunato A, Ribeiro A, Pawlowski K (2008) Chitinases in root nodules. Plant. Biotech 25:299–307

Sariah M, Cjoo CW, Zakaria H, Norihan MS (2005) Quantification and characterization of *Trichoderma spp.* from different ecosystems. Mycopathologia 159:113–117

Sela-Buurlage MB, Ponstein AS, Bres-Vloemans SA, Melchers LS, van den Elzen PJ, Cornelissen BJ (1993) Only specific tobacco (Nicotiana tabacum) chitinases and [beta]-1, 3-glucanases exhibit antifungal activity. Plant Physiol 101:857–863

Seneviratne DM, Prasannath K, De Costa DM (2014) Quantification of phenylalanine ammonia lyase activity in tomato and chili tissues grown under different crop management practices and agro-ecological regions in Sri Lanka. InProc 1st Fac of Agric Undergra res Symp, pp. 3

Shoresh M, Harman GE, Mastouri F (2010) Induced systemic resistance and plant responses to fungal biocontrol agents. Annu Rev Phytopathol 48(1):21–43

Sikorski P, Sørbotten A, Horn SJ, Eijsink VG, Vårum KM (2006) Serratia marcescens chitinases with tunnel-shaped substrate-binding grooves show endo activity and different degrees of processivity during enzymatic hydrolysis of chitosan. Biochemist 45:9566–9574

Silipo A, Erbs G, Shinya T, Dow JM, Parrilli M, Lanzetta R, Shibuya N, Newman MA, Molinaro A (2010) Glyco-conjugates as elicitors or suppressors of plant innate immunity. Glycobiol 20:406–419

Stintzi A, Heitz T, Prasad V, Wiedemann-Merdinoglu S, Kauffmann S, Geoffroy P, Legrand M, Fritig B (1993) Plant 'pathogenesis-related' proteins and their role in defense against pathogens. Biochimie 75:687–706

Verma JP, Yadav J, Tiwari KN, Lavakush SV (2010) Impact of plant growth promoting rhizobacteria on crop production. Inter J Agricul Res 5(11):954–983

Viterbo A, Montero M, Ramot O, Friesem D, Monte E, Llobell A, Chet I (2002) Expression regulation of the endochitinase chit36 from *Trichoderma asperellum* (*T. harzianum* T-203). Curr Genet 42(2):114–122

Wang C, Knill E, Glick BR, Défago G (2000) Effect of transferring 1-aminocyclopropane-1-carboxylic acid (ACC) deaminase genes into Pseudomonas fluorescens strain CHA0 and its gac a derivative CHA96 on their growth-promoting and disease-suppressive capacities. Can J Microbiol 46:898–907

Watanabe T, Kasahara NO, Aida KA, Tanaka HI (1992) Three N-terminal domains of beta-1, 3-glucanase A1 are involved in binding to insoluble beta-1, 3-glucan. J Bacteriol 174:186–190

Woo SL, Ruocco M, Vinale F, Nigro M, Marra R, Lombardi N, Pascale A, Lanzuise S, Manganiello G, Lorito MV (2014) *Trichoderma*-based products and their widespread use in agriculture. Open Mycol J 8:71–126

Yang CA, Cheng CH, Lo CT, Liu SY, Lee JW, Peng KC (2011) A novel L-amino acid oxidase from *Trichoderma harzianum* ETS 323 associated with antagonism of *Rhizoctonia solani*. J Agric Food Chem 59(9):4519–4526

Yeoh KA, Othman A, Meon S, Abdullah F, Ho CL (2012) Sequence analysis and gene expression of putative exo-and endo-glucanases from oil palm (*Elaeis guineensis*) during fungal infection. J Plant Physiol 169:1565–1570

Zaiton S, Sariah M, Zainal Abidin MA (2008) Effect of endophytic bacteria on growth and suppression of *Ganoderma* infection in oil palm. Inter J Agric Biol 10:127–132

Zakry FAA, Shamsuddin ZH, Rahim KA, Zakaria ZZ, Rahim AA (2012) Inoculation of *Bacillus sphaericus* UPMB-10 to young oil palm and measurement of its uptake of fixed nitrogen using the 15N isotope dilution technique. Microbes Environ 27(3):257–262

# Part III
# Plant-Microbe Interactions and Agriculture Sustenance

# Chapter 12
# *Pseudomonas* as Biocontrol Agent for Fungal Disease Management in Rice Crop

**Kausik Basu, Amit C. Kharkwal, and Ajit Varma**

## Contents

12.1 Introduction ........ 253
12.2 Rice Crop Diseases, Economic Losses, and Its Management ........ 254
12.3 *Pseudomonas*. as a Biocontrol Agent for Rice Disease Management with Effective Secondary Metabolite Production ........ 260
12.4 Industrial Perspective ........ 262
12.5 Research and Manufacturing Insights ........ 262
12.6 Future Prospect ........ 264
References ........ 265

## 12.1 Introduction

With increasing global population, demand for food crops is also equally in need at current scenario. Most of the underdeveloped and developing countries depend largely on food grains like rice, wheat, millets, and pulses for survival. Rice is the most staple food crop in a country like India (FAO 2015). More than 65% of the population depend on rice. India is the second largest producer of rice in the world after China. Rice is cultivated in more than 100 countries worldwide (Peng et al. 2009). Cultivating and growing rice are also a challenge considering several

K. Basu (✉)
Biofertilizer and Biologicals, Former Manager R&D, IPL Biologicals, Dehradun, Uttarakhand, India

A. C. Kharkwal · A. Varma
Amity Institute of Microbial Technology (AIMT), Amity University Uttar Pradesh (AUUP), Noida, Uttar Pradesh, India

 
R. Z. Sayyed et al. (eds.), *Antifungal Metabolites of Rhizobacteria for Sustainable Agriculture*, Fungal Biology, https://doi.org/10.1007/978-3-031-04805-0_12

limiting factors like unfavorable climate, soil health, and chemistry. Disease and pest management is another area of concern with increasing resistance of synthetic chemical fertilizer. Chemicals for years have been used in rice cultivation as fertilizer and for disease control. But uncontrolled use of synthetic chemicals has given rise to a serious concern of health hazards to all life forms including animals and human (Nicolopoulou-Stamati et al. 2016). With increased awareness and concern for safe food, there is a drift for using ecofriendly biological inputs that can replace the chemicals. Biological inputs for rice cultivation have attracted the attention of the growers due to its effectiveness in controlling diseases and pests and soil health management. Bacteria like *Pseudomonas fluorescens* are effective in controlling different soil-borne disease in rice crops. *P. fluorescens* is commercially used as biocontrol formulation due to its natural ability to survive in soil and root ecosystem. It produces a wide range of secondary metabolites which control different soil-borne diseases in rice. *P. fluorescens* control 70–85% of diseases like bakane and sheath blight in rice and help in increasing the yield and grain quality. *P. fluorescens* produces array of antimicrobial secondary metabolites like pyoluteorin, pyrrolnitrin, phenazine 1-carboxylic acid, and 2,4-diacetylphloroglucinol, cyclic lipopeptides (CLPs), etc. Disease-causing agents are suppressed by these metabolites and also induce systemic resistance to rice crops (Yasmin et al. 2017). In this chapter, we will provide an insight of biocontrol activities, mode of action, secreted antifungal metabolites, and commercial formulation of *Pseudomonas* in fungal disease management in rice.

## 12.2 Rice Crop Diseases, Economic Losses, and Its Management

Being the world's most staple food and highest consumption in India, rice crop is of utmost importance. There is serious need for increase in production of rice crop for feeding the ever-growing population of the country. An estimated 140 million tons of rice will be required for feeding the population of 1.6 billion by 2050 in India alone (FAO 2015). The severity of diseases caused by pathogens has evidential proof from the nineteenth century. Huge loss of potato crops in Europe by a fungal pathogen has affected the entire civilization and made us to understand the nature (Zadoks 2008).

### *12.2.1 Economic Losses Annually*

In 2019, India produced an estimated 2.6 thousand kg of rice per hectare (Statista 2020). Almost 44 million hectares of land in India is under cultivation of rice. Basmati rice with an estimated cost of 4.3 billion USD was exported in FY 2020 from India, but there is a gradual fall in the export quantity compared to the last 2 years (Keelery 2020).

### 12.2.2 Major Limiting Factors in Yield Loss for Rice

A major loss in rice yield is due to pests and diseases. Adequate production of rice is a challenge considering different factors. There is a huge competition of land for agriculture versus urbanization. With increase in industrialization, agricultural land is under threat of loss. Water scarcity and continuously decreasing ground water level are major concerns for shifting rice-growing farmers to other agricultural crops which require less irrigation (Pradhan and Parija 2020). There is also a diverse area of other biotic stresses responsible for this shift like climate changes, less rainfall, high temperature, soil fertility, etc. The years of heavy usage of chemical fertilizers for several decades have left many hectares of land into nutrient-deficient uncultivable land. With extensive usage of chemical pesticides, pests severely resistant to these chemicals have emerged. Cultivation of high-yielding varieties has created an uneven distribution of type of rice grown in country. This has resulted in disturbing the natural balance and also distributing different rice diseases to all parts. Few diseases which were earlier restricted to certain regions of the country are now present in different regions. A report from IRRI suggested that there is almost a 37% yield loss for rice due to diseases and pests (Rice Knowledge Bank, IRRI).

### 12.2.3 Different Rice Crop Diseases and Its Current Management

Diseases caused by pathogenic plant microorganism are almost 16% of which 70–80% is caused by fungi (Mooreet al. 2002). Other pathogenic agents that cause severe damage to crops are virus, bacteria, pests, etc. But in this chapter, we will discuss more of the losses caused due to fungal pathogens for the cultivation of rice. There are so many reported diseases caused to rice plant due to fungal pathogens. Few of the prominent disease and their current prevention methods are summarized below. *Pseudomonas* as a biocontrol agent was found effective in controlling all the diseases and is reported in several research articles. The mode of action of control of disease is described in subsequent sections.

#### 12.2.3.1 Blast Disease of Paddy

The blast is caused by the fungus *Magnaporthe grisea*. Appearance of lesions throughout the crop, including leaves, left neck, nodes, base, panicle, and grains, is the indications of the infections. The blast diseases are further classified on the basis of location and rise of the symptoms (TeBeest et al. 2007).

#### 12.2.3.2 Leaf Blast

The disease occurs in the seedling stage and the late stage of tillering. The fungus mainly affects the young leaves. In the later stages, the spots are transformed into spindles by the introduction of the yellow zone, from purple to brown borders. Brown spots occur only on the older leaves (Elazegui and Islam 2003).

#### 12.2.3.3 Neck Rot and Panicle Blast

They develop triangular lesions when the younger neck nodes get the infection. They stretch around the neck causing panicle damage. The infected panicles turn white in color. With severity of infection, the panicles may break and fall to the ground (Hayashi et al. 2019).

#### 12.2.3.4 Node Blast

Lesions appear on the infected stem nodes and thereby change the color of the rice tissue from brown to black. Leaves which are infected turn straw colored and finally drops (Sethy et al. 2020).

#### 12.2.3.5 Sheath Blight of Paddy Disease

The causative agent for the sheath blight of rice is *Rhizoctonia solani*. The fungus survives in the soil with the formation of a resting structure known as sclerotia or a dominant mycelia structure during unfavorable conditions for more than 2 years. The resting structure revives during land preparation and irrigations. The sclerotia bind with the seedling under favorable conditions and germinate. In such condition, the pathogens are very difficult to eradicate and can cause losses from 15% to 50% (Uppala and Zhou 2018).

Initially oval or ellipsoidal lesions appear with greenish gray spots measuring 1–2 cm in length (Fig. 12.1). The disease spreads rapidly in case of stagnant water. Later, the lesions enlarge turning into great white center surrounded by the dark brown margin. It spreads from one plant to another in this state (Manandhar 2016).

**Fig. 12.1** Prominent symptoms of paddy affected by sheath blight

Control of the blight is carried out by growing more tolerant rice varieties, avoiding the use of infected seeds, and treating the seeds with fungicide before sowing both by chemicals and biologicals like *Pseudomonas*, *Trichoderma*, etc. (Pal et al. 2015).

Also, avoiding the excessive use of nitrogen fertilizer can help in reduction of disease.Draining the excess water from the rice field early in the cropping season, during the sheath formation, controls the spread of an epidemic (Krishnamurthy and Gnanamanickam 1997).

#### 12.2.3.6 False Smut Disease of Rice

False smut is caused by fungus known as *Ustilaginoidea virens*. The fungus infects the spikes of young ovary and turns it into big balls known as smut balls (Fig. 12.2). The smut ball is yellow in color and has a membrane initially. After busting of the membrane, color changes to yellowish-green and at the later stages turns greenish-black. General practices for control of diseases are by avoiding the sowing of infected seeds and scheduling the sowing date properly to avoid the damage caused during the time of moving, using chemical spray and chemical soil application (post- and pre-sowing application) and seedling treatment with chemicals and biological products like *Bacillus*, *Pseudomonas*, etc. (Gupta et al. 2019).

#### 12.2.3.7 Bakane Disease or Foot Rot

Bakane disease of rice is caused by a fungus *Gibberella fujikuroi*. Infected plant reaches several inches taller than normal plants in seedbed and field. The infected plants are thin with yellowish green leaves and pale green flag leaves (Fig. 12.3). Seedlings dry at early tillering, while at late infection, tillering reduce and leaves

**Fig. 12.2** Prominent symptoms of paddy affected by false smut

**Fig. 12.3** Paddy showing parts of plants affected by bakane

dry. At maturity, the survived plants have grains or sterile or empty grains. In the seedbed, infected seedlings with lesions on roots die before or after transplanting. *Bakane* is a serious infection causing damages and crop losses mainly in parts of North India. Presently, few measures are followed to control the diseases like use of clean seeds which minimizes the occurrence of the disease. Salt water can be used to separate lightweight, infected seeds from seed lots and thereby reduce seed-borne inoculum. Chemical treatment like thiram, thiophanate-methyl, or benomyl is effective before planting. Seed treatment with organomercurials – Agrosan GN and Ceresan @ 2 g a.i. /kg seed – is highly effective; steeping seeds in 1% CuSO4 solution or 2% formalin is also recommended. Biological treatment of *Bakane* has been successful with *Pseudomonas* formulation and is extensively recommended by State Agricultural University (SAU) and nodal institutes (TNAU).

**Fig. 12.4** Prominent symptoms of paddy affected by brown spot

#### 12.2.3.8 Brown Spot

Brown spot of paddy is caused by *Cochliobolus miyabeanus* (earlier known as *Helminthosporium oryzae*). The fungus affects all the stages of growth from seedling to final stages. Yellow or brown lesions occur on the coleoptile at the seedling stage. First, small circular brown to purple-brown lesions occur on the leaves at the tillering point. The fully developed lesions are mostly circular to oval, light brown to gray clustered around the reddish-brown margin which is 5–14 mm long in some cases, causing the leaves to wax. On infected glumes and panicles, the dark brown to black oval spots are evident (Fig. 12.4). The grain value decreases as a result of incomplete grain filling when the florets are contaminated. By improving the soil fertility through regular monitoring of soil nutrients and by applying fertilizers, the disease can be controlled to some extent. The seeds are treated with fungicides such as propiconazole, trifloxystrobin, carbendazim, and prodione for 10–12 minutes in hot water at 53–54 degrees Celsius before planting to control the primary infection that may occur at the seedling stage (TNAU).

#### 12.2.3.9 Bacterial Blight

The causative agent of this disorder is *Xanthomonas oryzae pathovar* (PV). The species reach through the leaves and root pores of the plant. They grow within the plant and penetrate the veins of the leaf as well as block the roots' water-conducting tissues. After the initial lesions emerge, the bacteria ooze and spread by rain or wind to other crops. And the 25–34 degree Celsius atmospheric temperature and above 70% relative humidity are perfect conditions for the growth of these bacteria. Bacterial blight can cause a severe yield loss of paddy crops (Gnanamanickam et al. 1999).

At the initial stage of inferior operation, water-soaked streaks appear near the tip and margins of the leaf. Gradually, it transforms into yellow- and straw-colored stripes with wavy edges. At the creek phase, the leaves turn grayish-green, wilt, role up, and later wither (Fig. 12.5). Finally, the entire plant will dry or die. Management of bacterial blight in paddy is by developing resistant varieties that are able to withstand bacterial infection, through harvesting seeds from disease-free crops and disinfecting seeds before sowing. Avoiding transplantation of saplings in waterlogged

**Fig. 12.5** Prominent symptoms of paddy affected by bacterial blight

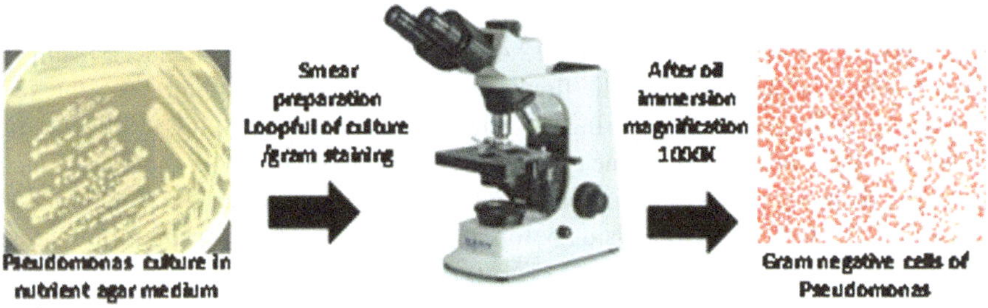


**Fig. 12.6** *Pseudomonas* spp. in cultured media and microscopic view after Gram staining

areas that are susceptible to flooding and using the appropriate dosage of fertilizers containing nitrogen and potassium can indirectly control the disease. Though bacterial blight is more efficiently controlled by *Bacillus subtilis* formulation, *Pseudomonas* has sometime proved to be effective when treated in combination with other biocontrol agents (Yasmin et al. 2017).

## 12.3 *Pseudomonas*. as a Biocontrol Agent for Rice Disease Management with Effective Secondary Metabolite Production

*Pseudomonas* spp. is a Gram-negative bacteria (Fig. 12.6) found in nature and is used extensively as biocontrol agent for crop disease management. *Pseudomonas* along with its biocontrol activity also acts as plant growth-promoting rhizobacteria and can indirectly increase plant resistance by improving the plant growth.

Interaction of *Pseudomonas* with host plant is through root colonization and thereby plays an important role in disease suppressions. The antagonistic bacteria suppress fungal and bacterial plant pathogens by producing different antimicrobial metabolites like antibiotics and enzymes like chitinase, glucanase, proteases, siderophores, etc. (Bardin et al. 2015).

*Pseudomonas* is also reported to produce hydrogen cyanide (HCN) which can significantly inhibit the growth of *Magnaporthe oryzae* (rice blast-causing pathogen) and show biocontrol activity (Spence et al. 2014). Diseases caused by rice pathogens like *Xanthomonas oryzae* and *Rhizoctonia solani* can be controlled up to 90% using *Pseudomonas* (Montano et al. 2014). Induced systemic resistance (ISR) is beneficial in disease control and developed a plant defense mechanism against different pathogens. The ISR is enhanced by different enzymes produced by *Pseudomonas* in association with plant host (Chithrashree et al. 2011).

*Pseudomonas* is used as broad spectrum biofertilizer and biofungicide in organic agriculture. Uncontrolled use of chemical fungicides has created a menace for environment. There is a need of the hour to identify an alternative strategy to control plant disease (Bashan 1997). *Pseudomonas* produces different extracellular secondary metabolites. These metabolites provide diverse properties like virulence factors, biosurfactants, and microbial agents for cell-to-cell signaling, etc. The metabolites enhance the activities of *Pseudomonas* and help it to get established in different environments (Novik et al. 2015). Some of the prominent metabolites are as follows:

- N-acylhomoserine lactone (Cataldi et al. 2009).
- Phenazine (Kumar et al. 2005).
- Pyochellin (Youard et al. 2007).
- Phloroglucinol (Kidarsa et al. 2011).
- Lahorenoic acid and diketopiperazines (Mehnaz et al. 2013).
- 4-Hydroxy-2-alkylquinolines (Lepine et al. 2004)
- Rhamnolipids (Soberon-Chavez et al. 2005).
- Cyclic lipopeptides (Mehnaz et al. 2013).
- 2, 4-Diacetylphloroglucinol (DAPG) (Reddy et al. 2007).

In a study, more than 500 rhizobacteria were isolated from different fields (Yasmin et al. 2017). The isolates were characterized for plant growth-promoting activities and biocontrol activities. Mass spectroscopy analysis has revealed that the isolates were capable of production of siderophores, rhamnolipids, etc. (Yasmin et al. 2017). The study also indicated that the biocontrol isolates were also capable of phosphate solubilization and indole acetic acid production (IAA). It was also reported that the solubilization of insoluble phosphates by microbial activity usually induces the secretion of certain metabolites mainly siderophores, lytic enzymes, and phytohormones that are involved in the control of phytopathogens (Vassilev et al. 2006).

In another study, 30 isolates of *Pseudomonas* were isolated from rice rhizosphere, and the antifungal activities were tested against *Magnaporthe grisea*, *Rhizoctonia solani*, and *Sarocladium oryzae*. Isolates were confirmed to produce 2,4-diacetylphloroglucinol (DAPG) and other antimicrobial compounds (Reddy et al. 2007).

## 12.4 Industrial Perspective

*Pseudomonas* spp. is used commercially as biofertilizers and biofungicides. In India, manufacturing and commercial activities of these agents are regulated by different authorities depending upon their categories and area of applications (Sekar et al. 2016). Isolation and selection of strains are of utmost importance for developing an efficient formulation for commercial use. Microbes are well diverse in their activities, and finding suitable isolates from nature is a challenging area. It is not correct to forecast a plant growth-promoting *Pseudomonas* to have a biocontrol activity. Strains are different in their traits and characteristics. Developing a suitable formulation of *Pseudomonas* is the secret for the success of the product.

## 12.5 Research and Manufacturing Insights

*Pseudomonas* strains having the efficiency of nutrient solubilization like phosphate, potassium, zinc, and other essential elements required for the growth of plants are selected strategically. The strains are developed into different formulations like liquid, powder, granular, and water soluble (Bejarano and Puopolo 2020).

Each formulation is developed after an intensive research for identifying compatible polymers which support the cell viability throughout the storage period along with *Pseudomonas*. *Pseudomonas* are reported as commercial products available in market for use (Mishra et al. 2018). The potent strains are developed using submerged fermentation technology. The strains are developed using statistical approaches like DOE (design of experiments). Growth parameter optimizations like growth medium, pH, oxygen concentration, temperature, and formulations are developed in research lab. After the initial parameter is optimized, the process is scaled up sequentially using sophisticated bioreactors. Bioreactors are vessels where microbes are cultivated under controlled condition for optimum yield of product using mass multiplication process. After the laboratory study, the technology is transferred to manufacturing facilities for mass production using large-scale commercial fermenters or bioreactors. The production process is controlled and regulated in such a way to avoid any failure or deviation in the process. Quality check at each place ensures a high value formulation of *Pseudomonas* that is produced. The cells thus manufactured are formulated for different types of products like liquid and carrier based. For liquid, the fermented broth is directly filled in the bottle under control environment (Clean air environment of Class 100) following the standard operating procedure. For carrier-based products, the liquid is mixed in an appropriate ration with a carrier like silicate powder, talcum powder, or granules like bentonite, gypsum, etc. (Bashan 1998) (Fig. 12.7).

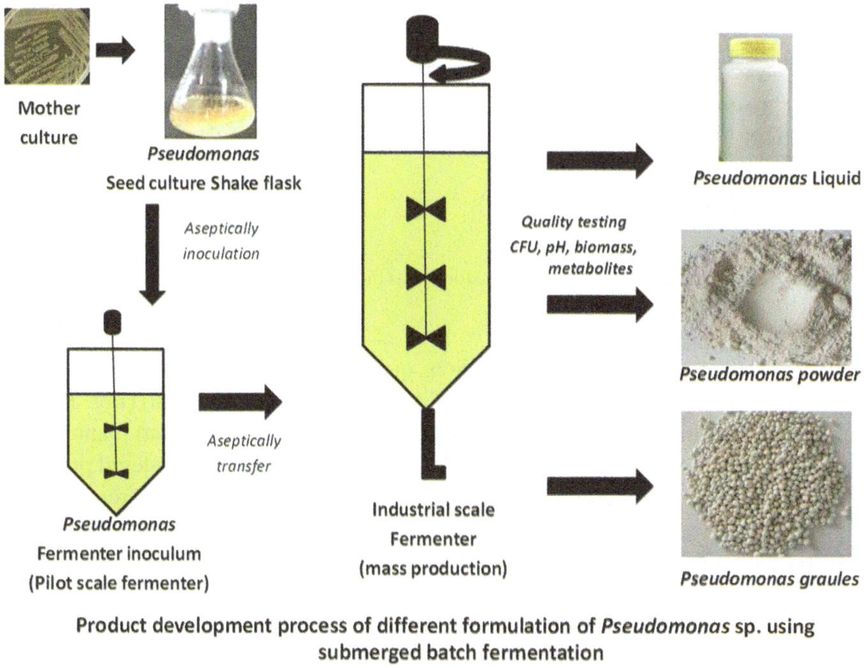


**Fig. 12.7** Product development process of different formulations of *Pseudomonas* sp. using submerged batch fermentation

## *12.5.1 Regulatory Procedure for Commercialization*

These products if intended to be used as biofertilizer need to follow the guidelines led by Biofertilizer and Organic Fertilizers in Fertilizer Control Order 1985, by the National Centre of Organic Farming (Department of Agriculture and Cooperation, Ministry of Agriculture, Government. of India) (FCO 1985) (Fig. 12.3). The products hence developed as biofertilizers are mandatory to follow the guidelines, but so far there is no requirement of registration. A separate category might be introduced by the Ministry of Agriculture for regulating biostimulants which will be different from existing biofertilizers. In the guidelines, it also helps scientist and industries to have an insight for process for development, formulations, recipe, method of analysis, sale of sampling, and quality check in Schedule III Clause 2(h) and 2(q) (FCO 1985).

While on the other hand, *Pseudomonas* with biocontrol activity for disease management are developed as products and are sold in Indian market. The products are categorized as plant protection products (biofungicides) and regulated strictly in India. Any new *Pseudomonas* formulation intended to use as a biocontrol need to be registered under the Central Insecticides Board and Registration Committee,

**Fig. 12.8** Schematic for *Pseudomonas* spp. for use as biofertilizer or biopesticides under different regulatory guidelines in India

CIB-RC (Directorate of Plant Protection, quarantine and storage, Department of Agriculture and Cooperation, Ministry of Agriculture, Govt. of India) (Fig. 12.8).

The license thus approved from CIB-RC permits an industry to start manufacturing and trading of product in Indian market under CIB-RC 9 (3) registered products with a valid registration number. For export and import, there is a different process for registration.

*Pseudomonas* products can thus be sold in the market through proper distributor network. The industry also applies for organic input approval for products under NPOP guidelines. Farmers and crop producers following organic agriculture practices approve their crops under guidelines led by APEDA (Ministry of Commerce and Industry). Thus, the product sold in the market is completely approved as organic. The inputs from nutrition to crop protection in such practices also need to be approved to use in organic agriculture under the guidelines of NPOP (National Program for organic Production) by Director General of Foreign Trade Development and Regulation Act 1992 (apeda.gov.in/apedawebsite). NPOP launched during 2001 laid the foundation for systematic development of organic agriculture sector in India.

## 12.6 Future Prospect

Biocontrol with microorganism is a novel approach for disease control as a substitute to chemicals. *Pseudomonas* is a prominent candidate that is being used for several years in organic agriculture having plant growth-promoting activities. There are several challenges when whole microbes are used as products like efficacy, viability, sustainability, etc. There are several biotic and abiotic stresses that restrict the growth of microbes. There is a less trust for its bio-efficacy among the cultivars compared to chemical inputs. The performance of biological products depends upon so many factors compared to chemical products that show immediate effects. Microbial metabolites are a new approach and area of research, where microbial cell will be processed and purified effective metabolites will be the products. The purified active metabolites will not depend upon any other factors to perform its desired biocontrol activity. This type of product will have head-to-head competition with established chemical products. To promote such type of novel biocontrol product development from microbes like *Pseudomonas*, new regulations are in process to be rolled out by the Government of India. As discussed earlier, there are so many factors that affect the production of rice. There is a serious urge to increase production of rice irrespective of these challenges. Disease management through integrated approach can be a novel approach for the future. *Pseudomonas* and its metabolite are ideal solutions for crop disease management of rice along with other conventional and modern techniques.

## References


Asibi AE, Chai Q, Coulter JA (2019) Rice blast: a disease with implications for global food security. Agronomy 9:451

Balgude YS, Gaikwad AP, Kshirsagar CR (2017) *Pseudomonas fluorescens* a potential bio agents for effective management of diseases in organic rice production. J Rice Res 10(2):80–84

Bardin M, Ajouz S, Comby M, Lopez-Ferber M, Graillot B, Siegwart M (2015) Is the efficacy of biological control against plant diseases likely to be more durable than that of chemical pesticides. Front Plant Sci 6:566

Bashan Y (1997) Alternative strategies for controlling plant diseases caused by pseudomonas syringae. Pseudomonas syringae Pathovars and Related Pathogens 9:575–583

Bashan Y (1998) Inoculants of plant growth-promoting bacteria for use in agriculture. Biotechnol Adv 16(4):729–770

Bejarano A, Puopolo G (2020) Bioformulation of microbial biocontrol agents for a sustainable agriculture. In: De Cal A, Melgarejo P, Magan N (eds) How research can stimulate the development of commercial biological control against plant diseases. Progress in Biological Control, vol 21. Springer, Cham, pp 275–293

Cataldi TR, Giuliana B, Salvatore A (2009) Accurate mass analysis of Nacyl-homoserine-lactones and cognate lactone-opened compounds in bacterial isolates of *Pseudomonas aeruginosa* PAO1 by LC-ESI-LTQ-FTICR-MS. J Mass Spectrom 44:182–192

Chithrashree UAC, Chandra NS, Reddy MS, Srinivas C (2011) Plant growth-promoting rhizobacteria mediate induced systemic resistance in rice against bacterial leaf blight caused by *Xanthomonas oryzae* pv. *Oryzae*. Biol Control 59:114–122

Elazegui F, Islam Z (2003) Diagnosis of common diseases of rice. International Rice Research Institute. Available online: http://www.knowledgebank.irri.org/images/docs/diagnostic-of-common-diseases-of-rice.pdf. Accessed 2 Mar 2021

FAO Food and agriculture organization, FAOSTAT Database (2015) Available online: http://www.faostat.fao.org. Accessed 20 Feb 2021

FCO (1985) (Fertilizer Control Order), Ministry of Agriculture and Rural Development, Government of India. Available online: https://agriodisha.nic.in/content/pdf/Fertiliser_Control_Order_1985.pdf. Accessed 10 Mar 2021

Gnanamanickam SS, BrindhaPriyadarisini V, Narayanan NN, Vasudevan P, Kavitha S (1999) An overview of bacterial blight disease of rice and strategies for its management. Curr Sci 77(11):1435–1444

Gupta AK, Paudel J, Yadav D, Kumar D, Yadav R, Kumar S (2019) Eco-friendly management of false smut (Ustilaginoideavirens) of rice. Int J Curr Microbiol App Sci 8(11):388–396

Hayashi K, Yoshida T, Hayano-Saito Y (2019) Detection of white head symptoms of panicle blast caused by *Pyricularia oryzae* using cut-flower dye. Plant Methods 6(15):159

India Rice Market - Growth, Trends, Covid-19 Impact, and Forecasts (2021–2026). Modor Intelligence (2020) Available online: https://www.mordorintelligence.com/industry-reports/rice-market. Accessed 2 Mar 2021

Keelery S (2020). Export value of rice from India FY 2015-2020 by type, Statista Research Department, Statista. Available online: https://www.statista.com/statistics/624674/export-value-of-rice-by-country-india/. Accessed 20 Mar 2021

Kidarsa TA, Goebel NC, Zabriskie TM, Loper JE (2011) Phloroglucinol mediates cross-talk between the pyoluteorin and 2, 4- diacetylphloroglucinol biosynthetic pathways in *Pseudomonas fluorescens* Pf-5. Mol Microbiol 81:395–414

Krishnamurthy K, Gnanamanickam SS (1997) Biological control of sheath blight of rice: induction of systemic resistance in rice by plant-associated pseudomonas spp. Current Science72 (5):331–334

Kumar RS, Ayyadurai N, Pandiaraja P, Reddy AV, Venkateswarlu Y, Prakash O (2005) Characterization of antifungal metabolite produced by a new strain *Pseudomonas aeruginosa* PUPa3 that exhibits broad-spectrum antifungal activity and biofertilizing traits. J Appl Microbiol 98:145–154

Kumar A, Dasgupta P, Kumar R (2014) Emerging opportunities and challenges in rice production. Popular Kheti 2(2):6–11

Lepine F, Milot S, Deziel E, He J, Rahme LG (2004) Electrospray/mass spectrometric identification and analysis of 4-hydroxy-2-alkylquinolines (HAQs) produced by *Pseudomonas aeruginosa*. J Am Soc Mass Spectrom 15:862–869

Manandhar HK (2016) A field guide for identification and scoring methods of. Bioversity International. Available online: https://www.bioversityinternational.org/fileadmin/user_upload/A_field_guide_Manandhar.pdf. Accessed 2 Mar 2021

Mehnaz S, Saleem RS, Yameen B, Pianet I, Schnakenburg G, Pietraszkiewicz H, Valeriote F, Josten M, Sahl HG, Franzblau SG, Gross H (2013) Lahorenoic acids A-C, ortho-dialkyl-substituted aromatic acids from the biocontrol strain Pseudomonas aurantiaca PB-St2. J Nat Prod 76(2):135–141

Mishra RK, Bohra A, Kamaal N, Kumar K, Gandhi K, Sujayanand GK, Saabale PR, Naik SJS, Sarma BK, Kumar D, Mishra M, Srivastava DK, Singh NP (2018) Utilization of biopesticides as sustainable solutions for management of pests in legume crops: achievements and prospects. Egypt J Biol Pest Control 28(3):1–11

Montano PF, Alias-Villegas C, Bellogin RA, Del Cerro P, Espuny MR, Jimenez-Guerrero I et al (2014) Plant growth promotion in cereal and leguminous agricultural important plants: from microorganism capacities to crop production. Microbiol Res 169:325–336

Moore D, Robson GD, Trinci APJ (2002) 21st century guide book to fungi, 2nd EDITION. Available online: https://www.researchgate.net/publication/342145987_21st_Century_Guidebook_to_Fungi_-_SECOND_EDITION_-_Reviews_and_Contents/link/5ee4a85692851ce9e7e05963/download. Accessed 2 Mar 2021

Nicolopoulou-Stamati P, Maipas S, Kotampasi C, Stamatis P, Hens L (2016) Chemical pesticides and human health: the urgent need for a new concept in agriculture. Frontier Public Health 4:148

Novik G, Savich V, Kiseleva E (2015) An insight into beneficial pseudomonas bacteria. Microbiology in agriculture and human health, Intech open. Available online: https://www.intechopen.com/books/microbiology-in-agriculture-and-human-health/an-insight-into-beneficial-pseudomonas-bacteria. Accessed 10 Mar 2021

Oerke EC (2019) Crop losses to pests. J Agric Sci 144:31–43

Pal P, Kumar S, Singh A (2009) Rice production in India — implications of land inequity and market imperfections. Agric Econ Res Rev 22:431–442

Pal R, Biswas MK, Mandal D, Naik B (2015) Management of sheath blight disease of rice through bio control agents in west central table land zone of Odisha. Int J Adv Res 3(11):747–753

Peng S, Tang Q, Zou Y (2009) Current status and challenges of rice production in China. Plant Product Sci 12:3–8

Pradhan B, Parija P (2020) Rising water crisis forces Indian farmers to rethink their crops selection, Bloomberg. Available online https://www.bloomberg.com/news/features/2020-07-21/india-water-crisis-forces-farmers-to-rethink-their-crops. Accessed 20 Feb 2021

Reddy KRN, Anithachoudary K, Reddy MS (2007) Antifungal metabolites of Pseudomonas fluorescens isolated from rhizosphere of rice crop. J Mycol Pl Pathol 37(2):1–5

Rice knowledge bank IRRI., http://www.knowledgebank.irri.org/step-by-step-production/growth/pests-and-diseases. Accessed 2 Mar 2021

Satterthwaite D, McGranahan G, Tacoli C (2007) Urbanization and its implications for food and farming. Philos Trans R SocLond B Biol Sci 365(1554):2809–2820

Sekar J, Raj R, Prabavathy VR (2016) Microbial consortial products for sustainable agriculture: commercialization and regulatory issues in India. Agriculturally Important Microorganisms:107–132

Sethy PK, Barpandaa NK, Rathb AK, Behera SK (2020) Image processing techniques for diagnosing Rice Plant disease: a survey. Procedia Computer Science 167:516–530

Sharma A, Diwevidi VD, Singh S, Pawar KK, Jerman M, Singh LB, Singh S, Srivastawa D (2013) International Journal of Biotechnology and Bioengineering Research 4(3):175–180

Soberon-Chavez G, Lepine F, Deziel E (2005) Production of rhamnolipids by *Pseudomonas aeruginosa*. Appl Microb Biotechnol 68:718–725

Spence CA, Raman V, Donofrio NM, Bais HP (2014) Global gene expression in rice blast pathogen *Magnaporthe oryzae* treated with a natural rice soil isolate. Planta 239:171–185

TeBeest DO, Guerber C, Ditmore M (2007) Rice blast. The plant health instructor. Available online: https://www.apsnet.org/edcenter/disandpath/fungalasco/pdlessons/Pages/RiceBlast.aspx. Accessed 2 Mar 2021

TNAU, Tamil Nadu Agricultural University official website. Available online: https://agritech.tnau.ac.in/crop_protection/crop_prot_crop%20diseases_cereals_paddy.html#2. Accessed 2 Mar 2021

Uppala S, Zhou XG (2018) Rice sheath blight. The plant health instructor. Available online: https://www.apsnet.org/edcenter/disandpath/fungalasco/pdlessons/Pages/RiceSheath.aspx. Accessed 6 Feb 2020

Vassilev N, Vassileva M, Nikolaeva I (2006) Simultaneous P-solubilizing and biocontrol activity of microorganisms: potentials and future trends. Appl Microbiol Biotechnol 71:137–144

World population prospects (2019) Ten key findings. United Nations Department of Economic and Social Affairs. Available online: www.population.un.org/wpp/Publications/Files/WPP2019_10KeyFindings.pdf. Accessed 6 Feb 2020

World-grain.com (2020) India's rice production on the rise, available online: https://www.world-grain.com/articles/14450-indias-rice-production-on-the-rise. Accessed 20 Feb 2021

Yasmin S, Hafeez FY, Mirza MS, Rasul M, Arshad HMI, Zubair M, Iqbal M (2017) Biocontrol of bacterial leaf blight of Rice and profiling of secondary metabolites produced by Rhizospheric *Pseudomonas aeruginosa* BRp3. Front Microbiol 8(1895):1–23

Youard ZA, Mislin GL, Majcherczyk PA, Schalk IJ, Reimmann C (2007) *Pseudomonas fluorescens* CHA0 produces enantio-pyochelin, the optical antipode of the *Pseudomonas aeruginosa* siderophore pyochelin. J Biol Chem 282:35546–35553

Zadoks, J. C. (2008). On the political economy of plant disease epidemics : capita selecta in historical epidemiology. Wageningen Academic Publishers. https://edepot.wur.nl/141867

# Chapter 13
# Interaction of Mycorrhizal Fungi with Rhizospheric Microbes and Their Mode of Action

**Jay Kant Raut, Kuber Baral, Mahesh Kumar Adhikari, and Pramod Kumar Jha**

**Contents**

13.1 Introduction .......... 269
13.2 Major Role of Mycorrhiza .......... 271
13.3 Interactions with Plant Growth-Promoting Rhizobacteria .......... 272
13.4 Interaction with *Rhizobia* (Rh) .......... 274
13.5 Interactions with Root Pathogens .......... 276
13.6 Interaction with Nematodes .......... 276
13.7 Interaction with Fungal Plant Pathogens .......... 277
13.8 Interaction with Other Plant Pathogens .......... 278
13.9 Interactions for Improving Soil Quality .......... 280
13.10 Interaction for Phyllospheric Protection .......... 281
13.11 Conclusion .......... 282
References .......... 282

## 13.1 Introduction

The rhizosphere is a site which is colonized by bacteria, fungi, and other microorganisms forming a living interface between roots and soil (Manoharachary and Mukerji 2006). It is also known as a hotspot of microbial activities as the plant roots influence microbial communities by depositing photosynthates (amino acids, low

J. K. Raut (✉) · K. Baral · M. K. Adhikari
Nepal Academy of Science and Technology, Khumaltar, Lalitpur, Nepal
e-mail: jaykant.raut@nast.gov.np

P. K. Jha
Nepal Academy of Science and Technology, Lalitpur, Nepal

Central Department of Botany, Tribhuvan University, Kathmandu, Nepal

 
R. Z. Sayyed et al. (eds.), *Antifungal Metabolites of Rhizobacteria for Sustainable Agriculture*, Fungal Biology, https://doi.org/10.1007/978-3-031-04805-0_13

molecular weight organic acids, or phytosiderophores) (Sasse et al. 2018). Interactions between these microorganisms in the rhizosphere influence plant health in both agro and natural ecosystems directly by providing nutrition and/or indirectly by protecting against biotic and abiotic stresses (Philippot et al. 2013). The microorganisms and their products interact with plant roots in a variety of ways such as positive, negative, and neutral (Kuzyakov and Xu 2013). The interactions among the microorganisms operate through a variety of modes of action. Understanding microbial interactions within the rhizosphere is urgent toward farming practices that are less reliant on conventional chemical fertilizers, which have known negative impacts on environmental receptors (Rascovan et al. 2016). Rhizospheric microbial interactions have been extensively studied especially in terms of plant health and nutrition (Whipps 2001; Johansson et al. 2004; Barea et al. 2005; Morgan et al. 2005; Lambers et al. 2009; Ryan et al. 2009; Philippot et al. 2013; De-la-Peña and

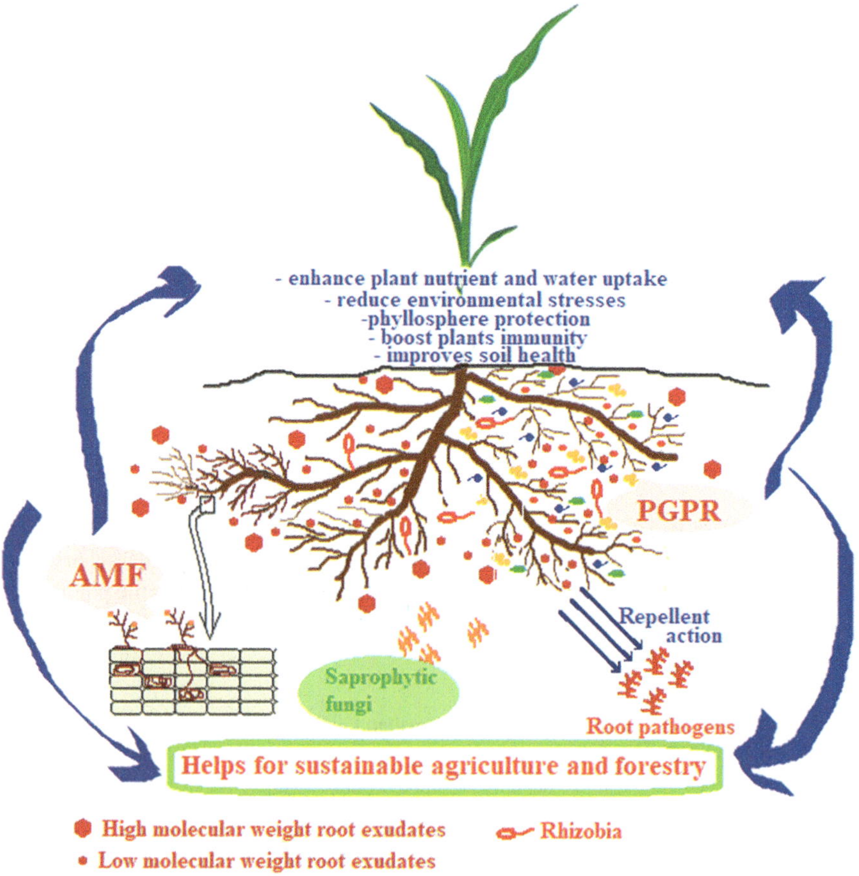


**Fig. 13.1** Schematic drawing representing the impacts of mycorrhizal interactions with other rhizospheric microbes on plant performance and soil fertility

Loyola-Vargas 2014; Raut 2014; Larsen et al. 2015; Choudhary et al. 2016; Jambon et al. 2018; Henkes et al. 2018; Nanjundappa et al. 2019; Odelade 2019; Pascale et al. 2020; Nacoon et al. 2020; Baral et al. 2021; Kannepalli et al. 2021; Kapadia et al. 2021; Khan et al. 2021; Manasa et al. 2021; Najafi et al. 2021). Mycorrhizal fungi are the most important rhizospheric microbe influencing plant growth and ubiquitous in the terrestrial ecosystem. The reported microbial interactions in the mycorrhizosphere may involve a variety of bacteria and fungi with specific functional capabilities that may influence plant growth (Fig. 13.1). However, most of the studies are focused on rhizospheric bacteria than fungi (Mendes et al. 2013). The purpose of this chapter is to outline the current knowledge on mycorrhizal interactions in the mycorrhizosphere and the practical implications that these interactions may have in sustainable agriculture and forestry. The information derived from recent studies will enable us to explore and gather more information at the plant and microbiome level.

## 13.2 Major Role of Mycorrhiza

Mycorrhizae establish symbiotic relationships with plants and play an essential role in plant growth, disease protection, and overall soil quality (Siddiqui and Pichtel 2008). Of the seven types of mycorrhizae described in current scientific literature (arbuscular, ecto, ectendo, arbutoid, monotropoid, ericoid, and orchidaceous mycorrhizae), the arbuscular and ectomycorrhizae are ubiquitous components of most ecosystems throughout the world influencing plant productivity (Brundrett 2002; Smith and Read 2008). Arbuscular mycorrhizal (AM) fungi are obligate root symbionts inhabiting almost all terrestrial ecosystems forming a symbiotic association with around 80% of vascular plants and with approximately 90% of agricultural plants (Smith and Read 2008). The distribution of ectomycorrhizal (ECM) fungi is also widespread, but they form associations with only 3% of terrestrial plant families (Smith and Read 2008). Mycorrhizal symbiosis, in particular arbuscular mycorrhizal fungi (AMF), to which this chapter mostly refers, is actively studied for its beneficial effect on plant growth, disease protection, and soil quality and for its potential use in sustainable agriculture and forestry. AMF are the most common fungi in soils and represent 9–55% of the soil microbe biomass and 5–36% of the total soil biomass (Olsson et al. 1999). AM fungus receives 10–20% of total photosynthates (Van Ooijen et al. 2008) and lipids (Bravo et al. 2017) from the host plant, whereas the host plant is enhanced through uptake of water and mineral nutrient by the mycorrhizal partner.

Besides their major role in nutrient uptake and translocation, extraradical mycelium (ERM) produced by AMF may also provide a large surface for interactions with other microorganisms, particularly soil bacteria (de Novais et al. 2020). ERM has the capacity to form extensive networks of subterranean hyphae which can comprise up to 25 m $g^{-1}$ soil (Mikkelsen et al. 2008) and due to the absence of host specificity may simultaneously connect the roots of different plant species (Pepe et al. 2018). Mycorrhizal fungi are also known to alter soil texture by increasing the extent of soil particle aggregation and thus stability (Sutton and Sheppard 1976). Symbiotic association of some ectomycorrhizal fungi produce edible mushrooms and truffles. In addition, mycorrhizae play a key role by influencing regulatory systems in plants by stimulating phytohormone production (i.e., cytokinins, gibberellins, and ethylene) (Barea and Azcon-Aguilar 1982). Furthermore, mycorrhizal fungi can reduce plant response to soil stresses such as high level of salt; toxicities associated with mine spoils or landfills and heavy metals; or toxicities due to minor element imbalance such as manganese toxicity (Tinker 1984). Some mycorrhizal fungi produce metabolites that can alter the plant's ability to produce roots from cuttings or to alter root regeneration and root morphology. Overall, these fungi play a vital role in agricultural ecosystems enhancing survival rate and development of seedlings, crop uniformity, and reproductive capacity (Lu and Koide 1994), decrease the input of P and N fertilizer, and increase resistance or tolerance to environmental adversities (Birhane et al. 2012; Hong and Posta 2018).

## 13.3 Interactions with Plant Growth-Promoting Rhizobacteria

Many beneficial soil bacteria are associated with plant roots called plant growth-promoting rhizobacteria (PGPRs). They belong to a wide range of bacterial genera including *Agrobacterium*, *Arthrobacter*, *Azotobacter*, *Azospirillum*, *Bacillus*, *Burkholderia*, *Caulobacter*, *Chromobacterium*, *Erwinia*, *Flavobacterium*, *Micrococcus*, *Pseudomonas*, and *Serratia* (Bhattacharyya and Jha 2012; Akinrinlola et al. 2018). PGPR play a significant role in enhancing plant growth and development both under non-stress and stress conditions by a number of direct and indirect mechanisms (Zahir et al. 2003; Glick et al. 2007; Nadeem et al. 2010; Shaikh et al. 2016; Fazell-Nasab and Sayyed 2019; Khan and Sayyed 2019; Yadav and Sayyed 2019; Zope et al. 2019; Basu et al. 2021). The mechanisms that promote plant growth include nitrogen fixation, phosphorus solubilization, production of siderophores, plant growth regulators and organic acids, as well as protection by enzymes like ACC-deaminase, chitinase, and glucanase (Glick et al. 2007; Berg 2009; Hayat et al. 2010; Kour and Sayyed 2019; Zakaria et al. 2019). Mycorrhizal fungi share the common habitat with PGPRs and interact with each other in the mycorrhizosphere. The interaction may be beneficial or neutral or detrimental.

Arbuscular mycorrhizal (AM) are probably the dominant mycorrhizal fungi found in agricultural soils (Willis et al. 2013). Co-inoculation of AM fungi and PGPR has been proposed as an efficient method to increase plant growth by many workers. Simultaneous inoculation of *Azospirillum* and AM fungi significantly increased the growth and chlorophyll content and mycorrhizal infection of the roots in sorghum, pearl millet, and wheat (Walley and Germida 1997). Synergism of *Azotobacter* and AM fungi have also been cited in the literature (Toro et al. 1997,

1998). Dual inoculation of *A. chroococcum* and *Glomus fasciculatum* enhanced root infection of AM fungi, stimulated the plant growth, and increased shoot N, Ca, Mg, and K in luxor tomatoes (El-Shanshoury et al. 1989). Improved growth and yield of lettuce plants in both sterile and non-sterile soils due to simultaneous inoculation with AM fungi and *A. chroococcum* have also been reported (Brown and Carr 1984). These synergisms could be mainly due to the production of phytohormones or growth regulators by these symbionts rather than the small increases in N and P availability (Barea et al. 1975). Inoculation of neem (*Azadirachta indica*) seedlings with *Azospirillum brasilense* and phosphate-solubilizing bacteria (PSB) in presence of AM fungi *Glomus intraradices* and *G. geosporum* resulted in increased mycorrhizal colonization, greater plant height, leaf area and number, root collar diameter, biomass, phosphorus, nitrogen and potassium content, and seedling quality (Saxena et al. 2006). In fruit crops like papaya, dual inoculation of *Azospirillum* and AM fungi enhanced the total dry matter and leaf area than non-colonized plants (Alarcón et al. 2002). Co-inoculation effects were significant even in field-established papaya and mulberry. *Bacillus coagulans* when co-inoculated with a mixture of *Glomus fasciculatum* and *G. caledonium* increase the leaf yield in 10-year-old mulberry and fruit yield in 1.5-year-old papaya (Mamatha et al. 2002).

The synergistic interaction of *Bacillus subtilis* and AM fungi altered *Acacia gerrardii* plant fitness under salt stress, significantly increasing plant biomass, nodulation, leghemoglobin, and crude protein content compared with untreated plants (Hashem et al. 2016). Synergistic effects on growth and systemic priming of wheat were reported with multilateral interactions between roots, mycorrhizal fungi, and PGPR. They can improve plant health via enhanced nutrient acquisition and priming of the plant immune system (Pérez-De-Luque et al. 2017).

In a recent field study, the inoculation with the consortium of PGPR-rhizobia-AM fungi (PRM) induced the greatest positive effect on the growth parameters (dry weight of shoots and roots) of faba bean (*Vicia faba* L.) and wheat (*Triticum durum* L.) compared to uninoculated plants. In addition, the consortium also improved N, P, Ca, K, and Na shoot contents, as well as the sugar and protein contents (Raklami et al. 2019). Arbuscular mycorrhizal fungi (AMF) and plant growth-promoting rhizobacteria (PGPR) are considered as highly efficient biological agents for conferring salt tolerance in host plants and improving soil fertility (Pan et al. 2020). Pan et al. (2020) found that *Elaeagnus angustifolia* L. was more responsive to specific inoculation by AMF than by a combination of AM fungi and PGPR or by solely PGPR in saline soils. AMF significantly promoted biomass accumulation in aboveground organs, increased the numbers of leaves and branches, and improved the leaf area, stem diameter, and plant height. AMF-mediated morphological characteristics of aboveground organs favored light interception and absorption and maximized the capacities for photosynthesis, transpiration, carbon dioxide assimilation, and gas exchange of *Elaeagnus angustifolia* L. seedlings in saline soil. AMF also promoted root growth, modified root architecture, and enhanced soil enzyme activities. In this way PGPR and mycorrhiza hold huge potential to substitute agrochemicals for restoring soil sustainability and agricultural productivity. The benefits of synergistic interactions among these efficient eco-friendly microbes have been studied and demonstrated generally under laboratory/greenhouse conditions (Singh et al. 2020). Their impact on plant growth and nutrition in the open field is scant and need more research (Hart et al. 2017; Ryan and Graham 2018; Raklami et al. 2019).

## 13.4 Interaction with *Rhizobia* (Rh)

The role of mycorrhizal fungi, especially the arbuscular mycorrhizae (AM fungi) in phosphorous mobilization in soils having a relatively low level of available phosphorous, is well established for cereals as well as legumes. Associative actions of mycorrhizal fungi in legumes have a great impact on root and shoot development and phosphorous uptake which results in the enhancement of nodulation and nitrogen fixation. Interaction between AM fungi and *Rhizobia* has been reported for several crop legumes like soybean (Zhao et al. 1997; Wang et al. 2011; Gao et al. 2012), common bean (Tajini et al. 2012; Razakatiana et al. 2020), faba beans (Ishac et al. 1994), medicago (Azcon et al. 1991; Ashrafi et al. 2014), green gram (Saxena et al. 1997), white clover (Tang et al. 2019; Xie et al. 2020), and tree legumes like *Anthyllis cytisoides* (Requena et al. 1997), *Leucaena leucocephala* (Dixon et al. 1993), *Sesbania* (Sengupta and Choudhuri 1995), and *Prosopis* (Dixon et al. 1993). The dual inoculation with AM fungi and *Rhizobium* bio-fertilizers is more effective in increasing growth, nutrition, chlorophyll content, and biomass production of legumes (Havugimana et al. 2016; Barea et al. 2017). Integration of AM fungi with *Rhizobium* spp. thus appears to be a promising approach for sustainable agriculture especially in legume crops where the net influence of this combination is supposed to be very high compared to other types of crop families.

The co-inoculation with AM fungi and *Rhizobium* increased nodule number and biomass significantly (Saxena et al. 1997; Zhao et al. 1997). The impact of such co-inoculation in improving nodulation and $N_2$ fixation is well established (Barea and Azcón-Aguilar 1983). The dual inoculation of *Rhizobium* and *Glomus mosseae* increased 176% and 305% of nodule biomass and number of nodules, respectively (Badji et al. 1989; Saxena et al. 2006). Nitrogenase is an important component in root nodule formation, and root nitrogenase activity was notably stimulated by dual inoculation with AMF and *Rhizobia* (Xie et al. 2020). Nitrogenase activity was closely associated with Fe levels, and AMF or Rh inoculation increased Fe contents of plants (Ibiang et al. 2017), thereby stimulating nitrogenase activity.

Dual inoculation with AMF and Rh exhibited higher leaf and root N levels in white clover (Xie et al. 2020), green grams (Musyoka et al. 2020), and soybean (Aliasgharzad et al. 2006). The effective association between AMF and Rh affects N accumulation to increase plant production and $N_2$ fixation (Herridge et al. 2008). Synergetic effects of dual inoculation of *Paraglomus occultum* and *Rhizobia* (*Rhizobium trifolii*) on accelerating N acquisition and plant growth of white clover was reported (Xie et al. 2020). An additive effect of co-infection by AM fungi and *Rhizobia* was found on plant growth and on the total N content of the plants, along with a synergistic effect on the total P content, the number of nodules, and the mycorrhizal rate of the plants. Dual symbioses with AM fungus and *Rhizobia* contribute to the success of legumes, especially in a harsh environment and on low-fertility soils (Franco and De Faria 1997; van der Heijden et al. 2016; Razakatiana et al. 2020). AMF and *Rhizobia* jointly stimulated plant growth, root soluble sugar, and soluble protein accumulation in white clover, while AMF did not show a cooperative effect on N fixation of *Rhizobia* (Tang et al. 2019). The AMF and *Rhizobia* collectively had the capability to increase nutrient acquisition and to protect plants from the infection of root pathogens (Akhtar and Siddiqui 2008a). However, it is unclear whether soil beneficial both microorganisms and plants co-work together (Tang et al. 2019).

High yield in legumes co-inoculated with AM and *Rhizobia* has been reported in comparison to uninoculated or inoculated with either microsymbiont (Corbera and Hernandez 1997). Antagonism between *Rhizobia* and AM fungi has also been reported due to competition for nutrients between them (Bethlenfalvay 1992; Schreiner et al. 1997). Such negative interaction could be due to specific compatibilities between the two microsymbionts (Saxena et al. 1997). Selective interactions between the two endosymbionts are important aspects for the yield of crop plants. Specific compatibility between microsymbionts can exist both in terms of competitive ability of introduced *Rhizobia* and percent root colonization by introduced AM fungi (Saxena et al. 1997). Introduced strains of *Bradyrhizobium* become more competitive than native *Rhizobia* to nodulate cowpea in the presence of AM fungus (*Glomus pallidum*) (Thiagarajan and Ahmad 1993). Nodulation competitiveness of introduced *Bradyrhizobium* spp. (Vigna) strain S24 was significantly higher (60–65%) in the presence of *Glomus mosseae*, *G. fasciculatum*, and *Scutellospora calospora* when compared to treatment with single inoculation of S24 (50%) (Saxena et al. 1997). They further reported that percentage AM colonization was higher in treatments having higher nodule occupancy of introduced strain (Saxena et al. 2006).

In a study the growth and yield response of pea to co-inoculation with AMF and *Rhizobium* strains depended on the particular AMF-*Rhizobium* combination. Careful co-selection of AMF species and *Rhizobium* strains enhanced pea yield and nutrition (Xavier and Germida 2003). Specific compatibilities have also been reported for other hosts like *Anthyllis cytisoides* (Requena et al. 1997), *Medicago arborea* (Valdenegro et al. 2001), kidney bean (Daniels-Hylton and Ahmad, 1994), green gram (Saxena et al. 1997), and *Acacia tortilis* (André et al. 2003). In stressed soils, the symbiosis with both AM fungi and *Rhizobium* enhances the ability of the plant to become established and cope with stress situations (nutrient deficiency, drought, trace element imbalance, soil disturbance) (Saxena et al. 2006). The use of both endophytes for legumes has been recommended for reclamation or for rehabilitation of desertified ecosystems (Requena et al. 1997), wastelands (Bhatia et al. 1998), saline soils (Azcón and El-Atrash 1997), and drought-prone soils (Goicoechea et al. 1998).

## 13.5 Interactions with Root Pathogens

Mycorrhizal fungi (both AM and ECM) have been known to protect the plants against several soil-borne pathogens like *Fusarium*, *Verticillium*, *Ralstonia*, *Macrophomina*, *Aphanomyces*, *Cylindrocladium*, *Macropomina*, *Phytophthora*, *Pythium*, *Rhizoctonia*, *Sclerotium*, etc. (Davis and Menge 1980). The negative-antagonistic interaction of mycorrhizal fungi with various soil-borne plant pathogens is the reason for their use as biocontrol agents (Singh et al. 2019). The protective effect of mycorrhizal symbioses against root pathogens has been reported in a number of literature (Dehne 1982; Barea and Azcón-Aguilar 1983; Smith 1988; Caron 1989; Paulitz and Linderman 1991; Fitter and Garbaye 1994; Siddiqui and Mahmood 1995; Azcón-Aguilar and Barea 1997; Mukerji 1999; Barea et al. 2005; Siddiqui and Akhtar 2006; Akhatar and Siddiqui 2008a, b; Singh et al. 2010; Banuelos et al. 2014; Schouteden et al. 2015; Olowe et al. 2018; Poveda et al. 2019; Swandi et al. 2020). Based on various observations, the mechanisms for the negative-antagonistic interaction of mycorrhizal symbiosis can be explained as the improvement of plant nutrition, damage compensation, competition for photosynthates or for colonization/infection sites, induction of changes in the morphology/anatomy of the root system, induction of changes in mycorrhizosphere populations, and activation of plant defense mechanisms.

Better control of pathogens can be achieved by using the combination of mycorrhizal fungi rather than individual use (Requena et al. 2001). In few cases some soil-borne diseases have been found to be increased with the mycorrhizal infection (Ross 1972; Davis and Menge 1980; Garmendia et al. 2004) suggesting that in some instances, a healthy mycorrhizal plant may be more susceptible than a poorly developed non-mycorrhizal one (Dehne 1982; Bending et al. 2006). Therefore, the interactions between different mycorrhizal fungi and plant pathogens vary with the host plant and the cultural system. Moreover, the protective effect of mycorrhizal inoculation may be both systemic and localized. Interactions of mycorrhizal fungi against phytopathogens can be categorized into interaction with nematodes, interaction with fungal pathogens, and interaction with other plant pathogens.

## 13.6 Interaction with Nematodes

A diverse group of nematodes like free-living nematodes as well as plant and animal parasites are ubiquitous in nature (Ferraz and Brown 2002). Plant-parasitic nematodes (PPN) of various species can act as pest on a wide range of important agricultural crops. They mainly reside in the soil, but some of them such as several species of *Ditylenchus* can act as aboveground pests. Nematode problems are more severe and complicated in warmer than cooler areas, in horticultural than field crops, mono-culture than multi-culture, and plantation crops than natural forests and vegetation. Given the high economic impact caused by parasitic nematodes, a large number of strategies have been developed for nematode control in agriculture, including the use of chemical nematicides. The nematicides are harmful to human health and contaminant for the environment (Zhang et al. 2014, 2017). One of the proposed environment-friendly options to manage PPN is the use of biological control organisms, such as AMF.

Normally, AMF act against nematodes through the host plant either by providing them with higher nutrient and water uptake, making the plants more competitive for nutrients and space with other plants, or altering root morphology by increasing root growth and branching, or altered rhizosphere interaction, but precise direct mechanisms of AMF against nematodes are not yet described (Schouteden et al. 2015; Wani et al. 2017; Pawlowski and Hartman 2020). Recent research has confirmed these mechanisms, for example, AMF (*Funneliformis mosseae*, *Rhizophagus intraradices*, and *G. etunicatum*) decrease peach root penetration by false root-knot nematode *M. javanica* in peach trees (Calvet et al. 2001). Similarly, the application of *Funneliformis mosseae* and *Rhizophagus intraradices* against *Nacobbus aberrans* was reported (Marro et al. 2018). On the other hand, the increment in nematode (*Pratylenchus brachyurus*) population in maize plant was due to the increase in root colonization by mycorrhizae (*Gigaspora rosea*, *G. margarita*, *Claroideoglomus etunicatum*, *Rhizophagus clarus*, *Scutellospora calospora*, and *S. heterogama*) which is opposite to the effect in cotton (Ferreira et al. 2018). The soybean cyst nematode (SCN) egg hatching, SCN juveniles, and cyst in roots were inhibited when plants were inoculated with *Funneliformis mosseae* compared with non-inoculated plants. To allow manipulation of AMF to be used more effectively as a biological control, a better understanding of the mechanisms involved is required specifically what antimicrobial exudates are involved in reducing egg hatch (Pawlowski and Hartman 2020).

The capacity of infection, their effectiveness, and all mechanisms of the phytoparasitic nematodes depend closely on the local environmental conditions. In this matter, the beneficial effect caused by mycorrhizal fungi negatively relates with the factors like variation in moisture, phosphorous, and potassium (Ferreira et al. 2018). The application of mycorrhizal fungi for the biological control of plant-parasitic nematodes has been widely studied in numerous crops like energy crops such as switchgrass (*Panicum virgatum*) and miscanthus (*Miscanthus giganteus*) (Emery et al. 2017) and maize (Alvarado-Herrejón et al. 2019) in spite of contradictory reports.

## 13.7 Interaction with Fungal Plant Pathogens

Fungal plant pathogens have been devastating threats throughout agricultural and forestry systems. More than 10,000 species of fungi are known to cause diseases of plants and are common in soil, in air (spores), and on plant surfaces in arid, tropical, temperate, and alpine regions of the entire globe (Agrios 2005). The occurrence of AM fungi and plant pathogenic fungi in roots of different crops and their dependence for nutrition on the host generally result in the interaction of AM fungi, plant pathogenic fungi, and host plant.

Association of these organisms generally exert opposite effects on the host. Thus, it is desirable to test the mutual effects of these organisms on plant growth and yield. The interaction of mycorrhizal fungi and plant pathogenic fungi has received considerable attention, and a large number of research papers have been published (Duchesne et al. 1989; Sreeramulu et al. 1998 Becker et al. 1999; Azcón-Aguilar et al. 2002; Bødker et al. 2002; Kasiamdari et al. 2002; Garmendia et al. 2004; Akhtar and Siddiqui 2008a, b; Saldajeno et al. 2008; Tahat et al. 2010; Tayal 2011; Elwakil et al. 2013; Eke et al. 2016; Olowe et al. 2018; Omomowo et al. 2018; Doley et al. 2019; Swandi et al. 2020).

Tobacco inoculated with a mixture containing *Gigaspora fasciculatum* and *Trichoderma harzianum* effectively controlled damping-off caused by *Pythium aphanidermatum* and black shank disease caused by *Phytophthora parasitica* var. nicotianae (Sreeramulu et al. 1998). The consortium of AM fungi (*G. fasciculatum*), *T. viride*, and *Pseudomonas fluorescens* reduced the *Fusarium* wilt disease in tomato plants by 94%, and in this association high level of phosphorus uptake was facilitated by AM fungi and *T. viride* assisted in the induction of antioxidant enzymes, viz., catalase and peroxidase (Tayal 2011). AM fungi in consortia rather than employing single mycorrhizal species increased growth parameters and significantly reduced disease incidence and severity due to *F. solani* on common bean seedlings (Eke et al. 2016). Presoaked seeds of cucumber plant with antioxidant formulations along with AM fungal spores against pathogen *F. oxysporum* resulted in significant control of pathogen and enhanced plant growth (Elwakil et al. 2013). Co-inoculation of AM fungi and *Trichoderma* spp. significantly reduced severity and incidence of powdery mildew disease in cowpea plant along with a significant plant growth (Omomowo et al. 2018). AM fungi was found very effective to control stem rot disease caused by *Sclerotium rolfsii* in the peanut plants along with increased salicylic acid content in the plants (Swandi et al. 2020). Currently, many companies are commercially producing mycorrhizal inoculants singly or in combinations with other biocontrol species with several constraints for actual field application (Gianinazzi and Vosátka 2004; Doley et al. 2019). Identifying an effective combination of mycorrhizal fungi might be valuable for transposition of this knowledge not only to increase crops production but also in pathogen management.

## 13.8 Interaction with Other Plant Pathogens

Besides nematodes and fungi, some other plant pathogens including plant pathogenic bacteria, phytoplasma, and plant viruses also interact with mycorrhizal fungi on various plants. Mycorrhizal fungi interact with a diverse group of bacteria and significantly affect plant growth and yield. Damage due to *Pseudomonas syringae* on tomato is significantly reduced when the plants are endomycorrhizal (Garcia-Garrido and Ocampo 1989). *Glomus mosseae* prevented the infection of soybean plants caused by *P. syringae* (Shalaby and Hanna 1998), by suppressing the population density of the pathogen in soybean rhizosphere. Li et al. (1997) also found that *G. macrocarpum* reduced the infection caused by *P. lacrymans* in eggplant and cucumber, although no positive growth or yield effect was noted, indicating tolerance to the pathogen as a possible mode of action.

The presence of phytoplasmas and their associated diseases is an emerging threat to vegetable production which leads to severe yield losses worldwide (Kumari et al. 2019). Phytoplasmas are associated with over 600 diverse plant diseases worldwide (Bertaccini et al. 2014). The effect of AM fungi on diseases caused by phytoplasmas has been reported in several works. Lingua et al. (2002) have reported the protective effect of *Glomus mosseae* on the stolbur phytoplasma which infects tomatoes. Mycorrhizal pear trees infected with pear decline (PD) phytoplasma had significantly increased shoot length compared to non-mycorrhizal trees, but stem diameter showed no difference between treatments (García-Chapa et al. 2004). According to D'Amelio et al. (2007), AM fungi slightly reduced the number of chrysanthemum yellows phytoplasma-infected plants and extended the life span of the affected plants. A positive effect of AM fungi inoculation on aster yellows phytoplasma-infected tobacco plants was reported by Kamińska et al. (2010).

Likewise, viral diseases cause huge losses in production and quality for all major agronomical crops, representing a serious threat to global food security (Varma and Malathi 2003). Mycorrhizal fungal inoculation has been proposed as a cost-effective and sustainable solution for plant virus control (Miozzi et al. 2019). However, the effects of AMF on plant-virus interactions are surprisingly low and contradictory. The impact of the AM symbiosis on infection by viral pathogens is still largely uncertain and barely explored. Stolyarchuk et al. (2009) found that AM fungi inhibit the development plant virus infection in *Nicotiana tabacum* and *Cucumis sativus* plants but not in *Lycopersicon esculentum*. Mycorrhizal symbiosis provided some protection against tomato yellow leaf curl Sardinia virus (TYLCSV) in tomato (Maffei et al. 2014). It is hypothesized that high jasmonic acid (JA) level in mycorrhizal plants creates an unfavorable environment for TYLCSV, limiting its replication and reducing symptom severity (Cameron et al. 2013). A symbiotic AM fungus, *Glomus intraradices*, significantly stimulated the growth of potato plants infected with potato virus Y (PVY) (Thiem et al. 2014). In contrast, enhanced viral multiplication and activity in mycorrhizal plants was reported (Daft and Okusanya 1973; Nemec, 1984). Similar virus titer increments in infected non-mycorrhizal plants grown with increased concentration of soluble phosphate were observed which is speculated to be attributed to higher P levels compared to non-mycorrhizal plants (Xavier and Boyetchko 2003). The inoculation of AMF reduced vegetative development in potato plants along with PVY infection, but it increased viral activity considerably (Sipahioglu et al. 2009).

More recently, Miozzi et al. (2011) observed increased tomato spotted wilt virus (TSWV) titer in infected mycorrhizal tomato plants at 34 and 56 days post-viral inoculation (dpi), but not at 14 dpi, compared to non-mycorrhizal controls. In this way several studies revealed increased virus multiplication and/or symptom severity in infected mycorrhizal plants from Solanaceae, Rosaceae, and Poaceae families. Such detrimental effect of mycorrhizal fungi on plant virus infection is termed as mycorrhiza-induced susceptibility (MIS) (Miozzi et al. 2019). The MIS outcome was also confirmed by Jabaji-Hare (1984) and Shaul et al. (1999), respectively, in TMV-infected tomato and tobacco plants colonized by *Glomus* spp. According to Shaul et al. (1999), MIS is not only associated with improved phosphorus nutrition but linked the increased plant susceptibility to viral infection with the delay in pathogenesis-related (PR) protein induction in mycorrhizal plants.

Most studies on the plant-virus-mycorrhizal interaction are focused on the effect of mycorrhiza on virus infection; however, viral infection can impact mycorrhization as well. The number of fungal spores and the percentage of mycorrhization were generally found higher in non-infected plants than infected ones (Nemec 1984). Information on the effect of the mycorrhizal symbiosis on viral infection as a biocontrol agent in agricultural environments is limited and not conclusive. More studies on different mycorrhizal species and even mycorrhizal consortia for detecting possible synergistic/antagonistic effects should be considered.

## 13.9 Interactions for Improving Soil Quality

Miller and Jastrow (2000) have demonstrated that the interaction of rhizospheric microorganisms contributes to the formation and stabilization of soil aggregate. Stable microaggregates of soil (2–20 μm in diameter) are formed due to binding of soil particles together with hyphae of AM and saprophytic fungi and with bacterial products. Then, they unite with bacterial polysaccharides to form larger microaggregates (20–250 μm in diameter). Thus, formed microaggregates incorporate to form macroaggregates (>250 μm in diameter) with bacterial polysaccharide as the binding agent. The size of microaggregates also increases due to AMF mycelia. The function of AMF is determined by the size of inoculums, branching habits, and three-dimensional structure of the external mycelium colonizing the rhizospheric soil (Miller and Jastrow 2000). Water-stable soil aggregates are formed due to the microbial cooperation of AMF with other rhizospheric microorganisms in different ecological situations (Bethlenfalvay 1992; Andrade et al. 1995, 1998; Bethlenfalvay et al. 1999; Requena et al. 2001). Glomalin, a glycoprotein produced from AMF hyphae, is mainly involved for the formation of such aggregates. It is glue-like hydrophobic in nature and has a major role in the binding of soil particles (Wright and Upadhyaya 1998) and hence participates in initiation and stabilization of soil aggregates (Miller and Jastrow 2000).

## 13.10 Interaction for Phyllospheric Protection

The part of plants above the ground level mainly around the leaves is called phyllosphere. Several researches showed how rhizospheric interaction can affect phyllospheric interaction with herbivores and carnivores (Philippot et al. 2013). Some microorganisms like *Pseudomonas*, *Bacillus*, and AMF are capable to evoke a systemic resistance response in plants which is detrimental to numerous disease-causing microorganisms (Zamioudis and Pieterse 2012; Philippot et al. 2013). A research about association between the plants through AMF common mycelial system indicates that transfer of fungal disease protective signals (volatile organic compounds) possibly occurs resulting in high level of resistance to disease, protective enzyme response, and expression of genes that encodes defensive proteins in healthy tomato plants (*Lycopersicon esculentum* Mill) linked to disease plants (*Alternaria solani*) with leaf early blight (Babikova et al. 2013). Many non-pathogenic rhizobacteria evoke defense reaction upon attack by pathogens by inducing systemic resistance (regulated by the host plant hormones). The possibility of AMF facilitating plant defense to herbivores is due to the common mycelial networks that could facilitate signaling compound's availability (Philippot et al. 2013). Similar effects on insects' enemies are also possible as the insects and their enemies react differently to volatile organic compounds (VOCs) released by plants. Herbivores like aphids use VOCs as signals for detecting host plants, but when they are attacked, the composition of subsequent VOCs liberated would change to repel other herbivores as well as attract parasitoid wasps (enemies of herbivores) which destroy the insects (De la Peña and Pueyo 2012; Dwivedi et al. 2015). The attacked plants release VOCs always systemically and transferred aerially between plants and transported to the rhizosphere through roots. Common AMF mycelial network then enhances

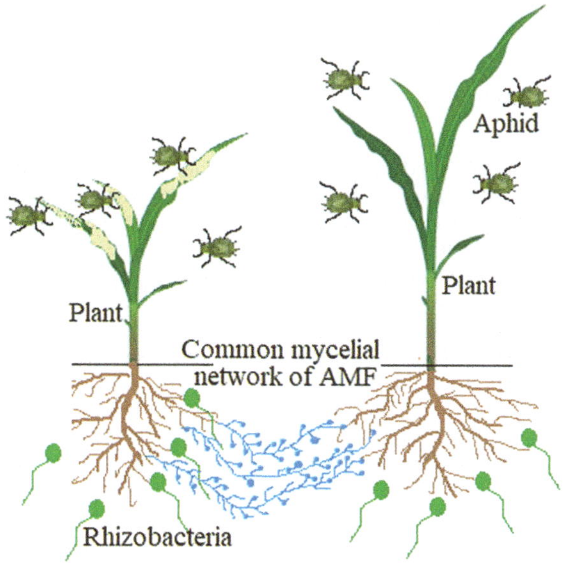


**Fig. 13.2** Rhizospheric interaction for phyllospheric protection through AMF common mycelial network. The attack of aphids on one plant (left) caused to release volatile organic compound through common mycelial system to prevent their attack on another plant (right)

transfer of signaling compounds from herbivore-attacked plants to herbivore-un-attacked plants leading to protection from insects (Fig. 13.2). The use of expensive and less eco-friendly conventional chemical insecticides as microbially induced plant protection against herbivore attack that guarantees to enhance the aboveground biomass can be minimized. Due to this, rhizospheric interaction of AMF functions in plant defense immunity for enhancing food security has clear understanding on a practical view (Babikova et al. 2013).

## 13.11 Conclusion

Mycorrhizal fungi are key components of sustainable plant-soil ecosystems. They interact with a wide array of rhizospheric microbes. These interactions are crucial in low-input cropping systems that rely on biological processes rather than agrochemicals to maintain soil fertility and plant health. Knowledge of such interactions and identification of an effective combination of mycorrhiza are useful in the development of mycorrhizal inoculants with bioprotectant and growth-enhancing potentials. Furthermore, interactions with mycorrhizal fungi and the underlying mechanisms involved will be helpful in the development of strategies to employ them as efficient biocontrol agents as well as crop improvers. Although there are a substantial number of studies on interactions between mycorrhizal fungi and rhizospheric microbes, the underlying mechanisms of these associations are not very well understood, and the proposed mechanisms still need further experimental confirmation. More insight into these mechanisms will enable optimization of the effective use of mycorrhizal fungi in combination with other rhizospheric microbes as a tool for increasing crop yields. In addition, more detailed and well-planned studies also need to address existing problems of inconsistent and unreliable results.

## References

Agrios GN (2005) Plant pathology, 5th edn. Elsevier-Academic, San Diego

Akhtar MS, Siddiqui ZA (2008a) Arbuscular mycorrhizal fungi as potential bioprotectants against plant pathogens. In: Siddiqui ZA, Akhtar S, Futai K (eds) Mycorrhizae: sustainable agriculture and forestry. Springer, pp 61–97. https://doi.org/10.1007/978-1-4020-8770-7_3

Akhtar MS, Siddiqui ZA (2008b) Biocontrol of a root-rot disease complex of chickpea by *Glomus intraradices*, *Rhizobium* spp. and *Pseudomonas straita*. Crop Prot 27(3–5):410–417. https://doi.org/10.1016/j.cropro.2007.07.009

Akinrinlola RJ, Yuen GY, Drijber RA, Adesemoye AO (2018) Evaluation of *Bacillus* strains for plant growth promotion and predictability of efficacy by in vitro physiological traits. Int J Microbiol. https://doi.org/10.1155/2018/5686874

Alarcón A, Davies FT, Egilla JN, Fox TC, Estrada-Luna AA, Ferrera-Cerrato R (2002) Short term effects of *Glomus claroideum* and *Azospirillum brasilense* on growth and root acid phosphatase activity of Carica papaya L. under phosphorus stress. Rev Latinoam Microbiol 44(1):31–37

Aliasgharzad N, Neyshabouri MR, Salimi G (2006) Effects of arbuscular mycorrhizal fungi and *Bradyrhizobium japonicum* on drought stress of soybean. Biologia 61(19):S324–S328. https://doi.org/10.2478/s11756-006-0182-x

Alvarado-Herrejón M, Larsen J, Gavito ME, Jaramillo-López PF, Vestberg M, Martínez-Trujillo M et al (2019) Relation between arbuscular mycorrhizal fungi, root-lesion nematodes and soil characteristics in maize agroecosystems. Appl Soil Ecol 135:1–8. https://doi.org/10.1016/j.apsoil.2018.10.019

Andrade G, Azcón R, Bethlenfalvay GJ (1995) A rhizobacterium modifies plant and soil responses to the mycorrhizal fungus *Glomus mosseae*. Appl Soil Ecol 2:195–202. https://doi.org/10.1016/0929-1393(95)00054-o

Andrade G, Mihara KL, Linderman RG, Bethlenfalvay GJ (1998) Soil aggregation status and rhizobacteria in the mycorrhizosphere. Plant Soil 202:89–96. https://doi.org/10.1023/A:1004301423150

André S, Neyra M, Duponnois R (2003) Arbuscular mycorrhizal symbiosis changes the colonization pattern of *Acacia tortilis* spp. Raddiana rhizosphere by two strains of *Rhizobia*. Microb Ecol 45(2):137–144. https://doi.org/10.1007/s00248-002-1022-3

Ashrafi E, Zahedi M, Razmjoo J (2014) Co-inoculations of arbuscular mycorrhizal fungi and *Rhizobia* under salinity in alfalfa. Soil Sci Plant Nutr 60(5):619–629. https://doi.org/10.1080/00380768.2014.936037

Azcón R, El-Atrash F (1997) Influence of arbuscular mycorrhizae and phosphorus fertilization on growth, nodulation and $N_2$ fixation ($^{15}N$) in *Medicago sativa* at four salinity levels. Biol Fertil Soils 24(1):81–86. https://doi.org/10.1007/s003740050212

Azcón R, Rubio R, Barea JM (1991) Selective interactions between different species of mycorrhizal fungi and *Rhizobium meliloti* strains, and their effects on growth, $N_2$ fixation ($^{15}N$) and nutrition of *Medicago sativa* L. New Phytol 117(3):399–404. https://doi.org/10.1111/j.1469-8137.1991.tb00003.x

Azcón-Aguilar C, Barea JM (1997) Arbuscular mycorrhizas and biological control of soil-borne plant pathogens - an overview of the mechanisms involved. Mycorrhiza 6(6):457–464. https://doi.org/10.1007/s005720050147

Azcón-Aguilar C, Jaizme-Vega MC, Calvet C (2002) The contribution of arbuscular mycorrhizal fungi to the control of soil-borne plant pathogens. Mycorrhizal Technol Agric 187–197. https://doi.org/10.1007/978-3-0348-8117-3_15

Babikova Z, Johnson D, Bruce T, Pickett J, Gilbert L (2013) How rapid is aphid-induced signal transfer between plants via common mycelial networks? Commun Integr Biol 6(6):1–3. https://doi.org/10.4161/cib.25904

Badji S, Ducousso M, Thoen D, Colonna JP (1989) Effects of double inoculation with *Rhizobium-Glomus mosseae* on the nodulation and growth of young *Acacia laetea*. In: Tress for development in sub-Saharan Africa. The International Foundation of Science (IFS), ICRAF House, Nairobi, pp 323–329

Banuelos J, Alarcon A, Larsen J, Trejo D (2014) Interactions between arbuscular mycorrhizal fungi and *Meloidogyne incognita* in the ornamental plant *Impatiens balsamina*. J Soil Sci Plant Nutr 14(1):63–74. https://doi.org/10.4067/S0718-95162014005000005

Baral K, Giri A, Shah PK, Kemmelmeier K, Stürmer SL, Gyawali S, Raut JK (2021) Diversity of arbuscular mycorrhizal fungi (Glomeromycota) in adjacent areas of different land use in Nepal. GSC Biological and Pharmaceutical Sciences 15(01):141–150. https://doi.org/10.30574/gscbps.2021.15.1.0098

Barea JM, Azcon-Aguilar C (1982) Production of plant growth-regulating substances by the vesicular-Arbuscular Mycorrhizal fungus *Glomus mosseae*. Appl Environ Microbiol 43(4):810–813

Barea JM, Azcón-Aguilar C (1983) Mycorrhizas and their significance in nodulating nitrogen- fixing plants. In: Brady N (ed) Advances in agronomy. Academic, New York, pp 1–54. https://doi.org/10.1016/S0065-2113(08)60351-X

Barea JM, Azcon R, Hayman DS (1975) Possible synergistic interactions between Endogyne and phosphate-solubilizing bacteria in low phosphate soils. In: Sanders FE, Mose B, Tinker PB (eds) Endomycorrhizas. Academic Press, London, pp 409–417

Barea JM, Pozo MJ, Azcon R, Azcon-Aguilar (2005) Microbial co-operation in the rhizosphere. J Exp Bot 56(417): 1761–1778. doi: https://doi.org/10.1093/jxb/eri197

Barea JM, Azcón R, Azcón-Aguilar C (2017) Mycorrhizosphere interactions to improve a sustainable production of legumes. In: Zaidi A, Khan MS, Musarrat J (eds) Microbes for legume improvement, 2nd edn. Springer, Cham, pp 199–225. https://doi.org/10.1007/978-3-319-59174-2

Basu A, Prasad P, Das SN, Kalam S, Sayyed RZ, Reddy MS, El Enshasy H (2021) Plant growth promoting Rhizobacteria (PGPR) as green bioinoculants: recent developments, constraints, and prospects. Sustainability 13(3):1140. https://doi.org/10.3390/su13031140

Becker DM, Bagley ST, Podila GK (1999) Effects of mycorrhizal-associated streptomycetes on growth of *Laccaria bicolor*, *Cenococcum geophilum*, and *Armillaria* species and on gene expression in *Laccaria bicolor*. Mycologia 91(1):33–40. https://doi.org/10.1080/00275514.1999.12060991

Bending GD, Aspray TJ, Whipps JM (2006) Significance of microbial interactions in the Mycorrhizosphere. Adv Appl Microbiol 60:97–132. https://doi.org/10.1016/S0065-2164(06)60004-X

Berg G (2009) Plant-microbe interactions promoting plant growth and health: perspectives for controlled use of microorganisms in agriculture. Appl Microbiol Biotechnol 84(1):11–18. https://doi.org/10.1007/s00253-009-2092-7

Bertaccini A, Duduk B, Paltrinieri S, Contaldo N (2014) Phytoplasmas and Phytoplasma diseases: a severe threat to agriculture. Am J Plant Sci 05(12):1763–1788. https://doi.org/10.4236/ajps.2014.512191

Bethlenfalvay GJ (1992) Mycorrhizal fungi in nitrogen-fixing legumes: problems and prospects. Method Microbiol 24:375–389. https://doi.org/10.1016/S0580-9517(08)70103-X

Bethlenfalvay GJ, Cantrell IC, Mihara KL, Schreiner RP (1999) Relationships between soil aggregation and mycorrhizae as influenced by soil biota and nitrogen nutrition. Biol Fertil Soils 28:356–363. https://doi.org/10.1007/s003740050504

Bhatia NP, Adholeya A, Sharma A (1998) Biomass production and changes in soil productivity during long term cultivation of *Prosopis juliflora* (Swartz) DC inoculated with VA mycorrhiza and *Rhizobium* spp. in a semi-arid wasteland. Biol Fertil Soils 26(3):208–214. https://doi.org/10.1007/s003740050369

Bhattacharyya PN, Jha DK (2012) Plant growth-promoting rhizobacteria (PGPR): emergence in agriculture. World J Microbiol Biotechnol 28(4):1327–1350. https://doi.org/10.1007/s11274-011-0979-9

Birhane E, Sterck FJ, Fetene M, Bongers F, Kuyper TW (2012) Arbuscular mycorrhizal fungi enhance photosynthesis, water use efficiency, and growth of frankincense seedlings under pulsed water availability conditions. Oecologia 169(4):895–904. https://doi.org/10.1007/s00442-012-2258-3

Bødker L, Kjøller R, Kristensen K, Rosendahl S (2002) Interactions between indigenous arbuscular mycorrhizal fungi and *Aphanomyces euteiches* in field-grown pea. Mycorrhiza 12(1):7–12. https://doi.org/10.1007/s00572-001-0139-4

Bravo A, Brands M, Wewer V, Dormann P, Harrison MJ (2017) Arbuscular mycorrhiza-specific enzymes FatM and RAM2 fine-tune lipid biosynthesis to promote development of arbuscular mycorrhiza. New Phytol 214(4):1631–1645. https://doi.org/10.1111/nph.14533

Brown ME, Carr GR (1984) Interactions between *Azotobacter chroococcum* and vesicular arbuscular mycorrhiza and their effects on plant growth. J Appl Bacteriol 56(3):429–437. https://doi.org/10.1111/j.1365-2672.1984.tb01371.x

Brundrett MC (2002) Coevolution of roots and mycorrhizas of land plants. New Phytol 154(2):275–304. https://doi.org/10.1046/j.1469-8137.2002.00397.x

Calvet C, Pinochet J, Hernández-Dorrego A, Estaún V, Camprubí A (2001) Field microplot performance of the peach-almond hybrid GF-677 after inoculation with arbuscular mycorrhizal fungi in a replant soil infested with root-knot nematodes. Mycorrhiza 10:295–300. https://doi.org/10.1007/pl00009998

Cameron DD, Neil AL, Wees S, Ton J (2013) Mycorrhiza-induced resistance: more than the sum of its parts? Trends Plant Sci 18(10):539–545. https://doi.org/10.1016/j.tplants.2013.06.004

Caron M (1989) Potential use of mycorrhizae in control of soil-borne diseases. Can J Plant Pathol 11(2):1–179. https://doi.org/10.1080/07060668909501135

Choudhary DK, Varma A, Tuteja N (2016) Plant-microbe interaction: an approach to sustainable agriculture. Springer, Singapore. https://doi.org/10.1007/978-981-10-2854-0_4

Corbera J, Hernandez A (1997) Assessment of the effects of the *Rhizobia*-vesicular arbuscular mycorrhiza association on growth and development of soybeans (*Glycine max*. L. Merrill). Cultivos-Tropicales 18:10–12

D'Amelio R, Massa N, Gamalero E, Dagostino G, Sampo S, Berta G, Faoro F, Iriti M, Bosco D, Marzachi C (2007) Preliminary results on the evaluation of the effects of elicitors of plant resistance on chrysanthemum yellows phytoplasma infection. Bull Insectology 60(2):317–318

Daft MJ, Okusanya BO (1973) Effect of Endogone mycorrhiza on plant growth. V. Influence of infection on the multiplication of viruses in tomato, petunia and strawberry. New Phytol 72:975–983

Daniels-Hylton KDM, Ahmad MH (1994) Inoculation response in kidney beans (*Phaseolus vulgaris*, L.) to vesicular-arbuscular mycorrhizal fungi and *Rhizobia* in non-sterilized soil. Biol Fertil Soil 18(2):95–98. https://doi.org/10.1007/BF00336452

Davis RM, Menge JA (1980) Influence of *Glomus fasciculatus* and soil phosphorus on Phytophthora root rot of citrus. Phytopathology. https://doi.org/10.1094/phyto-70-447

De la Peña TC, Pueyo JJ (2012) Legumes in the reclamation of marginal soils, from cultivar and inoculant selection to transgenic approaches. Agron Sustain Dev 32(1):65–91. https://doi.org/10.1007/s13593-011-0024-2

De Novais CB, Sbrana C, da Conceição JE, Rouws LFM, Giovannetti M, Avio L, Siqueira JO, Saggin Júnior OJ, da Silva EMR, de Faria SM (2020) Mycorrhizal networks facilitate the colonization of legume roots by a symbiotic nitrogen-fixing bacterium. Mycorrhiza 30(2–3):389–396. https://doi.org/10.1007/s00572-020-00948-w

Dehne HW (1982) Interaction between vesicular-arbuscular mycorrhizal fungi and plant pathogens. Phytopathology 72(8):1115–1119

De-la-Peña C, Loyola-Vargas VM (2014) Biotic interactions in the rhizosphere: a diverse cooperative enterprise for plant productivity. Plant Physiol 166(2):701–719. https://doi.org/10.1104/pp.114.241810

Dixon RK, Garg VK, Rao MV (1993) Inoculation of *Leucaena* and *Prosopis* seedlings with *glomus* and *rhizobium* species in saline soil: rhizosphere relations and seedling growth. Arid Soil Res Rehab 7(2):133–144. https://doi.org/10.1080/15324989309381343

Doley K, Borde M, Kulkarni M (2019) AM fungi and *Trichoderma* interaction for biological control of Soilborne plant pathogen *Fusarium oxysporum*. In: Varma A, Tripathi S, Prasad R (eds) Plant microbe interface. Springer, Cham, pp 95–128. https://doi.org/10.1007/978-3-030-19831-2_4

Duchesne LC, Peterson RL, Ellis BE (1989) The time course of disease suppression and antibiosis by the ectomycorrhizal fungus *Paxillus involutus*. New Phytol 111(4):693–698. https://doi.org/10.1111/j.1469-8137.1989.tb02364.x

Dwivedi SL, Sahrawat KL, Upadhyaya HD (2015) Chapter one-advances in host plant and *rhizobium* genomics to enhance symbiotic nitrogen fixation in grain legumes. Adv Agron 129:1–116. https://doi.org/10.1016/bs.agron.2014.09.001

Eke P, Chatue G, Wakam LN, Kouipou RMT, Fokou PVT, Boyom FF (2016) Mycorrhiza consortia suppress the fusarium root rot (*Fusarium solani* f. spp. *Phaseoli*) in common bean (*Phaseolus vulgaris* L.). Biol Control 103:240–250. https://doi.org/10.1016/j.biocontrol.2016.10.001

El-Shanshoury AR, Hassan MA, Abdel-Ghaffar BA (1989) Synergistic effects of vesicular-arbuscular-mycorrhizas and *Azotobacter chroococcum* on the growth and the nutrient contents of tomato plants. Phyton (Horn) 29(2):203–212

Elwakil MA, Baka ZA, Soliman HM, Sadek MS (2013) A modern tactic for reducing the biotic stress on cucumber plants caused by *Fusarium oxysporum*. Plant Pathol J 12:26–31

Emery SM, Reid ML, Bell-Dereske L, Gross KL (2017) Soil mycorrhizal and nematode diversity vary in response to bioenergy crop identity and fertilization. GCB Bioenergy 9:1644–1656. https://doi.org/10.1111/gcbb.12460

Fazeli-Nasab B, Sayyed RZ (2019) Plant growth promoting Rhizobacteria and salinity stress: a journey into the soil. In: Sayyed, Arora, Reddy (eds) Plant growth promoting Rhizobacteria for sustainable stress management Vol 1 abiotic stress management. Springer, Singapore, pp 21–34

Ferraz L, Brown D (2002) An introduction to nematodes - plant nematology. Pensoft, Sofia. https://doi.org/10.1163/156854105774355536

Ferreira BS, Santana MV, Macedo RS, Silva JO, Carneiro MA, Rocha MR (2018) Co-occurrence patterns between plant-parasitic nematodes and arbuscular mycorrhizal fungi are driven by environmental factors. Agric Ecosyst Environ 265:54–61. https://doi.org/10.1016/j.agee.2018.05.020

Fitter AH, Garbaye J (1994) Interactions between mycorrhizal fungi and other soil organisms. Plant Soil 159(1):123–132. https://doi.org/10.1007/BF00000101

Franco AA, De Faria SM (1997) The contribution of $N_2$-fixing tree legumes to land reclamation and sustainability in the tropics. Soil Biol Biochem 29(5–6):897–903. https://doi.org/10.1016/S0038-0717(96)00229-5

Gao X, Lu X, Wu M, Zhang H, Pan R, Tian J, Li S, Liao H (2012) Co-inoculation with *Rhizobia* and AMF inhibited soybean red crown rot: from field study to plant defense-related gene expression analysis. PLoS One 7(3):1–10. https://doi.org/10.1371/journal.pone.0033977

García-Chapa M, Batlle A, Laviña A, Camprubí A, Estaún V, Calvet C (2004) Tolerance increase to pear decline phytoplasma in mycorrhizal OHF-333 pear rootstock. Acta Hortic 657:437–441. https://doi.org/10.17660/ActaHortic.2004.657.70

Garcia-Garrido JM, Ocampo JA (1989) Effect of VA mycorrhizal infection of tomato on damage caused by *pseudomonas syringae*. Soil Biol Biochem 21(1):165–167. https://doi.org/10.1016/0038-0717(89)90027-8

Garmendia I, Goicoechea N, Aguirreolea J (2004) Effectiveness of three *Glomus* species in protecting pepper (*Capsicum annuum* L.) against verticillium wilt. Biol Control 31(3):296–305. https://doi.org/10.1016/j.biocontrol.2004.04.015

Gianinazzi S, Vosátka M (2004) Inoculum of arbuscular mycorrhizal fungi for production systems: science meets business. Can J Bot 82(8):1264–1271. https://doi.org/10.1139/B04-072

Glick BR, Todorovic B, Czarny J, Cheng Z, Duan J, McConkey B (2007) Promotion of plant growth by bacterial ACC deaminase. CRC Crit Rev Plant Sci 26(5–6):227–242. https://doi.org/10.1080/07352680701572966

Goicoechea N, Szalai G, Antolin MC, Sanchez-Diaz M, Paldi E (1998) Influence of arbuscular mycorrhizae and *Rhizobium* on free polyamines and proline levels in water-stressed alfalfa. J Plant Physiol 153(5–6):706–711. https://doi.org/10.1016/S0176-1617(98)80224-1

Hart MM, Antunes PM, Chaudhary VB, Abbott LK (2017) Fungal inoculants in the field: is the reward greater than the risk? Funct Ecol 32(1):126–135. https://doi.org/10.1111/1365-2435.12976

Hashem A, Allah EF, Alqarawi AA, Al-Huqail AA, Wirth S, Egamberdieva D (2016) The interaction between arbuscular mycorrhizal fungi and endophytic bacteria enhances plant growth of *Acacia gerrardii* under salt stress. Front Microbiol 7:1–15. https://doi.org/10.3389/fmicb.2016.01089

Havugimana E, Bhople BS, Byiringiro E, Mugabo JP (2016) Role of dual inoculation of *rhizobium* and arbuscular mycorrhizal (AM) fungi on pulse crops production. Walailak J Sci Technol 13(1):1–7. https://doi.org/10.14456/vol13iss4pp

Hayat R, Ali S, Amara U, Khalid R, Ahmed I (2010) Soil beneficial bacteria and their role in plant growth promotion: a review. Ann Microbiol 60(4):579–598. https://doi.org/10.1007/s13213-010-0117-1

Henkes GJ, Kandeler E, Marhan S, Scheu S, Bonkowski M (2018) Interactions of mycorrhiza and protists in the rhizosphere systemically alter microbial community composition, plant shoot-to-root ratio and within-root system nitrogen allocation. Front Environ Sci. https://doi.org/10.3389/fenvs.2018.00117

Herridge DF, Peoples MB, Boddey RM (2008) Global inputs of biological nitrogen fixation in agricultural systems. Plant Soil 311(1–2):1–18. https://doi.org/10.1007/s11104-008-9668-3

Hong Duc NN, Posta K (2018) Mycorrhiza-induced alleviation of plant disease caused by *Clavibacter michiganensis* subsp. *michiganensis* and role of ethylene in mycorrhiza-induced resistance in tomato. Acta Biol Hung 69(2):170–181. https://doi.org/10.1556/018.69.2018.2.6

Ibiang YB, Mitsumoto H, Sakamoto K (2017) *Bradyrhizobia* and arbuscular mycorrhizal fungi modulate manganese, iron, phosphorus, and polyphenols in soybean (*Glycine max* (L.) Merr.) under excess zinc. Environ Exp Bot 137:1–13. https://doi.org/10.1016/j.envexpbot.2017.01.011

Ishac YZ, Angle JS, EL-Borollosy ME, EL-Demerdas ME, Mostafa MI, Fares CN (1994) Growth of *Vicia faba* as affected by inoculation with vesicular-arbuscular mycorrhizae and *rhizobium leguminosarum* bv. *viceae* in two soils. Biol Fertil Soils 17(1):27–31. https://doi.org/10.1007/BF00418668

Jabaji-Hare SH (1984) Electron microscopic examination of tomato roots coinfected with *Glomus* spp. and tobacco mosaic virus. Phytopathology 74(3):277. https://doi.org/10.1094/phyto-74-277

Jambon I, Thijs S, Weyens N, Vangronsveld J (2018) Harnessing plant-bacteria-fungi interactions to improve plant growth and degradation of organic pollutants. J Plant Interact 13(1):119–130. https://doi.org/10.1080/17429145.2018.1441450

Johansson JF, Paul LR, Finlay RD (2004) Microbial interactions in the mycorrhizosphere and their significance for sustainable agriculture. FEMS Microbiol Ecol 48(1):1–13. https://doi.org/10.1016/j.femsec.2003.11.012

Kamińska M, Klamkowski K, Berniaka H, Treder W (2010) Effect of arbuscular mycorrhizal fungi inoculation on aster yellows phytoplasma-infected tobacco plants. Sci Hortic 125(3):500–503. https://doi.org/10.1016/j.scienta.2010.04.027

Kannepalli A, Davranov K, Narimanov A, Enakiev Y, Syed A, Elgorban AM, Bahkali AH, Wirth S, Sayyed RZ, Gafur A (2021) Co-inoculation of rhizobacteria promotes growth, yield, and nutrient contents in soybean and improves soil enzymes and nutrients under drought conditions. Sci Rep 11:22081. https://doi.org/10.1038/s41598-021-01337-9

Kapadia C, Sayyed RZ, Enshasy HEE, Vaidya H, Sharma D, Patel V, Malek RA, Syed A, Elgorban AM, Ahmad K, Zuan ATK (2021) Halotolerant microbial consortia for sustainable mitigation of salinity stress, growth promotion, and mineral uptake in tomato plant and soil nutrient enrichment. Sustainability 13(15):8369. https://doi.org/10.3390/su13158369

Kasiamdari RS, Smith SE, Smith FA, Scott ES (2002) Costs of secondary parasitism in the facultative hyperparasitoid Pachycrepoideus dubius: does host size matter? Plant Soil 238:235–244. doi: 10.1023/A

Khan A, Sayyed RZ (2019) Sonia Seifi Rhizobacteria : legendary soil guards in abiotic stress management. In: Sayyed, Arora, Reddy (eds) Plant growth promoting Rhizobacteria for sustainable stress management Vol 1 abiotic stress management. Springer, Singapore, pp 27–342

Khan N, Ali S, Shahi MA, Mustafa A, Sayyed RZ, Curaá JA (2021) Insights into the interactions among roots, rhizosphere and rhizobacteria for improving plant growth and tolerance to abiotic stresses: a review. Cell 10(6):1551. https://doi.org/10.3390/cells10061551

Kour D, Sayyed RZ (2019) Drought tolerant phosphorus solubilizing microbes: biodiversity and biotechnological applications for alleviation of drought stress in plant. In: Sayyed, Arora, Reddy (eds) Plant growth promoting Rhizobacteria for sustainable stress management Vol 1 abiotic stress management. Springer, Singapore, pp 255–308

Kumari S, Nagendran K, Rai AB, Singh B, Rao GB, Bertaccini A (2019) Global status of phytoplasma diseases in vegetable crops. Front Microbiol 10:1–15. https://doi.org/10.3389/fmicb.2019.01349

Kuzyakov Y, Xu X (2013) Competition between roots and microorganisms for nitrogen: mechanisms and ecological relevance. New Phytol 198(3):656–669. https://doi.org/10.1111/nph.12235

Lambers H, Mougel C, Jaillard B, Hinsinger P (2009) Plant-microbe-soil interactions in the rhizosphere: an evolutionary perspective. Plant Soil 321(1–2):83–115. https://doi.org/10.1007/s11104-009-0042-x

Larsen J, Jaramillo-López P, Nájera-Rincon M, Gonzaléz-Esquivel CE (2015) Biotic interactions in the rhizosphere in relation to plant and soil nutrient dynamics. J Soil Sci Plant Nutr 15(2):449–463. https://doi.org/10.4067/s0718-95162015005000039

Li SL, Zhao SJ, Zhao LZ, Li SL, Zhao SJ, Zhao LZ (1997) Effects of VA mycorrhizae on the growth of eggplant and cucumber and control of diseases. Acta Phytophylac Sin 24:117–120

Lingua G, D'Agostino G, Massa N, Antosiano M, Berta G (2002) Mycorrhiza-induced differential response to a yellows disease in tomato. Mycorrhiza 12(4):191–198. https://doi.org/10.1007/s00572-002-0171-z

Lu X, Koide RT (1994) The effects of mycorrhizal infection on components of plant growth and reproduction. New Phytol 128(2):211–218. https://doi.org/10.1111/j.1469-8137.1994.tb04004.x

Maffei G, Miozzi L, Fiorilli V, Novero M, Lanfranco L, Accotto GP (2014) The arbuscular mycorrhizal symbiosis attenuates symptom severity and reduces virus concentration in tomato infected by tomato yellow leaf curl Sardinia virus (TYLCSV). Mycorrhiza 24(3):179–186. https://doi.org/10.1007/s00572-013-0527-6

Mamatha G, Bagyaraj DJ, Jaganath S (2002) Inoculation of field-established mulberry and papaya with arbuscular mycorrhizal fungi and a mycorrhiza helper bacterium. Mycorrhiza 12(6):313–316. https://doi.org/10.1007/s00572-002-0200-y

Manasa M, Ravinder P, Gopalakrishnan S, Srinivas V, Sayyed RZ, Enshasy HE, Yahayu M, Zuan ATK, Kassem HS, Hameeda B (2021) Co-inoculation of bacillus spp. for growth promotion and iron fortification in sorghum. Sustainability 13(21):12091. https://doi.org/10.3390/su132112091

Manoharachary C, Mukerji KG (2006) In: Mukerji KG, Manoharachary J, Singh C (eds) Rhizosphere biology – an overview, microbial activity in the rhizosphere. Springer-Verlag, Berlin Heidelberg. https://doi.org/10.1007/3-540-29420-1_1

Marro N, Caccia M, Doucet ME, Cabello M, Becerra A, Lax P (2018) Mycorrhizas reduce tomato root penetration by false root-knot nematode *Nacobbus aberrans*. Appl Soil Ecol 124:262–265. https://doi.org/10.1016/j.apsoil.2017.11.011

Mendes R, Garbeva P, Raaijmakers JM (2013) The rhizosphere microbiome: significance of plant beneficial, plant pathogenic, and human pathogenic microorganisms. FEMS Microbiol Rev 37(5):634–663. https://doi.org/10.1111/1574-6976.12028

Mikkelsen BL, Rosendahl S, Jakobsen I (2008) Underground resource allocation between individual networks of mycorrhizal fungi. New Phytol 180(4):890–898. https://doi.org/10.1111/j.1469-8137.2008.02623.x

Miller RM, Jastrow JD (2000) Mycorrhizal fungi influence soil structure. In: Kapulnik Y, Douds DD (eds) Arbuscular mycorrhizas: physiology and functions. Kluwer Academic Publishers, Dordrecht, pp 3–18

Miozzi L, Catoni M, Fiorilli V, Mullineaux PM, Accotto GP, Lanfranco L (2011) Arbuscular mycorrhizal symbiosis limits foliar transcriptional responses to viral infection and favors long-term virus accumulation. Mol Plant-Microbe Interact 24(12):1562–1572. https://doi.org/10.1094/MPMI-05-11-0116

Miozzi L, Vaira AM, Catoni M, Fiorilli V, Accotto GP, Lanfranco L (2019) Arbuscular mycorrhizal symbiosis: plant friend or foe in the fight against viruses? Front Microbiol. https://doi.org/10.3389/fmicb.2019.01238

Morgan JAW, Bending GD, White PJ (2005) Biological costs and benefits to plant-microbe interactions in the rhizosphere. J Exp Bot 56(417):1729–1739. https://doi.org/10.1093/jxb/eri205

Mukerji KG (1999) Mycorrhiza in control of plant pathogens : molecular approaches. In: Mukerji KG, Chamola BP, Upadhyay RK (eds) Biotechnological approaches in biocontrol of plant pathogens. Springer, Boston, pp 135–155. https://doi.org/10.1007/978-1-4615-4745-7_7

Musyoka DM, Njeru EM, Nyamwange MM, Maingi JM (2020) Arbuscular mycorrhizal fungi and *Bradyrhizobium* co-inoculation enhances nitrogen fixation and growth of green grams (*Vigna radiata* L.) under water stress. J Plant Nutr 43(7):1036–1047. https://doi.org/10.1080/01904167.2020.1711940

Nacoon S, Jogloy S, Riddech N, Mongkolthanaruk W, Kuyper TW, Boonlue S (2020) Interaction between phosphate solubilizing bacteria and Arbuscular Mycorrhizal fungi on growth promotion and tuber inulin content of *Helianthus tuberosus* L. Sci Rep 10(1):1–10. https://doi.org/10.1038/s41598-020-61846-x

Nadeem SM, Ahmad M, Zahir ZA, Javaid A, Ashraf M (2010) Microbial ACC-deaminase: prospects and applications for inducing salt tolerance in plants. CRC Crit Rev Plant Sci 29(6):360–393. https://doi.org/10.1080/07352689.2010.524518

Najafi V, Sohrabi Y, Sayyed RZ, Suriani NL, Datta R (2021) Effects of combinations of Rhizobacteria, mycorrhizae, and seaweeds on growth and yields in wheat cultivars under the influence of supplementary irrigation. Plan Theory 10(3):811. https://doi.org/10.3390/plants10040811

Nanjundappa A, Bagyaraj DJ, Saxena AK, Kumar M, Chakdar H (2019) Interaction between arbuscular mycorrhizal fungi and *Bacillus* spp. in soil enhancing growth of crop plants. Fungal Biol Biotechnol 6(1):1–10. https://doi.org/10.1186/s40694-019-0086-5

Nemec S (1984) Virus- *Glomus etunicatum* interactions in citrus rootstocks. Plant Dis. https://doi.org/10.1094/pd-69-311

Odelade KA (2019) Bacteria, fungi and archaea domains in Rhizospheric soil and their effects in enhancing agricultural productivity. Int J Environ Res Public Health 16:3873

Olowe OM, Olawuyia OJ, Sobowalea AA, Odebode AC (2018) Role of arbuscular mycorrhizal fungi as biocontrol agents against *Fusarium verticillioides* causing ear rot of *Zea mays* L. (Maize). Curr Plant Biol 15:30–37. https://doi.org/10.1016/j.cpb.2018.11.005

Olsson PA, Thingstrup I, Jakobsenb I, Baath E (1999) Estimation of the biomass of arbuscular mycorrhizal fungi in a linseed field. Soil Biol Biochem 31(13):1879–1887. https://doi.org/10.1016/S0038-0717(99)00119-4

Omomowo IO, Fadiji AE, Omomowo OI (2018) Assessment of bio-efficacy of *glomus versiforme* and *Trichoderma harzianum* in inhibiting powdery mildew disease and enhancing the growth of cowpea. Ann Agric Sci 63(1):9–17. https://doi.org/10.1016/j.aoas.2018.03.001

Pan J, Huang C, Peng F, Zhang W, Luo J, Ma S, Xue X (2020) Effect of arbuscular mycorrhizal fungi (AMF) and plant growth-promoting bacteria (PGPR) inoculations on *Elaeagnus angustifolia* L. in saline soil. Appl Sci 10(3):945. https://doi.org/10.3390/app10030945

Pascale A, Proietti S, Pantelides IS, Stringlis IA (2020) Modulation of the root microbiome by plant molecules: the basis for targeted disease suppression and plant growth promotion. Front Plant Sci 10:1–23. https://doi.org/10.3389/fpls.2019.01741

Paulitz TC, Linderman RG (1991) Lack of antagonism between the biocontrol agent *Gliocladium virens* and vesicular arbuscular mycorrhizal fungi. New Phytol 117(2):303–308. https://doi.org/10.1111/j.1469-8137.1991.tb04911.x

Pawlowski ML, Hartman GL (2020) Impact of arbuscular mycorrhizal species on *Heterodera glycines*. Plant Dis 104(9):2406–2410. https://doi.org/10.1094/PDIS-01-20-0102-RE

Pepe A, Giovannetti M, Sbrana C (2018) Lifespan and functionality of mycorrhizal fungal mycelium are uncoupled from host plant lifespan. Sci Rep 8(1):2–11. https://doi.org/10.1038/s41598-018-28354-5

Pérez-De-Luque A, Tille S, Johnson I, Pascual-Pardo D, Ton J, Cameron DD (2017) The interactive effects of arbuscular mycorrhiza and plant growth-promoting rhizobacteria synergistically enhance host plant defences against pathogen. Sci Rep 7(1):1–10. https://doi.org/10.1038/s41598-017-16697-4

Philippot L, Raaijmakers JM, Lemanceau P, Van Der Putten WH (2013) Going back to the roots: the microbial ecology of the rhizosphere. Nat Rev Microbiol 11(11):789–799. https://doi.org/10.1038/nrmicro3109

Poveda J, Hermosa R, Monte E, Nicolás C (2019) *Trichoderma harzianum* favours the access of arbuscular mycorrhizal fungi to non-host Brassicaceae roots and increases plant productivity. Sci Rep 9(1):1–11. https://doi.org/10.1038/s41598-019-48269-z

Raklami A, Bechtaoui N, Tahiri A, Anli M, Meddich A, Oufdou K (2019) Use of rhizobacteria and mycorrhizae consortium in the open field as a strategy for improving crop nutrition, productivity and soil fertility. Front Microbiol 10:1–11. https://doi.org/10.3389/fmicb.2019.01106

Rascovan N, Carbonetto B, Perrig D, Díaz M, Canciani W, Abalo M, Alloati J, González-Anta G, Vazquez MP (2016) Integrated analysis of root microbiomes of soybean and wheat from agricultural fields. Sci Rep 6(1150):1–12. https://doi.org/10.1038/srep28084

Raut JK (2014) Spores of native ECM fungi as inoculant for *Pinus roxburghii* seedlings. In Abstracts of the 10th International Mycological Congress, Bangkok, Thailand

Razakatiana ATE, Trap J, Baohanta RH, Raherimandimby M, Le Roux C, Duponnois R, Ramanankierana H, Becquer T (2020) Benefits of dual inoculation with arbuscular mycorrhizal fungi and *rhizobia* on *Phaseolus vulgaris* planted in a low-fertility tropical soil. Pedobiologia 83:150685. https://doi.org/10.1016/j.pedobi.2020.150685

Requena N, Jimenez I, Toro M, Barea J (1997) Interactions between plant-growth-promoting rhizobacteria (PGPR), arbuscular mycorrhizal fungi and *rhizobium* spp. in the rhizosphere of *Anthyllis cytisoides*, a model legume for revegetation in mediterranean semi-arid ecosystems. New Phytol 136(4):667–677. https://doi.org/10.1046/j.1469-8137.1997.00786.x

Requena N, Perez-Solis E, Azcón-Aguilar C, Jeffries P, Barea JM (2001) Management of indigenous plant-microbe symbioses aids restoration of desertified ecosystems. Appl Environ Microbiol 67(2):495–498. https://doi.org/10.1128/AEM.67.2.495-498.2001

Ross JP (1972) Influence of Endogone mycorrhiza on Phytophthora rot of soybean. Phytopathology 62:896–897

Ryan MH, Graham JH (2018) Little evidence that farmers should consider abundance or diversity of arbuscular mycorrhizal fungi when managing crops. New Phytol 220(4):1092–1107. https://doi.org/10.1111/nph.15308

Ryan PR, Dessaux Y, Thomashow LS, Weller DM (2009) Rhizosphere engineering and management for sustainable agriculture. Plant Soil 321(1–2):363–383. https://doi.org/10.1007/s11104-009-0001-6

Saldajeno MGB, Chandanie WA, Kubota M, Hyakumachi M (2008) Effects of interactions of arbuscular mycorrhizal fungi and beneficial saprophytic mycoflora on plant growth and disease protection. In: Siddiqui Z, Akhatar M, Futai K (eds) Mycorrhizae: sustainable agriculture and forestry. Springer, pp 211–226. https://doi.org/10.1007/978-1-4020-8770-7_9

Sasse J, Martinoia E, Northen T (2018) Feed your friends: do Plant exudates shape the root microbiome? Trends Plant Sci 23(1):25–41. https://doi.org/10.1016/j.tplants.2017.09.003

Saxena AK, Rathi SK, Tilak KVBR (1997) Differential effect of various endomycorrhizal fungi on nodulating ability of green gram by *Bradyrhizobium* spp. (Vigna) strain S24. Biol Fertil Soils 24(2):175–178. https://doi.org/10.1007/s003740050227

Saxena AK, Shende R, Grover M (2006) Interaction among beneficial microorganisms. In: Mukerji KG, Singh C, Manoharachary J (eds) Microbial activity in the rhizosphere. Springer, Berlin Heidelberg, p 356

Schouteden N, De Waele D, Panis B, Vos CM (2015) Arbuscular mycorrhizal fungi for the biocontrol of plant-parasitic nematodes: a review of the mechanisms involved. Front Microbiol 6:1–12. https://doi.org/10.3389/fmicb.2015.01280

Schreiner RP, Mihara KL, McDaniel H, Bethlenfalvay GJ (1997) Mycorrhizal fungi influence plant and soil functions and interactions. Plant Soil 188(2):199–209. https://doi.org/10.1023/A:1004271525014

Sengupta A, Choudhuri S (1995) Effect of dual inoculation of *Rhizobium* and mycorrhiza on growth response of *Sesbania grandiflora* L. in coastal saline and sand dune soil. Ind J Forestry 18:35–37

Shaikh SS, Reddy MS, Sayyed RZ (2016) Plant growth promoting rhizobacteria: an eco-friendly approach for sustainable agro-ecosystem plant soil-microbes. Springer, Cham, pp 182–201

Shalaby AM, Hanna MM (1998) Preliminary studies on interactions between VA mycorrhizal fungus *Glomus mosseae*, *Bradyrhizobium japonicum* and *Pseudomonas syringae* in soybean plants. Acta Microbiol Pol 47:385–391

Shaul O, Galili S, Volpin H, Ginzberg I, Elad Y, Chet I, Kapulnik Y (1999) Mycorrhiza-induced changes in disease severity and PR protein expression in tobacco leaves. Mol Plant-Microbe Interact 12(11):1000–1007. https://doi.org/10.1094/MPMI.1999.12.11.1000

Siddiqui ZA, Akhtar MS (2006) Biological control of root-rot disease complex of chickpea by AM fungi. Arch Phytopathol Plant Protection 39(5):389–395. https://doi.org/10.1080/03235400500320117

Siddiqui ZA, Mahmood I (1995) Some observations on the management of the wilt disease complex of pigeonpea by treatment with a vesicular arbuscular fungus and biocontrol agents for nematodes. Bioresour Technol 54(3):227–230. https://doi.org/10.1016/0960-8524(95)00135-2

Siddiqui Z, Pichtel J (2008) Mycorrhizae: an overview. In: Siddiqui Z, Akhtar S, Futai K (eds) Mycorrhizae: sustainable agriculture and forestry. Springer, pp 1–35

Singh PK, Singh M, Vyas D (2010) Biocontrol of fusarium wilt of chickpea using arbuscular mycorrhizal fungi and *Rhizobium leguminosorum* biovar. Caryologia 63(4):349–353. https://doi.org/10.1080/00087114.2010.10589745

Singh A, Kumar R, Singh D (2019) Mycorrhizal fungi as biocontrol agent for soil borne pathogens : a review. J Pharmacogn Phytochem:281–284. Available at: http://www.phytojournal.com/archives/2019/vol8issue1S/PartG/Sp-8-1-84-415.pdf

Singh V, Sharma S, Gosal SK, Choudhary R, Singh R, Adholeya A, Singh B (2020) Synergistic use of plant growth-promoting Rhizobacteria, Arbuscular Mycorrhizal fungi, and spectral properties for improving nutrient use efficiencies in wheat (*Triticum aestivum* L.). Commun Soil Sci Plant Anal 51(1):14–27. https://doi.org/10.1080/00103624.2019.1689259

Sipahioglu MH, Demir S, Usta M, Akkopru A (2009) Biological relationship of potato virus Y and Arbuscular Mycorrhizal fungus *Glomus intraradices* in potato. Pest Technol 3(3):63–66

Smith GS (1988) The role of phosphorous nutrition in interaction of vesicular-arbuscular mycorrhizal fungi wih soilborne nematodes and fungi. Phytopathology 78(3):371–374

Smith SE, Read DJ (2008) Mycorrhizal symbiosis. Academic Press, London

Sreeramulu KR, Onkarappa T, Swamy HN (1998) Biocontrol of damping off and black shank disease in tobacco nursery. Tob Res 24:1–4

Stolyarchuk M, Shevchenko P, Polischuk P, Kripka V (2009) Virus infection course in different plant species under influence, pp 70–75

Sutton JC, Sheppard BR (1976) Aggregation of sand-dune soil by endomycorrhizal fungi. Can J Bot 54(3–4):326–333. https://doi.org/10.1139/b76-030

Swandi F, Sulyanti E, Reflin D (2020) The potential of Arbuscular Mycorrhizal fungi (AMF) as biocontrol agent against stem rot diseases caused *Sclerotium rolfsii* of peanut (*Arachis hypogaea* L). JERAMI Indones J Crop Sci 2(2):65–71. https://doi.org/10.25077/jijcs.2.2.65-71.2020

Tahat MM, Sijam K, Othman R (2010) Mycorrhizal Fungi as a biocontrol agent. Plant Pathol J 9(4):198–207. https://doi.org/10.3923/ppj.2010.198.207

Tajini F, Trabelsi M, Drevon JJ (2012) Combined inoculation with *Glomus intraradices* and *Rhizobium tropici* CIAT899 increases phosphorus use efficiency for symbiotic nitrogen fixation in common bean (*Phaseolus vulgaris* L.). Saudi. J Biol Sci 19(2):157–163. https://doi.org/10.1016/j.sjbs.2011.11.003

Tang T-T, Xie M-M, Chen S-M, Zhang S-M, Wu Q-S (2019) Effects of Arbuscular Mycorrhizal fungi and *Rhizobia* on physiological activities in White clover (*Trifolium repens*). Biotechnology 18(2):49–54. https://doi.org/10.3923/biotech.2019.49.54

Tayal P, Kapoor R. Bhatnagar AK (2011) Functional synergism among *Glomus fasciculatum*, *Trichoderma viride* and *Pseudomonas fluorescens* on Fusarium wilt in tomato. J Plant Pathol 93: 745–750

Thiagarajan TR, Ahmad MH (1993) Influence of a vesicular-arbuscular mycorrhizal fungus on the competitive ability of *Bradyrhizobium* spp. for nodulation of cowpea *Vigna unguiculata*

(L.) Walp in non-sterilized soil. Biol Fertil Soils 15(4):294–296. https://doi.org/10.1007/BF00337215

Thiem D, Szmidt-Jaworska A, Baum C, Muders K, Niedojadło K, Hrynkiewicz K (2014) Interactive physiological response of potato (*Solanum tuberosum* L.) plants to fungal colonization and Potato virus Y (PVY) infection. Acta Mycol 1(1):291–303. https://doi.org/10.5586/am.2014.015

Tinker PB (1984) The role of microorganisms in mediating and facilitating the uptake of plant nutrients from soil. Plant Soil 76(1–3):77–91. https://doi.org/10.1007/BF02205569

Toro M, Azcón R, Barea JM (1997) Improvement of arbuscular mycorrhiza development by inoculation of soil with phosphate-solubilizing rhizobacteria to improve rock phosphate bioavailability (32P) and nutrient cycling. Appl Environ Microbiol 63(11):4408–4412. https://doi.org/10.1128/aem.63.11.4408-4412.1997

Toro M, Azcón R, Barea JM (1998) The use of isotopic dilution techniques to evaluate the interactive effects of *Rhizobium* genotype, mycorrhizal fungi, phosphate-solubilizing rhizobacteria and rock phosphate on nitrogen and phosphorus acquisition *Medicago sativa*. New Phytol 138(2):265–273. https://doi.org/10.1046/j.1469-8137.1998.00108.x

Valdenegro M, Barea JM, Azcón R (2001) Influence of arbuscular-mycorrhizal fungi, *Rhizobium meliloti* strains and PGPR inoculation on the growth of *Medicago arborea* used as model legume for re-vegetation and biological reactivation in a semi-arid mediterranean area. Plant Growth Regul 34(2):233–240. https://doi.org/10.1023/A:1013323529603

Van Der Heijden MGA, De BS, Luckerhoff L, Van Logtestijn RSP, Schlaeppi K (2016) A widespread plant-fungal-bacterial symbiosis promotes plant biodiversity, plant nutrition and seedling recruitment. ISME J 10(2):389–399. https://doi.org/10.1038/ismej.2015.120

Van Ooijen G, Van Den Burg HA, Cornelissen BJC, Takken FLW (2008) Structure and function of resistance proteins in solanaceous plants. Annu Rev Phytopathol 45:43–72. https://doi.org/10.1146/annurev.phyto.45.062806.094430

Varma A, Malathi VG (2003) Emerging geminivirus problems: a serious threat to crop production. Ann Appl Biol 142(2):145–164. https://doi.org/10.1111/j.1744-7348.2003.tb00240.x

Walley FL, Germida JJ (1997) Response of spring wheat (*Triticum aestivum*) to interactions between *pseudomonas* species and *Glomus clarum* NT4. Biol Fertil Soils 24(4):365–371. https://doi.org/10.1007/s003740050259

Wang X, Pan Q, Chen F, Yan X, Liao H (2011) Effects of co-inoculation with arbuscular mycorrhizal fungi and *rhizobia* on soybean growth as related to root architecture and availability of N and P. Mycorrhiza 21(3):173–181. https://doi.org/10.1007/s00572-010-0319-1

Wani KA, Manzoor J, Shuab R, Lone R (2017) Arbuscular mycorrhizal fungi as biocontrol agents for parasitic nematodes in plants. In: Varma A, Prasad R, Tuteja N (eds) Mycorrhiza-Nutrient Uptake, Biocontrol, Ecorestoration. Springer, Cham, pp 195–210. https://doi.org/10.1007/978-3-319-68867-1_10

Whipps JM (2001) Microbial interactions and biocontrol in the rhizosphere. J Exp Bot 52:487–511. https://doi.org/10.1093/jxb/52.suppl_1.487

Willis A, Rodrigues BF, Harris JC (2013) The ecology of Arbuscular Mycorrhizal fungi. CRC Crit Rev Plant Sci 32(1):1–20. https://doi.org/10.1080/07352689.2012.683375

Wright SF, Upadhyaya A (1998) A survey of soils for aggregate stability and glomalin, a glycoprotein produced by hyphae of arbuscular mycorrhizal fungi. Plant Soil 198:97–107. https://doi.org/10.1023/a:1004347701584

Xavier L, Boyetchko S (2003) Arbuscular Mycorrhizal Fungi In Plant disease control. https://doi.org/10.1201/9780203913369.ch16

Xavier LJC, Germida JJ (2003) Selective interactions between arbuscular mycorrhizal fungi and *Rhizobium leguminosarum* bv. *viceae* enhance pea yield and nutrition. Biol Fertil Soils 37(5):261–267. https://doi.org/10.1007/s00374-003-0605-6

Xie MM, Zou YN, Wu QS, Zhang ZZ, Kuča K (2020) Single or dual inoculation of arbuscular mycorrhizal fungi and *Rhizobia* regulates plant growth and nitrogen acquisition in white clover. Plant Soil Environ 66(6):287–294. https://doi.org/10.17221/234/2020-PSE

Yadav AN, Sayyed RZ (2019) Psychrotrophic microbes: biodiversity, mechanisms of adaptation and biotechnological implications in alleviation of cold stress in plant. In: Sayyed, Arora, Reddy (eds) Plant growth promoting Rhizobacteria for sustainable stress management Vol 1 abiotic stress management. Springer, Singapore, pp 219–253

Zahir ZA, Arshad M, Frankenberger WT (2003) Plant growth promoting Rhizobacteria: applications and perspectives in agriculture. Adv Agron 81(03):97–168. https://doi.org/10.1016/S0065-2113(03)81003-9

Zakaria AK, Sayyed RZ, Enshasy HE (2019) Biosynthesis of antibiotics by PGPR and their roles in biocontrol of plant diseases. In: Plant growth promoting Rhizobacteria for sustainable stress management Vol II Rhizobacteria in biotic stress management. Springer, Singapore, pp 1–36

Zamioudis C, Pieterse CM (2012) Modulation of host immunity by beneficial microbes. Mol Plant-Microbe Interact 25(2):139–150. https://doi.org/10.1094/mpmi-06-11-0179

Zhang S, Gan Y, Xu B, Xue Y (2014) The parasitic and lethal effects of *Trichoderma longibrachiatum* against *Heterodera avenae*. Biol Control 72:1–8. https://doi.org/10.1016/j.biocontrol.2014.01.009

Zhang S, Gan Y, Ji W, Xu B, Hou B, Liu J (2017) Mechanisms and characterization of *Trichoderma longibrachiatum* T6 in suppressing nematodes (*Heterodera avenae*) in wheat. Front Plant Sci 8:1491. https://doi.org/10.3389/fpls.2017.01491

Zhao B, Trouvelot A, Gianinazzi S, Gianinazzi-Pearson V (1997) Influence of two legume species on hyphal production and activity of two arbuscular mycorrhizal fungi. Mycorrhiza 7(4):179–185. https://doi.org/10.1007/s005720050179

Zope VP, El Enshasy H, Sayyed RZ (2019) Plant growth promoting Rhizobacteria: an overview in agricultural perspectives. In: Sayyed RZ (ed) Plant growth promoting Rhizobacteria for sustainable stress management Vol II Rhizobacteria in biotic stress management. Springer, Singapore, pp 345–362

# Chapter 14
# The Potential of Plant Growth-Promoting Rhizobacteria (PGPR) as Biological Tools in Enhancing Agricultural Sustainability

**Garima Malik, Samira Chugh, and Akanksha Madan**

**Contents**

14.1 Introduction .......... 295
14.2 PGPRs and Their Role in Sustainable Agriculture .......... 296
14.3 The Rhizosphere .......... 297
14.4 Rhizobacteria .......... 298
14.5 Approaches for Enhancing Rhizosphere Colonization By PGPR Inoculants .......... 299
14.6 Commercialization of PGPR as Biofertilizers .......... 302
14.7 Future Prospects .......... 304
References .......... 305

## 14.1 Introduction

The United Nations' forecast is that by 2050 the human populace will raise to 9.3 billion (United Nations Population Division 2001). The total populace is likewise speedily getting more urbanized. In 1975, about 33% of the world's population lived in urban communities; this number is estimated to ascend to more than 60% by 2030 (United Nations Population Division 1999). Rapid growth in population without a corresponding increase in global food production along with reduction in arable land is a major problem the world is facing (Goswami et al. 2016). Climate change, increase in population, and burning of fossil fuels have further resulted in significant loss of areas of agricultural land to rising seas, erosion, salinization, and desertification (Pachauri et al. 2014). These problems may soon lead to food

G. Malik (✉)
Raghunath Girls' Post Graduate College, Chaudhary Charan Singh University, Meerut, India

S. Chugh · A. Madan
Gargi College, University of Delhi, New Delhi, India



R. Z. Sayyed et al. (eds.), *Antifungal Metabolites of Rhizobacteria for Sustainable Agriculture*, Fungal Biology, https://doi.org/10.1007/978-3-031-04805-0_14

insecurity especially in developing and underdeveloped countries (Goswami et al. 2016). New agricultural strategies to increase agricultural productivity to feed all are the need of the hour in the twenty-first century (Glick 2014). The Green Revolution that took place in twentieth century was successful in solving the worldwide food problem. Increase in synthetic fertilizers, herbicides, and pesticides and use of superior crop varieties were the main strategies employed to increase crop yield during the Green Revolution. However, indiscriminate use of pesticides and fertilizers had high environmental costs.

The large-scale agriculture system adopted at present consumes excessive amounts of water, fossil fuel, and topsoil at untenable rates. It adds to ecological dilapidation, as well as pollution, soil depletion, and loss of biodiversity (Horrigan et al. 2002). The pesticides used in current food production are now known to be carcinogenic for workers and consumers and may cause reproductive dysfunction and endocrine disruption (Horrigan et al. 2002). Once again, a fresh agricultural revolution with reduced negative environmental impact is required to withstand the needs of the growing population and the changing climate (Backer et al. 2018). Efficient and extensive term answers to supply food for the world must be supportable and with eco-accommodating natural arrangements (Goswami et al. 2016).

Many human activities contribute to increasing amounts of chemical pollutants in the atmosphere, one of the foremost being agriculture which adds synthetic chemical fertilizers and pesticides to the environment. These cause environmental damage and harm human health. Over time the employment of synthetic pesticides, fertilizers, and herbicides has augmented manifold because of their convincing results and ease of use. However, they cause a lot of environmental and health complications like contamination of crop products and groundwater by heavy metals, physical and chemical corrosion of farming soil, disruption of the innate biogeochemical nutrient cycle, and damage to soil microbiota specially the useful rhizobacteria (Noumavo et al. 2016). The continued application of mineral fertilizers devoid of natural supplements results in moisture- and nutrient-poor soil which is also more prone to erosion via wind and rain (Alalaoui 2007). The heavy metal pollution in groundwater and food crops is one of the leading causes of cancer in the world today (Koo and Kyung-Suk 2009). Excessive use of nitrogen fertilizers in fields produces nitrous oxide ($N_2O$) which is a greenhouse gas, triggering global warming (Vejan et al. 2016).

## 14.2 PGPRs and Their Role in Sustainable Agriculture

In intensive unsustainable agriculture, the soil microbial diversity gets reduced owing to cultivation and lofty addition of chemical fertilizers, pesticides, and herbicides in these systems (Johansson et al. 2004). In low-input sustainable cropping systems, the high density of useful microorganisms contributes to the biocontrol of harmful microbes and improves the inventory of supplements (Johansson et al. 2004). Hence, it is indispensable to sustainably enhance agrarian efficiency, and one of the ways to do this is through the routine usage of plant growth-promoting bacteria (Glick 2014). To accomplish infection resistance, drought resilience, salt resilience, heavy metal stress resilience, and improved nutritional value in crops, earth microbes such as bacteria, fungi, algae, etc. that enhance the nutrient uptake capability and water use competence can be used (Khan et al. 2021; Patel et al. 2016). PGPRs can be used for sustainable agriculture to encourage plant growth without causing ecological degradation (Vejan et al. 2016; Basu et al. 2021).

There are now many reports on how inoculating plants with PGPR or treating plants with microbe-to-plant signal compounds can stimulate crop growth. These strategies are relied upon to turn out to be more accepted notwithstanding the unavoidable environmental change as they confer tolerance to crops for abiotic stresses like heat, drought, and salinity (Ilyas et al. 2020; Kusale et al. 2021; Sagar et al. 2022). Multifunctional PGPR-based compositions for profitable farming are being used to diminish the utilization of agrochemicals and fertilizers (Backer et al. 2018; Hamid et al. 2021).

## 14.3 The Rhizosphere

It has now been established for a few decades that plants are associated with a microbial community in the endosphere (internal tissues), phyllosphere (aerial surfaces), and rhizosphere (root surface) (Turner et al. 2013). Important exchanges transpire between the plant, soil, soil microfauna, and microbes in the rhizosphere (Antoun and Prevost 2005). The phyllosphere is not as abundant as the rhizosphere in terms of associated microbial communities as it is comparatively deprived of nutrients and exposed to varying moisture content, sunlight, and temperature (Turner et al. 2013). Soil rhizosphere is the intricate zone around the plant roots with a varied population of microorganisms like bacteria, protists, invertebrates, nematodes, and fungi; however, bacteria are the most copious microbes in the rhizosphere (Kaymak 2010; Venturi and Keel 2016). These bacterial communities in the rhizosphere are called rhizobacteria. The rhizosphere has a higher number and diversity of bacteria than the rest of the bulk soil due to the presence of root exudates containing sugars and amino acids. These secretions called rhizodeposition provide energy and nutrients to the soil microbes (Novo et al. 2018). Rhizodeposition may comprise of nearly 10% of photosynthetically fixed carbon (C) and 15% of plant total nitrogen (N) (Venturi and Keel 2016).

The rhizosphere of the soil has a rich deposition of plant root exudates, mucilage, and dry sloughed root cells and varies between different species and cultivars, age, and developmental stage of the host plant (Turner et al. 2013). Root exudates mainly comprise of amino acids, organic acids, fatty acids, sugars, growth factors, hormones, vitamins, and antimicrobial compounds (Bertin et al. 2003). The organization of the rhizosphere microbiome is determined by the root exudates (Turner et al. 2013). Also, the microbiome influences the exudates of the roots (Micallef et al. 2009). Along with root exudates, the rhizosphere is rich in cellulose and pectin, plant cell wall polymers acquired by the sloughing of root cells (Dennis et al. 2010). Methanol released due to decomposition of pectin may also be utilized as a source of C by other microbes (Galbally and Kirstine 2002). Apart being a C source and providing nutrients, plant roots offer an attachment framework to rhizosphere microbes (Bulgarelli et al. 2012).

## 14.4 Rhizobacteria

Kloepper et al. (1980a, b) proposed the term PGPR (plant growth-promoting rhizobacteria). A bacterium qualifies as a PGPR when it produces a positive outcome on the host plant (Kloepper and Schroth 1978). PGPRs include the bacteria which can keenly colonize plant root and enhance its overall growth and development (Wu et al. 2005). Rhizobacteria are grouped as beneficial, neutral, and harmful, contingent to their influence on the growth of plant (Huang et al. 2014). According to Antoun and Kloepper (2001), PGPRs amount to nearly 2–5% of overall rhizospheric bacteria. These soil microorganisms significantly influence soil fertility and plant health (Johansson et al. 2004). Most PGPRs belong to *Acinetobacter*, *Agrobacterium*, *Arthrobacter*, *Azotobacter*, *Azospirillum*, *Bacillus*, *Bradyrhizobium*, *Burkholderia*, *Frankia*, *Pseudomonads*, *Rhizobium*, *Serratia*, and *Thiobacillus* genera (Glick 1995; Vessey 2003). Various researches are being conducted to study the PGPR–plant relationship (Zhuang et al. 2007). PGPRs can improve host plant growth by generating and oozing plant growth regulators, viz., auxins, cytokinins, and gibberellins (GAs), facilitating the uptake of minerals (such as phosphorus) and nutrients from land, thereby reducing the reliance on harmful fertilizers and thwarting various diseases prone to infect the plant (Kaymak 2010; Ahemad and Kibret 2014).

Depending on their beneficial effects on plants, PGPRs are categorized as *phytostimulators* (release plant growth regulators and enhance plant overall development), *biofertilizers* (increase the accessibility of minerals and nutrients to the plant), *biopesticides* (control diseases by producing antibiotics and antifungal metabolites), and *rhizoremediators* (break down organic contaminants) (Somers et al. 2004). Useful rhizobacteria can influence plant growth both directly and indirectly. The direct mechanism involves the synthesis of phytohormones and resource accumulation including N fixation, bioavailability of phosphorus (P), and iron sequestration. PGPR have different roles, e.g., free-living N-fixing bacteria (viz., *Azotobacter* spp.) and symbiotic N-fixing bacteria (viz., root-nodulating *Rhizobium* spp.) make available N to the host plant, while some bacteria increase P bioavailability by solubilizing minerals containing P (Turner et al. 2013). Indirect mechanism includes biocontrol which is the reduction of the effects of phytopathogens by production of antibiotics and antifungal compounds (Glick 2012; Nova et al. 2018; Zakaria et al. 2019). The role of diacetylphloroglucinol (DAPG), an antifungal compound produced by *Pseudomonas fluorescens*, has been reported in the decline of "take-all" disease in cereals caused by the fungus *Gaeumannomyces graminis* (Combes-Meynet et al. 2012). The disease harshness declines with recurring cultivation of a plant such as wheat in case of "take-all" disease as the soil itself becomes disease suppressive due to the establishment of antagonistic microbial communities (Raaijmakers and Weller 1998). According to Turner et al. (2013), a rich and diverse presence of rhizobacteria may be quite sufficient to avert infection in host plants. The list of few PGPRs which may play crucial role in sustainable agriculture is displayed in Table 14.1.

## 14.5 Approaches for Enhancing Rhizosphere Colonization By PGPR Inoculants

The prosperous range of root exudates and plant cell fragments is the controlling factor that attracts a unique diversity and pattern of microbes for rhizosphere colonization. The identity of this rhizomicrobiome decides the growth and development of the plant. This relationship is very well known to be harnessed in order to stimulate crop growth clearly indicating toward the huge benefits of these PGPRs in stimulating the agricultural outputs at global level.

It is not only the diversity and amount of PGPR inoculum that is important but also the strategy of inoculation which results in enhanced crop outputs. Multiple studies have listed numerous strategies of inoculation of PGPRs which have worked in different ways for different groups of plants/crops. Sometimes, only inoculation is not found to be sufficient; however, it has to be supplemented by treating the plants with microbe-to-plant signal compounds.

PGPR inoculums and formulations have been proven to be an effective strategy due to their multifunctionality and multipurpose utilization in agricultural sector and utilizing bio-based fertilizers to gain commercial benefits instead of depending on synthetic fertilizers and agrochemicals. PGPRs have indeed proven to be low-cost commercial agricultural inputs, with multilevel benefits, the most important being improvement in rhizosphere colonization.

While considering field conditions, in addition to the internal factors (root exudate composition, microbiont diversity, and so on), external factors (C and N availability, organic matter content, moisture, pH, soil type, climate) also come into play. These external stimuli can modify the plant-microbe relationships and plant growth (the main purpose of exploiting PGPRs) might get impaired. Therefore, it is imperative to develop effective strategies, high-quality products, suitable formulations, and inoculation methods that are applicable to a wide range of these parameters. Some of these methods of inoculum which are currently used on regular basis include the following.

**Table 14.1** List of PGPRs that may play crucial roles in sustainable agriculture

| PGPR | Plant | Role of PGPR (phytostimulator/biofertilizer/biocontrol/rhizomediator) | References |
|---|---|---|---|
| *Azoarcus* | Rice | Biofertilizer: Nitrogen fixation | Reinhold-Hurek and Hurek (1998) |
| *Azotobacter Vinelandii (Az3)* | Rice | Biofertilizer: Nitrogen fixation | Sahoo et al. (2014) |
| *Azotobacter* | Cucumber | Phytostimulator: Synthesis of phytohormone cytokinin | Aloni et al. (2006) |
| *Azorhizobium caulinodans* | Wheat | Biofertilizer: Nitrogen fixation | Sabry et al. (1997) |
| *Azospirillum brasilense ab-V5* | Maize | Biofertilizer: Nitrogen fixation | Oliveira et al. (2017) |
| *Bacillus* | Potato | Phytostimulator: Synthesis of phytohormone auxin | Ahmed and Hasnain (2010) |
| *Bacillus cereus* UW85 | Alfalfa | Biocontrol of damping-off of alfalfa caused by *Phytophthora medicaginis* | Silo-Suh et al. (1994) |
| *Bacillus* | Pepper | Phytostimulator: Synthesis of phytohormone gibberelin | Joo et al. (2005) |
| *Bacillus* | Cucumber, pepper | Biofertilizer: Solubilization of phosphate | Han et al. (2006) |
| *Frankia* | Alnus | Biofertilizer: Nitrogen fixation | Simonet et al. (1990) |
| *Herbaspirillum Seropedicae* DX35 | Rice | Biofertilizer: Nitrogen fixation and phosphorus solubilization, production of siderophore | Wang et al. (2014) |
| *Pseudomonas fluorescens* | Cereals specially wheat | Biocontrol of "take-all" disease caused by the fungus *Gaeumannomyces graminis* | Combes-Meynet et al. (2012) |
| *Rhizobium leguminosarumbv. Trifolii* | Rice | Phytostimulator: Auxin synthesis<br>Biofertilizer: Acquisition of N, P, K etc. | Yanni et al. (2001) |
| *Pseudomonas aeruginosa* | Chick pea | Biofertilizer: N fixation | Verma et al. (2013) |
| *Mesorhizobium* sp. | Chick pea | Biofertilizer: Solubilization of phosphate | Verma et al. (2013) |

### 14.5.1 Biofilm Inoculum Method

Biofilms are small or expansive and structured aggregates of complex multicellular populations made by harmless or beneficial microbial symbiont populations that have multiplied and colonized by adhering to inert plant surfaces, in response to conducive environment and plant-based surroundings. These microcolonies that get encased in a polymer matrix, when used as BBs (biofilm biofertilizers), have impact on overall plant health because they are not only physically and metabolically linked to the plant and each other but also communicate with the plant via molecular signals. This phenomenon is exploited to incorporate the establishment of certain plant-based biofilms on various plant parts, viz., leaves, stems, roots, seeds, etc., in order to enhance the survival of beneficial bacteria that aid in improving plant growth. This biofilm method is proven beneficial as compared to planktonic counterparts in benefitting the plant health. In addition, these biofilms are resistant to antibiotics, making them more competitive for survival under harsh conditions. These traits may come in handy when applying such microbial films to the plants where these microbes have to compete with pre-existing and indigenous soil microbes (Baudoin et al. 2010). This way, these biofilms also help in controlling the propagation of disease-causing microbes in the vicinity (Chen et al. 2013). PGPR activity of biofilms has utilized single-species and multi-species biofilms of microbes like *Pseudomonas*, *Trichoderma*, *Penicillium*, and so on which had been seen to improve the auxin production, nitrogenase activity, and a diverse range of improved growth.

### 14.5.2 Planktonic Inoculum Method

Utility of planktonic inoculants has been widely studied especially pertaining to applications in PGPR-based improvement of plant health. These are the initial, therefore, reversible stages of smaller group of cells that can be, if needed, easily detached from the plant surface because the stage is too young to be encased in a protective matrix. Therefore, these planktonic inocula (of same microorganisms especially bacteria when in biofilms) are readily prone to antibiotics and do not manifest spectacular results under harsher conditions (Guilhen et al. 2017).

### 14.5.3 Biochar Inoculum

Biomass-based organic matter produces biochar upon its irreversible pyrolysis (high-temperature and pressure-mediated chemical decomposition in the absence of air). This change is physical state, and chemical composition of organic matter gives rise to a stable, C-rich solid which has tremendous potential of C sequestration (capture and storage), altering soil pH and increasing soil fertility in itself. Biochar not only enhances the survival and multiplication of microbial inoculants but also is a proven carrier material because it is an eco-friendlier substitute to peat carriers. It is known to provide a protective niche to the microbial strain and promote colonization process of the beneficial microbes when applied as a seed coating. Biochar source, origin, and method of production decide its physical, chemical, and biological properties. Therefore, care must be taken while selecting a biochar as a microbial inoculant carrier as some may prove poisonous to microorganisms and plants as well (Sun et al. 2016; Liu et al. 2018).

Developing such strategies includes helping the bacterial microbes gain advantage during colonization as compared to the efficiency of other non-relevant/redundant microbes. Development of commercia-level strategies also includes creating a high-quality bio-product which is stable and ensures selective but effective colonization of only the desirable microbial strains and whose inoculum is consistent in performance under a broad array of field conditions (Li et al. 2020).

## 14.6 Commercialization of PGPR as Biofertilizers

Scientific community and agri-based industries constantly work together to come up with commercially viable and agriculturally sustainable formulations of PGPRs which could efficiently replace the use of chemical-based synthetic fertilizers. Commercial success of any product, including PGPR strains, is a result of various hurdles being crossed. It has to be economical, viable, consistent, effective on a broad spectrum, safe, stable, and easily mingled with carrier material. Various stages and trials are needed for development of a successful market product, viz., isolation, screening, fermentation, mass production, viability tests, toxicology tests, quality check, and field efficacy (Gupta et al. 2015). During fermentation, identity (phenotype testing, genome identification) and density of the strains must be managed and ensured to be contaminant-free. The carrier must be sterile, aseptically mixed with the desired strains, and stored under specific temperature and humidity conditions for better shelf life (checked by CFU in plate counting). The final product again goes through identity, viability, and stability checks (Fig. 14.1).

PGPR-based bioformulations that are meant for field application comprise one or more strains of biologically active rhizomicroorganism along with an economical and suitable carrier material (as discussed in biochar) which maintains the cell viability even in harsh environmental conditions. Diverse materials could be used as delivery system, viz., clay, talc, diatomaceous earth, rice and wheat bran, etc. on the basis of the microbial viability and the mode of application (liquid, powder, or granules). There are certain polymer-based carriers as well, e.g., alginate beads that could be used (natural polymer). Sterilized carriers are mixed with the desired strain of rhizobacterial inoculum (pure liquid cultures), where more than one bacterial strain could also be used. Sometimes, genetically engineered bacteria could be used as well. The shelf life of this concoction could be increased either by

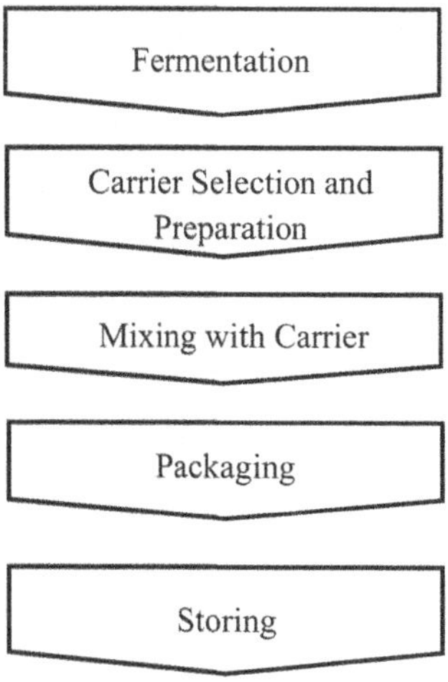


**Fig. 14.1** Flow diagram representing various steps taken to ensure quality and functionality maintenance

**Table 14.2** Quality measures taken during biofertilizer manufacture process

| Step | Quality checks |
|---|---|
| Fermentation | Purity, identity, viability |
| Carrier selection and preparation | Aseptic, pH, texture |
| Mixing with carrier | Sterile conditions |
| Packaging | Sterile conditions, label information |
| Storing | Viability |

cryopreservation or addition of suitable carriers (C sources and protectants like sugars). Due to this reason, the cost of liquid formulations becomes higher as compared to the solid or dry formulations.

Once the formulation is made, it undergoes various trials and is approved by passing all the quality control measures (Table 14.2), it is packaged, and the packaging must have information, viz., name of the product, bacterial strains used, crops for which the formulation is effective on, date of manufacture and expiry, and instructions or directions to use.

Applications of commercially available PGPR formulations:

- Liquid: addition in soil using misters or irrigation systems. Seeds or roots of seedlings could also be dipped in these formulations. Water-in-oil emulsions could also be used as inoculation method during irrigation of the field.
- Powdered/encapsulated: sprayed with centrifuges and spreaders on the fields.

Commercialization of PGPRs is indeed an eco-friendly way of progression toward sustainability, increased crop productivity, and improved crop health (Das and Singh 2014). Recent studies and data collected for commercially used biofertilizers in Canada reveal that companies like Novozymes, BASF Canada Inc., Biomax, etc. have created and patented many PGPR-based biofertilizer products (Table 14.3).

**Table 14.3** A list of common commercially available PGPR-based biofertilizer products in Canada

| Company | Product® |
|---|---|
| Novozymes | Cell-tech, Nitragin gold (rhizobia) |
| | TagTeam |
| BASF Canada Inc. | Nodulator N/T, pro, XL |
| BrettYoung seeds | Bioboots |
| Biomax | Life |
| | Biomix, Biozink, biodine |
| Symborg | VitaSoil |
| TFAC (Tokachi Federation of Agricultural Cooperatives) | Mamezo |
| | R-processing seeds |
| | Hyper coating seeds |

**Table 14.4** Some of the inoculants that have potential use or are being utilized in commercial formulations in India (Reference: Garcia de Salamone et al. 2019)

| Microorganism (inoculum) | Assayed crop | Additional data | References |
|---|---|---|---|
| *Azotobacter chroococcum, Bacillus coagulans, bacillus* spp. | *Cicer arietinum* | Commercial name: Anubhav Pravahi | Ansari et al. (2015) |
| | | Application: $10^8$–$10^9$ CFU /mL | |
| *A. chroococcum, B. megaterium, glomus intraradices, pseudomonas monteilii* | *Andrographis paniculata* (creat/kalmegh) | Vermicompost along with bacterial inoculums give better impact | Khan et al. (2015) |
| *Trichoderma* spp. | *G. hirsutum* (cotton) | Application: 300 g seeds needed per acre | Prasanna et al. (2016) |
| *Pseudomonas fluorescens* AMB-8 | Chili and tomato | Shelf life: Half year at RT ($10^6$ CFU/mL) | Anith et al. (2016) |
| *Azotobacter*, *Azospirillum*, *Rhizobium*, and P-, K-, and Zn-solubilizing bacteria, arbuscular mycorrhiza, and *Acetobacter* | *O. sativa, Z. mays, G. hirsutum, P. sativum, C. arietinum, S. officinarum, I. batatas, Rubus, P. granatum*, and *C. roseus* | Commercial name: Buchneri 500 inoculant | Sruthilaxmi and Babu (2017) |

A good number of bio-inoculants that have potential use as PGPR are being utilized in a variety of commercial formulations in India (Table 14.4).

## 14.7 Future Prospects

The proper use of PGPRs has a huge potential for sustainable crop management practices (Backer et al. 2018). Agricultural products like bioformulations of alive solitary or multi-species inoculum or formed on sequestered signal molecules for soil fertility, plant growth support, and inhibition of plant pathogens offer environment-friendly options to synthetic agriculture chemicals (Arora et al. 2016). Few PGPRs like *Azospirillum*, *Azotobacter*, *Bacillus*, *Enterobacter*, *Pseudomonas*, *Klebsiella*, and *Serratia* species have been commercialized (Glick 2012). There is an increase in the numeral of reports focusing to separate, pick out, and assess plant-growth promoting bacteria and their association with wide range of crops so that they can be altered into innovative microbial formulations in the coming time. Microbial inoculants composed of living microorganisms competent of promoting the growth of crop plant are being designed as biofertilizers and biopesticides. Mixed formulations of different strains of PGPRs which can promote plant growth can have great potential. Effective signal compounds from strains showing potential to function as biostimulants can be isolated and developed into agro-products, e.g., thuricin 17, a bacteriocin produced by *Bacillus thuringiensis* NEB17, and lipochitooligosaccharides (LCOs), produced by rhizobia. To isolate, PGPR strains are cultured in broth medium and then by centrifugation and filtration process cells are removed (Backer et al. 2018; Gray et al. 2006).

However, the current application of PGPR in agriculture sector amounts to only a small section of agricultural setup. The thriving use of PGPR depends on its endurance in soil, its affinity with the host crop on which it is inoculated, how well it can interact with the indigenous soil microbiome, and environmental conditions (Vejan et al. 2016). There are certain factors like biotic and abiotic stresses, nutrient-poor soils, pests, etc. that hamper the utilization of inoculants in several areas. These limitations restrict the widespread application of PGPRs. Attempts are being made to improve the efficacy of rhizobacteria already investigated and reported and in the discovery of novel strains to be utilized as inoculants which are also effective under adverse and demanding environmental circumstances. These inoculants will become widespread only if they are cost-effective, easy to use, and viable for a long time. PGPRs can help to drive the "next Green Revolution" for global food security in a climate change-resilient and sustainable manner. This can be achieved by collaboration between industry, academia, and the government for a cost-effective, easy to apply, and successful agro-product development based on PGPR inoculants or their signal compounds.

## References

Ahemad M, Kibret M (2014) Mechanisms and applications of plant growth promoting rhizobacteria: current perspective. J King saud Univ Sci 26(1):1–20

Ahmed A, Hasnain S (2010) Auxin producing *bacillus* sp.: auxin quantification and effect on the growth *Solanum tuberosum*. Pure Appl Chem 82:313–319

Alalaoui AC (2007) Fertilisation minérale des cultures: les elements fertilisantsmajeurs (Azote, Potassium et Phosphore). Bull Mens Inform Liaison 155:1–4

Aloni R, Aloni E, Langhans M (2006) Role of cytokinin and auxin in shaping root architecture: regulating vascular differentiation, lateral root initiation, root apical dominance and root gravitropism. Ann Bot 97:883–893

Anith KN, Vaishakhi AS, Viswanathan A, Varkey S, Aswini S (2016) Population dynamics and efficiency of coconut water based liquid formulation of *Pseudomonas fluorescens* AMB-8. J Trop Agric 54:184–189

Ansari MF, Tipre DR, Dave SR (2015) Efficiency evaluation of commercial liquid biofertilizers for growth of *Cicer aeritinum* (chickpea) in pot and field study. Biocatal Agric Biotechnol 4(1):17–24

Antoun H, Kloepper JW (2001) Plant growth promoting rhizobacteria. In: Brenner S, Miller JH (eds) Encyclopedia of genetics. Nova Science Publishers, New York, pp 1477–1480

Antoun H, Prevost D (2005) Ecology of plant growth promoting rhizobacteria. In: Siddiqui ZA (ed) PGPR: biocontrol and biofertilization. Springer, Dordrecht, Netherlands, pp 1–38

Arora NK, Mehnaz S, Balestrini R (2016) Bioformulations: for sustainable agriculture. Springer, Berlin. https://doi.org/10.1016/j.mib.2017.03.011

Backer R, Rokem JS, Ilangumaran G, Lamont J, Praslickova D, Ricci E, Subramanian S, Smith DL (2018) Plant growth-promoting Rhizobacteria: context, mechanisms of action, and roadmap to commercialization of biostimulants for sustainable agriculture. Front Plant Sci 9:1473. https://doi.org/10.3389/fpls.2018.01473. PMID: 30405652; PMCID: PMC6206271

Basu A, Prasad P, Das SN, Kalam S, Sayyed RZ, Reddy MS, El Enshasy H (2021) Plant growth promoting Rhizobacteria (PGPR) as green bioinoculants: recent developments, constraints, and prospects. Sustainability 13:1140. https://doi.org/10.3390/su13031140

Baudoin E, Lerner A, Mirza MS, El Zemrany H, Prigent-Combaret C, Jurkevich E, Spaepen S, Vanderleyden J, Nazaret S, Okon Y Moënne-Loccoz Y (2010) Effects of *Azospirillum brasilense* with genetically modified auxin biosynthesis gene ipdC upon the diversity of the indigenous microbiota of the wheat rhizosphere. Res Microbiol, 161(3):219–226

Bertin C, Yang XH, Weston LA (2003) The role of root exudates and allelochemicals in the rhizosphere. Plant Soil 256:67–83

Bulgarelli D, Rott M, Schlaeppi K, Loren V, van Themaat E, Ahmadinejad N, Assenza F, Rauf P, Huettel B, Reinhardt R, Schmelzer E, Peplies J, Gloeckner FO, Amann R, Eickhorst T, Schulze-Lefert P (2012) Revealing structure and assembly cues for *Arabidopsis* root-inhabiting bacterial microbiota. Nature 488:91–95

Chen Y, Yan F, Chai Y, Liu H, Kolter R, Losick R, Guo JH (2013) Biocontrol of tomato wilt disease by *Bacillus subtilis* isolates from natural environments depends on conserved genes mediating biofilm formation. Environ Microbiol 15(3):848–864

Combes-Meynet E, Pothier JF, Moenne-Loccoz Y, Prigent-Combaret C (2012) The *Pseudomonas* secondary metabolite 2,4-diacetylphloroglucinol is a signal inducing rhizoplane expression of *Azospirillum* genes involved in plant-growth promotion. Mol Plant-Microbe Interact 24:271–284

Das I, Singh AP (2014) Effect of PGPR and organic manures on soil properties of organically cultivated mung bean. The Bioscan 9(1):27–29

Dennis PG, Miller AJ, Hirsch PR (2010) Are root exudates more important than other sources of rhizodeposits in structuring rhizosphere bacterial communities? FEMS Microbiol Ecol 72:313–327

Galbally IE, Kirstine W (2002) The production of methanol by flowering plants and the global cycle of methanol. J Atmos Chem 43:195–229

García de Salamone IE, Esquivel-Cote R, Hernández-Melchor DJ, Alarcón A (2019) Manufacturing and quality control of inoculants from the paradigm of circular agriculture. In: Singh D, Gupta V, Prabha R (eds) Microbial interventions in agriculture and environment. Springer, Singapore

Glick BR (1995) The enhancement of plant growth by free-living bacteria. Can J Microbiol 41:109–117

Glick BR (2012) Plant growth-promoting bacteria: mechanisms and applications, Scientifica (Cairo), pp 1–15

Glick BR (2014) Bacteria with ACC deaminase can promote plant growth and help to feed the world. Microbiol Res 169:30–39

Goswami D, Thakker JN, Dhandhukia PC (2016) Portraying mechanics of plant growth promoting rhizobacteria (PGPR): a review. Cogent Food Agric 2:1. https://doi.org/10.1080/23311932.2015.1127500

Gray EJ, Di Falco M, Souleimanov A, Smith DL (2006) Proteomic analysis of the bacteriocin thuricin 17 produced by *Bacillus thuringiensis* NEB17. FEMS Microbiol Lett 255(1):27–32

Guilhen C, Forestier C, Balestrino D (2017) Biofilm dispersal: multiple elaborate strategies for dissemination of bacteria with unique properties. Mol Microbiol 105(2):188–210

Gupta G, Parihar SS, Ahirwar NK, Snehi SK, Singh V (2015) Plant growth promoting Rhizobacteria (PGPR): current and future prospects for development of sustainable agriculture. J Microb Biochem Technol 7(2):096–102. https://doi.org/10.4172/1948-5948.1000188

Hamid B, Zaman M, Farooq S, Fatima S, Sayyed RZ, Baba ZA, Sheikh TA, Reddy MS, El Enshasy H, Gafur A, Suriani NL (2021) Bacterial plant biostimulants: a sustainable way towards improving growth, productivity, and health of crops. Sustainability 13(5):2856. https://doi.org/10.3390/su13052856

Han HS, Supanjani S, Lee KD (2006) Effect of co-inoculation with phosphate and potassium solubilizing bacteria on mineral uptake and growth of pepper and cucumber. Plant Soil Environ 52:130–136

Horrigan L, Lawrence RS, Walker P (2002) How sustainable agriculture can address the environmental and human health harms of industrial agriculture. Environ Health Perspect 110(5):445–456

Huang X-F, Chaparro JM, Reardon KF, Zhang R, Shen Q, Vivanco JM (2014) Rhizosphere interactions: root exudates, microbes, and microbial communities 1. Botany 92(4):267–275

Ilyas N, Mumtaz K, Akhtar N, Yasmin H, Sayyed RZ, Khan W, El Enshasy HA, Dailin DJ, Elsayed EA, Ali (2020) Exopolysaccharides producing bacteria for the amelioration of drought stress in wheat. Sustainability,12:8876. Doi:https://doi.org/10.3390/su12218876

Johansson JF, Paul LR, Finlay RD (2004) Microbial interactions in the mycorrhizosphere and their significance for sustainable agriculture. FEMS Microbiol Ecol 48(1):1–13

Joo GJ, Kim YM, Kim JT (2005) Gibberellins-producing rhizobacteria increase endogenous gibberellins content and promote growth of red peppers. J Microbiol 43:510–515

Kaymak DC (2010) Potential of PGPR in agricultural innovations. In: Maheshwari DK (ed) Plant growth and health promoting bacteria. Springer, Berlin/Heidelberg

Khan K, Pankaj U, Verma SK, Gupta AK, Singh RP, Verma RK (2015) Bio-inoculants and vermicompost influence on yield, quality of *Andrographis paniculata*, and soil properties. Ind Crop Prod 70:404–409

Khan N, Ali S, Shahid MA, Mustafa A, Sayyed RZ, Curá JA (2021) Insights into the interactions among roots, rhizosphere and Rhizobacteria for improving plant growth and tolerance to abiotic stresses: a review. Cell 10(6):1551. https://doi.org/10.3390/cells10061551

Kloepper JW, Schroth MN (1978) Plant growth promoting rhizobacteria on radishes. Proceedings of 4th International Conference Plant pathogenic bacteria, Angers. pp 879–882

Kloepper J, Leong J, Teintze M et al (1980a) Enhanced plant growth by siderophores produced by plant growth-promoting rhizobacteria. Nature 286:885–886

Kloepper JW, Schroth MN, Miller TD (1980b) Effects of rhizosphere colonization by plant growth promoting rhizobacteria on potato plant development and yield. Ecol Epidemiol 70:1078–1082

Koo SY, Kyung-Suk C (2009) Isolation and characterization of a plant growth-promoting rhizobacterium, *Serratia* sp. SY5. J Microbiol Biotechnol 19:1431–1438

Kusale SP, Attar YC, Sayyed RZ, Malek RA, Ilyas N, Suriani NL, Khan N, El Enshasy H (2021) Production of plant beneficial and antioxidants metabolites by *Klebsiella variicola* under salinity stress. Molecules 26:1894. https://doi.org/10.3390/molecules26071894

Li K, Cao R, Mo S, Yao R, Ren Z, Wu J (2020) Swine manure composting with compound microbial inoculants: removal of antibiotic resistance genes and their associations with microbial community. Front Microbiol 11:2686

Liu L, Li J, Yue F, Yan X, Wang F, Bloszies S, Wang Y (2018) Effects of arbuscular mycorrhizal inoculation and biochar amendment on maize growth, cadmium uptake and soil cadmium speciation in cd-contaminated soil. Chemosphere 194:495–503

Micallef SA, Channer S, Shiaris MP, Colon-Carmona A (2009) Plant age and genotype impact the progression of bacterial community succession in the *Arabidopsis* rhizosphere. Plant Signal Behav 4:777–780

Noumavo PA, Agbodjato NA, Baba-Moussa F, Adjanohoun A, Baba-Moussa L (2016) Plant growth promoting rhizobacteria: beneficial effects for healthy and sustainable agriculture. Afr J Biotechnol 15(27):1452–1463

Novo LAB, Castro PML, Alvarenga P, da Silva EF (2018) Plant Growth–Promoting Rhizobacteria-Assisted Phytoremediation of Mine Soils. Bio-Geotechnologies for Mine Site Rehabilitation 281–295

Oliveira ALM, Santos OJAP, Marcelino PRF, Milani KML, Zuluaga MYA, Zucareli C, Gonçalves LSA (2017) Maize inoculation with *Azospirillum brasilense* Ab-V5 cells enriched with exopolysaccharides and Polyhydroxybutyrate results in high productivity under low N fertilizer input. Front Microbiol 26(8):1873

Pachauri RK, Allen MR, Barros VR, Broome J, Cramer W, Christ R et al (2014) Climate change 2014: synthesis report. In: Pachauri R, Meyer L (eds) Contribution of working groups I, II and III to the fifth assessment Report of the intergovernmental panel on climate change. IPCC, Geneva

Patel PR, Shaikh SS, Sayyed RZ (2016) Dynamism of PGPR in bioremediation and plant growth promotion in heavy metal contaminated soil. Indian J of Exp Biol 54:286–290

Prasanna R, Ramakrishnan B, Ranjan K, Venkatachalam S, Kanchan A, Solanki P, Monga D, Shivay YS, Kranthi S (2016) Microbial inoculants with multifaceted traits suppress *Rhizoctonia* populations and promote plant growth in cotton. J Phytopathol 164(11–12):1030–1042

Raaijmakers JM, Weller DM (1998) Natural plant protection by 2,4-diacetylphloroglucinol-producing Pseudomonas spp. in take-all decline soils. Mol Plant-Microbe Interact 11:144–152

Reinhold-Hurek B, Hurek T (1998) Interactions of gramineous plants with *Azoarcus* spp. and other diazotrophs: identification, localization, and perspectives to study their function. Crit Rev Plant Sci 17:29–54

Sabry SRS, Saleh SA, Batchelor CA (1997) Endophytic establishment of *Azorhizobium caulinodans* in wheat. Proc Biol Sci 264:341–346

Sagar A, Rai S, Ilyas N, Sayyed RZ, Al-Turki AI, El Enshasy HA, Simarmata T (2022) Halotolerant Rhizobacteria for salinity stress mitigation: diversity. Mechanism Mol Approaches Sustain 14:490. https://doi.org/10.3390/su14010490

Sahoo RK, Ansari MW, Dangar TK, Mohanty S, Tuteja N (2014) Phenotypic and molecular characterisation of efficient nitrogen-fixing *Azotobacter* strains from rice fields for crop improvement. Protoplasma 251(3):511–523

Silo-Suh LA, Lethbridge BJ, Raffel SJ (1994) Biological activities of two fungistatic antibiotics produced by Bacillus cereus UW85. Appl Environ Microbiol 60:2023–2030

Simonet P, Normand P, Moiroud A (1990) Identification of *Frankia* strains in nodules by hybridization of polymerase chain reaction products with strain-specific oligonucleotide probes. Arch Microb 153:235–240

Somers E, Vanderleyden J, Srinivasan M (2004) Rhizosphere bacterial signaling: a love parade beneath our feet. Crit Rev Microbiol 30:205–240

Sruthilaxmi CB, Babu S (2017) Microbial bio-inoculants in Indian agriculture: ecological perspectives for a more optimized use. Agric Ecosyst Environ 242:23–25

Sun D, Hale L, Crowley D (2016) Nutrient supplementation of pinewood biochar for use as a bacterial inoculum carrier. Biol Fertil Soils 52(4):515–522

Turner TR, James EK, Poole PS (2013) The plant microbiome. Genome Biol 14:209. https://doi.org/10.1186/gb-2013-14-6-209

United Nations Population Division (1999) World urbanization prospects: the 1999 revision. United Nations, New York

United Nations Population Division (2001) World population prospects: the 2000 revision. United Nations, New York

Vejan P, Abdullah R, Khadiran T, Ismail S, Nasrulhaq Boyce A (2016) Role of plant growth promoting Rhizobacteria in agricultural sustainability-a review. Molecules (Basel, Switzerland) 21(5):573. https://doi.org/10.3390/molecules21050573

Venturi V, Keel C (2016) Signaling in the rhizosphere. Trends Plant Sci 21(3):187–198

Verma JP, Yadav J, Tiwari KN, Kumar A (2013) Effect of indigenous *Mesorhizobium* spp. and plant growth promoting rhizobacteria on yields and nutrients uptake of chickpea (*Cicer arietinum* L.) under sustainable agriculture. Ecol Eng 51:282–286. https://doi.org/10.1016/j.ecoleng.2012.12.022

Vessey JK (2003) Plant growth promoting rhizobacteria as biofertilizers. Plant Soil 255:571–586

Wang X, Cao Y, Tang X, Ma X, Gao J, Zhang X (2014) Rice endogenous nitrogen fixing and growth promoting bacterium *Herbaspirillum seropedicae* DX35. Wei Sheng Wu Xue Bao 54(3):292–298

Wu SC, Cao ZH, Li ZG, Cheung KC, Wong MH (2005) Effects of biofertilizer containing N-fixer, P and K solubilizers and AM fungi on maize growth: a greenhouse trial. Geoderma 125:155–166

Yanni Y, Rizk R, Abd-El Fattah F (2001) The beneficial plant growth-promoting association of *Rhizobium leguminosarum* bv. *trifolii* with rice roots. Aust J Plant Physiol 28:845–870

Zakaria AK, Sayyed RZ, Enshasy HE (2019) Biosynthesis of antibiotics by PGPR and their roles in biocontrol of plant diseases. In: Plant growth promoting Rhizobacteria for sustainable stress management Vol II Rhizobacteria in biotic stress management. Springer, Singapore, pp 1–36

Zhuang X, Chen J, Shim H, Bai Z (2007) New advances in plant growth-promoting rhizobacteria for bioremediation. Environ Int 33:406–413

# Chapter 15
# Role of Plant Growth-Promoting Rhizobacteria (PGPR) in Food Security

**Ram Awadh Ram, Ajit Kumar Dubedi Anal, G. S. Tiwari, Vishal Nath, and Ram Kripal Pathak**

**Contents**


15.1 Introduction .......... 311
15.2 Need of Crop Productivity Improvement .......... 314
15.3 PGPR for Enhancing Crop Productivity .......... 315
15.4 PGPRs in Bio-Enhancers .......... 316
15.5 Biodynamic Liquid Manure/Pesticides .......... 321
15.6 Future Prospects .......... 330
References .......... 332


## 15.1 Introduction

Worldwide food and nutritional security is a necessity. Agriculture, thereby, is the most important priority. Earth has biodiversity that has an ability to provide enough food and nutrition to the natural habitat. In present times, certain issues related to soil biology, environmental ecology, and quality of food have emerged. These issues are relevant as production has taken over productivity of our natural environment. In addition, inorganic practices have been introduced in the farming system resulting in short-term improvement in production but depletion of soil health, contamination of water, and adverse effects on the ecosystem. This has attracted to reduce the application of agrochemicals and conceptualizing the organic farming to promote PGPRs.

R. A. Ram (✉) · R. K. Pathak
ICAR-Central Institute for Subtropical Horticulture, Lucknow, India

A. K. D. Anal · V. Nath
ICAR-National Research Centre on Litchi, Muzaffarpur, India

G. S. Tiwari
Teerthanker Mahaveer University, Moradabad, India



R. Z. Sayyed et al. (eds.), *Antifungal Metabolites of Rhizobacteria for Sustainable Agriculture*, Fungal Biology, https://doi.org/10.1007/978-3-031-04805-0_15

Indiscriminate use of agrochemicals, coupled with monoculture of rice and wheat system, and use of heavy machineries have resulted short-term yield enhancement. But it has disturbed the soil biology and environmental ecology. Somehow during this period, importance of soil biology and ecology has been ignored. The Green Revolution is exhibiting second-generation's problem owing to over-exploitation and mismanagement of natural resources. Soil organic carbon content in most of the soils has been reduced to >0.5 percent. Excessive mining of micronutrients and ignorance of minor foods have led to malnutrition, under-nutrition, and many other ill impacts in the society.

Researchers are ever-involved to find out the solutions to overcome on these hurdles, by boosting soil fertility and crop productivity. In the recent decades, role of soil microbiome emerged as important tools for management of soil fertility as well as quality crop production. The microbial life plays a major role in enriching the soil by enhancing the breakdown of organic matter that can be recycled into new life, if only they could be unlocked from that organic matter and also from the mineral matter. The microbes, especially plant growth-promoting rhizobacteria (PGPR), are incredibly important. The plant growth-promoting rhizobacteria (PGPR) are versatile candidates providing its services in plant growth and health promotion tirelessly. The microorganisms are powered with super abilities to produce antimicrobial metabolites, volatile compounds, induced systemic resistance (ISR), etc. These defense mechanisms can cause substantial changes in plants' structural and functional activities, that is, pathogen resistance, which in turn enhance productivity and yield of crops (Fig. 15.1). We have tried to review the role of plant growth-promoting rhizobacteria (PGPR) to ensure food security through high-quality production.

Presently, the agricultural fields are being fertilized abundantly with agrochemicals that were begun in the nineteenth and twentieth centuries. Initially, the application of agrochemicals had shown quantum jump in crop productivity but, later, with the rapid technological evolution and un-precedent application of higher dose of fertilizers declined the fertility indices of soil. This not only affected flora and fauna

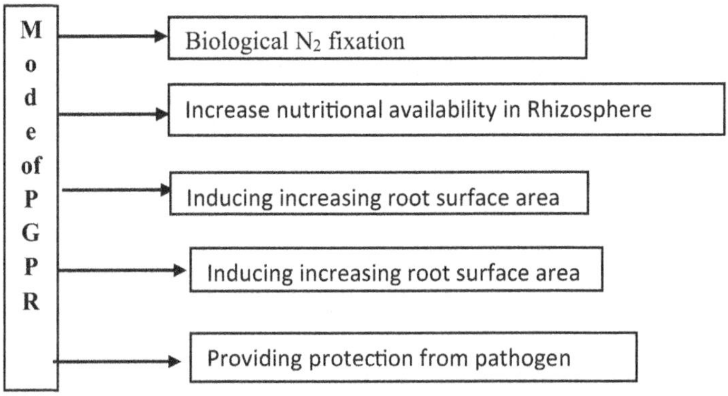


**Fig. 15.1** Modes of action of PGPR

of the soil but also increased the financial unsustainability of farmers especially in developing countries. Hence, application of agrochemicals is no more viable solution for sustainable crop production. It has resulted around 870 million populations put under hunger-stricken slab due to severe scarcities of accessible food in 2012 (FAO 2012). A number of reports have emphasized the need to improve agriculture production by 60% to plateauing food production corresponding to population blast, which is roughly estimated to reach about 9 billion by 2050. In the present scenario, farmers have blind reliance on the use of various agrochemicals, which led in depleting nutrition index in crops and also affecting microscopic life of soil. Across the globe, governments are now discouraging uses of chemical fertilizers and pesticides in agriculture production to store biological properties of the soil. To sort this matter, farmers have to change trend of organic crop production and their alternatives. Living soil is one of the basic components of organic farming. Life, in the basic form of microbes, has been the Earth' companion since its inception. It has close bond between the environment of the Earth and the microscopic organisms thriving upon it. Within soil, procreating in high concentrations, the bacteria, fungi, and other bio-life ensure soil fertility by recycling the elements through the chemical laboratory that constitutes their bodies, making them available to plants. Nitrogen and carbon alone aren't requiring the help of microbes before they can be rendered fit for plants. As nitrogen is converted into nitrate, phosphorus is turned into phosphate, sulfur to sulphate, chlorine to chloride, boron to borate, molybdenum to molybdate, and so on through elements, thanks to microbes.

High soil biological activities, particularly beneficial microorganisms, offer great potential. In fact, with all advancement, systematic research is lacking. The role of microbial lives is immense for the soil which can naturally regulate the ecology of soil and provide plenty and sustained nutrients for crop health thereby quality production. But, now at this diaspora, to obviate these problems, it has become an indispensable need to administer such methods of farming which are ecologically compatible, holistic, and organic in nature. Sustainable approaches in agricultural sector are of utmost importance to improve the food security and nutrition problem around the globe (Arora 2018).

The microbe-based inoculants are cheaper and helpful in protection of indigenous flora and fauna and other natural resources, have less or no harmful by-products, improve soil organic matter, and above all, maintain sustainability of ecosystems (Arora 2018). At present, the biological alternatives appear to be one of the ways to save planet Earth from the abusive agrochemicals entering in the food chain and causing serious health hazards to mankind.

The beneficial bacteria in the rhizosphere are recognized as plant growth-promoting rhizobacteria (PGPR) that influence plant growth and health promotion by several mechanisms. These can influence plant traits under fluctuating environmental conditions and improve yield and productivity in a sustainable way. The use of PGPR in field crops has attracted attention of many researches to enhance crop production and productivity in a sustainable manner. The inoculation strategies such as co-inoculation of two or more beneficial rhizobacteria have provided greater phytostimulation because of the synergistic and multifarious effects with field crops. PGPR are emerging tools of food security, also providing strategic avenues to combat biotic and abiotic stresses in crops.

Food security brought about emerging issues all over the world and consequently resulting serious influence on crop production in general and pressure for higher production of major crops. To meet the demand of food, emphasis has been given on few selected crops and thereby cultivation of other minor crops bearing rich amount of nutrients, i.e., essential amino acids, vitamins, minerals, etc. have been ignored. This has ultimately resulted in nutrient deficiency and malnutrition especially in children (Gundersen and Ziliak 2015).

The review is outlined to explore the current prospects and scenario of using beneficial PGPR to enhance quality food production in a sustainable way.

## 15.2 Need of Crop Productivity Improvement

One of the major problems of agriculture is facing to ensure availability of safe quality food to meet the challenges of food requirements of the burgeoning population coupled with deterioration of natural resources by assertive promotion of agrochemical-led farming. This has forced to conceive technology which can ensure "ever-sustained Green Revolution." The major crop such as rice and wheat were largely cultivated on the call of green revolution, but now, it has been observed that sustainability in production has become problems in many regions. Since these crops are staple food, researchers from across the world have paid ultimate attention to enhance the production of rice wheat system. Assefa and Tadesse (2019) cautioned that most of these crops possess less amounts of essential nutrients such as amino acids, vitamins, minerals, etc.; thus, they do not meet out the complete nutrition requirement for a healthy life. On the other hand, Ram and Pathak (2016a, b) mentioned that the agrochemicals such as pesticides and fertilizers got accumulated in the ecosystem and consequently enter in the food chain, when used in high doses in order to raise disease-free crops. The pesticides, once escaped in the environment, influence soil and climatic factors according to their half-life, amounts applied, and physicochemical reactions with the existing minerals of the soil (Carpio et al. 2021). The harmful effect of such chemicals in ground water and agricultural products was discussed in various studies elaborating agriculture-environment relationship (Meena et al. 2016a, b; Chandra et al. 2018). This calls for a possibility of reverting back to encourage those time-tested crops rich in essential nutrients and bioactive components for ensuring availability of nutritive and therapeutic food. The crop is enriched with plenty of essential nutrients so as to complete a balanced diet and can be used to accommodate the escalating food requirement.

## 15.3 PGPR for Enhancing Crop Productivity

The harmful effects and ill impacts of agrochemicals have posed serious problems in people involved in research, farmers, and policy makers and call for a viable alternative. Out of a number of alternative approaches, the use of beneficial group of bacteria such as plant growth-promoting rhizobacteria (PGPR) as bio-fertilizers and biocontrol agents to reduce crop losses, enhance crop productivity, and soil fertility has emerged as a viable alternative in recent decades (Table 15.1).

Along with plant growth promotion, they reforest eroded areas, restore the contaminated sites, and thereby cause positive effects on the degraded ecosystem (Gupta et al. 2015). These PGPR were seen as successful bacteria to be established in soil ecosystem due to their high adaptability in a wide variety of environments,

**Table 15.1** Multifarious plant growth-promoting traits of bacterial cultures isolated from biodynamic preparations

| Culture no. | HCN production | Ammonia production | Siderophore production | IAA production |
|---|---|---|---|---|
| CISH-PGPR-BD-502 I | – | ++ | + | + |
| CISH-PGPR-BD-503 4MR | – | ++ | – | – |
| CISH-PGPR-BD-503 5MR | – | ++ | – | – |
| CISH-PGPR-BD-503 6MR | – | ++ | – | – |
| CISH-PGPR-BD-503 6 | – | ++ | – | + |
| CISH-PGPR-BD-503 N7 | – | ++ | – | – |
| CISH-PGPR-BD-503 C8 | – | ++ | – | – |
| CISH-PGPR-BD-503 8 | – | ++ | | + |
| CISH-PGPR-BD-503 9 | – | ++ | – | + |
| CISH-PGPR-BD-503 10 | – | ++ | – | + |
| CISH-PGPR-BD-503 11 | – | ++ | – | + |
| CISH-PGPR-BD-504 C1 | – | ++ | – | – |
| CISH-PGPR-BD-506 I | – | ++ | – | + |
| CISH-PGPR-BD-507 CR1 | – | ++ | – | – |
| CISH-PGPR-BD-507 CR2 | – | + | + | – |
| CISH-PGPR-BD-507 CR3 | – | ++ | – | – |
| CISH-PGPR-BD-507 CR4 | – | ++ | + | – |

faster growth rate, and biochemical versatility to metabolize a wide range of natural and xenobiotic compounds. Several workers have found that in the rhizosphere ecology, plant growth-promoting rhizobacteria (PGPR) enhance plant growth by a wide variety of functions like biological nitrogen fixation, phytohormone production, phosphate solubilization, siderophore production, and 1-aminocyclopropane-1-carboxylate deaminase production (ACC), exhibiting antifungal activity due to the secretion of chitinolytic enzymes and production of volatile organic compounds (VOCs) and promoting beneficial plant-microbe symbiosis, interference with pathogen toxin production, etc. (Kundan et al. 2015; Prasad et al. 2015). Successful studies using PGPR on the growth enhancement of various crops were achieved in laboratory and field conditions (Agarwal et al. 2017). To overcome these situations, there is a need to develop eco-friendly and cost-effective microbial inoculants which can withstand the extreme environmental conditions. Plant growth-promoting bacteria (PGPB) are known to improve crop health even at low soil temperature, and the survival of the PGPR at different temperatures makes them suitable candidates for enhancement of vegetative and reproductive plant growth even in adverse conditions. These bacteria are well known for their beneficial effect on different host plants (Gouda et al. 2017).

As stated, PGPR ameliorate plant health and productivity by enhancing the nutrient status of soil and host plants subsequently (Dos Santos et al. 2020). The bioavailability of nutrients and their increased uptake may significantly enhance the nutrient use efficiency of plants. Increased solubilization of nutrients by PGPBs enhanced their uptake and accumulation in grain of amaranth (Pandey et al. 2021). Nutrient availability has been influenced by solubilization, chelation, and oxidation-reduction reaction in soil (Alori et al. 2017). The plant growth significantly influenced by the micronutrients along with the macronutrients supported metabolic and enzymatic activities in plant. Application of PGPR has been reported as an effective approach under field conditions in the management of plant diseases (Swarnalakshmi et al. 2020).

## 15.4 PGPRs in Bio-Enhancers

"Bio-enhancers are organic preparations, obtained by active fermentation of animal & plant residues over specific duration. These are rich source of microbial consortia, macro, micronutrients and plant growth promoting substances including immunity enhancers. In general, these are utilized to treat seeds/seedlings, enhance decomposition of organic materials thereby enrich soil and induce better plant vigour" (Pathak and Ram 2013). These are concentrated manures and bio-products in powder or in liquid form, henceforth termed as bio-enhancers. These could be a potent tool to utilize these in fertigation in various crops (Pathak and Ram 2013).

### 15.4.1 Characteristics of Bio-Enhancers

- Potent source for macro- and micronutrients.
- Contains plant growth-promoting factors.
- Immunity enhancer.
- Pesticide and fungicidal property.
- Its efficacy is influenced by inputs used and method of preparation.
- Used for seed/seedling treatment, enhancing decomposition and improving soil fertility and productivity.
- An effective and potent tool for fertigation.

These preparations can be applied with irrigation water, drenched on organic mulches, diluted and sprayed as foliar fertilizers (Pathak and Ram 2013). On the basis of materials used in the preparation and impact on crops, these organic fertilizers/organic liquid manures have varying response. In general, these play an important role in quick decomposition of organic wastes: improve humus content of the soil which is essential to maintain the activity of microorganisms and other life forms in the soil. Bio-enhancers are useful in suppressing the growth of pathogen also. In an experiment Ram et al. (2016) have studied the antimicrobial properties of some bio-enhancers. They have tested bio-enhancers, viz., amritpani, panchagavya, cow pat pit, and jeevamrita, against some selected pathogens in laboratory conditions. Growth of *Aspergillus fumigatus* was significantly reduced, while complete growth inhibition was observed for *Colletotrichum gloeosporioides* and *Fusarium solani*. These are prepared locally and can resolve number of apprehensions, helpful in boosting production and mitigating number of nutritional disorders including effective management of soil-borne diseases in crops. These products belong to the Ayurvedic medical tradition, where indigenous cow products (dung, urine, milk, ghee, and curd) are critical central ingredients in addition to few selected medicinal herbs. These are known to supplement major and minor nutrients, growth stimulants, and other beneficial substances to the plants (Ram and Pathak 2017). Salient features of bioenhancers are as follows.

### 15.4.2 Types of Enhancers

In general bio-enhancers are of the following types.

#### 15.4.2.1 Bio-Enhancers

These are prepared with cattle dung, sheep and goat droppings, and fish manures. Combinations of plant and animal byproducts have better impacts on crop production. Liquid manure and liquid fertilizer preparations obtained by active fermentation of animal and plant residues over specific duration are important. Organic liquid manures play a key role in promoting growth and providing immunity to plant system (Ram et al. 2020a, b). Effects of foliar sprays of few of the formulations have been observed to increase the plant growth, yield, and quality of several crops. In fact, liquid manures are rich source of microbial consortia, macro- and micronutrients, and plant growth-promoting substances. These are used to treat seeds/seedlings, enrich soil, and induce better plant vigor.

There is increasing trend for naturally derived formulations for sustainable production in organic farming system (Ram et al. 2018a, b). Many formulations of liquid manures are being used by the farmers of different regions. Few important and widely used formulations are discussed in the following. On the basis of preparation, bio-enhancers can be grouped as simple and special preparations. Brief account of these has been dealt below.

#### 15.4.2.2 Special Bio-Enhancers

A number of cow byproduct-based bio-enhancer alone or in combination with a few other products have been developed in different organic farming systems, and their impact has been recorded. Salient features of few of the selected bio-enhancers and their impact have been discussed below.

#### 15.4.2.3 Biodynamic Preparation (502–507)

One group of preparation is known as compost preparation (BD-502-BD-507), and the other group is field (BD-500 and BD-501) sprays. In biodynamic compost, BD preparations (502–507) are used to enhance decomposition of materials and help in mediating cosmic energy in the soil. These preparations are rich in microbial load as well as mediate cosmic energy in the heap. These compost preparations are produced with use of different herbs. Herbal plants used in preparation have been presented in Fig. 15.2.

Plant growth-promoting traits of bacterial cultures isolated from biodynamic preparations are presented in Table 15.1.

Biodynamic agriculture principals are based on aligning with cosmic energy preponderant in the universe which is responsible for creation and preservation of all life. While practicing biodynamic agriculture, more physical evidences of impact are observed in the drastic improvement of soil structures, increased carbon content, balanced pH, availability of trace elements, increased production cycle of crops, suppressed diseases and pest incidence, and high nutrition content in the food products (Pathak and Ram 2013).

Brief accounts of biodynamic preparations are enumerated as follows.

**Fig. 15.2** (Left to right) Plants and flowers of yarrow grass (*Achillea millefolium*), chamomile flower (*Matricaria reticulata*), stinging nettle (*Urtica dioica*), oak (*Quercus robur*), (*Taraxacum officinale*), and valerian (*Valeriana officinalis*)

**Fig. 15.3** Preparation of cow horn manure (BD-500) and stirring in water for spray

### 15.4.2.3.1 Cow Horn Manure (BD-500)

It is basically fermented cow dung and is the basis for soil fertility and renewal of degraded soils. This is a fundamental biodynamic field spray preparation.

*Stirring*– The reason for stirring of BD-500 and BD-501 in water for an hour before spraying it onto the land or plant is to release an "immense ethereal and astral force" preserved in cow horn, a force that is useless to the plant until released by such stirring (Fig. 15.3). These are released by being placed into a rhythmic sun movement, a spiraling in and out, forming a vortex. Seed treatment of *Vigna* sp. seed with BD-500 has improved the root system for better growth and development (Ram and Pathak 2019) (Fig. 15.4).

Microbial analysis of BD-500 was done and PGPR of some isolated bacteria were carried out by standard methods (Table 15.2).

#### 15.4.2.3.2 Cow Horn Silica (BD-501)

It is another biodynamic preparation (Fig. 15.5) that is very effective in enhancing photosynthetic activity through increase in chlorophyll content in the leaves and thereby the plant immune system. It helps in improvement in protein and sugar (Brix) level, metabolism, mechanical rigidity, growth, tolerance to environmental stresses (frost, drought, salinity, mineral toxicity, and deficiency), and resistance to fungal attack. It also improves the taste, color, aroma, and shelf-life of produce.

Using cow horn for making horn manure and horn silica employs on ideal focal device to concentrate the earthly or cosmic stream of forces on the material filled in the horn. Microbes isolated from cow horn and PGP tests are presented in Table 15.3.

The primary function of BD-500 is to promote root activity, especially the fine root hairs, to stimulate soil micro-life and to increase beneficial bacterial growth. During incubation microflora, humus-forming bacteria accumulated, like castings of earthworms. Five hundred millions of aerobic bacteria per gram were calculated in the finished product. The preparation also serves to regulate the lime and nitrogen content of the soil and to help release trace elements (Ram and Kumar 2019).

**Fig. 15.4** Root system development after seed treatment with BD-500

**Table 15.2** Identification and PGP test of microorganisms isolated from BD-500

| Code | Zinc solubilization | Phosphate solubilization | Siderophore production |
|---|---|---|---|
| BD-500(5) | – | ++++ | ++++ |
| BD-500(6) | – | ++++ | ++++ |
| BD-500(7) | – | +++ | + |
| BD-500(9) | – | ++++ | ++++ |
| BD-500(1) | – | ++++ | ++++ |
| BD-500(18) | – | ++++ | ++++ |
| BD-500(19) | – | ++++ | ++++ |
| BD-500(21) | – | +++ | ++ |

**Fig. 15.5** Preparation of cow horn silica (BD-501)

#### 15.4.2.3.3 Cow Pat Pit (CPP)

It is a special biodynamic field preparation also called as "soil shampoo." Fresh cow dung obtained from lactating and pasture-going cows is used for preparation. Cow dung is fermented along with crushed egg shells (calcium) and basalt/bentonite (clay) dust duly mixed and placed in a pit size of 3 × 2 × 1.5′. Two sets of BD-502-507 preparations are incorporated for catalyzing the composting process (Fig. 15.6). It is a concentrated source of beneficial organisms. It also contained the highest amount of *B. subtilis* ($1.9 \times 10^6$) responsible for disease tolerance (Ram and Kumar 2019) in plants. PGPR isolated in cow pat pit are presented in Table 15.4 which showed that it's a consortium of beneficial microbes.

## 15.5 Biodynamic Liquid Manure/Pesticides

Biodynamic liquid manures/pesticides are prepared from cow dung, urine, and leaves of leguminous tree, neem leaves, fish waste, castor leaves, and other medicinal plant parts. Besides cow dung, cow urine and one set of biodynamic preparations (502–507) are also incorporated (Fig. 15.7). The liquid manures are used to promote the vigor and quality production. On an average, preparation of liquid

**Table 15.3** PGP test of microorganisms isolated from BD-501

| Culture name | Zinc solubilization | Potassium solubilization | HCN production | Siderophore production | IAA production | Amylase production | Potassium solubilization |
|---|---|---|---|---|---|---|---|
| 501–7 | +++ | _ | – | +++ | + | + | _ |
| 501–11 | ++ | +++ | _ | _ | + | _ | _ |
| 501–15 | +++ | – | – | – | – | – | – |
| 501–25 | ++++ | ++ | – | – | + | – | – |
| 501–26 | ++++ | _ | – | – | – | – | – |
| 501–32 | ++ | – | – | – | – | + | – |

**Fig. 15.6** Preparation of cow pat pit

**Fig. 15.7** Preparation of biodynamic liquid pesticides

**Table 15.4** Identification and PGP test of microorganisms isolated from cow pat pit

| Microbial culture | Zn solubilization | P solubilization | HCN production | Siderophore production |
|---|---|---|---|---|
| C1 | + | – | – | ++ |
| C2 | ++++ | ++++ | – | ++ |
| C3 | +++ | – | – | – |
| C7 | ++++ | ++++ | – | ++ |
| C8 | ++++ | ++++ | – | +++ |
| C14 | ++++ | +++ | – | ++ |
| C15 | – | – | – | ++++ |
| C16 | ++++ | – | – | +++ |

manure takes 10–12 days. Liquid manure can also be prepared with *Pongamia*, *Calotropis*, and nettle leaves, which also have insecticidal and fungicidal properties. Microbial analysis of biodynamic liquid pesticide was done from 0 to 20 days after preparation and PGP test of isolated microbes was done and observations are presented in Table 15.5.

**Table 15.5** PGP test of microorganisms isolated from biodynamic liquid manure/pesticide

| Microbial culture | Zn solubilization | P solubilization | HCN production | Siderophore production | IAA production | Amylase production | Potassium solubilization |
|---|---|---|---|---|---|---|---|
| BL2 | ++ | + – | – | ++ | – | – | – |
| BL 5 | ++++ | +++ | – | + | +++ | – | ++ |
| BL 8 | +++ | +++ | – | +++ | ++++++++ | – | ++++ |

### 15.5.1 Panchagavya

It is a special bio-enhancer prepared from five products obtained from cow, i.e., dung, urine, milk, curd, and ghee (Fig. 15.8). When these are properly mixed, incubated for recommended period, and ready fermented, the solution has miraculous effect on all crops. Preparation is rich in nutrients, auxins, gibberellins, and microbial fauna and acts as tonic to enrich soil and induce plant vigor with quality production. Panchagavya is now gaining attention as an efficient organic growth promoter.

Composition of panchagavya was investigated by Rakesh et al. (2017), and they observed the presence of aerobic heterotrophic bacteria, lactic acid bacteria, yeast, fungi, and anaerobic bacteria. In a study, the highest microbial load was recorded in a 7-day-old preparation. However, a gradual reduction in the microbial load was observed up to 50 days and population reduced significantly after 30 days.

The effective microorganisms in panchagavya were the mixed culture of naturally occurring, beneficial microbes, mostly *Actinomycetes*, P-solubilizing bacteria, *Rhizobium*, *Azospirillum*, and certain fungi. In fact, panchagavya contains naturally occurring beneficial microorganisms, some of which are nitrogen fixers and P solubilizers (Ram et al. 2019), which can be considered as an ideal organic growth promoter. However, it is advisable to be used within 30 days of its preparation to achieve better success. Application of panchagavya has been found more effective than recommended chemical fertilizers and chemical sprays. Kumar et al. (2015)

**Fig. 15.8** Preparation of panchagavya

reported that among the different treatments tested against the management of insect pest in teak (*Tectona grandis*), 7 percent and 5 percent panchagavya solution were found to be more effective in controlling the pests. However, cost-effective analysis showed 7 percent panchagavya was found to be cheaper and affordable to the tree growers.

Nutritional analysis of panchagavya revealed that it possesses almost all macro- and micronutrients and growth-promoting hormones (IAA, GA) required for plant growth (Ram et al. 2017). Predominance of fermentative microorganisms like yeast and *Lactobacillus* are due to combined effect of low pH milk products and addition of jaggery/sugarcane juice as substrate for their growth. In general, 3 percent (3 kg/100 liters) solution of panchagavya has been found effective for most of the crops. PGP test of microbes isolated from panchagavya is presented in Table 15.6.

### 15.5.2 *Jeevamrita*

Jeevamrita is prepared by fermenting cow dung, urine, jaggery, pulse flour, and virgin soil by simple facilities created in the village with minimum expenditure. Jeevamrita is a rich bioformulation that contains consortia of beneficial microbes. Systematic microbial analysis of jeevamrita from the day 0 to day 20 of preparation suggests that formulation showed maximum population of microbial consortia within 6–9 days, hence should be used for better response (Ram et al. 2017). PGPR test of bacteria isolated from jeevamrita has been given in Table 15.7.

### 15.5.3 *Beejamrita*

It is special bio-enhancer prepared with locally available materials for seed and seedling treatment. As preparation is very cheap and cost-effective, it can easily be prepared and used by small and marginal farmers (Fig. 15.9).

Microbial analysis of Beejamrita showed that it should be used on the seventh day of preparation (Table 15.8). In case the crop is transplanted, it is advisable that seeds are treated with Beejamrita before sowing, and while planting the root portion of the seedlings is dipped in the solution. PGP tests of isolated bacteria from Beejamrita are presented in Table 15.8.

**Table 15.6** Identification and PGP test of microorganisms isolated from panchagavya

| Culture name | Zn solubilization | P solubilization | HCN production | Siderophore production |
|---|---|---|---|---|
| P3 | – | + | – | +++ |
| P4 | – | +++ | – | – |
| P8 | ++++ | ++++ | – | ++ |
| P10 | – | – | – | + |

**Table 15.7** Identification and PGP test of microorganisms isolated from jeevamrita

| Culture Name | Zn solubilization | P solubilization | HCN production | Siderophore production |
|---|---|---|---|---|
| J2 | +++ | ++++ | _ | ++++ |
| J4 | +++ | +++ | _ | +++ |
| J6 | ++++ | +++ | _ | + |

**Fig. 15.9** Preparation of Beejamrita

**Table 15.8** Identification and PGP TEST of microorganisms isolated from Beejamrita

| Code | Zinc production | Phosphate solubilization | Siderophore production |
|---|---|---|---|
| BJ10 | _ | ++ | +++ |
| BJ13 | ++++ | + | ++++ |
| BJ14 | ++++ | ++ | + |
| BJ23 | ++++ | +++ | ++ |

## *15.5.4 Amritpani*

This formulation is rich in nutrients (Fig. 15.10) and beneficial microbes (Table 15.9). Ingredients for preparation of amritpani and its intensive use are suggested. It is used to improve seed germination, soil fertility, and plant vigor. The use of amritpani along with 250 g rhizospheric soil of *Ficus benghalensis* tree and organic mulching in organic guava production gives a maximum return of INR 1,27,746/ha and benefit-cost ratio of 4.4 compared to INR of 1,20,820 and benefit-cost ratio of 3.7 with application of 350 g N, 150 g $P_2O_5$, and 350 g $K_2O$ tree$^{-1}$ (Ram and Verma 2017).

Systematic microbial profiling analysis of amritpani from 0 to 20 days suggests that amritpani should be utilized from the sixth to ninth day of preparation for

**Fig. 15.10** Preparation of amritpani

maximum effectiveness. Ram et al. (2018a, b) have estimated microbial load and nutrient status of few selected bioenhancers. This indicates that regular use of these formulations can resolve many problems associated with soil fertility and crop productivity.

### *15.5.5 Vermiwash*

Vermiwash is a liquid leachate obtained by excess water to saturate the vermicomposting substrate. It is a collection of excretory products and mucus recreations of earthworms along with nutrients from the soil organic molecules. In fact, vermiwash is an enriched bioformulation prepared from the heavy population of earthworms reared in earthen pots/plastic or cement containers (Fig. 15.11). Vermiwash was found to contain enzyme cocktail of proteases, amylases, urease, and phosphatase.

Laboratory scale trial showed effectiveness of vermiwash on cowpea plant growth (Asha et al. 2016). Vermi wash contains *Azotobacter*, *Agrobacterium*, and *Rhizobium* and phosphate-solubilizing microbes. The presence of these microbes makes available inorganic nitrogen, amino acids, and inorganic phosphate to plants through ammonification and nitrification process.

Vermi wash can be used for better growth, yield, and quality production. Recently microbial study of vermiwash revealed that it contains nitrogen-fixing bacteria like *Azotobacter* spp., *Agrobacterium* sp., *Rhizobium* spp., and phosphate-solubilizing bacteria (Table 15.10). Protease in soils helps in seed germination, while amylases help in the availability of simple carbon source for enhancement of plant vigor and productivity as well. Soil-born microflora is essential for growth of plants because organic nitrogenous compounds and phosphorus are decomposed and mineralized by fixing and phosphate-solubilizing bacteria. The presence of a large number of

**Table 15.9** Identification and PGP test of microorganisms isolated from amritpani

| Isolates | IAA production | Amylase production | Phosphate-solubilizing microbes | HCN production | Siderophore production | Potassium-solubilizing microbes | Zinc-solubilizing microbes | Catalase production |
|---|---|---|---|---|---|---|---|---|
| A8 | +++ | + | ++ | – | +++ | ++++ | ++++ | + |
| A10 | ++++ | +++ | ++ | – | ++++ | +++ | ++++ | ++ |

**Fig. 15.11** Preparation of vermiwash

**Table 15.10** Identification and PGP test of microorganisms isolated from vermiwash

| Culture name | Zn solubilization | P solubilization | HCN production | Siderophore production |
|---|---|---|---|---|
| V9 | – | +++ | – | +++ |
| V10 | +++ | ++++ | ++++ | ++++ |
| V11 | ++++ | ++ | – | ++++ |
| V12 | ++++ | ++++ | +++++ | ++++ |

beneficial microorganisms helps in plant growth and protects from a number of pathogens in the field (MacHfudz et al. 2020). Repeated spray of vermiwash has been found effective for the control of two-spotted spider mite, *Tetranychus urticae Koch*, on bean plant (*Phaseolus vulgaris* L.) (Aghamohammadi et al. 2016).

## 15.6 Future Prospects

Overuse of chemical fertilizers and insect pest management through pesticides have become a threat to the soil and environment and human health; hence, these practices are now being discouraged worldwide. This has led to conceive viable alternatives to manage soil fertility and manage insect pests. Composts, bio-enhancers, and biopesticides, particularly PGPR, for a wide range of crops are effective tools for bioremediation of soil and biocontrol. PGPR-mediated agriculture is now gaining worldwide importance and acceptance for a large number of crops, to improve soil biology and ecosystems associated with quality production without use of agrochemicals as the safe method of insect pest management. The limited understanding of soil micro-life often hinders attempts, e.g., employing tailored application strategies, to optimize the biological activity. One can envision a number of different ways in which biocontrol efficacy of PGPR might be improved. Biocontrol efficacy of PGPR may be improved by genetically engineering them to overexpress one or more of these traits so that strains with several different anti-phytopathogen traits can act synergistically. Systematic studies are needed on the composition of the rhizosphere population, the effect of cultivar on bacterial population dynamics, the influence of inoculum density on antagonistic activity, the survival of inoculum under adverse conditions, and the role of environmental conditions in altering the activity of rhizobacteria. An attempt to overcome problems of varying efficacy may be attained by strain mixing, improved inoculation techniques, or gene transfer of active genetic source of antagonists to the host plant. The soil microbes are active elements for soil development and the basis of sustainable agriculture. From the point of sustainable agricultural development and sound environment establishment, one should encourage organic manures coupled with bio-enhancers and biopesticides in a rational way to ensure high soil biological activities to improve rhizobacteria, responsible for sustainable quality production (Table 15.11).

**Table 15.11** Beneficial effects of some PGPRs on various crops

| PGPR | Name of plant | Scientific name | Effect | References |
|---|---|---|---|---|
| *Azospirillum brasilense* | Strawberry | *Fragaria*× *ananassa* | Stimulate plant growth, increase plant resistance to abiotic and biotic stresses, and might thus have a positive effect on fruit quality | Pii et al. (2018) |
| *Phyllobacterium endophyticum strain PEPV15* | Strawberry | *Fragaria* × *ananassa* | Siderophore production | Flores-Felix et al. (2015) |
| *Alcaligenes* 637Ca, *agrobacterium* A18, *staphylococcus* MFDCa1, *staphylococcus* MFDCa2, *bacillus* M3, and *Pantoea* FF1 | Apple | *Malus domestica* | Bacterial strains used in the present study have an important potential to increase plant growth | İpek et al. (2021) |
| *Brevibacillus laterosporus* AMCC100017 | Apple | *Malus robusta* | Increased IAA content in *M. robusta* and promoted root secretion of tryptophan | Wang et al. (2022) |

(continued)

**Table 15.11** (continued)

| PGPR | Name of plant | Scientific name | Effect | References |
|---|---|---|---|---|
| *Bacillus amyloliquefaciens, bacillus megaterium* M3, *and* mix *(Bacillus subtilis* OSU142, *B. megaterium* M3, *Azospirillum brasilense* Sp245) | Barley | *Hordeum vulgare* | Increased yield, growth of seedlings, salinity tolerance | Kasim et al. (2016) |
| *Bacillus amyloliquefaciens* | Mango seedlings | *Mangifera indica* | Showed maximum PGPR properties including phosphorus, zinc solubilization, siderophore production | Ram et al. (2020a, b) |
| *B. pumilus, Alcaligenes piechaudii* | Cucumber | *Cucumis sativus* L. | Seedling growth and quality | Yıldırım et al. (2015) |
| *Paenibacillus polymyxa* (SN-22), *Sinomonas atrocyanea* (NSB-27) | Watermelon | *Acidovorax citrulli* | Total chlorophyll content, plant height, total fresh weight, total dry weight | Adhikari et al. (2017) |
| *Neurospora* sp. SR8 | Sorghum | *Sorghum bicolor* (L.) Moench | Potent phosphate solubilizer in stressed condition | Sharma et al. (2016) |
| *Bradyrhizobium japonicum* USDA 110 and *pseudomonas putida* TSAU1 | Soybean | *Glycine max* L. Merr. | Produced good amounts of indole-acetic acid, siderophores, and phosphate solubilization which also significantly improved the growth and nutrient content of soybean and the level of nutrients and enzymes in the soil | Jabborova et al. (2020) |
| *Enterobacter* sp. PR14 | Rice | *Oryza sativa* | Promotes the growth of under salinity stress | Sagar et al. (2020) |

# References


Adhikari M, Yadav DR, Kim SW et al (2017) Biological control of bacterial fruit blotch of watermelon pathogen (*Acidovorax citrulli*) with rhizosphere associated bacteria. Plant Patholog J 33:170–183. https://doi.org/10.5423/PPJ.OA.09.2016.0187

Agarwal M, Dheeman S, Dubey RC, Kumar P, Maheshwari DK, Bajpai VK (2017) Differential antagonistic responses of *Bacillus pumilus* MSUA3 against *Rhizoctonia solani* and *Fusarium oxysporum* causing fungal diseases in *Fagopyrum esculentum* Moench. Microbiol Res 205:40–47

Aghamohammadi Z, Etesami H, Alikhani HA (2016) Vermiwash allows reduced application rates of *acaricide azocyclotin* for the control of two spotted spider mite, *Tetranychus urticae Koch*, on bean plant (*Phaseolus vulgaris* L.). Ecol Eng 93:234–241

Alori ET, Glick BR, Babalola OO (2017) Microbial phosphorus Solubilization and its potential for use in sustainable agriculture. Front Microbiol 8(971)

Arora NK (2018) Environmental sustainability—necessary for survival. Environ Sustain 1:2
Asha V Pillai, Aswathy, K K, Preethy, T T and Renisha, Mannambeth (2016). Effect of organic liquid manures on crop growth and productivity. International Journal of Current Research, 8(04): 29023–29029.
Assefa S, Tadesse S (2019).The principal role of organic fertilizer on soil properties and agricultural productivity -a review.Agri Res& Tech: Open Access J. 22(2): 556192. DOI: https://doi.org/10.19080/ARTOAJ.2019.22.556192
Carpio MJ, Sánchez-Martín MJ, Rodríguez-Cruz MS, Marín-Benito JM (2021) Effect of organic residues on pesticide behavior in soils: a review of laboratory research. Environments 8:32. https://doi.org/10.3390/environments8040032
Chandra S, Askari K, Kumari M (2018) Optimization of indole acetic acid production by isolated bacteria from Stevia rebaudiana rhizosphere and its effects on plant growth. J Gene Eng Biotechnol 16(2):581–586
Dos Santos RM, Diaz PAE, Lobo LLB, Rigobelo EC (2020) Use of plant growth-promoting Rhizobacteria in maize and sugarcane: characteristics and applications. Front Sustain Food Syst 4:136. https://doi.org/10.3389/fsufs.2020.00136
Flores-Felix JD, Silva LR, Rivera LP (2015) Plants probiotics as a tool to produce highly functional fruits: the case of Phyllobacterium and vitamin C in strawberries. PLoS One 10:e0122281
FAO (2012) Current world fertilizer trends and outlook to 2016. Food and Agriculture Organization of the United Nations (FAO), Rome.
Gouda S, Nayak S, Bishwakarma S, Kerry RG, Das G, Patra JK (2017) Role of microbial technology in agricultural sustainability. Microbial biotechnology. Springer, Singapore, pp 181–202
Gundersen C, Ziliak JP (2015) Food insecurity and health outcomes. Health Aff 34(11):1830–1839
Gupta G, Parihar SS, Ahirwar NK, Snehi SK, Singh V (2015) Plant growth promoting rhizobacteria (PGPR): current and future prospects for development of sustainable agriculture. J Microb Biochem Technol 7(2):096–102
İpek M, Arıkan Ş, Eşitken A, Pırlak L, Turan M, Dönmez M (2021) Effects of some plant growth-promoting Rhizobacteria (PGPR) on growth and nutrition of apple Cv. "Braeburn" under high lime soil condition. Commun Soil Sci Plant Anal 52(5):432–442
Jabborova D, Wirth S, Kannepalli A, Narimanov A, Desouky S, Davranov K, Sayyed RZ, El Enshasy H, Malek RA, Syed A, Bahkali AH (2020) Co-inoculation of Rhizobacteria and biochar application improves growth and nutrients in soybean and enriches soil nutrients and enzymes. Agronomy 10:1142. https://doi.org/10.3390/agronomy10081142
Kasim WA, Gaafar RM, Abou-Ali RM et al (2016) Effect of biofilm forming plant growth promoting rhizobacteria on salinity tolerance in barley. Ann Agric Sci 61:217–227. https://doi.org/10.1016/j.aoas.2016.07.003
Kumar MS, Bharath M, Josmin LL, Nisha L, Basavaraju H (2015) Field efficacy of Panchagavya on insect pests recorded during the study in *Tectona Grandis*. Int J Res Agric For 2(7):1–8
Kundan R, Pant G, Jadon N, Agrawal PK (2015) Plant growth promoting rhizobacteria: mechanism and current prospective. J Fertil Pest 6:155. https://doi.org/10.4172/2155-6202.1000155
MacHfudz M, Basit A, Handoko RNS (2020) Effectiveness of vermicompost with additives of various botanical pesticides in controlling *Plutella xylostella* and their effects on the yield of cabbage (*Brassica oleracea L. var. Capitata*). Asian J Agricult 8(3):223–232
Meena SK, Rakshit A, Meena VS (2016a) Effect of seed bio-priming and N doses under varied soil type on nitrogen use efficiency (NUE) of wheat (*Triticum aestivum* L.) under greenhouse conditions. BiocatalAgricBiotechnol 6:68–75
Meena VS, Bahadur I, Maurya BR, Kumar A, Meena RK, Meena SK, Verma JP (2016b) Potassium-solubilizing microorganism in evergreen agriculture: an overview. In: Potassium solubilizing microorganisms for sustainable agriculture. Springer, New Delhi, pp 1–20
Pandey C ,Dheeman, S, Prabha D, Negi, Y, Maheshwari, D (2021) Plant growth-promoting bacteria: effective tools for increasing nutrient use efficiency and yield of crops. https://doi.org/10.1007/978-3-030-65447-4_13
Pathak R K and Ram R A (2013). Bio-enhancers:A potential tool to improve soil fertility and plant health in organic production of horticultural crops, Progress Hortic, 45(2):237–254

Pii Y, Graf H, Valentinuzzi F, Cesco S, Mimmo T (2018) The effects of plant growth-promoting rhizobacteria (PGPR) on the growth and quality of strawberries. Acta Hortic 1217:231–238. https://doi.org/10.17660/ActaHortic.2018.1217.29

Pillai AV, Aswathy KK, Preethy TT, Renisha M (2016) Effect of organic liquid manures on crop growth and productivity. Int J Curr Res 8(04):29023–29029

Prasad R, Kumar M, Varma A (2015) Role of PGPR in soil fertility and plant health. In: Plant-growth-promoting rhizobacteria (PGPR) and medicinal plants. Springer, pp 247–260

Rakesh S, Poonguzhali S, Saranya B, Suguna S, Jothibasu K (2017) Effect of panchagavya on growth and yield of okra (*Abelmoschus esculentus* L. cv. Arka Anamika). Int J Curr Microbiol Appl Sci 6(9):3090–3097

Ram RA, Kumar A (2019) Bio-dynamic agriculture: An advance stage of organic farming. J Eco-Friendly Agric 14(1):34–37

Ram RA, Pathak RK (2016a) Organic approaches for sustainable production of horticultural crops: a review. Progress Hortic 48(1):1–16

Ram RA, Pathak RK (2016b) Organic approaches for sustainable production of horticultural crops: a review. Progress Hortic 48(1):1–16

Ram RA, Pathak RK (2017) Bioenhancers. LAP Lambert Academy Publishing, Germany, ISBN 978-3-330-33128-0, P30

Ram RA, Pathak RK (2019) Indigenous technologies of organic agriculture-a review. Progress Hortic 50(1&2):70–81

Ram RA, Verma AK (2017) Energy input, output and economic analysis in organic production of guava (*Psidium guajava*) cv. Allahabad Safeda Ind J Agric Sci 87(6):462–467

Ram RA, Garg N, Preeti (2016) Study on antimicrobial property of amritpani, cow pat pit, jeevamrita and panchagavya on pathogens. ICAR network project report pp 12-14

Ram RA, Singha A, Verma AK (2017) Annual report. Development of organic package of practice in mango cv. Mallika. ICAR network project report. pp 2–8

Ram RA, Singha A, Vaish S (2018a) Microbial characterization of on-farm produced bio-enhancers used in organic farming. Indian J Agric Sci 88(1):35–40

Ram RA, Singha A, Vaish S (2018b) Microbial characterization of on-farm produced bio-enhancers used in organic farming. Indian J Agric Sci 88(1):35–40

Ram RA, Ahmad I, Kumar A (2019) Isolation and characterization of PGPRs from organic preparations. J Eco-Friendly Agric 14(2):18–22

Ram R A, Govind Kumar, Shailendra Kumar Maurya, S. Rajan and Israr Ahmad (2020a). Plant growth promoting, biochemical and antifungal properties of microbes isolated from cow dung, Progress Hortic, 52(2):111–115

Ram RA, Ahamad I, Kumar G, Maurya S (2020b) Uttar Pradesh Council of Science and Technolgy, Lucknow. Annu Rep pp50

Sagar A, Sayyed RZ, Ramteke PW, Sharma S, Marraiki N, Elgorban AM, Syed A (2020) ACC deaminase and antioxidant enzymes producing halophilic Enterobacter sp. PR14 promotes the growth of rice and millets under salinity stress. Physiol Mol Biol Plants 26:1847–1854

Sharma S, Sayyed RZ, Sonawane M, Trivedi MH, Thivakaran GA (2016) *Neurospora* sp SR8, a novel phosphate solubiliser from rhizosphere of soil of Sorghum in Kachh, Gujarat. Indian J Exp Biol 54:644–649

Swarnalakshmi K, Yadav V, Tyagi D, Dhar DW, Kannepalli A, Kumar S (2020) Significance of plant growth promoting Rhizobacteria in grain legumes: growth promotion and crop production. Plan Theory 9(11):1596. https://doi.org/10.3390/plants9111596

Tomkins P, Bird C (2004) Secrets of the soil. Rupa Publications, New Delhi, p 444

Wang X, Zhang J, Wang X, An J, You C, Zhou B, Hao Y (2022) The growth-promoting mechanism of *Brevibacillus laterosporus* AMCC100017 on apple rootstock *Malus robusta*. Hortic Plant J 8(1):22–34

Yıldırım E, Ekinci M, Dursun A, Karagöz K (2015) Plant growth-promoting rhizobacteria improved seedling growth and quality of cucumber (*Cucumis sativus* L.), International conference on chemical, food and environment engineering (ICCFEE'15) Jan 11–12, Dubai, pp 6–8

# Chapter 16
# Plant-Microbe-Insect Interactions: Concepts and Applications for Agricultural Sustainability

**V. Prithiv Raj, S. Thangavel, N. Ponnusamy, and V. Prabakaran**

## Contents

16.1 Introduction ........ 335
16.2 Plant-Microbe Interactions ........ 337
16.3 Plant-Insect Interactions ........ 339
16.4 Tritrophic Interaction ........ 342
16.5 Influence of Insect-Associated Microbes on Plants ........ 342
16.6 Contribution of Microbes on Detoxifying Plant Defense ........ 343
16.7 Natural Enemies Attracted by Microbes ........ 343
16.8 Effect of Phytopathogen on Herbivore Fitness ........ 344
16.9 Benefits of Plant-Microbe-Insect Interaction in Sustainable Agriculture ........ 344
16.10 Conclusion ........ 345
References ........ 346

## 16.1 Introduction

The globe has been facing a challenge to produce adequate food in a sustainable manner for the increasing global population, as this has been decreasing food resources. Here, stresses (abiotic and biotic) were involved in sinking the plant food production. Among them, the over-abundance of biotic stresses like pathogens and

V. Prithiv Raj (✉)
Department of Agricultural Entomology, Sardarkrushinagar Dantiwada Agricultural University, Dantiwada, Gujarat, India

S. Thangavel · V. Prabakaran
Department of Agricultural Entomology, B. A. College of Agriculture, Anand Agricultural University, Anand, Gujarat, India

N. Ponnusamy
Department of Agricultural Entomology, Dr. Rajendra Prasad Central Agricultural University, Pusa, Bihar, India

 
R. Z. Sayyed et al. (eds.), *Antifungal Metabolites of Rhizobacteria for Sustainable Agriculture*, Fungal Biology, https://doi.org/10.1007/978-3-031-04805-0_16

herbivores predominantly hampers the food production (Thakur and Sohal 2013). Generally insects need food sources, which provide nutrients for their metabolism. The herbivores as arthropods entirely depend on plant source for their growth, development, and reproduction. To identify the potential host plants, phytophagy habitually relies on cues that give the relevant information about the particular plant traits (Bruce et al. 2005) like location (Couty et al. 2006), nutritional quality (Bruce and Pickett 2011), and defensive status (De Moraes et al. 2001).

Cues are significant for the selection process of plant for oviposition and foraging. In another case, most of the plants are associated with beneficial and pathogenic microorganism, often alter the plant traits, and indirectly affect the cues as well as herbivores. The pathogenic and beneficial plant associated-microbes interfere the plant-herbivore interaction by modifying the olfactory, visual, and gustatory cues in plants (Fig. 16.1), thereby affecting the selection process of herbivores (Grunseich et al. 2019). There are several microorganisms lodge on the roots as well as shoots that provide good support to the plants against various biotic and abiotic stresses. Among them, bacteria habituated on below ground like plant growth-promoting rhizobacteria (PGPR) which include *Azotobacter*, *Arthrobacter*, *Azospirillum*, *Burkholderia*, *Pseudomonas*, *Serratia*, *Rhizobium*, and *Klebsiella* help in enhancing the plant health by producing various metabolites (Basu et al. 2021).

Several endophytic bacteria like *Burkholderia solanacearum*, *Phyllobacterium rubiacearum*, and *Sphingomonas trueperi* were colonizing the internal tissues and

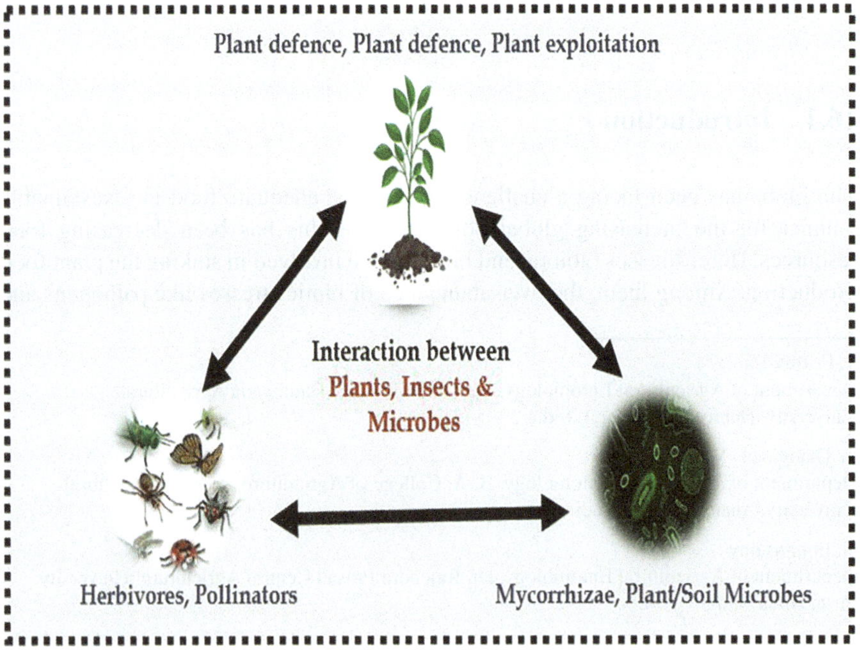


**Fig. 16.1** Interactions between plants, insects, and microbes

support plant health. These endophytic bacteria are used for fixing nitrogen and also degrade the organic pollutants. There are specific hormones like Indole-3-acetic acid (IAA) and pyocyanin which are doing multiple functions in the rhizosphere root system. Same compound used as growth regulator, biocontrol agent, induces nodulation and also acts as a stress regulator in the case of IAA (Zope et al. 2019a, b).

Plant chemicals have a critical role in an ecosystem and they provisioned the insect to identify the specific potential host for its own development. Certain activities like feeding, oviposition, and reproduction may depend on the plant cues as well as metabolites. These cues and metabolites are influenced by several microorganisms, and the third trophic level has the capacity to alter the chemical metabolite which is beneficial and but also deleterious sometimes. These plant chemicals also give chemical defense to certain insects like lemon butterfly against its natural enemies. Beneficial microbes stimulate the ISR in plants against fungi, virus, insect, and nematodes. *Beauveria bassiana* and *Lecanicillium lecanii* provide induced systemic resistance (ISR) in plants and the ability to withstand against various stresses. These tritrophic interactions have crucial effects on plant health and make provisions for the eco-friendly crop management in sustainable agriculture (Patel et al. 2016a, b).

## 16.2 Plant-Microbe Interactions

There is this famous saying: don't take more than you need, give as much as you get. This saying is well suited for plants and microbes which maintain the sustainability. All plants are associated with a kind of microorganism, some may be beneficial and some may be harmful. Harmful microorganisms always lead to the diseases which induce great crop losses. Beneficial microbes like AM fungi, root-inhabiting bacteria, and several endophytes promote the crop growth and yield (Bastami et al. 2021; Ayshasiddeka et al. 2021). The association between the plant and microbe is essential to promote plant growth, protecting plant health and preventing the attack from pathogens through biological control. Several functions accomplished by the microbes in the locality of plants (phyllosphere, rhizosphere, etc.) are exceptionally interlinked and inseparable. Besides harmful pathogen several beneficial microbes are enhancing the growth and development as discussed below.

Bacteria which are present on the rhizosphere zone and give beneficial effects to the plants are called PGPR (plant growth-promoting rhizobacteria) (Patel et al. 2016a, b). PGPR comprise *Azotobacter*, *Arthrobacter*, *Azospirillum*, *Alcaligenes*, *Bacillus*, *Enterobacter*, *Burkholderia*, *Pseudomonas*, *Serratia*, *Rhizobium*, and *Klebsiella* which improve the crop growth (Jadhav et al. 2020; Manasa et al. 2021). The comparison of PGPR with 2,4-D, which acts as a growth regulator at certain concentration and acts as a herbicide in higher doses like that leaving doses, PGPRs act as a growth promoters and bio-control agents by direct and indirect mechanism (Hamid et al. 2021).

Direct mechanisms preferred by PGPR consist of enhancing the nutrient (phosphorous) uptake by plants; iron sequestration through siderophores; nitrogen fixation; synthesis of plant hormones, viz., auxins, cytokinins, and gibberellins; and reducing the ethylene level in plants by using ACC deaminase, which accumulates during unfavorable conditions (Mayak et al. 2004). Likewise, some indirect mechanisms executed by PGPR were antibiotic production as production of 2,4-diacetylphloroglucinol (DAPG), pyrrolnitrin, phenazine, and pyoluteorin were used against the pathogenic fungi and bacteria; in the rhizosphere, PGPR reduce the available iron to the phytopathogens by producing siderophores; produce fungal cell-wall lytic enzymes; create competition with harmful microorganisms for space requirement on roots of the plants; and induce systemic resistance not in favor of several pathogens of plants (Nithyapriya et al. 2021).

Mutualistic, commensalistic, parasitic, predatory, and neutral relationships were showed by the microbes in rhizosphere region. These relationships are not stable and continuously changing with respect to time and climatic conditions in that region. Here, PGPRs show mutualism with plants and work in diverse aspects (Khan et al. 2021; Baba et al. 2021). Even particular mechanism works on visible and invisible ways on plants. For an instance, in providing nutrients to the plants, PGPR directly regulate and enhance the growth which is visible through increasing crop yield, but in providing nutrients to the host at deficient conditions, the functions from PGPRs are unnoticeable. Once the host becomes healthy, its ability against pathogens might increase (Suriani et al. 2020). In N- and P-deficient soils, the nitrogen-fixing bacteria and phosphate-solubilizing bacteria promote the plant growth. Thus, in several ways microbes sustain the plant growth that helps in enhancing the yield.

### *16.2.1 Plant Ecosphere*

With the presence of microbial loads, the regions surrounding the plants have been named as rhizosphere (Barea et al. 2005), phyllosphere (Whipps et al. 2008), and spermosphere (Liu et al. 2012), and the whole region is called plant ecosphere. There is no even distribution of microbes in ecosphere; all are unstable in nature (Whipps et al. 2008). Different studies demonstrated that *Pseudomonas* is the most dominant and common population inhabiting the plant rhizosphere, and its abundance was noticed by Liu et al. (2012) in sunflower (*Helianthus annuus* L.), finger millet (Pennisetum glaucum L.), and maize (*Zea mays* L.). The diversity and distribution of microorganism may differ from host species to species and niche to niche, for instance, Nelson and Mele (2007) said that the wheat rhizosphere is majorly occupied by *Klebsiella pneumonia*, *Vibrio diazotrophicus*, *Azospirillum*, *Rhodobacter*, *Pseudomonas*, and *Azotobacter chroococcum* and *Beijerincka*. Likewise in ryegrass, *Flavobacterium* and *Cytophaga* constitute the major fraction (Marilley and Aragno 1999). Bacteroidetes, *Chryseobacterium*, and *Oxalobacteraceae* dominate the canola (Kaiser et al. 2001). *Variovorax*,

*Acinetobacter calcoaceticus*, and *Arthrobacter ramosus* dominate the rhizosphere of alfa-alfa (*Medicago sativa*) and white goosefoot (*Chenopodium album*). Rhizosphere regions in legumes are primarily characterized by symbionts, like *Rhizobium*, *Mesorhizobium*, *Sinorhizobium*, and *Bradyrhizobium* (Tan et al. 2001). *Klebsiella* is a non-rhizobial symbiont, and several cases have been recognized constantly on the root surfaces and nodules of alfalfa, clover, and soybeans. *Azolla-Anabaena* symbiosis is a unique association for nitrogen fixation. *Pseudomonas*, *Azospirillum*, and *Bacillus* are root-colonizing bacteria and exist as endophytes by colonizing internal plant tissues (Berg and Hallmann 2006).

Apart from root-inhabiting bacteria, some bacterial species improve plant development as endophytes like *Burkholderia solanacearum*, *Phyllobacterium rubiacearum*, *Sphingomonas trueperi*, *Serratia plymuthica*, etc. (Kalam et al. 2020). *Burkholderia*, endophytic bacteria, present in most of the plants, having the capacity to fixing nitrogen and also degrades the organic pollutants in greatly contaminated soils (Kour et al. 2021). *Methylobacterium*, an endophyte extracted from aspen roots, has the capacity to degrade the nitro-substituted explosives (Jabborova et al. 2020). *Neotyphodium* sp., *Pestalotiopsis microspora*, *Guignardia* sp., *Acremonium* spp., and *Epichloe typhina* are endophytic fungi, responsible for resistance against biotic and abiotic stresses. Most of the higher plants, 80%, are able to form mycorrhizal symbiosis; among them, AM is the most regular association (Brundrett 2009).

IAA and pyocyanin are metabolites secreted by microbes and some plants in the rhizosphere, performing multiple roles. IAA, a metabolite able to serve as a plant growth stimulator, biocontrol agent, and stress regulator, induces nodulation in legumes. Pyocyanin acts as an iron chelator. Some *Pseudomonas* strains were reported that they have *hcn* genes which are responsible for the production of biocide HCN (Duffy et al. 2004), and the expression and production of HCN depend on availability of iron. Several examples are there for the mechanism and metabolite secreted in the rhizosphere by microbes, doing multiple tasks concurrently under various conditions may help in reclaiming agricultural lands, eco-friendly management and crop production.

## 16.3 Plant-Insect Interactions

Insect is an organism that always depends on other food resources for their metabolism. Though there are several categories of insects, herbivores play a detrimental role in today's agriculture. Herbivores fuel their metabolic activity by feeding plenty of plant matters for their nutritional requirement. Besides herbivores, several beneficial insects like predators (generally includes parasitic wasps), scavengers, and pollinators are enhancing the crop yield in direct and indirect ways (Fig. 16.2). Majority of the insects belong to herbivores (plant-dependent). Both plants and insects produce certain chemicals which have a significant role in their lives. Phytochemicals, specifically a secondary metabolite, play a critical role in several herbivores and

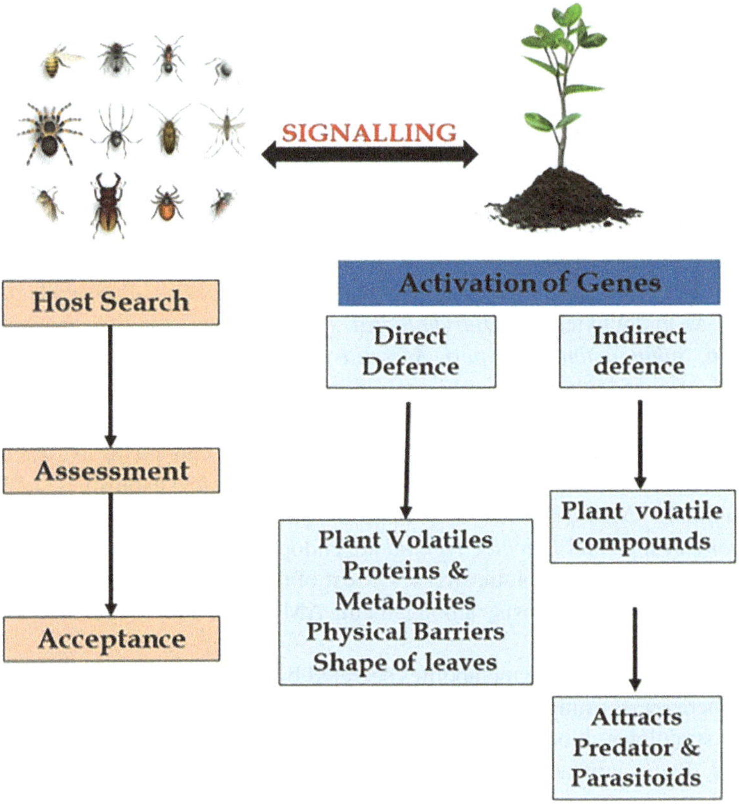


**Fig. 16.2** Overview of plant-insect interactions: challenges and promises

also affect the metabolism and reproduction in several ways and thus affect the fitness (Honda et al. 2010). The action and reaction of interacting organisms lead to develop a co-evolution which helps the insects to counteract the plants' chemical barriers. Later, insects use certain metabolites as cues to place and identify appropriate host. Several metabolites act as attractants, repellents, and deterrents, involved in this practice.

Host selection is an important process for survival of an insect herbivore. This selection process could be done by the various plant cues which could be identified by various olfactory and gustatory receptors. Generally, the selection process consists of two phases: food choice during preimaginal feeding and seeking host plants by egg-laying females. Besides plant chemical components, several visual and mechanical stimuli decide whether the particular host is a potential or not (Honda et al. 2010).

### 16.3.1 Oviposition Preference

There are several phytochemicals that act as a feeding and/or oviposition stimulant or deterrent to various specific insect species, for example, swallowtail butterfly mostly prefer rutaceous plants as hosts. *Eurytides marcellus*, commonly known as zebra swallowtail butterfly, is the feeding specialist of Annonaceae family. *Papilio polyxenes*, specialist member of Apiaceae, insect used to lay eggs with the response to two chemicals namely trans-chlorogenic acid and luteolin, known from *Daucus carota* (wild carrot). Pyrrolizidine alkaloids (PAs) are an important class of secondary metabolites, which certain lepidopteran in specific groups, that is, Danainae, Ithomiinae, and Arctiidae, sequester as larvae or adults and exploit as chemical protective (defense) substances against its natural enemies, possibly due to their bitter taste and hepatotoxicity (Honda and Maes 2008). In PA-storing lepidopterans, PAs work as precursors of male pheromones. *Idea leuconoe* (present in the Old World tropics) feeds on *Parsonsia* (Apocynaceae), which include PAs, and uses the alkaloids as typical cues in host recognition. The cinnabar moth (*Tyria jacobaeae*) exclusively depends on *Senecio jacobaea*, which belongs to Asteraceae that contain PAs. In a similar manner, several crops have specific chemical compounds responsible for herbivore attraction.

### 16.3.2 Floral Attraction

Most of the insect species visit angiosperm flowers for nutrition intake or reproductive reward and accidentally involved in pollination. The floral scents and colors of a flower act as cues for a long-range pull of insects (Knudsen et al. 2006). The floral volatile may differ from plant to plant, and more than 1700 compounds were identified in 1000 floral plants (Table 16.1). Among them, 12 compounds, namely, limonene, (E)-$\beta$-ocimene, $\beta$-myrcene, $\beta$-pinene, $\beta$-pinene, linalool, 6-methyl-5-hepten-2-one, $\beta$-caryophyllene, benzyl alcohol, 2-phenylethanol, benzaldehyde, and methyl salicylate, were commonly present at the rate of more than 50% as a floral volatile (Knudsen et al. 2006).

**Table 16.1** Generalized plant volatiles for insect pollinators

| Pollinator type (insects) | Plant volatile traits |
|---|---|
| Various bees | Terpenes, phenylpropenoids/benzenoids |
| Butterflies | Terpenes, benzenoids, and volatile organic compounds |
| Syrphid flies | Terpenes, phenylpropenoids (phenylethanol) |
| Hymenopteran insects | Blend of volatiles |
| Tropical beetles | Methoxylated benzenoids |
| Temperate beetles | Nitrogen compounds |
| Dipteran insects | Amine/sulfur-containing volatile organic compounds |

## 16.4 Tritrophic Interaction

Price et al. (1980) argued a thing that "theory on insect-plant interactions cannot progress realistically without consideration of the third trophic level." They included natural enemies and microorganisms in the discussion of third trophic level. Some research finding says that natural enemies used to locate the herbivore for predation with the help of herbivore-induced volatiles (HIVs) (Turlings et al. 1993). Identification of HIVs by natural enemies may be indirectly advantageous to plants. The plants trigger the HIVs through receiving the fatty acid-amino acid conjugates from herbivores' secretion (Alborn et al. 1997). In the earlier days, researchers focused on predator- and parasitoid-mediated plant-herbivore interaction but not microbes due to lack of knowledge. However, recent technologies bring out the microbe-mediated plant-herbivore interaction.

## 16.5 Influence of Insect-Associated Microbes on Plants

Recent research findings say that the herbivore perception and their HAMPs (herbivore-associated molecular patterns) by plants may be altered by the higher trophic levels which is advantageous to the herbivore, viz., cracking the direct plant defense may drive the herbivore to grow faster and act as a healthier host for the third trophic level. Altering the herbivore perception is beneficial for natural enemies by suppressing the HIV release which attracts competitors. A naturally occurring parasitic intracellular bacterium, *Wolbachia*, infects a broad range of phytophagous insects. A western corn worm, *Diabrotica virgifera virgifera*, which is affected by the *Wolbachia*, reduces or removes the genes which is responsible for defense but the process is unknown (Barr et al. 2010).

The *Wolbachia*, present on the oral secretions or regurgitant of root worm, produces non-coding RNAs which can normalize the plant as well as bacterial genes (Mayoral et al. 2014) and thus may indirectly affect HAMPs, present in rootworms. Generally, while feeding, many insects like larva of lepidoptera and some beetle larva used to regurgitate and thus may get out the microbes from digestive tract. The source of regurgitant is foregut anterior midgut, which has plenty of symbionts. Research findings said that Hessian fly (*Mayetiola destructor*)-infested wheat contains a variety of bacterial genera which are present on the fly's body (Bansal et al. 2011). Nearly 70% of bacterial genera were cultured from a fly's body, also present on infested wheat. Hence, the bacteria are part of Hessian fly oral secretion, secreted over the plants which could affect the plant defense.

Likewise, large amounts of bacteria were present on Colorado potato beetle's oral secretions which influence the jasmonic acid present in tomato. Jasmonic acid (JA) maintains the defense transcripts (PPO, protein inhibitors) in tomato (Chung et al. 2013a).

Without antibiotic treatments, plants which were fed by larvae showed reduced JA production, reasonable for anti-herbivore defense, but salicylic acid (SA) accumulation and expression was increased. JA mitigation helps the larvae to attain enormous growth, and SA-deficient plants not showed any suppression. This indicates that plant defense regulated by JA depends on SA signaling pathway. The bacteria responsible for suppressing plant defense are *Pseudomonas*, *Stenotrophomonas*, and *Enterobacter* which showed result while applying on wounded plants (Chung et al. 2013b). Reinsertion of bacteria to a larvae treated with antibiotic treatments restores its ability to reduce defense. Therefore, herbivores use symbiotic bacteria as a trick to distract plants into mistakenly perceiving the menace as microbial. The oral secretion of fall army worm (*Spodoptera frugiperda*) contains *Pantoea* and mitigate the antiherbivore defense in tomato, but in the case of maize, same bacteria in fall army worm provoke the plant defense (Acevedo et al. 2017).

## 16.6 Contribution of Microbes on Detoxifying Plant Defense

Insect-microbe association may be beneficial for insects in many aspects. Microbes play a vital role from enzyme synthesis to polymer degradation and chemical detoxification (Douglas 2015). Symbionts associated with poor feeders (phloem feeder) like aphids get their nutrients from microbial association for their metabolism. In detoxification, *Symbiotaphrina kochi*, a yeast-like symbiont, helps the *Lasioderma serricorne* against compounds such as tannic acid, resorcinol, and flavones in artificial diets. Hydrolases, phosphatases, glucosidases, and glutathione transferases are detoxifying enzymes synthesized by symbionts (Dowd and Shen 1990). In pine beetle *Dendroctonus ponderosae*, gut bacteria (*Pseudomonas*, *Serratia*, *Rahnella*, and *Burkholderia*) are involved in the terpenes' detoxification process (Adams et al. 2013). However, single bacteria may not be enough for detoxification which enhance larval development, and it may be done by the bacterial community as per several experiments.

## 16.7 Natural Enemies Attracted by Microbes

Biological control is a part of IPM as well as an important alternative method for managing harmful insect pests, and this could be done with the help of predators and parasitoids. But the effectiveness of the control method is affected by the lack of nutritional imbalance like lack of carbohydrate-rich food in modern agriculture. Most of them require additional food sources like nectar or sugar solution for their metabolism and reproduction in adult stage. Therefore, while growing natural enemies in laboratories, supplementary sugar solution is necessary for maintenance.

However, retaining the natural enemies on the crop for effective control is still a challenging one.

Recent studies have showed that the specific plant-insect-associated microbes can retain the natural enemies on crop, thus improving the efficacy of biocontrol programs. For instance, *Staphylococcus sciuri*, a bacterium isolated from *Acyrthosiphon pisum* (aphid) honey dew, has the kairomone effect, thus increasing the efficacy of natural enemies (Leroy et al. 2011). Here, host-associated bacterium steering prey locality and egg-laying preference for the natural enemy. *Staphylococcus sciuri* has been producing specific semiochemical which has the attractant and oviposition stimulant activity.

## 16.8 Effect of Phytopathogen on Herbivore Fitness

The plants which are infected by pathogens may influence the herbivore performance, and the temporal and spatial distribution also influenced on a single leaf, whole plant, and the entire landscape level (Tack and Dicke 2013). The behavior and distribution of insects affected by pathogens either improve or worsen the plant health. In one case, rice tungro virus alters the chlorophyll, phenol, and carbohydrate contents in rice phloem. The effect on herbivore may vary depend on time and space. Due to the infection of *Melampsora allii-fragilis*, rust fungi on willow trees affect the willow leaf beetle, scientifically *Plagiodera versicolora*, by reducing the weight and increased the mortality and development time (Simon and Hilker 2003). In other cases, the infected plants may show some phenotypic variance like alteration in color and altered odor which influence the colonizing tendency of vectors (Fereres and Moreno 2009) and non-vectors (Purcell 1988). The common symptom for virus-infected plants is yellowing of the leaves, which are the landing sites for most of the vectors like leafhoppers, aphids, and whiteflies. Some fungus attracts the pollinators for further infection by phenotypic appearance as a flower. The abovementioned case occurred in mummy berry fungus (Batra and Batra 1985).

## 16.9 Benefits of Plant-Microbe-Insect Interaction in Sustainable Agriculture

In insects all the mechanisms include host selection, feeding, reproduction, and oviposition driven by various plant metabolites which act as cues. All these metabolites are changeable in nature through many factors such as climatic conditions, several soil-inhabiting microorganism, endophytes, other trophic level organisms, etc. All the processes in nature are interconnected. Here, microbes which initiate certain alteration in plants have both negative and positive effects. Microbes would be promising organisms for the sustainable agriculture through the production of various enzymes which enhances the plant growth. For instance, endophytic bacteria *Acidovorax radicis* N35 promotes the growth of barley through the production of *N*-acyl homoserine lactone. Various fungal and bacterial endophytes reduce the herbivore performance by altering the plant traits.

Apart from deleterious microbes, beneficial microbes are coupled with the plants at below-ground as well as above-ground plant parts. In below-ground mycorrhizal fungi, plant growth-promoting fungi (PGPF), endophytic fungi, and plant growth-promoting rhizobacteria (PGPR) play a major role in supporting plant growth (Shaikh et al. 2016). Nowadays, beneficial microbe has been available in the commercial scale as EM-1, having phototropic bacteria, yeast, lactic acid bacteria, and actinomycetes (Najafi et al. 2021). Also microbes provoke the induced systemic resistance engaged by increasing the sensitivity to JA and ethylene done through priming (Pineda et al. 2010).

Though ISR may be effective for a broad range of pests which include virus, fungi, nematode, and insects, in insects and nematodes, it depends on feeding specialization. There are several theories on mycorrhizal fungi that have neutral effects on specialist feeders, but for phloem feeders, it may give positive effects. In the case of generalized and mesophyll feeders, fungi have negative effects (Hartley and Gange 2009). Studies show that fungal entomopathogens protect the plants from phytophagous insects. These beneficial insect pathogenic fungi colonize the plant system and give protection against insects for a period of time which ultimately supports the crop yield (Biswas et al. 2012). The 04/01-Tip strain of *Beauveria bassiana* colonizes *Papaver somniferum* (opium) and gave protection against *Iraella luteipes* (stem gall wasp) (Moraga et al. 2006).

The fungal entomopathogens build a direct contact with herbivore insects and protect the plants by producing a secondary metabolite which is toxic to insects (Enshasy et al. 2020). The entomopathogenic fungi such as *Beauveria bassiana* when treated with groundnut (*Arachis hypogea*), synthesize higher number of defence chemicals while fed up by *Aproaerema modicella* (Senthilraja et al. 2013).

## 16.10 Conclusion

Microbes and insects have a common trait that is ubiquitous in nature, but microbes are ubiquitous in insects. Of course, microbes were mediating the plant-insect interaction in both positive and negative ways. Microbes like PGPRs directly as well as indirectly facilitate the plant by providing nutrients and synthesis of various enzymes that are detrimental to herbivores (Khan and Sayyed 2019). Several cues are involving in all cases of interaction. Herbivore, generalist and specialist response differently based on the plant traits and microbes. Plant-microbe-insect interactions have major effect on plant health and may give suggestions for the design of environment-friendly crop management strategies in agriculture (Nasab and Sayyed 2019; Zope et al. 2019a, b). It is crucial in eco-physiological perspective and integrated pest management in relation to assessing the contribution to sustainable agriculture.

Identification and function of microbial communities are providing prosperity of appropriate chances to control insect pests. Probiotic approach directly induces plant resistance or dysbiosis approach which introduces microbes that disrupt the microbial function in the gut region of insect pests. Additionally, the application of antibiotic approach supressed the key or symbiotic gut microbes that are crucial to insect fitness. Finally, we propound that entomopathogens have the capability to supremacy plant defenses explicitly as endophytes or by altering insect physiology which is happening indirectly.

## References

Acevedo FE, Peiffer M, Tan CW, Stanley BA, Stanley A, Wang J, Felton G (2017) Fall armyworm-associated gut bacteria modulate plant defense responses. MOI 30(2):127–137

Adams AS, Aylward FO, Adams SM, Erbilgin N, Aukema BH, Currie CR, Raffa KF (2013) Mountain pine beetles colonizing historical and naive host trees are associated with a bacterial community highly enriched in genes contributing to terpene metabolism. Appl Environ Microbiol 79(11):3468–3475

Alborn HT, Turlings TCJ, Jones T, Stenhagen G, Loughrin JH, Tumlinson JH (1997) An elicitor of plant volatiles from beet armyworm oral secretion. Science 276(5314):945–949

Ayshasiddeka M, Islam M, Akter S, Ali H, Yeasmin T, Enshasy HAE, Yahayu M, Gafur A, Syed A, Bahkali AH, Sayyed RZ (2021) Biodiversity analysis of Arbuscular Mycorrhizal fungal association with Euphorbiaceae Family. PLoS ONE 17(3):e0266403

Baba ZA, Hamid B, Sheikh TA, Alotaibi S, Enshasy HE, Ansari JA, Zuan ATK, Sayyed RZ (2021) Psychrotolerant *Mesorhizobium* sp. isolated from temperate and cold desert regions solubilize potassium and produces multiple plant growth promoting metabolites. Molecules 26:5758

Bansal R, Hulbert S, Schemerhorn B, Reese JC, Whitworth RJ, Stuart JJ, Chen MS (2011) Hessian fly-associated bacteria: transmission, essentiality, and composition. PLoS One 6(8):e23170

Barea JM, Pozo MJ, Azcon R, Azcon-Aguilar C (2005) Microbial co-operation in the rhizosphere. J Exp Bot 56(417):1761–1778

Barr KL, Hearne LB, Briesacher S, Clark TL, Davis GE (2010) Microbial symbionts in insects influence down-regulation of defense genes in maize. PLoS One 5(6):e11339

Bastami A, Amirnia R, Sayyed RZ, Enshasy HE (2021) The effect of mycorrhizal fungi and organic fertilizers on quantitative and qualitative traits of two important Satureja species. Agronomy 11:1285

Basu A, Prasad P, Das SN, Kalam S, Sayyed RZ, Reddy MS, Enshasy H (2021) Plant Growth Promoting Rhizobacteria (PGPR) as green bioinoculants: recent developments, constraints, and prospects. Sustainability 13:1140

Batra LR, Batra SW (1985) Floral mimicry induced by mummy-berry fungus exploits host's pollinators as vectors. Science 228(4702):1011–1013

Berg G, Hallmann J (2006) Control of plant pathogenic fungi with bacterial endophytes. In: Microbial root endophytes. Springer, Berlin, Heidelberg, pp 53–69

Biswas C, Dey P, Satpathy S, Satya P (2012) Establishment of the fungal entomopathogen *Beauveria bassiana* as a season long endophyte in jute *(Corchorus olitorius)* and its rapid detection using SCAR marker. BioControl 57(4):565–571

Bruce TJ, Pickett JA (2011) Perception of plant volatile blends by herbivorous insects–finding the right mix. Phytochemistry 72(13):1605–1611

Bruce TJ, Wadhams LJ, Woodcock CM (2005) Insect host location: a volatile situation. Trends Plant Sci 10(6):269–274

Brundrett MC (2009) Mycorrhizal associations and other means of nutrition of vascular plants: understanding the global diversity of host plants by resolving conflicting information and developing reliable means of diagnosis. Plant Soil 320(1):37–77

Chung SH, Rosa C, Hoover K, Luthe DS, Felton GW (2013a) Colorado potato beetle manipulates plant defenses in local and systemic leaves. Plant Signal Behav 8:e27592

Chung SH, Rosa C, Scully ED, Peiffer M, Tooker JF (2013b) Herbivore exploits orally secreted bacteria to suppress plant defenses. PNAS 110:15728–15733

Couty A, Van Emden H, Perry JN, Hardie J, Pickett JA, Wadhams LJ (2006) The roles of olfaction and vision in host-plant finding by the diamondback moth, *Plutella xylostella*. Physiol Entomol 31:134–145

De Moraes CM, Mescher MC, Tumlinson JH (2001) Caterpillar-induced nocturnal plant volatiles repel conspecific females. Nature 410:577–580

Douglas AE (2015) Multiorganismal insects: diversity and function of resident microorganisms. Annu Rev Entomol 60:17–34

Dowd PF, Shen SK (1990) The contribution of symbiotic yeast to toxin resistance of the cigarette beetle (*Lasioderma serricorne*). Entomol Exp Appl 56:241–248

Duffy B, Keel C, Defago G (2004) Potential role of pathogen signaling in multitrophic plant-microbe interactions involved in disease protection. Appl Environ Microbiol 70(3):1836–1842

Enshasy HEA, Ambehati KK, Ashraf F, Ramchuran S, Sayyed RZ, Amalin D, Dailin DJ, Hanapi SZ (2020) Trichoderma: biocontrol agents for promoting plant growth and soil health. In: proceedings of Functional annotation for crop protection, pp 239–259

Fereres A, Moreno A (2009) Behavioural aspects influencing plant virus transmission by homopteran insects. Virus Res 141:158–168

Grunseich JM, Thompson MN, Aguirre NM, Helms AM (2019) The role of plant-associated microbes in mediating host-plant selection by insect herbivores. Plan Theory 9(1):6

Hamid B, Zaman M, Farooq S, Fatima S, Sayyed RZ, Baba ZA, Sheikh TA, Reddy MS, Enshasy H, Gafur A, Suriani NL (2021) Bacterial plant biostimulants: a sustainable way towards improving growth, productivity, and health of crops. Sustainability 13:2856

Hartley SE, Gange AC (2009) Impacts of plant symbiotic fungi on insect herbivores: mutualism in a multitrophic context. Annu Rev Entomol 54:323–342

Honda K, Maes RP (2008) Addiction to pyrrolizidine alkaloids in male danaine butterflies: a quest for the evolutionary origin of pharmacophagy. In: proceedings of Insect physiology: new research, pp 73–118

Honda K, Omura H, Hori M, Kainoh Y (2010) Allelochemicals in plant-insect interactions. In: proceedings of Comprehensive natural products II: Chemical biology, pp 563–594

Jabborova D, Annapurna K, Fayzullaeva M, Sulaymonov K, Kadirova D, Jabbarova Z, Sayyed RZ (2020) Isolation and characterization of endophytic bacteria from ginger (*Zingiber officinale* Rosc.). Ann Phytomedicine 9(1):116–121

Jadhav HP, Sonawane MS, Khairnar MH, Sayyed RZ (2020) Production of alkaline protease by rhizospheric *Bacillus cereus* HP_RZ17 & *Paenibacillus xylanilyticus* HP_RZ19. Environ Sustainab 3:5–13

Kaiser O, Puhler A, Selbitschka W (2001) Phylogenetic analysis of microbial diversity in the rhizoplane of oilseed rape (*Brassica napus* cv *Westar*) employing cultivation dependent and cultivation- independent approaches. Microb Ecol 42:136–149

Kalam S, Basu A, Ahmad I, Sayyed RZ, Enshasy HA, Dailin DJ, Suriani NL (2020) Recent understanding of soil Acidobacteria and their ecological significance: a critical review. Front Microbiol 11:580024

Khan A, Sayyed RZ (2019) Rhizobacteria: legendary soil guards in abiotic stress management. In: Plant growth promoting Rhizobacteria for sustainable stress management, vol Vol. 1. Springer, Singapore, pp 27–342

Khan N, Al S, Shahi MA, Mustafa A, Sayyed RZ, Curaá JA (2021) Insights into the interactions among roots, rhizosphere and Rhizobacteria for improving plant growth and tolerance to abiotic stresses: a review. Cell 10(6):1551

Knudsen JT, Eriksson R, Gershenzon J, Stahl B (2006) Diversity and distribution of floral scent. Bot Rev 72(1):1

Kour D, Kaur T, Devi R, Yadav A, Singh M, Joshi D, Singh J, Suyal DC, Kumar A, Rajput VD, Yadav AN, Singh K, Singh J, Sayyed RZ, Arora NK, Saxena AK (2021) Beneficial microbiomes for bioremediation of diverse contaminated environments for environmental sustainability: present status and future challenges. Environ Sci Pollut Res 28:24917–24939

Leroy PD, Sabri A, Heuskin S, Thonart P, Lognay G, Verheggen FJ, Haubruge E (2011) Microorganisms from aphid honeydew attract and enhance the efficacy of natural enemies. Nat Commun 2(1):1–7

Liu Y, Zou S, Zou Y, Wang J, Song W (2012) Investigation on diversity and population succession dynamics of indigenous bacteria of the maize spermosphere. World J Microbiol Biotechnol 28(1):391–396

Manasa M, Ravinder P, Gopalakrishnan S, Srinivas V, Sayyed RZ, Enshasy HE, Yahayu M, Zuan ATK, Kassem HS, Hameeda B (2021) Co-inoculation of *bacillus* spp. for growth promotion and iron fortification in sorghum. Sustainability 13:12091

Marilley L, Aragno M (1999) Phytogenetic diversity of bacterial communities differing in degree of proximity of *Lolium perenne* and *Trifolium repens* roots. Agric Ecosyst Environ Appl Soil Ecol 13:127–136

Mayak S, Tirosh T, Glick BR (2004) Plant growth-promoting bacteria confer resistance in tomato plants to salt stress. Plant Physiol Biochem 42:565–572

Mayoral JG, Hussain M, Joubert DA, Iturbe-Ormaetxe I, O'Neill SL, Asgari S (2014) *Wolbachia* small non-coding RNAs and their role in cross-kingdom communications. PNAS 111:18721–18726

Moraga QE, Landa BB, Munoz-Ledesma J, Jimenez-Diáz RM, Santiago-Alvarez C (2006) Endophytic colonisation of opium poppy, *Papaver somniferum*, by an entomopathogenic *Beauveria bassiana* strain. Mycopathologia 161:323–329

Najafi V, Sohrabi Y, Sayyed RZ, Suriani NL, Datta R (2021) Effects of combinations of Rhizobacteria, mycorrhizae, and seaweeds on growth and yields in wheat cultivars under the influence of supplementary irrigation. Plan Theory 10:811

Nasab BF, Sayyed RZ (2019) Plant growth promoting Rhizobacteria and salinity stress: a journey into the soil. In: Plant growth promoting Rhizobacteria for sustainable stress management, vol Vol. 1. Springer, Singapore, pp 21–34

Nelson DR, Mele PM (2007) Subtle changes in rhizosphere microbial community structure in response to increased boron and sodium chloride concentrations. Soil Biol Biochem 39:340–351

Nithyapriya S, Lalitha S, Sayyed RZ, Reddy MS, Dailin EHA, Luh N, Herlambang S (2021) Production, purification, and characterization of bacillibactin siderophore of *Bacillus subtilis* and its application for improvement in plant growth and oil content in sesame. Sustainability 13:5394

Patel PR, Shaikh SS, Sayyed RZ (2016a) Dynamism of PGPR in bioremediation and plant growth promotion in heavy metal contaminated soil. Indian J Exp Biol 54:286–290

Patel S, Sayyed RZ, Saraf M (2016b) Bacterial determinants and plant defense induction: their role as bio-control agent in agriculture plant soil microbes. In: Plant soil microbes. Springer, Switzerland, pp 187–204

Pineda A, Zheng SJ, van Loon JJA, Pieterse CMJ, Dicke M (2010) Helping plants to deal with insects: the role of beneficial soil-borne microbes. Trends Plant Sci 15:507–514

Price PW, Bouton CE, Gross P, McPheron BA, Thompson JN, Weis AE (1980) Interactions among three trophic levels: influence of plants on interactions between insect herbivores and natural enemies. Annu Rev Ecol Evol Syst 11:41–65

Purcell AH (1988) Increased survival of *Dalbulus maidis*, a specialist on maize, on non-host plants infected with mollicute plant pathogens. Entomol Exp Appl 46:187–196

Senthilraja G, Anand T, Kennedy JS, Raguchander T, Samiyappan R (2013) Plant growth promoting rhizobacteria (PGPR) and entomopathogenic fungus bio-formulation enhance the expression of defense enzymes and pathogenesis-related proteins in groundnut plants against leaf miner insect and collar rot pathogen. Physiol Mol Plant Path 82:10–19

Shaikh SS, Reddy MS, Sayyed RZ (2016) Plant growth promoting rhizobacteria: an eco-friendly approach for sustainable agro-ecosystem. In: Plant soil-microbes. Springer, Switzerland, pp 182–201

Simon M, Hilker M (2003) Herbivores and pathogens on willow: do they affect each other? Agric For Entomol 5:275–284

Suriani N, Suprapta D, Novizar N, Parwanayoni N, Darmadi A, Dewi D, Sudatri N, Ahmad F, Sayyed RZ, Syed A, Elgorban A, Bahkali A, Enshasy H, Dalin DJ (2020) A mixture of piper leaves extracts and Rhizobacteria for sustainable plant growth promotion & biocontrol of blast pathogen of organic Bali Rice. Sustainability 12:8490

Tack AJM, Dicke M (2013) Plant pathogens structure arthropod communities across multiple spatial and temporal scales. Funct Ecol 27:633–645

Tan ZY, Kan FL, Peng GX, Wang ET, Reinhold-Hurek B, Chen WX (2001) *Rhizobium yanglingense* sp. nov isolated from arid and semi-arid regions in China. Int J Syst Evol Microbiol 51:909–914

Thakur M, Sohal BS (2013) Role of elicitors in inducing resistance in plants against pathogen infection: a review. ISRN Biochem 2013:762412

Turlings TCJ, McCall PJ, Alborn HT, Tumlinson JH (1993) An elicitor in caterpillar oral secretions that induces corn seedlings to emit chemical signals attractive to parasitic wasps. J Chem Ecol 19:411–425

Whipps JM, Hand P, Pink D, Bending GD (2008) Phyllosphere microbiology with special reference to diversity and plant genotype. J Appl Microbiol 105:1744–1175

Zope VP, Enshasy HE, Sayyed RZ (2019a) Plant growth promoting Rhizobacteria: an overview in agricultural perspectives. In: Plant growth promoting rhizobacteria for sustainable stress management, vol vol. 2. Springer, Singapore, pp 345–362

Zope VP, Jadhav HP, Sayyed RZ (2019b) Neem cake carrier prolongs shelf life of biocontrol fungus *Trichoderma viridae*. Indian J Exp Biol 57(5):372–375

# Chapter 17
# Exploring the Crucial Role of Plant Growth-Promoting Rhizobacteria (PGPR) in Plant Secondary Metabolite Production and Diabetes Management

**Deegendra Khadka and Kishor Pandey**

**Contents**

17.1 Introduction .... 351
17.2 Beneficial Plant Growth-Promoting Rhizobacteria (PGPR) .... 354
17.3 Plant Secondary Metabolites (SMs) .... 355
17.4 Role of PGPR in Secondary Metabolite Production/Plant Disease Suppression .... 357
17.5 Secondary Metabolites as Antidiabetic Medicine .... 358
17.6 Future Prospects .... 365
17.7 Conclusion .... 365
References .... 365

## 17.1 Introduction

In recent years, both quantity and quality of pharmaceutical plants, fruits, vegetables, and crops are deteriorating due to plant diseases induced by soil-dwelling microorganisms (Wang et al. 2015). In addition, a wide range of fertilizers and chemical pesticides have been utilized for better crop production but result in reduced plant product quality, development of chemical-resistant pathogens, and increased environmental issues (Wang et al. 2018). A large number of bacterial species inhabiting in plant rhizosphere are involved in plant growth called plant

D. Khadka (✉)
Molecular Biotechnology Unit, Faculty of Science, Nepal Academy of Science and Technology, Khumaltar, Lalitpur, Nepal
e-mail: deegendra.khadka@nast.gov.np

K. Pandey
Central Department of Zoology, Tribhuvan University, Kirtipur, Kathmandu, Nepal



R. Z. Sayyed et al. (eds.), *Antifungal Metabolites of Rhizobacteria for Sustainable Agriculture*, Fungal Biology, https://doi.org/10.1007/978-3-031-04805-0_17

growth-promoting rhizobacteria (PGPR). PGPR plays a crucial role in boosting plant growth, curbing soilborne diseases, ameliorating soil quality, augmenting crops yield, and so on (Shaikh et al. 2016). A PGPR named *Streptomyces* species, gram-positive actinomycetes, localized massively in the soil activates plants for achieving nutrients. *Streptomyces* develop a number of active compounds required for plant growth. Some of them are hydrolytic enzymes and antibiotic and volatile organic compounds (Kannepalli et al. 2021). PGPR can produce phytohormones like indoleacetic acid, cytokinins, gibberellins, and inhibitors of ethylene which are actively participating in coordinating cellular activities to grow a sound and healthy plant. Hence, PGPR acts as a crucial role in the growth of disease resistant plants by significantly increasing plant heightand root length reaching to the water surface (Backer et al. 2018).

Many research studies proved that PGPR inoculation augmented in the production of secondary metabolites of aromatic plant species (Grover et al. 2021; Cappellari et al. 2013). Plant seeds or roots with PGPR increase secondary metabolites such as terpenes, phenolics, flavonoids, and alkaloids. PGPR can improve the quality of medical plant products by eliciting medically active secondary metabolites. Phytochemicals produced by the medicinal plants have been related to the involvement of PGPR (Zope et al. 2019; Dojima and Craker 2016). Different strains of rhizobacteria work in the various regions. Rhizobacteria such as *Rhizobium* and *Bradyrhizobium* promote plant growth in the absence of pathogens by converting nitrogen to ammonia. *Azospirillum* is dedicated to increase root development so plants can uptake required amounts of water and minerals easily (Zakaria et al. 2019). *P. putida* PCL1444 acts effectively as a rhizoremediator by utilizing root exudates and degrading naphthalene around the root to protect seeds from being killed by naphthalene so that the plant becomes healthy (Bisht et al. 2015). *B. subtilis* GB03 works as a photo-stimulator by increasing the photosynthetic efficiency and chlorophyll content. Some stress-tolerant bacteria such as *Azospirillum lipoferum* and *Azospirillum brasilense* containing enzyme 1-aminocyclopropane-1-carboxylate (ACC) deaminase alleviate plant development by maintaining ethylene levels. Such bacteria convert ethylene precursor ACC into 2-oxobutanoate and $NH_3$. ACC combat with phytogenic bacteria to resist stress developed by polyaromatic hydrocarbon (Sagar et al. 2022).

A person becomes a diabetic patient when his/her blood glucose level is out of control due to occurrence of either poor insulin secretion or uncertain insulin action or both. Chronic hyperglycema is a symbol of developing severe diabetes that consequences in malfunctioning and longterm damaging of several organs, especially the heart, kidney, blood vessel and eyes (Khadka 2021). Generally, diabetes is categorized into three classes. Type 1 diabetes mellitus (T1DM) also called insulin-dependent diabetes (IDDM) is a condition developed by lack of insulin secretion when β-cells are deteriorating due to microorganisms or environmental and genetic factors (Boucher et al. 2014). Only below 10% of total diabetic patients are under IDDM. IDDM patients have been treated using insulin injection. Type 2 diabetes mellitus (T2DM) also known as insulin-independent diabetes mellitus (IIDM) occurs when β -cells can produce sufficient or less insulin but insulin doesn't work properly to insert glucose molecules into other cells rather than blood cells. So, the glucose level remains high in the blood cells. More than 90% of diabetic patients belong to this category. IIDM patients are being treated by providing oral medications which are abundantly available in the markets. Much research has been carried out on the title of T2DM but still remains unsolved. Such synthetic drug molecules develop severe side effects. Natural product-based medicine would be more lucrative to the diabetic patients. Gestational diabetes (GD) is the third category of diabetes formed from glucose intolerance during pregnancy period. Ignoring such diabetes increases the risk of complications in the pregnancy period and delivery time. However, GD is a temporary disorder and disappears after giving birth but carries a risk of T2DM in the future. Patients belonging to this category are treated regarding their health conditions either by insulin injection or oral medications (Khadka 2021).

The proper function of PGPR supports medicinal plants in the production of secondary metabolities quantitatively and qualitatively. Simply, it is acceptable that PGPR is one of the key factors in forming secondary metabolites (SMs). SMs have been used as a drug for a long time. More than 80% of the global population still has faith in natural healers (Adeniyi et al. 2021). Synthetic drugs with deleterious side effects are being used massively for the sake of treating diabetes. Chemically produced antidiabetic drugs are not only costly but also have a germ of producing deleterious effects. Regarding the disadvantages of synthetic drugs, nowadays intentions in searching for potential natural ingredients have increased exponentially (Karimi et al. 2015). The SMs of medicinal plants work through different mechanisms to treat diabetes. The principal working mechanisms are functioning as rejuvenating pancreatic ®-cell, inhibiting phosphatase enzymes, and improving insulin activity (Rorsman and Ashcroft 2018). For instance, *Vaccinium arctostaphylos* L. contains anthocyanoside myrtillin as a secondary metabolite that works similar to insulin, *Ginkgo biloba* L. produces flavonglycosides which is effective in wound healing caused by diabetes by increasing blood circulation, and *Trigonella foenum-graecum* L. seed contains bioactive compounds coumarin and nicotinic acid that are responsible for suppressing blood glucose level in type 2 diabetic patients (Przeor 2022).

This review paper reveals kinship of PGPR and its crucial role in type 2 diabetes management. Sufficient research articles have been published on the title relating PGPR and SMs and also enough articles unveiling SMs used as antidiabetic medicine, but no scientific paper has been published describing the role of PGPR in diabetes treatment. So, we focus on the comprehensive analysis that helps to show the role of PGPR in diabetes management through producing secondary metabolites as a diabetes healer.

## 17.2 Beneficial Plant Growth-Promoting Rhizobacteria (PGPR)

Several microbials play a pivotal role in plant growth, and showing their effects in the rhizosphere is significant to develop a healthy plant. Such bacteria are assigned as plant growth-promoting rhizobacteria. PGPR demonstrates beneficial effects by direct mechanism where the environment is free of pathogens and indirect mechanism where plants are protected against soilborne diseases (Wang et al. 2021). Rhizosphere, the layer of soil influenced by the root, contains more bacteria in the ratio of surrounding bulk soil. Bacteria concentrated in the rhizosphere are 10–100 times higher than the surrounding soil. Root-colonizing bacteria are needed for biocontrol mechanisms, such as antibiosis (Sayyed et al. 2019). PGPR is directed by two working mechanisms. The first mechanism is that PGPR directly participates in raising plant growth in the absence of any kinds of pathogens. The second mechanism is that PGPR works in direct mechanisms acting as biofertilizers (Yadav and Sayyed 2019), rhizoremediators, and photo-stimulators (Patel et al. 2016). Rhizobacteria protect plants by combating soilborne pathogen being a stress controller in which PGPR functions as antagonism, signal interference, predation and parasitism (Gouda et al. 2018), induced systemic resistance, competition for ferric iron ions, competition for nutrients and niches (Romera et al. 2019), and interference with activity, survival, germination, and sporulation of the pathogen.

PGPR is classified into two groups depending on their location, namely, extracellular PGPR (ePGPR) and intracellular PGPR (iPGPR). PGPR found in the rhizospheric soil are termed ePGPR, while PGPR inhabiting the inner side of roots are known as iPGPR (Zerihun et al. 2019). Several PGPR bacteria such as *Pseudomonas*,

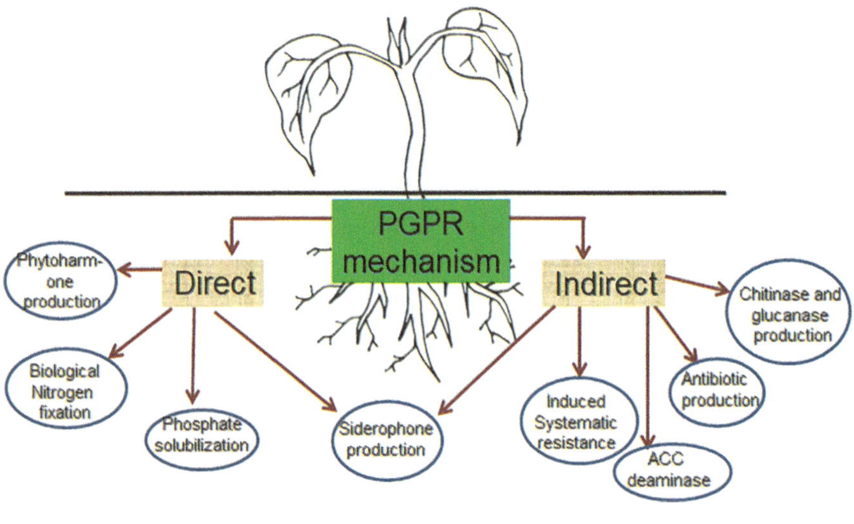


**Fig. 17.1** Direct and indirect working mechanism of PGPR

*Enterobacter*, *Bacillus*, *Variovorax*, *Klebsiella*, *Burkholderia*, *Azospirillum*, *Serratia*, and *Azotobacter* have been reported till now (Zaidi et al. 2015; Huang et al. 2016). How the PGPR works to assist in plant growth is shown in Fig. 17.1. Exudates secreted from the plant roots play a crucial role for survival rate in the root zone, bacterial flourishment, and growing microbial density rapidly (Canarini et al. 2019). The effective action of PGPR in the rhizosphere are known to enhance root shoot, root hair growth, root colonization, plant biomass, weight of panicles, and grain production, increase the number of tillers, and block pathogen destruction (Khan et al. 2021; Kapadia et al. 2021). Hence, PGPR protects plants from various deleterious factors by direct working mechanism (Fig. 17.1).

In indirect mechanism, PGPR acts as a biocontrol agent by producing antifungal metabolites such as HCN, phenazine, pyrrolnitrin, 2,4-diacetylphloroglucinol, pyoluteorin, viscosinamide, and tensin (Pourakbar et al. 2021). Plants can develop resistivity against pathogenic bacteria, fungi, and viruses when the root interacts with some rhizobacteria (Fig. 17.1). This plant action is called induced systemic resistance (IRS) which involves jasmonate and ethylene signaling within the plants. The plants harmone are responsible for combating with a variety of pathogens in host plants (Suriani et al. 2020).

## 17.3 Plant Secondary Metabolites (SMs)

The end products obtained after metabolism are categorized as SMs. Plants produce such vast and diverse organic compounds that appear indirectly in plants' growth and development, traditionally referred to as SMs. Plant SMs are one of the principal sources utilized in pharmaceuticals, food additives, flavors, and other industrial materials (Chiocchio et al. 2021). Humans have been applying SMs as medicines, flavorings, pigments, and recreational drugs in current trends. Primary metabolities are required to keep plants healthy and SMs protects plants from the problems of fertility, phenotype, and intermediate death. SMs are often found in a narrow spectrum of plant species belonging to a similar phylogenetics (Kang et al. 2017). One of the important roles of SMs is defending plants from herbivory and interspecies (Kang et al. 2017). We simply say, SMs develop new adaptive characters to the plants with altering environmental conditions. Many research articles reported that SM contains antiviral, antifungal, and antibiotic activities to save plants from toxic microbes (Tiwari and Rana 2015).

SMs are divided on the basis of various parameters depicted in Fig. 17.2. In the classification of chemical structure, molecules are grouped based on the number of rings with sugar molecules and in composition category. SMs are also divided regarding presence or absence of nitrogen atoms and on the basis of their solubility in several solvents. They are mainly classified based on their biosynthetic pathway (Jan et al. 2021). Generally, few large molecule families, namely flavonoids, alkaloids, phenolic, terpenes, and coumarin, are taken (Sharma et al. 2020). Some of them are described briefly below:

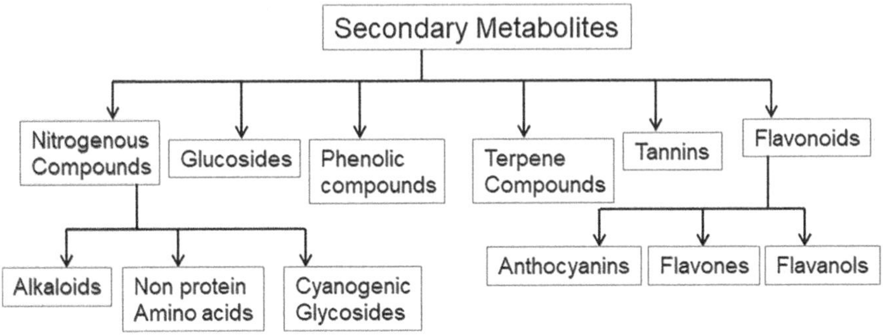


**Fig. 17.2** Broad classification of secondary metabolites

(i) **Flavonoids**

Flavonoids represent a diverse group of phenolic compounds which play a wide range of functions. More than 4500 flavonoids are reported. Specific flavonoids protect plants against UV-B radiation. Flavonoids consist of various groups of plant metabolites which include chalcones, aurones, flavanones, isoflavonoids, flavones, flavonols, leucoanthocyanidins, catechins, and anthocyanins.

(ii) **Alkaloids**

Alkaloids are well-known for their potent pharmacological activities like anti-malaria (e.g., quinine), anti-asthma (e.g., ephedrine), anticancer (e.g., homoharringtonine) (Singh and Sharma 2018), analgesic (e.g., morphine), antibacterial (e.g., chelerythrine), and antidiabetic (e.g., piperine) (Aryal et al. 2021).

(iii) **Phenolics**

These are aromatic benzene ring compounds containing one or more hydroxyl groups produced by plants as SMs for stress management. Phenolic compounds play significant roles in plant development, particularly in lignin and pigment biosynthesis and pathogen controls. For example, phenolic compound phytoalexins secreted by injured plants repel or kill many microorganisms. Estradiol, tyrosine, thymol, gallic acid, etc. are some examples of phenolics.

(iv) **Terpenes**

Terpenes are large and various groups of organic compounds with a strong odor. Terpenes are polymeric forms of the monomer isoprene unit. Most terpenes contain a unique aroma which is used in perfumery and food additives. Terpenes possessing desirable properties are utilized in food, cosmetics, pharmaceutical, and biotechnology industries (Jaegera and Cuny 2016). The presence of terpenes in plants functions both in ecological and physiological aspects such as allelopathy, insecticidal agents, insect pollinators, plant hormones (abscisic acid, gibberellin), and so on. ⟨ and ®-carotenes, steroid, natural rubber, retinol, etc. are representative compounds of terpenes.

(v) **Coumarins**

Coumarins are colorless crystalline solid with sweet odor like vanilla smell but the taste is caustic. The compound is found on a wide range of taxonomy functioning as a chemical weapon for predator destruction (Favaretto et al. 2015). Coumarins have been used as anticoagulant, in laser dyes and perfume, and as aromatizers. Warfarin, a kind of coumarin derivative, acts as anticoagulant. Coumarin tetramethyl laser dyes provide wide tunability and high laser power (Ibnaouf et al. 2017). Coumarins are present in more than 800 species with above 1500 representative compounds.

## 17.4 Role of PGPR in Secondary Metabolite Production/Plant Disease Suppression

Alternation of plant characters is made possible in the presence of rhizobacteria in the root zone. Roots released some exudates fascinate in aggregating rhizobacteria. After inhabiting in the root zone, rhizobacteria secrete allelochemical as SMs which improve plant growth and crop yield by providing sufficient nutrients and combating against diseases. Allochemicals such as siderophores, antibiotics, biocidal volatiles like hydrogen cyanide, and lytic enzymes (e.g., chitinase) restrict the maturation of phytopathogens (Chandran et al. 2021). Siderophore produced by *Alcaligenes* spp., *Serratia* spp., and *Acinetobacter* spp. of PGPPR has been reported in several articles as a bio-controller (Sukmawati et al. 2021). Heterocyclic SMs named phenazines produced by eubacteria and some archaea have been reported for their antibiotic activities (Zakaria et al. 2019). Auxin indoleacetic acid (IAA) as a phytohormone is secreted by rhizobacteria. Such hormones have the capacity to change various physiological progressions by increasing root surface area and length so that plants can absorb soil nutrients sufficiently (Fu et al. 2015). Ethylene, a kind of plant hormone, is also involved in plant growth and development as well as a crucial indicator responding toward stress caused by environmental factors (Iqbal et al. 2017).

*Azospirillum* strain of PGPR in maize plant has been reported for modifying SMs qualitatively and quantitatively by the root exudate activities (Walker et al. 2011). *Azospirillum* rhizobacteria have the ability to modify hydroxycinnamic derivatives and flavonoids (Panche et al. 2016). *Bacillus subtilis* stimulates in accumulating glycine betaine and choline (Nithyapriya et al. 2021). So, several such examples have been published describing the modification of SMs by PGPR to the plants' nourishment. Plants are known as a repository of biologically active compounds possessing therapeutic properties for healing many ailments. Such biologically active compounds are primarily SMs which are produced by PGPR (Chandaran et al. 2021). PGPR assists in modification and accumulations of many compounds which are inevitably required for plant growth and defy during stress periods created by environmental factors. In summary, it can be said that PGPR plays a vital role in producing SMs which assist in developing healthy plants by suppressing diseases.

## 17.5 Secondary Metabolites as Antidiabetic Medicine

Secondary metabolites have been used as a hypoglycemic agent through much of human history (Shehadeh et al. 2021). Plant products and their derivatives are abundantly available on the global market as food accessories (Ota and Ulrih 2017). Pharmaceutical companies rely on the structure of natural products to design a drug molecule. For instance, metformin, a hypoglycemic drug used by type 2 diabetic patients, is structurally designed taking the natural product of *Galega officinalis* as a reference. The SMs produced by a herd of medicinal plants have been used in the management of diabetes. There are several mechanisms behind how SMs are being exploited as hypoglycemic agents. Some of the significant mechanisms are stimulating pancreatic ®-cells to secrete insulin, decreasing glucose absorption by blood cells, improving insulin sensitivity, and inhibiting phosphatase enzymes. A number of plants contain some degree of hypoglycemic activity, but magnificent research along with conventional usage is limited to the useful plants. Some of them with active SMs participating in blood glucose level controls through different mechanistic ways are included in Table 17.1.

**Table 17.1** Working mechanism of secondary metabolites obtained from diverse plant species in diabetes management

| Plant species/family | Plant parts | Secondary metabolites | Mechanism of action |
|---|---|---|---|
| *Acacia arabica* (lam.) Willd/Fabaceae | Barks | Corosolic acid, 4-hydroxybenzoic acid, caffeic acid, kaempferol | Carbohydrate digestion and absorption and insulin secretion (Bharti et al. 2018) |
| *Achyranthes rubrofusca* (Wight) Hook.f./ Amaranthaceae | Leaves | Chlorogenic acid, rutin, astragalin | Increase in the pancreatic enzyme SOD, CAT, and glutathione expression (Geetha et al. 2011) |
| *Aegle marmelos* L./ Rutaceae | Stem barks | Aegelin, coumarins, alkaloids (Chauhan et al. 2010), umbelliferone β-D-galactopyranoside | Improve in pancreatic insulin secretion (Kumar et al. 2013) |
| *Albizia lebbeck* L. Benth/ Fabaceae | Steam barks | Kaempferol, ®-sitosterol, benzoic acid (Verma et al. 2013) | Protecting β-cell by free radical scavenging properties (Ahmed et al. 2014) |
| *Allium cepa* L./ Amaryllidaceae | Whole body parts | Allyl sulfide (Chauhan et al. 2010) | Stimulating insulin release and enhancing cellular uptake and utilization of glucose (Ozougwu 2011) |

(continued)

**Table 17.1** (continued)

| Plant species/family | Plant parts | Secondary metabolites | Mechanism of action |
|---|---|---|---|
| *Allium sativum* L./ Amaryllidaceae | Whole body parts | Allicin, alliin, diallyl sulfide, ajoene (Shang et al. 2019), s-allylcysteine sulfoxide | Increasing insulin secretion from the β-cell or releasing bound insulin (Ozougwu and Eyo 2010) |
| *Aloe vera*/Asphodelaceae | Leaves | Aloin, aloe emodin (Patel et al. 2012), pseudoprototinosaponin AIII, prototinosaponin AIII (Eddouks et al. 2014) | Increasing insulin secretion by interfering with ATP-sensitive potassium channels on pancreatic β-cell (Hammeso et al. 2019) |
| *Anacardium occidentale* Linn/Anacardiaceae | Fruits | Glucosides, flavonoids | Stimulating pancreatic functions (Jaiswal et al. 2016) |
| *Arctium lappa* L./ Asteraceae | Roots | Sitosterol-®-D-glucopyranoside | Restoring pancreatic tissue function and improving insulin production (Chan et al. 2011) |
| *Achyranthes aspera* L./ Amaranthaceae | Leaves | Oleanolic acid, ursolic acid, β-sitosterol | Increasing insulin sensitivity, decreasing glucose absorption on blood cells (Silva et al. 2016) |
| *Andrographis paniculata* (Burm f.) Nees/ Acanthaceae | Aerial parts | Andrographolide, flavonoids, polyphenols | Stimulating insulin production (Nugroho et al. 2012) |
| *Artemisia herba-alba* Asso/Asteraceae | Aerial parts | α-Thujone, α-chrysanthenone, camphor, β-thujone | Inhibition of the proximal tubular reabsorption mechanism for glucose in the kidneys (Boudjelal et al. 2015) |
| *Azadirachta indica* A. Juss./Meliaceae | Leaves | Rutin, azadirachtin | Improving the morphology of Langerhans islet and β- cells (Sunarwidhi et al. 2014) |
| *Bauhinia racemosa* lam./Fabaceae | Leaves | Saponins, sterols, flavonoids, triterpenes | Activation of pancreatic β-cells (Gawade and Farooqui 2018) |
| *Bauhinia thonningii* Schum./ Fabaceae | Fruits | Saponosides, catechin | Stimulating insulin secretion (Salehi et al. 2019) |

(continued)

**Table 17.1** (continued)

| Plant species/family | Plant parts | Secondary metabolites | Mechanism of action |
|---|---|---|---|
| *Caesalpinia bonducella* (L.) Fleming/Fabaceae | Seeds | Flavonoid, tannin, caesalpinianone, bonducellin, lignin | Protecting β-cells from oxidative damages (Subbiah et al. 2019) |
| *Caesalpinia ferrea*/ C. Mart./Caesalpiniaceae | Leaves | Di-O-glycosyl-C-glucosyl flavones, saponins, terpenoids, phenolic compounds | Stimulating insulin secretion (Hassan et al. 2015) |
| *Camellia sinensis* L. cv. Sofu /Theaceae | Leaves | Gallocatechin, epicatechin, epigallocatechin | Protecting pancreatic β- cells (Fu et al. 2017) |
| *Cannabis sativa* L./ Cannabaceae | Leaves | Cannabinoids, cannabinol | Protecting pancreatic β-cells from oxidative injuries (Chauhan et al. 2010) |
| *Carthamus tinctorius* L./ Asteraceae | Flowers | Triterpenoids, phenolic compounds, flavonoids, glycosides, anthraquinone organic acid | Regenerating and restoring the islets of Langerhans (Sabet et al. 2012) |
| *Casearia esculenta* (Roxb.)/Salicaceae | Roots | 3-Hydroxymethyl xylitol | Regenerating and restoring increased insulin secretion |
| *Carum carvi* L./Apiaceae | Seeds | Carvone, phenolic compounds | Protecting pancreatic β-cells (Hajlaoui et al. 2021) |
| *Cassia fistula* L./Fabaceae | Leaves | Amentoflavone | Protecting pancreatic β-cells from oxidative damages (Srividhya et al. 2017) |
| *Cassia grandis* L.f./ Fabaceae | Fruits | Alkaloids, coumarins, flavonoids, saponins, and triterpenes, | Inhibition of α-glucosidase (Prada et al. 2018) |
| *Catharanthus roseus* (L.)/ Apocynaceae | Leaves and flowers | Caffeoylquinic acids, flavonol glycosides | Protecting insulin cells from damages (Vega-Ávila et al. 2012) |
| *Cinnamomum cassia* (L.) J.Presl/ Lauraceae | Bark | Flavonoids, polyphenols | Regenerating damaged pancreatic β-cells (Singh et al. 2018) |
| *Cinnamomum zeylanicum* Blume/Lauraceae | Bark | Methyl hydroxy chalcone polymers (MHCP) | Insulin sensitizer (Beji et al. 2018) |
| *Cucumis prophetarum* L./ Cucurbitaceae | Roots | N-Trissacharide (Kavishankar and Lakshmidevi 2014), secoisolariciresinol diglucoside (Moree et al. 2013) | α-Amylase and α-glucosidase inhibitors (Gawli and Lakshmidevi 2015) |

(continued)

**Table 17.1** (continued)

| Plant species/family | Plant parts | Secondary metabolites | Mechanism of action |
|---|---|---|---|
| *Citrus aurantifolia* (Christm.)/Rutaceae | Leaves | Diosmin (Campanero et al. 2010), naringin, hesperidin, camphor | Increasing the number of β-cells in the islets of Langerhans (Mawarti 2018) |
| *Cyamopsis tetragonoloba* (L.) Taub.—Guar/Fabaceae | Beans | *p*-Coumaric, caffeic, gallic, and gentisic acids, quercetin, kaempferol | Glucose sensor of the β-cell (Kumareswari 2018) |
| *Dioscorea batatas*/Dioscoreaceae | Roots | Diosgenin (Pari et al. 2012), allantoin | Improving the function of β-cells, maintaining normal insulin and glucose levels (Go et al. 2015) |
| *Emblica officinalis*/Phyllanthaceae | Fruits | Ellagic acid (Fatima et al. 2017) | Stimulating insulin secretion from isolated islets (Fatima et al. 2017) |
| *Eugenia jambolana* lam./Myrtaceae | Fruits | 2-(4-[(2-Hydroxybenzyl) amino]-phenyl amino-methyl)-phenol (HBPMP) (Moree et al. 2013) | Stimulating the pancreatic β-cells to secrete insulin (Jana et al. 2015) |
| *Ferula asafoetida* H. karst./ Umbelliferae | Whole parts | Ferulsinaic acid umbelliferone, diallyl sulfide, dimethyl trisulfide | Preserving pancreatic β-cells' integrity (Akhlaghi et al. 2012) |
| *Gongronema latifolium* Benth./Apocynaceae | Leaves | Polyphenols, triterpenoid, sapogenin (Khadka 2021) | Glucose sensor in pancreatic β-cells (Ojo et al. 2020) |
| *Fraxinus rhynchophylla* L./Oleaceae | Dry barks | 5-Methoxyl aesculetin, fraxetin (Sarfraz et al. 2017) | Mechanism not known |
| *Ginkgo biloba*/ Ginkgoaceae | Leaves | Quercetin, kaempferol, isorhamnetin | Improving β-cells (Cheng et al. 2013) |
| *Gymnema sylvestre* (Retz.) Schult/Apocynaceae | Leaves | Gymnemic acids, gymnemagenin, gymnestrogenin | Delaying glucose absorption in the blood cells (Saneja and Sharma 2010) |
| *Heinsia crinita* (Afzel.) G. Taylor/Rubiaceae | Leaves | Saponin | Stimulating glycolytic enzyme activity for increasing flux of glucose for cellular energy (Akpan et al. 2020) |
| *Lagerstroemia speciosa* L./ Lythraceae | Leaves | Gallotannin-penta-O-galloyl-glucopyranose | Suppression of gluconeogenesis and stimulation of glucose oxidation (Sharmin et al. 2018) |
| *Lycium barbarum* L./ Solanaceae | Roots | Saponin, polysaccharide, anthraquinone, flavone | Improving insulin secretion by restoring the impaired pancreatic β-cells (Cai et al. 2015) |

(continued)

**Table 17.1** (continued)

| Plant species/family | Plant parts | Secondary metabolites | Mechanism of action |
|---|---|---|---|
| *Lycopersicon esculentum* mill./Solanaceae | Fruits | Lycopene, resveratrol (Banihani 2018) | Mechanism not known |
| *Mangifera indica* L./ Anacardiaceae | Fruits | Mangiferin, sesquiterpenes, monoterpenes | α-Amylase inhibition, glucose adsorption, and glucose uptake capacity (Ngo et al. 2019) |
| *Momordica charantia* L./Cucurbitaceae | Fruits | Charantin (Joseph and Jini 2013), momordicine II, saponins | Insulin-releasing activity in MIN6 β-cells (Poolperm and Jiraungkoorskul 2017) |
| *Morus alba* L./Moraceae | Leaves | Flavonoids, polysaccharides, amino acids, phenylpropanoids | Rejuvenating the β-cells (Tian et al. 2016) |
| *Moringa oleifera* lam./ Moringaceae | Leaves/ pods | Isothiocyanates, phenolic acids, flavonoids | Reducing insulin resistance and influencing β-cell mass and function (Vergara-Jimenez et al. 2017) |
| *Murraya koenigii* (L.) Spreng./Rutaceae | Leaves | Carbazole | Insulin secretagogue effect stimulation or regeneration of ®-cell functioning (Bindu and Narendhirakannan 2019) |
| *Musa paradisiaca* L./Musaceae | Fruits | Syringin, phenylpropanoid glycoside (Sundaram et al. 2014) | Restoring pancreatic function (Khizar et al. 2019) |
| *Nigella sativa* L./ Ranunculaceae | Seeds | β-Carotene (), γ-Sitosterol, Thymoquinone | Increasing glucose-induced insulin release in isolated pancreatic islets (Mathur et al. 2011) |
| *Ocimum sanctum* Linn./ Lamiaceae | Leaves | Eugenol | Increasing the insulin secretion from pancreatic β-cells (Antora and Salleh 2017) |
| *Ocimum tenuiflorum* L./Lamiaceae | Leaves | 3,4-Dimethoxycinnamic acid, caffeic acid, kaempferol, rosmarinic acid, apigenin | Maintaining glucose homeostasis (Mousavi et al. 2018) |
| *Olea europaea* L./Oleaceae | Leaves | Triterpenoids (Eddouks et al. 2014), oleuropein | Increasing insulin secretion by protecting β-cells (Qadir et al. 2016) |

(continued)

**Table 17.1** (continued)

| Plant species/family | Plant parts | Secondary metabolites | Mechanism of action |
|---|---|---|---|
| *Opuntia streptacantha* Lem./Cactaceae | Aerial parts | Betalains, phytosterol (Gouws et al. 2019) | Mechanism not known |
| *Opuntia ficus-indica*/Cactaceae | Aerial parts | Myricetin, rutin, kaempferol and ®-carotene | Increasing plasma insulin and facilitating disposal of an oral glucose load from the circulation (Habtemariam 2019) |
| *Origanum vulgare* L./ Lamiaceae | Leaves | Biochanin A | Improving insulin sensitivity (Morshedloo et al. 2018) |
| *Persea americana*/Lauraceae | Fruits/ leaves | Polyphenols | Repairing ®-cells to secrete insulin (Kouamé et al. 2019) |
| *Phyllanthus niruri* L./ Phyllanthaceae | Aerial parts | Flavonoids, glycosides, tannins | Suppressing postprandial rise in blood glucose levels (Okoli et al. 2010) |
| *Polygala senega* L./ Polygalaceae | Rhizomes | Triterpenoid glycoside-senegin II, saponins | Regulating insulin secretion (Kolhe and Rachh 2018) |
| *Panax quinquefolius* L./ Araliaceae | Roots | Ginsenoside | Insulin-sensitizing (Szczuka et al. 2019) |
| *Phyllanthus amarus*/ Phyllanthaceae | Leaves | Phyllanthin, hypophyllanthin, bioflavonoids | Regulating Akt/β-- catenin signaling (Nisar et al. 2018) |
| *Piper longum* Linn./ Piperaceae | Fruits | Alkaloids, glycoside | Utilizing glucose, so blood glucose levels were significantly balanced (Nabi et al. 2013) |
| *Poncirus trifoliata* (L.) Raf./Rutaceae | Peel oil | Tangeritin, neoeriocitrin, Hesperidin, chlorogenic acid | α-Amylase and α-glucosidase enzymes (Tundis et al. 2016) |
| *Psidium guajava* L./ Myrtaceae | Leaves | Vescalagin, strictinin (Patel et al. 2012), isostrictinin, pedunculagin (Eddouks et al. 2014), triterpenoids | Protecting β-cells (Bagri et al. 2016) |
| *Pterocarpus marsupium* Roxb/Fabaceae | Barks | Pterocarpol, pterostilbene (Hariharan et al. 2005), (−)-epicatechin, marsupsin | β-cell regeneration, insulin release (Perera 2016) |
| *Punica granatum* L./ Punicaceae (flower) | Fruits/ flowers | Gallic, ellagic acid | PPAR-γ gamma activators improve the sensitivity of insulin receptor (Hai et al. 2019) |

(continued)

**Table 17.1** (continued)

| Plant species/family | Plant parts | Secondary metabolites | Mechanism of action |
|---|---|---|---|
| *Salvia officinalis* L./ Lamiaceae | Leaves | Ellagic acid, camphor, apigenin 1, 8-cineole, β-thujone, β-caryophyllene (Khadka 2021) | α-Glucosidase, pancreatic lipase inhibition (Kianbakht et al. 2016) |
| *Stevia rebaudiana* Bertoni/ Asteraceae | Leaves | Stevioside | Revitalizing the β-cells of the pancreas (Patel et al. 2012) |
| *Swertia chirata* Buch.-ham. Ex Wall/ Gentianaceae | Aerial parts | Amarogentin, swertiamarin, mangiferin, swerchirin | Stimulating insulin release from islets of Langerhans (Dey et al. 2020) |
| *Syzygium cumini* (L.) Skeels/Myrtaceae | Fruits | Kaempferol 7-O-methylether, myristic acid, quercetin | Expressing glucose transporter (Glut-4) and activating receptor gamma (PPARγ) (Gajera et al. 2017) |
| *Syzygium jambolana* lam./ Myrtaceae | Fruits | Alkaloids, tannins, phenols | Repairing of pancreatic ®-cell caused by higher insulin levels in the serum (Baliga et al. 2013) |
| *Symplocos cochinchinensis* (Lour.) S. Moore/ Symplocaceae | Barks | ®-Sitosterol, phloretin 2′glucoside, oleanolic acid | Improved insulin sensitivity with moderate antiglycation (Antu et al. 2014) |
| *Tamarindus indica* L./Fabaceae | Seeds | Polyphenol | Protective effect on pancreatic β-cells (Sole and Srinivasan 2012) |
| *Terminalia catappa* Linn./ Combretaceae | Barks | Ursolic acid | Protecting pancreatic β-cells against immunoresponse (Behl and Kotwani 2017) |
| *Terminalia chebula* Retz./ Combretaceae | Leaves | Gallic acid, chebulic acid, ellagic acid, flavonol, glycosides, triterpenoids, coumarin (Khadka 2021) | Insulin-stimulating action from the remnant β-cells or from regenerated β-cells (Dutta and Kalita 2018) |
| *Tinospora cordifolia* (Willd.) Miers/ Menispermaceae | Leaves/ flowers | Tinosporin, berberine, and tinosporinone | Inducting insulin secretion from β-cells (Rafe 2017) |
| *Trigonella foenum-graecum* L./Fabaceae | Seeds/ green leaves | Trigonelline, choline, gentianine, flavonoids, quercetin, kaempferol, saponins (Khadka 2021) | β-Cell renewal and insulin secretion stimulation (Kadan et al. 2013) |
| *Urtica dioica* L./Urticaceae | Leaves | Oxalic acid, terpenes, histamine, acetylcholine, serotonin (Khadka 2021) | Modulation of GLUT4 and protection of pancreatic β-cells (Kadan et al. 2013) |
| *Ziziphus spina-christi* (L.) Willd/Rhamnaceae | Leaves | Christinin-A (Patel et al. 2012) | Stimulating insulin secretion from the pancreatic β-cell (Khaleel et al. 2020) |

## 17.6 Future Prospects

The research should focus on the new methodologies through which several biomolecules can be isolated from the medicinal plants, and the identification of PGPR based on specific SM productions would be more lucrative in the future. Studies relating the role of root-dwelling microbes in the manufacturing of bioactive compounds are limited, and strengthening this field would make future studies more accessible and reliable about improving the yield of SMs. Much more studies are indispensable to show the strong bonding between PGPR and SMs as diabetic herbal medicines.

## 17.7 Conclusion

This chapter aimed to establish the role of PGPR in diabetes management. No literature has been published describing their relationships. PGPR functions indirectly to ameliorate diabetes. The vital role played by PGPR in producing plant SMs is not a new topic, but explaining the PGPR in diabetes management would be totally a novel chapter. PGPRs are well-known as a producer of the plant SMs, and the SMs are ultimately used as a medicine for healing various ailments, of course, including diabetes. Hence, plant SMs produced by PGPR are used as treatment for diabetes. PGPR produces SMs, and SMs play more significant roles in the management of diabetes. Investigations on improving the growth of PGPR will be helpful for producing SMs both quantitatively and qualitatively and finally work out in diabetes improvement.

## References


Adeniyi O, Washington L, Glenn CJ, Franklin SG, Scott A, Aung M, Niranjan SA, Jolly PE (2021) The use of complementary and alternative medicine among hypertensive and type 2 diabetic patients in Western Jamaica: a mixed methods study. PLoS One 16(2):1–15. https://doi.org/10.1371/journal.pone.0245163

Ahmed D, Kumar V, Verma A, Gupta PS, Kumar H, Dhingra V, Mishra V, Sharma M (2014) Antidiabetic, renal/hepatic/pancreas/cardiac protective and antioxidant potential of methanol/dichloromethane extract of *Albizzia Lebbeck* Benth. Stem bark (ALEx) on streptozotocin induced diabetic rats. BMC Complement Altern Med 14(233):1–17. https://doi.org/10.1186/1472-6882-14-243

Akhlaghi F, Rajaei Z, Hadjzadeh MAR, Iranshahi M, Alizadeh M (2012) Antihyperglycemic effect of asafoetida (*Ferula Assa-foetida* oleo-gum-resin) in streptozotocin-induced diabetic rats. World Appl Sci J 17(2):157–162

Akpan UA, Eka OU, Ekanemesang UM, Akpanabiatu MI (2020) Effect of ethanolic leaf extract of *Heinsia crinita* (Atama) on the haematological indices and lipid profile in normoglycemic and diabetic albino Wistar rats. Niger J Pharma Appl Sci Res 5(1):48–53. https://nijophasr.net/index.php/nijophasr/article/view/110

Antora RA, Salleh RM (2017) Antihyperglycemic effect of Ocimum plants: a short review. Asian Pac J Trop Biomed 7(8):755–759. https://doi.org/10.1016/j.apjtb.2017.07.010

Antu KA, Riya MP, Mishra A, Anilkumar KS, Chandrakanth CK, Tamrakar AK, Srivastav AK, Raghu KG (2014) Antidiabetic property of *Symplocos cochinchinensis* is mediated by inhibition of alpha glucosidase and enhanced insulin sensitivity. PLoS One 9(9):1–13

Aryal B, Niraula P, Khadayat K, Adhikari B, Khatri Chhetri D, Sapkota BK, Bhattarai BR, Aryal N, Parajuli N (2021) Antidiabetic, antimicrobial, and molecular profiling of selected medicinal plants. Evid Based Complement Alternat Med 5510099:1–15. https://doi.org/10.1155/2021/5510099

Backer R, Rokem JS, Ilangumaran G, Lamont J, Praslickova D, Ricci E, Subramanian S, Smith DL (2018) Plant growth-promoting rhizobacteria: context, mechanisms of action, and roadmap to commercialization of biostimulants for sustainable agriculture. Front Plant Sci 9(1473):1–17. https://doi.org/10.3389/fpls.2018.01473

Bagri P, Ali M, Aeri V, Bhowmik M (2016) Isolation and antidiabetic activity of new lanostenoids from the leaves of *Psidium guajava* L. Int J Pharm Pharm Sci 8(9):14–18. https://doi.org/10.22159//ijpps.2016.v8i9.10425

Baliga MS, Fernandes S, Thilakchan KR, D'Souza P, Rao S (2013) Scientific validation of the antidiabetic effects of *Syzygium jambolanum* dc (black plum), a traditional medicinal plant of India. J Altern Complement Med 19(3):191–197. https://doi.org/10.1089/acm.2011.0752

Banihani SA (2018) Tomato (*Solanum lycopersicum* L.) and type 2 diabetes. Int J Food Prop 21(1):99–105. https://doi.org/10.1080/10942912.2018.1439959

Behl T, Kotwani A (2017) Proposed mechanisms of *Terminalia catappa* in hyperglycaemia and associated diabetic complications. J Pharm Pharmacol 69(2):123–134. https://doi.org/10.1111/jphp.12676

Beji RS, Khemir S, Wannes WA, Ayari K, Ksouri R (2018) Antidiabetic, antihyperlipidemic and antioxidant influences of the spice cinnamon (*Cinnamomum zeylanicumon*) in experimental rats. Braz J Pharm Sci 54(2):1–18. https://doi.org/10.1590/s2175-97902018000217576

Bharti SK, Krishnan S, Kumar A, Kumar A (2018) Antidiabetic phytoconstituents and their mode of action on metabolic pathways. Ther Adv Endocrinol Metab 9(3):81–100. https://doi.org/10.1177/2042018818755019

Bindu J, Narendhirakannan RT (2019) Role of medicinal plants in the management of diabetes mellitus: a review. 3. Biotech 9(1):1–17. https://doi.org/10.1007/s13205-018-1528-0

Bisht S, Pandey P, Bhargava B, Sharma S, Kumar V, Sharma KD (2015) Bioremediation of polyaromatic hydrocarbons (PAHs) using rhizosphere technology. Braz J Microbiol 46(1):7–21. https://doi.org/10.1590/S1517-838246120131354

Boucher J, Kleinridders A, Ronald KC (2014) Insulin receptor signaling in normal and insulin-resistant states. Cold Spring Harb Perspect Biol 6:1–24

Boudjelal A, Siracusa L, Henchiri C, Sarri M, Abderrahim B, Baali F, Ruberto G (2015) Antidiabetic effects of aqueous infusions of *Artemisia herba-alba* and *Ajuga Iva* in alloxan-induced diabetic rats. Planta Med 81(9):696–704. https://doi.org/10.1055/s-0035-1546006

Cai H, Liu F, Zuo P, Huang G, Song Z, Wang T, Lu H, Guo F, Han C, Sun G (2015) Practical application of antidiabetic efficacy of *Lycium barbarum* polysaccharide in patients with type 2 diabetes. Med Chem 11(4):383–390. https://doi.org/10.2174/1573406410666141110153858

Campanero MA, Escolar M, Perez G, Garcia-Quetglas E, Sadaba B, Azanza JR (2010) Simultaneous determination of diosmin and diosmetin in human plasma by ion trap liquid chromatography-atmospheric pressure chemical ionization tandem mass spectrometry: application to a clinical pharmacokinetic study. J Pharma and Biomed Anal 51(4):875–881. https://doi.org/10.1016/j.jpba.2009.09.012

Canarini A, Kaiser C, Merchant A, Richter A, Wanek W (2019) Root exudation of primary metabolites: mechanisms and their roles in plant responses to environmental stimuli. Front Plant Sci 10(157):1–19. https://doi.org/10.3389/fpls.2019.00157

Cappellari LDR, Santoro MV, Nievas F, Giordano W, Banchio E (2013) Increase of secondary metabolite content in marigold by inoculation with plant growth-promoting rhizobacteria. Appl Soil Ecol 70:16–22. https://doi.org/10.1016/j.apsoil.2013.04.001

Chan YS, Cheng LN, Wu JH, Chan E, Kwan YW, Lee SMY, Leung GPH, Yu PHF, Chan SW (2011) A review of the pharmacological effects of *Arctium lappa* (burdock). Inflammopharmacol 19(5):245–254. https://doi.org/10.1007/s10787-010-0062-4

Chandran H, Meena M, Swapnil P (2021) Plant growth-promoting rhizobacteria as a green alternative for sustainable agriculture. Sustainability 13(10986):1–30. https://doi.org/10.3390/su131910986

Chauhan A, Sharma P, Srivastava P, Kumar N, Dudhe R (2010) Plants having potential antidiabetic activity: a review. Der Pharm Lett 2(3):369–387. ISSN 0975-5071

Cheng D, Liang B, Li Y (2013) Antihyperglycemic effect of *Ginkgo biloba* extract in streptozotocin-induced diabetes in rats. Biomed Res Int 2013:1–8. https://doi.org/10.1155/2013/162724

Chiocchio I, Mandrone M, Tomasi P, Marincich L, Poli F (2021) Plant secondary metabolites: an opportunity for circular economy. Molecules 26(495):1–31. https://doi.org/10.3390/molecules26020495

Dey P, Singh J, Suluvoy JK, Dilip KJ, Nayak J (2020) Utilization of *Swertia chirayita* plant extracts for management of diabetes and associated disorders: present status, future prospects and limitations. Nat Prod Bioprospect 10:431–443. https://doi.org/10.1007/s13659-020-00277-7

Dojima T, Craker LE (2016) Potential benefits of soil microorganisms on medicinal and aromatic plants. ACS Symp Ser 1218:75–90

Dutta J, Kalita MC (2018) Anti-hyperglycemic evaluation of *Terminalia chebula* leaves. Int J Pharm Pharm 10(11):43–48. https://doi.org/10.22159/ijpps.2018v10i11.28167

Eddouks M, Bidi A, El Bouhali B, Hajji L, Zeggwagh NA (2014) Antidiabetic plants improving insulin sensitivity. J Pharm Pharmacol 66(9):1197–1214. https://doi.org/10.1111/jphp.12243

Fatima N, Hafizur RM, Hameed A, Ahmed S, Nisar M, Kabir N (2017) Ellagic acid in *Emblica officinalis* exerts antidiabetic activity through the action on β-cells of pancreas. Eur J Nutr 56(2):591–601. https://doi.org/10.1007/s00394-015-1103-y

Favaretto A, Chini SO, Scheffer-Basso SM, Sobottka AM, Bertol CD, Perez NB (2015) Pattern of allelochemical distribution in leaves and roots of tough lovegrass (*Eragrostis plana* Nees). Aust J Crop Sci 9(11):1119–1125. ISSN 1835-2693

Fu S-F, Wei J-Y, Chen H-W, Liu Y-Y, Lu H-Y, Chou J-Y (2015) Indole-3-acetic acid: a widespread physiological code in interactions of fungi with other organisms. Plant Signal Behav 10(8):1–9. https://doi.org/10.1080/15592324.2015.1048052

Fu QY, Li QS, Lin XM, Qiao RY, Yang R, Li XM, Dong ZB, Xiang LP, Zheng XQ, Lu JH, Yuan CB Ye JH, Liang YR (2017) Antidiabetic effects of tea. Molecules 22(5), 849:1-19. https://doi.org/10.3390/molecules22050849

Gajera HP, Gevariya SN, Hirpara DG, Patel SV, Golakiya BA (2017) Antidiabetic and antioxidant functionality associated with phenolic constituents from fruit parts of indigenous black jamun (*Syzygium cumini* L.) landraces. J Food Sci Technol 54(10):3180–3191. https://doi.org/10.1007/s13197-017-2756-8

Gawade B, Farooqui M (2018) Screening of phytochemicals and *in-vitro* antidiabetic activity of *Bauhinia racemosa* lam. Leaves. Asian J Pharm Clin Res 11(6):190–193. https://doi.org/10.22159/ajpcr.2018.v11i6.24038

Gawli K, Lakshmidevi N (2015) Antidiabetic and antioxidant potency evaluation of different fractions obtained from *Cucumis prophetarum* fruit. Pharm Biol 53(5):689–694. https://doi.org/10.3109/13880209.2014.937503

Geetha G, Gopinathapillai PK, Sankar V (2011) Antidiabetic effect of *Achyranthes rubrofusca* leaf extracts on alloxan induced diabetic rats. Pak J Pharm Sci 24(2):193–199. PMID: 21454169

Go HK, Rahman M, Kim GB, Na CS, Song CH, Kim JS, Kim SJ, Kang HS (2015) Antidiabetic effects of yam (*Dioscorea batatas*) and its active constituent, allantoin, in a rat model of streptozotocin-induced diabetes. Nutrients 7(10):8532–8544. https://doi.org/10.3390/nu7105411

Gouda S, Kerry RG, Das G, Spiros P, Shin HS (2018) Patra JK (2018) revitalization of plant growth promoting rhizobacteria for sustainable development in agriculture. Microbiol Res 206:131–140. https://doi.org/10.1016/j.micres.2017.08.016132

Gouws CA, Georgousopoulou EN, Mellor DD, McKune A, Naumovski N (2019) Effects of the consumption of prickly pear cacti (*Opuntia* spp.) and its products on blood glucose levels and insulin: a systematic review. Medicina (Lithuania) 55(138):1–18. https://doi.org/10.3390/medicina55050138

Grover M, Bodhankar S, Sharma A, Sharma P, Singh J, Nain L (2021) PGPR mediated alterations in root traits: way toward sustainable crop production. Front Sustain Food Syst 4:618230. https://doi.org/10.3389/fsufs.2020.618230

Habtemariam S (2019) The chemical and pharmacological basis of prickly pear cactus (*Opuntia species*) as potential therapy for type 2 diabetes and obesity. Medicinal foods as potential therapies for type-2 diabetes and associated diseases.Chapter13, pp 435–472

Hai NT, Nguyen TH, Nguyen VK, Dang KT, Bui TT (2019) Phytochemicals and antidiabetic activity of the aqueous extract of the *Punica granatum* fruit in streptozotocin-induced diabetic mice. J Basic Clin Physiol Pharmacol 30(4):1–7. https://doi.org/10.1515/jbcpp-2019-0061

Hajlaoui H, Arraouadi S, Noumi E, Aouadi K, Adnan M, Khan MA, Kadri A, Snoussi M (2021) Antimicrobial, antioxidant, anti-acetylcholinesterase, antidiabetic, and pharmacokinetic properties of Carum carvi L. and Coriandrum sativum L. essential oils alone and in combination. Molecules 26(3625):1–18. https://doi.org/10.3390/molecules26123625

Hammeso WW, Emiru YK, Getahun KA, Kahaliw W (2019) Antidiabetic and antihyperlipidemic activities of the leaf latex extract of *aloe megalacantha* baker (Aloaceae) in streptozotocin-induced diabetic model. Evid Based Complement Alternat Med 2019:1–9. https://doi.org/10.1155/2019/8263786

Hassan SK, El-Sammad NM, Mousa AM, Mohammed MH, Farrag ARH, Hashim ANE, Werner V, Lindequist V, Nawwar MAEM (2015) Hypoglycemic and antioxidant activities of *Caesalpinia ferrea* martius leaf extract in streptozotocin-induced diabetic rats. Asian Pac J Trop Biomed 5(6):462–471. https://doi.org/10.1016/j.apjtb.2015.03.004

Huang C, Zhou J, Jie Y, Xing H, Zhong Y, Yu W, She W, Ma Y, Liu Z, Zhang Y (2016) A ramie (*Boehmeria nivea*) bZIP transcription factor BnbZIP3 positively regulates drought, salinity and heavy metal tolerance. Mol Breed 35(12):776–786

Ibnaouf KH, Taha KK, Idriss H, Aldaghri O (2017) Amplified spontaneous emission (ASE) properties of a laser dye (LD-473) in solid state. J Eur Opt Soc Rapid Publ 13(20):1–5. https://doi.org/10.1186/s41476-017-0049-5

Iqbal N, Khan NA, Ferrante A, Trivellini A, Francini A, Khan MIR (2017) Ethylene role in plant growth, development and senescence: interaction with other phytohormones. Front Plant Sci 8(475):1–19. https://doi.org/10.3389/fpls.2017.00475

Jaegera R, Cuny E (2016) Terpenoids with special pharmacological significance: a review. Nat Prod Commun 11(9):1373–1390. PMID: 30807045

Jaiswal YS, Tatke PA, Gabhe SY, Vaidya AB (2016) Antidiabetic activity of extracts of *Anacardium occidentale* Linn. Leaves on n-streptozotocin diabetic rats. J Trad Complement Med 7(4):421–427. https://doi.org/10.1016/j.jtcme.2016.11.007

Jan R, Asaf S, Numan M, Lubna K, Kyung MK (2021) Plant secondary metabolite biosynthesis and transcriptional regulation in response to biotic and abiotic stress conditions. Agronomy 11(968):1–31. https://doi.org/10.3390/agronomy11050968

Jana K, Bera TK, Ghosh D (2015) Antidiabetic effects of *Eugenia jambolana* in the streptozotocin-induced diabetic male albino rat. Biomark Genom Med 7(3):116–124. https://doi.org/10.1016/j.bgm.2015.08.001

Joseph B, Jini D (2013) Antidiabetic effects of *Momordica charantia* (bitter melon) and its medicinal potency. Asian Pac J Trop Dis 3(2):93–102. https://doi.org/10.1016/S2222-1808(13)60052-3

Kadan S, Saad B, Sasson Y, Zaid H (2013) *In-vitro* evaluations of cytotoxicity of eight antidiabetic medicinal plants and their effect on GLUT4 translocation. Evid Based Complementary Altern Med 2013:1–10. https://doi.org/10.1155/2013/549345

Kang L, Azian AA, Ming H, Takaaki N, Md Altaf U-A, Shigehiko K (2017) Novel approach to classify plants based on metabolite-content similarity. BioMed Res Intern 2017(5296729):1–12. https://doi.org/10.1155/2017/5296729

Kannepalli A, Davranov K, Narimanov A, Enakiev Y, Syed AA, Elgorban AM, Bahkali AH, Wirth S, Sayyed RZ, Gafur A (2021) Co-inoculation of rhizobacteria promotes growth, yield, and nutrient contents in soybean and improves soil enzymes and nutrients under drought conditions. Scient Rep 11:22081. https://doi.org/10.1038/s41598-021-01337-9

Kapadia C, Sayyed RZ, Enshasy HEE, Vaidya H, Sharma D, Patel V, Malek RA, Syed A, Elgorban AM, Ahmad K, Zuan ATK (2021) Halotolerant microbial consortia for sustainable mitigation of salinity stress, growth promotion, and mineral uptake in tomato plant and soil nutrient enrichment. Sustainability 13(8369):1–14. https://doi.org/10.3390/su13158369

Karimi A, Majlesi M, Rafieian-Kopaei M (2015) Herbal versus synthetic drugs; beliefs and facts. J Nephropharmacol. 4(1): 27–30. PMC5297475

Kavishankar GB, Lakshmidevi N (2014) Antidiabetic effect of a novel N-Trisaccharide isolated from *Cucumis prophetarum* on streptozotocin-nicotinamide induced type 2 diabetic rats. Phytomedicine 21(5):624–630. https://doi.org/10.1016/j.phymed.2013.12.002

Khadka D (2021) Ethnopharmacology of wild plants. In: Rai M, Bhattarai S, Feitosa CM (eds) Ameliorating insulin signaling pathway by phytotherapy: a review. Taylor & Francis Group, CRC Press, pp 317–342

Khaleel AM, Almuhur RA, Al-Deeb TM, Jaran AS, Al-Jamal AA (2020) Antidiabetic and hypolipidemic effects of ethanolic leaf extract of *Ziziphus spina-christi* on normal and streptozotocin induced diabetic rats. Eurasia J Biosci 14:5865–5870

Khan N, Al S, Shahi MA, Mustafa A, Sayyed RZ, Curaá JA (2021) Insights into the interactions among roots, rhizosphere and rhizobacteria for improving plant growth and tolerance to abiotic stresses: a review. Cells 10(6):1551:1–1551:19. https://doi.org/10.3390/cells10061551

Khizar A, Rhizwania GH, Zahida H, Shareef H, Taqi MM (2019) *Musa paradisiaca* L. may restore pancreatic morphology and function to trigger its anti-diabetic and hypolipidemic activities in alloxon-induce diabetic rats. Medicinal Aromat Plants 08(333):1–8. https://doi.org/10.35248/2167-0412.19.8.333

Kianbakht S, Nabati F, Abasi B (2016) *Salvia officinalis* (Sage) leaf extract as add-on to statin therapy in hypercholesterolemic type 2 diabetic patients: A randomized clinical trial. Inter J Mole Cellul Med 5(3):141–148. PMC5125366

Kolhe SS, Rachh PR (2018) Review on potent antidiabetic plants or herbs from traditional medicine. J Drug Del Ther 8(5):92–98. https://doi.org/10.22270/jddt.v8i5.1856

Kouamé NM, Koffi C, N'zoué KS, Yao NAR, Doukouré B, Kamagate M (2019) Comparative antidiabetic activity of aqueous, ethanol, and methanol leaf extracts of persea Americana and their effectiveness in type 2 diabetic rats. Evid Based Complementary Alter Med 2019:1–14. https://doi.org/10.1155/2019/5984570

Kumar V, Ahmed D, Verma A, Anwar F, Ali M, Mujeeb M (2013) Umbelliferone β-D-galactopyranoside from *Aegle marmelos* (L.) corr. An ethnomedicinal plant with antidiabetic, antihyperlipidemic and antioxidative activity. BMC Complement Alter Med 13(273):1–19. https://doi.org/10.1186/1472-6882-13-273

Kumareswari T (2018) Antidiabetic activities of foliar application of extract of Ulva reticulata grown fruit of *Cyamopsis tetragonoloba* (L.) taub. In alloxan induced diabetic rats. J Appl Biotechnol Bioeng 5(3):146–149. https://doi.org/10.15406/jabb.2018.05.00130

Mathur ML, Gaur J, Sharma R, Haldiya KR (2011) Antidiabetic properties of a spice plant *Nigella sativa*. J Endocrinol Metab 1(1):1–8. https://doi.org/10.4021/jem12e

Mawarti H, Khotimah As'ad MZ, Rajin M (2018) Ameliorative effect of *Citrus aurantifolia* and *Cinnamomum burmannii* extracts on diabetic complications in a hyperglycemic rat model. Trop J Pharm Res 17(5):823–829. https://doi.org/10.4314/tjpr.v17i5.11

Moree SS, Kavishankar GB, Rajesha J (2013) Antidiabetic effect of secoisolariciresinol diglucoside in streptozotocin-induced diabetic rats. Phytomedicine 20(3–4):237–245. https://doi.org/10.1016/j.phymed.2012.11.011

Morshedloo MR, Ahmadi H, Pirali HM, Yazdani D (2018) An over review to *Origanum vulgare* L. and its pharmacological properties. J Med Plants 17(68):15–31. https://doi.org/10.22037/ijpr.2020.113874.14539

Mousavi L, Salleh RM, Murugaiyah V (2018) Phytochemical and bioactive compounds identification of *Ocimum tenuiflorum* leaves of methanol extract and its fraction with an antidiabetic potential. Inter J Food Prop 21(1):2390–2399. https://doi.org/10.1080/10942912.2018.1508161

Nabi SA, Kasetti RB, Sirasanagandla S, Tilak TK, Kumar MVJ, Rao CA (2013) Antidiabetic and antihyperlipidemic activity of *Piper longum* root aqueous extract in STZ induced diabetic rats. BMC Complement Altern Med 37:1–9. https://doi.org/10.1186/1472-6882-13-37

Ngo DH, Ngo DN, Vo TTN, Vo TS (2019) Mechanism of action of Mangifera indica leaves for anti-diabetic activity. Sci Pharm 87(13):1–12. https://doi.org/10.3390/scipharm8702001

Nisar F, He J, Ahmed A, Yang Y, Li M, Wan C (2018) Chemical components and biological activities of the genus phyllanthus: a review of the recent literature. Molecules 23(2567):1–25. https://doi.org/10.3390/molecules23102567

Nithyapriya S, Lalitha S, Sayyed RZ, Reddy MS, Dailin DJ, El Enshasy HA, Luh Suriani N, Herlambang S (2021) Production, purification, and characterization of Bacillibactin siderophore of Bacillus subtilis and its application for improvement in plant growth and oil content in sesame. Sustainability 13(5394):1–18. https://doi.org/10.3390/su131053

Nugroho AE, Andrie M, Warditiani NK, Siswanto E, Pramono S, Lukitaningish E (2012) Antidiabetic and antihyperlipidemic effect of *Andrographis paniculata* (Burm. F.) Nees and andrographolide in high-fructose-fat-fed rats. Indian J Pharmacol 44(3):377–381. https://doi.org/10.4103/0253-7613.96343

Ojo OA, Okesola MA, Lisa E, Ajiboye BO, Oyinloye BE, Agboinghale PE, Onikanni AS (2020) *Gongronema latifolium* Benth. Leaf extract attenuates diabetes induced neuropathy via inhibition of cognitive, oxidative stress and inflammatory response. Sci Food Agricl 00(12):4504–4511. https://doi.org/10.1002/jsfa.10491

Okoli CO, Ibiam AF, Ezike AC, Akah PA, Okoye TC (2010) Evaluation of antidiabetic potentials of *Phyllanthus niruri* in alloxan diabetic rats. Afr J Biotech 9(2):248–259. ISSN 1684–5315

Ota A, Ulrih NP (2017) An overview of herbal products and secondary metabolites used for management of type two diabetes. Front Pharmacol 8(436):1–14. https://doi.org/10.3389/fphar.2017.00436

Ozougwu JC (2011) Antidiabetic effects of *Allium cepa* (onions) aqueous extracts on alloxan-induced diabetic *Rattus norvegicus*. J Med Plants Res 5(7):1134–1139. ISSN 1996-0875

Ozougwu CJ, Eyo EJ (2010) Studies on the antidiabetic activity of *Allium sativum* (garlic) aqueous extracts on alloxan-induced diabetic albino rat. Pharmacologyonline 2:1079–1088

Panche AN, Diwan AD, Chandra SR (2016) Flavonoids: an overview. J Nutr Sci 5:1–15. https://doi.org/10.1017/jns.2016.41

Pari L, Monisha P, Jalaludeen MA (2012) Beneficial role of diosgenin on oxidative stress in aorta of streptozotocin induced diabetic rats. Eur J Pharmacol 691(1–3, 143):–150. https://doi.org/10.1016/j.ejphar.2012.06.038

Patel DK, Prasad SK, Kumar R, Hemalatha S (2012) An overview on antidiabetic medicinal plants having insulin mimetic property. Asian Pac Trop Biomed 2(4):320–330. https://doi.org/10.1016/s2221-1691(12)60032-x

Patel PR, Shaikh SS, Sayyed RZ (2016) Dynamism of PGPR in bioremediation and plant growth promotion in heavy metal contaminated soil. Indian J Exp Biology 54:286–290. PM 27295926

Perera H (2016) Antidiabetic effects of *Pterocarpus marsupium* (Gammalu). Eur J Med Plants 13(4):1–14. https://doi.org/10.9734/EJMP/2016/23930

Poolperm S, Jiraungkoorskul W (2017) An update review on the anthelmintic activity of bitter gourd, *Momordica charantia*. Phcog Rev 11(21):31–34. https://doi.org/10.4103/phrev.phrev_52_16

Pourakbar L, Moghaddama SS, Enshasy HE, Sayyed RZ (2021) Antifungal activity of the extract of a macroalgae, *Gracilariopsis persica*, against four plant pathogenic fungi *in-vitro*. Plants 10(1781):1–13. https://doi.org/10.3390/plants10091781

Prada AL, Amado JRR, Keita H, Zapata EP, Carvalho H, Lima ES, Sousa TP, Carvalho JCT (2018) *Cassia grandis* fruit extract reduces the blood glucose level in alloxan-induced diabetic rats. Biomed Pharmacother 103:421–428. https://doi.org/10.1016/j.biopha.2018.04.059

Przeor M (2022) Some common medicinal plants with antidiabetic activity, known and available in Europe (A Mini-Review). Pharma 15(65):1–25. https://doi.org/10.3390/ph15010065

Qadir NM, Ali KA, Qader SW (2016) Antidiabetic effect of oleuropein from *Olea europaea* leaf against alloxan induced type 1 diabetic in rats. Braz Arch Biol Technol 59:1–10. https://doi.org/10.1590/1678-4324-2016150116

Rafe MR (2017) A review of five traditionally used anti-diabetic plants of Bangladesh and their pharmacological activities. Asian Pac J Trop Dis 10(10):933–939. https://doi.org/10.1016/j.apjtm.2017.09.002

Romera FJ, García MJ, Lucena C, Martínez-Medina A, Aparicio MA, Ramos J, Alcántara E, Angulo M, Pérez-Vicente R (2019) Induced systemic resistance (ISR) and Fe deficiency responses in dicot plants. Front Plant Sci 10(287):1–17. https://doi.org/10.3389/fpls.2019.00287

Rorsman P, Ashcroft FM (2018) Pancreatic β-cell electrical activity and insulin secretion: of mice and men. Physiol Revi 98(1):117–214. https://doi.org/10.1152/physrev.00008.2017

Sabet F, Asgary S, Rahimi P, Mahzouni P, Madani H (2012) Antidiabetic effect of hydroalcoholic extract of *Carthamus tinctorius* L. in alloxan-induced diabetic rats. J Res Med Sci 17(6):386–392

Sagar A, Rai S, Ilyas N, Sayyed RZ, Al-Turki AI, El Enshasy HA, Simarmata T (2022) Halotolerant Rhizobacteria for salinity stress mitigation: diversity, mechanism and molecular approaches. Sustainability 14(490):1–26. https://doi.org/10.3390/su14010490

Salehi B, Ata A, Kumar NVA, Sharopov F, Ramírez-Alarcón K, Ruiz-Ortega A, Ayatollahi SA, Fokou PVT, Kobarfard F, Zaka ZA, Iriti M, Taheri Y, Martorell M, Sureda A, Setzer WN, Durazzo A, Lucarini M, Santini A, Capasso R, Ostrander EA, Rahman AU, Choudhary MI, Cho WC, Sharifi-Rad J (2019) Antidiabetic potential of medicinal plants and their active components. Biomol Ther 9(551):1–121. https://doi.org/10.3390/biom9100551

Saneja A, Sharma C (2010) *Gymnema sylvestre* (Gurmar): a review. Der Pharm Lett 2(1):275–284. ISSN 0975-5071

Sarfraz I, Rasul A, Jabeen F, Younis T, Zahoor MK, Arshad M, Muhammad Ali M (2017) Fraxinus: a plant with versatile pharmacological and biological activities. J Evid Based Complementary Altern Med 2017(4269868):1–12. https://doi.org/10.1155/2017/4269868

Sayyed RZ, Seifi S, Patel PR, Shaikh SS, Jadhav HP, El Enshasy H (2019) Siderophore production in groundnut rhizosphere isolate, *Achromobacter* sp. RZS2 influenced by physicochemical factors and metal ions. Environ Sustain 2:117–124. https://doi.org/10.1007/s42398-019-00070-4

Shaikh SS, Reddy MS, Sayyed RZ (2016) Plant growth promoting rhizobacteria: an eco-friendly approach for sustainable agro-ecosystem plant soil-microbes. Springer, Switzerland, Chapter 10:182–201

Shang A, Cao SY, Xu XY, Gan RY, Tang GY, Corke H, Mavumengwana V, Li HB (2019) Bioactive compounds and biological functions of garlic (*Allium sativum* L.). Foods 8(7):246:1–246:31. https://doi.org/10.3390/foods8070246

Sharma A, Gupta A, Dalela M, Sharma S, Sayyed RZ, Enshasy HA, El EEA (2020) Linking organic metabolites as produced by *Purpureocillium lilacinum* 6029 cultured on Karanja deoiled cake medium for the sustainable management of root-knot nematodes. Sustainability 12:8276:1–8276:9. https://doi.org/10.3390/su12198276

Sharmin T, Rahman S, Mohammadi H (2018) Investigation of biological activities of the flowers of Lagerstroemia speciosa, the Jarul flower of Bangladesh. BMC Complement Altern Med 18(231):1–10. https://doi.org/10.1186/s12906-018-2286-6

Shehadeh MB, Suaifan GARY, Abu-Odeh AM (2021) Plants secondary metabolites as blood glucose-lowering molecules. Molecules 26(4333):1–46. https://doi.org/10.3390/molecules26144333

Silva FSG, Oliveira PJ, Duarte MF (2016) Oleanolic, ursolic, and betulinic acids as food supplements or pharmaceutical agents for type 2 diabetes: promise or illusion? J Agric Food Chem 64(15):2991–3008. https://doi.org/10.1021/acs.jafc.5b06021

Singh N, Sharma B (2018) Toxicological effects of Berberine and Sanguinarine. Front Mol Biosci 5(21):1–7. https://doi.org/10.3389/fmolb.2018.0002

Singh J, Parasuraman S, Kathiresan S (2018) Antioxidant and antidiabetic activities of methanolic extract of *Cinnamomum cassia*. Phcog Res 10(3):237–242

Sole SS, Srinivasan BP (2012) Aqueous extract of tamarind seeds selectively increases glucose transporter-2, glucose transporter-4, and islets' intracellular calcium levels and stimulates β-cell proliferation resulting in improved glucose homeostasis in rats with streptozotocin-induce. Nutr Res 32(8):626–636. https://doi.org/10.1016/j.nutres.2012.06.015

Srividhya M, Hridya H, Shanthi V, Ramanathan K (2017) Bioactive amento flavone isolated from *Cassia fistula* L. leaves exhibits therapeutic efficacy. 3. Biotech 7(1):1–5. https://doi.org/10.1007/s13205-017-0599-7

Subbiah V, Nagaraja P, Narayan P, Nagendra HG (2019) Evaluation of pharmacological properties of Caesalpinia bonducella seed and shell extract. Pharmacog J 11(1):150–154. https://doi.org/10.5530/pj.2019.1.25

Sukmawati D, Family N, Hidayat I, Sayyed RZ, Elsayed EA, Dailin DJ, Hanapi SZ, Wadaan MA, Enshasy HE (2021) Biocontrol activity of *Aureobasidium pullulans* and *Candida orthopsilosis* isolated from *Tectona grandis* L. Phylloplane against Aspergillus sp in post-harvested citrus fruit. Sustainability 13(7479):1–15. https://doi.org/10.3390/su13137479

Sunarwidhi AL, Sudarsono S, Nugroho AE (2014) Hypoglycemic effect of combination of *Azadirachta indica* a. Juss. And *Gynura procumbens* (Lour.) Merr. Ethanolic extracts standardized by rutin and quercetin in alloxan-induced hyperglycemic rats. Adv Pharma Bull 4:613–618. https://doi.org/10.5681/apb.2014.090

Sundaram CKS, Pillai SI, Pillai SS (2014) Isolation, characterization of syringin, phenylpropanoid glycoside from *Musa paradisiaca* tepal extract and evaluation of its antidiabetic effect in streptozotocin-induced diabetic rats. Biomed Prev Nutr 4(2):105–111

Suriani N, Suprapta D, Novizar N, Parwanayoni N, Darmadi A, Dewi D, Sudatri N, Ahmad F, Sayyed RZ, Syed A, Elgorban A, Bahkali A, Enshasy H, Dalin DJ (2020) A mixture of piper leaves extracts and Rhizobacteria for sustainable plant growth promotion & biocontrol of blast pathogen of organic Bali Rice. Sustain 12(8490):1–18. https://doi.org/10.3390/su12208490

Szczuka D, Nowak A, Zakłos-Szyda M, Kochan E, Szymańska G, Motyl I, Blasiak J (2019) American ginseng (*Panax quinquefolium* L.) as a source of bioactive phytochemicals with pro-health properties. Nutrients 11(1041):1–27. https://doi.org/10.3390/nu11051041

Tian S, Tang M, Zhao B (2016) Current antidiabetes mechanisms and clinical trials using *Morus alba* L. J Tradit Chinese Medical Sci 3(1):3–8. https://doi.org/10.1016/j.jtcms.2016.04.001

Tiwari R, Rana CS (2015) Plant secondary metabolites: a review. Int J Eng Res Gener Sci 3(5):661–670. ISSN 2091-2730

Tundis R, Bonesi M, Sicari V, Pellicanò TM, Tenuta MC, Leporini M, Menichini F, Loizzo MR (2016) *Poncirus trifoliata* (L.) Raf.: chemical composition, antioxidant properties and hypoglycaemic activity via the inhibition of α-amylase and α-glucosidase enzymes. J Funct Foods 25:477–485. https://doi.org/10.1016/j.jff.2016.06.034

Vega-Ávila E, Cano-Velasco JL, Alarcón-Aguilar FJ, Fajardo Ortíz MDC, Almanza-Pérez JC, Romas RR (2012) Hypoglycemic activity of aqueous extracts from *Catharanthus roseus*. Evid Based Complementar Alter Med 934258:1–7. https://doi.org/10.1155/2012/934258

Vergara-Jimenez M, Almatrafi MM, Fernandez ML (2017) Bioactive components in *Moringa oleifera* leaves protect against chronic disease. Antioxidants 6(91):1–17. https://doi.org/10.3390/antiox6040091

Verma SC, Vashishth E, Singh R, Kumari A, Meena AK, Pant P, Bhuyan GC, Padhi MM (2013) A review on parts of *Albizia lebbeck* (L.) Benth. Used as ayurvedic drugs. Res J Pharm Technol 6(11):1307–1313. ISSN: 0974-360X

Walker V, Bertrand C, Bellvert F, Moënne-Loccoz Y, Bally R, Comte G (2011) Host plant secondary metabolite profiling shows a complex, strain-dependent response of maize to plant growth-promoting rhizobacteria of the genus Azospirillum. New Phytol 189(2):494–506. https://doi.org/10.1111/j.1469-8137.2010.03484.x

Wang LY, Xie YS, Cui YY, Xu J, He W, Chen HG, Guo JH (2015) Conjunctively screening of biocontrol agents (BCAs) against fusarium root rot and fusarium head blight caused by *fusarium graminearum*. Microbiol Res 177:34–42. https://doi.org/10.1016/j.micres.2015.05.005

Wang C, Wang Y, Ma J, Hou Q, Liu K, Ding Y, Du B (2018) Screening and whole-genome sequencing of two *Streptomyces species* from the rhizosphere soil of peony reveal their characteristics as plant growth-promoting rhizobacteria. Biomed Res Int: Article ID 2419686:1–11. https://doi.org/10.1155/2018/2419686

Wang H, Liu R, You MP, Barbetti MJ, Chen Y (2021) Pathogen biocontrol using plant growth-promoting bacteria (PGPR): role of bacterial diversity. Microorganisms 9(1988):1–18. https://doi.org/10.3390/microorganisms9091988

Yadav AN, Sayyed RZ (2019) Psychrotrophic microbes: biodiversity, mechanisms of adaptation and biotechnological implications in alleviation of cold stress in plant. In: Plant growth promoting rhizobacteria for sustainable stress management Vol 1 abiotic stress management. Sayyed, Arora & Reddy Eds. Springer, Singapore, pp 219–253. https://doi.org/10.1007/978-981-13-6536-2_12

Zaidi A, Ahmad E, Khan MS, Saif S, Rizvi A (2015) Role of plant growth promoting rhizobacteria in sustainable production of vegetables: current perspective. Sci Hortic 193:231–239

Zakaria AK, Sayyed RZ, Enshasy HE (2019) Biosynthesis of antibiotics by PGPR and their roles in biocontrol of plant diseases. In: Plant growth promoting rhizobacteria for sustainable stress management Vol II rhizobacteria in biotic stress management. Springer, Singapore, pp 1–36

Zerihun T, Birhanu G, Genene T, Adey F, Solomon C (2019) Isolation and biochemical characterization of plant growth promoting (PGP) bacteria colonizing the rhizosphere of tef crop during the seedling stage. J Plant Sci Phytopathol 3(1):13–27. https://doi.org/10.29328/journal.jpsp.1001027

Zope VP, Enshasy HE, Sayyed RZ (2019) Plant growth promoting Rhizobacteria: an overview in agricultural perspectives. In: Sayyed RZ (ed) Plant growth promoting Rhizobacteria for sustainable stress management Vol II Rhizobacteria in biotic stress management. Springer, Singapore, pp 345–362

# Index

**A**
ACC-deaminase, 13, 108, 116–119, 125, 173, 174, 214, 272, 338, 352
Acetyltransferase, 73
*Acinetobacter,* 13, 104, 298, 357
*Acinetobacter calcoaceticus,* 67, 112, 339
Actinomycetes, 3, 12, 15, 17, 24, 86, 158, 170, 195, 198, 217, 240, 325, 345, 352
Acylated homoserine lactones (AHL), 68, 74, 116, 125
Aerugine, 75, 86
Agrobacterium, 84, 104, 116, 272, 298, 328, 331
Agrochemicals, 5, 9, 10, 26, 85, 109, 125, 159, 160, 231, 273, 282, 297, 299, 311–315, 331
Agrosan GN, 258
AHL-dependent, 68
AHL synthase, 68
Aldehydes, 35, 69, 86, 223
Alkaloid, vii, 27, 42, 198, 341, 352, 355, 356, 358, 360, 363, 364
Allicin, 33, 35, 41–43, 359
Allochemical, 357
α-amylase, 363
Ammonification, 328
Amritpani, 317, 327–329
Antagonistic, vii, 4, 5, 16, 24, 27, 33, 38, 58, 72, 73, 84, 86, 109, 117, 124, 125, 137–138, 140, 144, 174, 184, 185, 188–190, 192–194, 198, 213, 217, 229, 240, 242, 243, 261, 276, 280, 299, 331
Antagonistic nature, 137–138
Antagonistic property, 84, 109
Antibiosis, 4, 5, 10, 86, 115, 121, 137, 144, 146, 187, 242, 354
Antidiabetic, 353, 356, 358–364
Antifungal, 12, 25, 57, 61, 85, 107, 136, 161, 186, 215, 244, 254, 298, 316, 353
Antihelminthic, 72
Antimicrobial, 4, 7, 9, 12, 15, 17, 33, 38, 42, 43, 72, 74, 75, 85, 88–91, 109, 120, 125, 136–141, 144–146, 148, 161, 165, 182–187, 189, 191, 192, 198, 213, 228, 241, 254, 261, 277, 297, 312, 317
Antimicrobial compounds, 12, 17, 138, 182, 183, 241, 261, 297
Antioxidant, 13, 72, 85, 109, 278
Antipenetrant, 42
Antitumor, 6, 72, 85, 86, 109, 140
Antiviral, 72, 85, 89, 92, 109, 140, 355
Aphids, 115, 281, 343, 344
Appressorium, 25, 29, 30, 33, 34, 42
Arbuscular, 271–274, 304
Arbutoid, 271
Arthrobacter, 13, 59, 84, 104, 272, 298, 336, 337, 339
Ascomycete, 23, 88, 122, 172
ATP-binding cassettes (ABC), 11
ATP synthase inhibitor, 39
Auxins, 12, 108, 110–112, 298, 325, 338
Azomycin, 71, 72, 86
*Azorhizobium caulinodans,* 67, 70, 300


R. Z. Sayyed et al. (eds.), *Antifungal Metabolites of Rhizobacteria for Sustainable Agriculture*, Fungal Biology,
https://doi.org/10.1007/978-3-031-04805-0

*Azospirillum,* 13, 84, 104, 107, 112, 116, 120, 121, 158, 212, 272, 273, 282, 298, 300, 304, 305, 325, 331, 332, 336–339, 352, 355, 357
*Azotobacter,* 14, 67, 84, 92, 104, 107, 112, 121, 212, 230, 272, 298, 300, 304, 305, 328, 336–338, 355

**B**
Bacillomycin, 27, 28, 36, 38, 59, 71, 75, 89, 93, 138, 140, 141, 144, 146, 162, 164
Bacillomycin D, 4, 75, 141, 144, 164
*Bacillus,* 4, 24, 57, 67, 84, 104, 136, 158, 186, 212, 240, 257, 272, 298, 326, 337, 353
*Bacillus cereus,* 57, 75, 90, 138, 163, 165, 170, 189, 217, 220, 222, 225, 240, 245–246, 300
*Bacillus subtilis,* 27, 57, 75, 88, 107, 138, 159, 186, 212, 260, 273, 321, 350
Bacterial blight, 191, 196, 259, 260
Bactericidal, 74, 87, 90, 185, 188
Bakane disease, 257–259
Beejamrita, 326–327
Beneficial insects, 339, 345
Benomyl, 57, 258
β-glucanase, 225, 244, 246
β-glucosidase, 173, 224, 225
Bioactive secondary metabolites, 27, 28, 33, 35–43, 115, 161, 163
Biochar, 301–302
Biochemical regulation, 54
Biocontrol, 4, 25, 57, 69, 84, 104, 136, 161, 182, 212, 242, 254, 276, 296, 315, 337, 352
Biocontrol agent, 4, 10, 14–17, 26, 43, 44, 57, 58, 69, 75, 76, 84–85, 90, 109, 110, 113–115, 120, 121, 125, 136, 146, 147, 161, 187, 200, 213, 214, 217, 218, 220, 222, 223, 227, 228, 230, 244, 253–265, 276, 280, 282, 315, 337, 339, 355
Biodiversity, 104–107, 114, 126, 158, 296, 311
Biodynamic preparations, 315, 318–321
Bio-enhancers, 316–321, 325, 326, 330, 331
Biofertilizers, 15, 70, 105, 120, 121, 125, 159, 261–264, 298, 300–305, 354
Biofilm, 12, 65–76, 91, 119–122, 126, 137, 143–144, 183, 192, 198, 301
Biofilmed biofertilizers, 120–121, 301
Bioformulation, 146, 302, 304, 326, 328
Biofungicides, 43, 146, 159, 160, 174, 229–230, 261–263
Biomass, 15, 27, 183, 271, 273, 274, 282, 301, 355
Biopesticides, 16, 17, 35, 264, 298, 305, 331
Bioprotectant, 282
Biosafety, 15–17, 60
Blast disease, 24–26, 28, 32, 42–44, 90, 91, 93, 94, 162, 256
*Bradyrhizobium,* 13, 57, 70, 84, 104, 113, 121, 158, 275, 298, 332, 339, 352
*Brevibacterium,* 75, 88, 107, 189
Brown spot, 114, 123, 256, 259
*Burkholderia,* 13, 59, 70, 75, 88, 92, 104, 107, 113, 116, 121, 158, 186, 187, 189–191, 212, 213, 272, 298, 336, 337, 339, 343, 355
Butyrolactones, 69, 86, 162

**C**
Carbendazim, 259
Carbon source, 66, 74, 194, 222, 328
Carnivores, 281
Cellobiohydrolases, 224
Cellulase, 11, 32, 85, 94, 108, 109, 139, 162, 170, 173, 187, 217, 220, 223–226, 243, 244
Cell wall degrading enzymes (CWDEs), 85, 94, 108, 125, 139, 227–230, 240, 243–246
Central Insecticide Board and Registration Committee (CIB-RC), 264
Ceresan, 258
Chalcone synthase (CHS), 73
Chitinases, 11, 30, 59, 69, 85, 105, 137, 161, 187, 213, 242, 261, 272
Chitinolytic enzymes, 137, 138, 228, 316
Chitin synthesis inhibitor, 27
Circulin, 72, 124
Cis-11-methyl-2-dodecanoic acid (DSF), 68
Climate change, 24, 255, 295, 305
*Cochliobolus miyabeanus,* 259
Co-inoculation, 13, 272–275, 278, 313
Colistin, 72, 124
Commercialization, 15–17, 68, 263–264, 302–304
Cow horn manure, 319–320
Cow horn silica, 320–321
Cow pat pit (CPP), 317, 321, 323
Crop diseases, 10–11, 158, 188, 239, 242, 254–260, 265
C sequestration, 301
$CuSO_4$, 258
Cyclic lipopeptides (CLPs), 29, 32, 38, 72, 86, 89–91, 95, 105, 107, 146, 165, 187, 254, 261

Cytokinins, 12, 54, 125, 272, 298, 300, 338, 352
Cytotoxic, 42, 72, 85, 91, 109, 186

**D**
Defense mechanisms, 124, 161, 227, 242, 245, 261, 276, 312
Detoxification, 343
Diketopiperazines, 261
Direct mechanism, 13, 55, 125, 277, 298, 338, 354

**E**
Eco-friendly, vii, viii, 3, 13, 15, 24, 57, 110, 158, 191, 230, 239, 273, 282, 303, 316, 337, 339
Ecomycins, 72, 86, 162
Ectomycorrhizal (ECM), 271, 272, 276
Endochitinases, 220, 228
Endogenous, 13, 111, 118
Endoglucanase, 173, 224, 225
Endophytes, 16, 54, 70, 76, 116, 191, 211, 225, 275, 337, 339, 344–346
Endosphere, 297
Environmental effects, 15–17
Ericoid, 271
Exochitinases, 220, 228
Exoglucanases, 224, 225
Extracellular metabolites, 4, 69, 73, 198, 244
Exudates, 66–68, 83, 105, 111, 115, 116, 118, 182–184, 193, 194, 211, 212, 277, 297, 299, 352, 355, 357

**F**
False smut, 257, 258
Fengycin, 4, 28, 32, 38, 59, 72, 75, 86, 88, 89, 91, 93, 94, 124, 138, 140, 141, 144–146, 148, 162–165, 187
Flavaglines, 33, 42
Flavonoid, 352, 355–357, 359, 360, 362–364
Food security, 23, 56, 135, 279, 282, 305, 311–332
Foot rot, 257–259
Forestry, 271, 277
Formulations, 5, 15–17, 35, 146, 229, 230, 254, 258, 260, 262–263, 278, 299, 302–305, 318, 326–328
*Frankia,* 84, 104, 113, 225, 298, 300
Fungal cell wall, 109, 137, 139, 164, 171, 215–228, 230, 240, 243–246, 338
Fungal pathogen, 11, 23, 88, 89, 91, 92, 108, 109, 115, 121, 122, 136, 140, 143, 148, 157, 158, 211–231, 239–246, 254, 255, 276
Fungal resistance, 57
*Fusarium udum,* 11, 213

**G**
*Gaeumannomyces graminis,* 146, 162, 220, 299, 300
*Gibberella fujikuroi,* 257
Gibberellins (GAs), 12, 54, 108, 112, 125, 272, 298, 325, 326, 338, 352, 356
Gliotoxin, 75
Glucanase, viii, 69, 85, 105, 108, 109, 139, 162, 172, 187, 213, 216, 217, 225–228, 243, 244, 261, 272
Glumes, 259
Green revolution, 296, 305, 312, 314

**H**
HCN production, 193, 315, 322, 323, 326, 327, 329, 330
*Helminthosporium oryzae,* 259
Herbicolin, 75
Herbivore-associated molecular patterns (HAMPs), 342
Herbivores, 281, 282, 336, 339–345
Heterocyclic group, 75
Heterothalic, 23
Hydrogen cyanide (HCN), vii, viii, 69, 72, 85, 86, 95, 105, 107–109, 121, 125, 169, 193, 194, 261, 315, 322–324, 326, 327, 329, 330, 339, 355, 357
Hydrolytic enzymes, vii, viii, 13, 69, 85, 93, 94, 109, 115, 158, 161, 170–174, 184, 187, 211–231, 239–246, 352
Hydrophobic, 138, 163, 215, 280
4-Hydroxy-2-alkylquinolines, 261
Hydroxyphenazines, 69
Hyperglycema, 352

**I**
Indirect mechanism, 10, 54, 55, 105, 125, 272, 298, 337, 338, 354, 355
Indole acetic acid (IAA), 13, 70, 108, 110–112, 118, 121, 241, 261, 315, 322, 324, 326, 329, 331, 332, 337, 339, 357
Induced systemic resistance (ISR), vii, 4, 27, 59, 73, 85, 86, 90, 91, 115, 124, 125, 138, 144–145, 159, 161, 163, 171, 190, 229, 261, 312, 337, 345, 354, 355
Infection mechanism, 25–26

Inoculants, vii, 14–17, 278, 282, 299–302, 304, 305, 313, 316
Inoculation methods, 16–17, 299, 303
Insulin, 352, 353, 358–364
Integrated pest management (IPM), viii, 105, 115, 182, 186, 343, 345
Islet of Langerhans, 359–361
Isoleucyl-tRNA synthetase, 74, 188
Iturin, 4, 38, 72, 86, 88, 89, 91, 93, 94, 124, 138, 140, 141, 143–146, 148, 159, 162–165, 187, 228, 229

J
Jeevamrita, 317, 326, 327

K
Kanosamine, 59, 71, 86
*Klebsiella pneumonia,* 67, 341

L
Lahorenoic acid, 261
Leaf blast, 256
Lipopeptides, vii, 32, 38, 71, 72, 86, 88–89, 91–93, 121, 136–138, 140, 141, 145, 146, 162–165, 174, 186–187, 198, 200
Liquid manure, 317, 318, 321–330
LuxI, 68, 116
LuxR, 68
Lytic enzymes, 10, 11, 59, 105, 108–109, 124, 139, 162, 217, 223, 225, 229, 242, 261, 338, 357

M
*Magnaporthe grisea,* 39, 162, 165, 186, 256, 261
*Magnaporthe oryzae,* 23–44, 90, 91, 94, 141, 261
Mannoproteins, 216, 221, 222
Metabolites, 4, 24, 59, 66, 85, 107, 136, 158, 181, 228, 240, 254, 272, 298, 312, 336, 352
Metformin, 358
Microaggregates, 280
Microbiome, viii, 67–68, 103, 271, 297, 305, 312
Microcolonies, 120, 301
Monoacetylphloroglucinol (MAPG), 73, 241
Mupirocin, 69, 72, 74, 188
Mycelial, 27–35, 39, 87, 107, 143, 146, 161, 171, 190, 281

N
N-acyl homoserine lactones, 116, 345
National Centre of Organic Farming (NCOF), 263
National Program for organic Production (NPOP), 264
Natural compounds, 25, 26, 36, 43
Natural enemies, vii, 337, 341–344
Natural product, vii, 25, 27–33, 41, 76, 199, 353, 358
Neck rot, 256
Nematicide, 161, 277
Nitrification, 328
Nitrogen fertilizer, 241, 257, 296
Nitrogen fixation, 13, 14, 55, 66, 70, 84, 113, 118, 121, 125, 174, 184, 241, 272, 274, 300, 316, 338, 339
Node blast, 256
Nodulation, 14, 111, 118, 273–275, 337, 339
Non-ribosomal peptides (NRPs), 140–142, 168, 186–187, 199, 243
Non-ribosomal peptide synthetases (NRPSs), 85, 88, 142, 168, 186, 189, 199
Non-volatile, 69, 86–87, 90, 196, 244
Non-volatile antibiotics, 86–90, 244

O
Oligomycin A, 7, 28, 29, 32, 39, 69
Oligopeptides, 11, 68, 88, 186
Oomycetes, 42, 87, 145, 165, 167, 187–189, 191, 214, 215, 223, 225, 226
Oomycin, 4, 86, 162
Orchidaceous, 271
Oxidative stress, 28, 55, 91

P
Paddy, 256–260
Panchagavya, 317, 325–326
Pancreatic β cells, 359–362, 364
Panicle blast, 256
Panicle damage, 256
Panicles, 25, 26, 256, 259, 355
*Pantonea agglomerans,* 67, 75
Pathogenic, vii, 4, 7, 23, 27, 56, 59, 69, 72, 85, 87, 90, 108, 109, 122, 124, 135, 140, 142, 161–168, 172–174, 182, 186, 189, 190, 193, 212, 214, 215, 217, 222, 223, 228, 229, 240, 242–246, 255, 277, 278, 336, 338, 345, 355
Peptidyl nucleoside antibiotic, 39
Phenazine, 4, 59, 69, 72, 75, 88, 90–91, 94, 95, 105, 107, 115, 124, 125, 189–190, 197, 241, 254, 261, 338, 355, 357

Phenazine-1-carboxamide (PCN), 72, 86, 88, 90, 162
Phenazine derivatives, 69, 88, 90, 190
Phenolic, vii, 74, 87, 90, 145, 188, 352, 355, 356, 360, 362
Phenylpyrrole, 69, 88, 91, 114
Phenylpyrrole antibiotics, 69
Phloroglucinol (PHL), 73, 87, 95, 115, 261
Phosphate solubilization, 13, 15, 84, 105, 112–113, 121, 136, 261, 273, 316, 321, 327–329, 332, 338
Photosynthates, 269, 271, 276
Phyllosphere, 281, 297, 337, 338
Physiological shift, 66
Phytochemicals, 33, 42, 339, 341, 352
Phytohormone, 12–13, 15, 105, 136, 158, 184, 212, 213, 261, 272, 273, 298, 300, 316, 352, 357
Phytomicrobiome, vii, 53, 60, 212
Phytopathogenic fungi, 7–11, 87, 90, 109, 121, 122, 136, 141, 146, 184, 186, 188, 189, 218–223, 225–228, 241, 244
Phytopathogens, vii, 3, 5, 6, 8, 9, 24, 32, 43, 53–60, 84, 87, 92, 106, 109, 110, 114–116, 122, 124, 125, 135–137, 139, 141, 148, 166, 170, 181–201, 213, 214, 216–218, 241, 261, 276, 298, 338, 344, 357
Phytoplasmas, 278, 279
Phytostimulators, 298, 300
Phytotoxic, 72, 85, 109
PKC inhibitor, 42
Planktonic form, 67, 120
Plant defense mechanism, 124, 261, 276
Plant growth-promoting actinomycetes (PGPA), 3, 10, 11
Plant-growth promoting rhizobacteria (PGPR), 3, 54, 66, 84, 103, 158, 182, 212, 260, 272, 297, 311, 336, 352
Plant growth promotion, 3, 10, 13–15, 54, 55, 103–126, 136, 168, 174, 213, 315
Plant hormones, 66, 108, 112–113, 281, 338, 356, 357
Plant pathogen, vii, 3–17, 25–27, 33, 43, 55, 67, 69, 71, 73, 74, 85–87, 90, 92, 95, 104, 109, 120, 124, 125, 136, 142, 158, 161, 163, 170–174, 183, 184, 190, 213, 214, 216, 217, 220, 222, 227, 230, 239, 240, 261, 276–280, 304
Plant resistance, 4, 168, 240, 260, 331, 346
Plant-virus, 278–280
Plipastatins A and B, 71, 165
Polyketides (PKs), 69, 72–74, 85–87, 90, 140, 158, 165, 187–188, 199
Postprandial, 363
Preparations, 15, 115, 196, 198, 229, 230, 256, 303, 315–321, 323, 325–328, 330
Prodione, 259
Propiconazole, 259
Protease, viii, 7, 11, 32, 59, 69, 85, 94, 105, 108, 109, 139, 158, 162, 172, 173, 213, 216, 217, 219–222, 225, 227–229, 243, 244, 261, 328
Proteobacteria, 68, 104, 106
Pseudomonads, 4, 67, 69, 73, 74, 86, 90, 109, 188, 298
*Pseudomonas aeruginosa,* 14, 59, 74, 86, 87, 94, 110, 115, 167, 196, 198, 220, 225, 241, 243, 245, 246, 300
*Pseudomonas fluorescens,* 32, 57, 59, 67, 68, 70, 72, 74, 75, 86, 87, 89, 90, 95, 115, 120, 124, 167, 171, 172, 187, 188, 193, 195, 198, 212, 228, 229, 254, 278, 299, 300, 304
*Pseudomonas putida,* 29, 57, 67, 68, 87, 90, 95, 107, 120, 167, 186, 332, 352
Pseudomonic acid, 71, 72, 74, 86, 188
P-solubilizers, 112, 325, 332
Pyochellin, 261
Pyocyanin, 69, 90, 91, 162, 189, 337, 339
Pyoluteorin (Plt), 4, 72, 74, 75, 86, 87, 90, 95, 105, 107, 109, 115, 162, 188, 241, 254, 338, 355
Pyrrolnitrin, 30, 59, 72, 75, 95, 107, 124
*Pythium* spp., 57, 72, 75, 87, 90, 92, 95, 109, 124, 141, 144, 159, 161, 162, 167, 188, 215, 217, 223, 225–227, 229, 276, 278

## Q

Quorum sensing, 6, 66, 116, 119, 125, 137, 142, 183, 192, 193

## R

Regulatory procedure, 263–264
Rhamnolipids, 86, 125, 163, 241, 261
Rhizobacteria, 3, 66, 83, 103, 158, 184, 212, 260, 272, 296, 312, 336, 352
Rhizobia, 10, 112, 113, 116, 212, 274–275, 304, 305
*Rhizobium leguminosarum,* 67, 70, 167, 300
*Rhizoctonia solani,* 13, 57–59, 72, 75, 87–90, 93–95, 109, 121, 124, 138, 140, 141, 143, 144, 147, 159, 162, 164, 171, 187, 191, 217, 219, 220, 227, 256, 261
Rhizomicrobiome, 66, 212, 299
Rhizoremediator, 298, 352, 354

Rhizosphere, 13, 54, 83, 103, 137, 158, 182, 211, 240, 261, 269, 297, 313, 337, 351
Ribosomal peptides (RPs), 139, 142

**S**
Safe food, 254
Salinity stress, 117, 332
Saponin, 33, 42, 359–364
*Sarocladium oryzae,* 261
Sclerotia, 171, 190, 191, 227, 228, 256
Secondary metabolites, vii, viii, 3–17, 23–44, 59, 66, 73, 75, 88, 107, 115, 138, 140, 145, 158, 161, 163, 164, 174, 181–201, 241, 243, 254, 260–261, 339, 341, 345, 351–365
Seed-borne, 190, 258
Semiochemicals, 344
*Serratia,* 14, 57, 70, 84, 104, 107, 124, 172, 186, 195, 220, 228, 272, 298, 305, 336, 337, 339, 343, 355, 357
Sheath blight, 88, 95, 213, 254, 256–257
Siderophore, 13, 59, 84, 105, 137, 158, 183, 214, 241, 261, 272, 300, 315, 338, 357
Signal compounds, 54, 66, 212, 297, 299, 305
Signal transduction, 15, 107, 145, 227
Soil microbiota, 68, 120, 296
*Streptomyces,* 3–17, 26, 28, 29, 31, 32, 39, 41, 42, 59, 68, 70, 75, 88, 89, 104, 172, 186–189, 191, 192, 195, 197, 198, 217, 220, 227, 230, 352
Streptomycin, 28, 42, 56
Stress-tolerant bacteria, 352
Surfactin, 4, 28, 32, 38, 72, 86, 88, 89, 93, 94, 125, 138, 140–142, 144–146, 148, 162–164, 187, 228
Susceptibility, 223, 228, 280
Sustainable agriculture, 104, 119, 125, 136, 157, 168, 174, 214, 271, 274, 296–297, 299, 300, 331, 337, 344–345
Symbiont, 106, 195, 242, 271, 273, 301, 339, 342, 343
Symbiotic, 17, 65, 105, 113, 120, 183, 212, 271, 272, 279, 298, 343, 346
Synergism, 228–229, 272, 273
Synergistic, 137, 273, 275, 280, 314
Synthetic chemical fertilizer, 168, 254, 296

**T**
*Thiobacillus,* 104, 298
Thiophanate-methyl, 28, 258
Thiram, 258
Tillering, 256, 257, 259
*Trichoderma asperellum,* 242, 243, 245, 246
*Trichoderma harzianum,* 227, 242–246, 278
Trifloxystrobin, 259
2,4-diacetylphloroglucinol (2,4, DAPG), 4, 69, 72, 86, 95, 105–107, 115, 163, 187, 188, 241, 254, 261, 338, 355
Type 2 diabetes mellitus (T2DM), 352, 353

**U**
Ubiquitous, 106, 271, 276, 345
*Ustilaginoidea virens,* 257

**V**
Vermi wash, 328
Viscosinamide, 69, 86, 162, 355
Volatile, 7, 27, 69, 86, 108, 136, 158, 183, 212, 244, 312, 341, 357
Volatile metabolites, 138, 184
Volatile organic compounds (VOCs), 4, 5, 12, 93, 107, 142, 143, 168–171, 184, 190–192, 197, 199, 281, 316, 341, 352

**X**
Xanthobaccin, 69, 75
Xanthobaccin A, 75
*Xanthomonas,* 9, 68, 84
*Xanthomonas oryzae pathovar* (PV), 259
*Xylella,* 68

**Z**
Zwittermicin A, 4, 69, 71, 72, 75, 86, 90, 92, 138, 165, 189

The manufacturer's authorised representative in the EU is Springer Nature Customer Service Centre GmbH, Europaplatz 3, 69115 Heidelberg, Germany. If you have any concerns regarding our products, please contact ProductSafety@springernature.com

Printed and bound by CPI Group (UK) Ltd, Croydon, CR0 4YY
30/09/2026
02225239-0015